Electronics

ALL-IN-ONE

FOR

DUMMIES®

Electronics
ALL-IN-ONE
FOR
DUMMIES®

by Doug Lowe

WILEY

John Wiley & Sons, Inc.

Electronics All-in-One For Dummies®

Published by
John Wiley & Sons, Inc.
111 River Street
Hoboken, NJ 07030-5774

www.wiley.com

About the Author

When **Doug Lowe** was 10 years old, his father gave him an electronics experimenter's kit that started a life-long love of all things electronic. As a child, he dreamed of becoming an electrical engineer. But he soon discovered that he loved writing as much as he loved electronics, and his first technical book was published in 1981. Since then, he has written more than 100. All of those books have been about one specific corner of the electronics world: computers. But now he has finally written a book about his first love, the joy of designing and building electronic circuits from scratch.

Doug lives in sunny Fresno, California, where the motto is "Fres-YES!," (unfortunately, I'm not making that up). The fruits of his electronic passions can often be seen around the Halloween season, in the form of animated computer-controlled Halloween decorations that rival Disney's Haunted Mansion.

Dedication

My grandpa had a wonderful workshop behind his home. It was cramped and cold in the winter, and smelled of machine oil, stale coffee, aged rubber, and damp wood — like an ancient hardware store clinging to life down the road from a giant Big Box store. It was there that I began a lifelong love of electronics, as my grandpa tried to explain the mystery of electricity and taught me how to use a soldering iron, read a voltmeter, and test a questionable vacuum tube.

In the spring of 1970, he showed up at my house with a trailer full of everything an 11-year-old boy could possibly need to create his own electronics lab. The trailer was stacked high with broken radios and box after box filled with tools and spare parts: vacuum tubes, soldering irons, condensers, volt meters, resistors, tube testers, and tuning capacitors.

I can still remember the look on my dad's face when he saw that trailer full of junk pull up in front of our house, hitched to the back of grandpa's old Buick.

I was in heaven. I think my dad may have been somewhere else.

But he helped unload the trailer, and then he built me an amazing workbench with enough shelves and drawers to store everything and more.

This book is dedicated to the memory of my grandpa, Kenneth D. Lowe Sr., and to my dad, Kenneth D. Lowe, Jr.

To my grandfather, for giving me an amazing gift buried in a trailer full of junk. And to my dad, for letting me keep it.

Author's Acknowledgments

First, I'd like to thank Katie Feltman for giving me the opportunity to write this book. We've talked about doing something like this for years, and I was so delighted when she called me to discuss this project. I am especially grateful for her patience as deadlines came and went.

Next, I'd like to thank Christopher Morris, not only for doing such an incredible job as project editor on this long project, but also giving me the encouragement and support I needed along the way.

Thanks also to Kirk Kleinschmidt, who gave the book a great technical review and offered many excellent suggestions not only on purely technical considerations but also on the presentation of complex material. And thanks to Heidi Unger, who as usual did a great job whipping my prose into shape, crossing all the i's and dotting all the t's, or something like that. And, of course, thanks to all the behind-the-scenes people who chipped in with help I'm not even aware of.

I'd also like to thank my niece, Miranda Chapman, for allowing me to jumpstart her modeling career by using her beautiful image to illustrate the internal structure of the atom.

Finally, I'd like to thank a few others who made important contributions: Jon Williams and John Barrowman of EFX-TEK for providing the prop controllers used in several of the chapters; Light-O-Rama for providing the lighting controller used in Chapter 3 of Book VIII; and my good friend Dave Youngs, who parted with a very fine 100mm macro lens for the duration of the project.

Publisher's Acknowledgments

We're proud of this book; please send us your comments at http://dummies.custhelp.com. For other comments, please contact our Customer Care Department within the U.S. at 877-762-2974, outside the U.S. at 317-572-3993, or fax 317-572-4002.

Some of the people who helped bring this book to market include the following:

Acquisitions, Editorial, and Vertical Websites

Senior Project Editor: Christopher Morris

Senior Acquisitions Editor: Katie Feltman

Copy Editors: Heidi Unger, Kathy Simpson

Technical Editor: Kirk Kleinschmidt

Editorial Manager: Kevin Kirschner

Vertical Websites Project Manager: Laura Moss-Hollister

Vertical Websites Assistant Project Manager: Jenny Swisher

Vertical Websites Associate Producers: Josh Frank, Marilyn Hummel, Douglas Kuhn, Shawn Patrick

Editorial Assistant: Amanda Graham

Sr. Editorial Assistant: Cherie Case

Cover Photos: © istockphoto.com / Rudyanto Wijaya

Cartoons: Rich Tennant (www.the5thwave.com)

Composition Services

Project Coordinator: Katherine Crocker

Layout and Graphics: Carrie A. Cesavice, Cheryl Grubbs, Joyce Haughey, Mark Pinto, Corrie Socolovitch

Proofreaders: Lauren Mandelbaum, Evelyn Wellborn

Indexer: BIM Indexing & Proofreading Services

Publishing and Editorial for Technology Dummies

Richard Swadley, Vice President and Executive Group Publisher

Andy Cummings, Vice President and Publisher

Mary Bednarek, Executive Acquisitions Director

Mary C. Corder, Editorial Director

Publishing for Consumer Dummies

Kathleen Nebenhaus, Vice President and Executive Publisher

Composition Services

Debbie Stailey, Director of Composition Services

Contents at a Glance

Table of Contents

Introduction

*W*elcome to the amazing world of electronics!

Ever since I was a kid, I've been fascinated with electronics. When I was about 10 years old, my dad bought me an electronic experimenter's kit from the local RadioShack store. I still have it; it's pictured in Figure IN-1. I have incredible memories of evenings spent with my dad, wiring together the sample circuits to make squawking police sirens, flashing lights, a radio receiver, and even a telegraph machine.

The best part was dreaming that when I grew up, I'd have a job in the field of electronics, that someday I'd understand exactly how those resistors, capacitors, inductors, transistors, and integrated circuits actually worked and I'd use that knowledge to design televisions or computers or communication satellites.

Well, that dream didn't come true. Instead, I went into a closely related field: computer programming. But my love of electronics never died, and I've spent the last 40 years or so experimenting with electronics as a hobbyist.

This book is an introduction to electronics for people who have always been fascinated by electronics but didn't make a career out of it. In these pages, you'll find clear and concise explanations of the most important concepts that form the basis of all electronic devices, concepts such as the nature of electricity (if you think you really know what it is, you're kidding yourself); the difference between voltage, amperage, and wattage; and how basic components such as resistors, capacitors, diodes, and transistors work.

Not only will you gain an appreciation for the electronic devices that are a part of everyday life, you'll also learn how to build simple circuits that will not only impress your friends but may actually be useful!

Figure IN-1:
My first
electronics
kit.

About This Book

Electronics All-in-One For Dummies is intended to be a reference for the most important topics you need to know when you dabble in building your own electronic circuits. It's a big book made up of eight smaller books, which we at the home office like to call *minibooks*. Each of these minibooks covers the basics of one key topic for working with electronics, such as circuit building techniques, how electronic components like diodes and transistors work, or using integrated circuits.

This book doesn't pretend to be a comprehensive reference for every detail on every possible topic related to electronics. Instead, it shows you how to get up and running fast so that you have more time to do the things you really want to do. Designed using the easy-to-follow For Dummies format, this book helps you get the information you need without laboring to find it.

Whenever one big thing is made up of several smaller things, confusion is always a possibility. That's why this book is designed with multiple access points to help you find what you want. At the beginning of the book is a detailed table of contents that covers the entire book. Then each minibook begins with a minitable of contents that shows you at a miniglance what chapters are included in that minibook. Useful running heads appear at the top of each page to point out the topic discussed on that page, and handy thumbtabs run down the side of the pages to help you find each minibook quickly. Finally, a comprehensive index lets you find information anywhere in the entire book.

This isn't the kind of book you pick up and read from start to finish, as if it were a cheap novel. If I ever see you reading it at the beach, I'll kick sand in your face. Beaches are for reading romance novels or murder mysteries, not electronics books. Although you could read this book straight through from start to finish, this book is designed like a reference book, the kind of book you can pick up, open to just about any page, and start reading.

You don't have to memorize anything in this book. It's a "need-to-know" book: You pick it up when you need to know something. Need a reminder on how to calculate the correct load resistor for an LED circuit? Pick up the book. Can't remember the pinouts for a 555 timer IC? Pick up the book. After you find what you need, put the book down and get on with your life.

How to Use This Book

This book works like a reference. Start with the topic you want to find out about, and look for it in the table of contents or index to get going. The table of contents is detailed enough that you can find most of the topics you're looking for. If not, turn to the index, where you can find even more detail.

Of course, the book is loaded with information — so if you want to take a brief excursion into your topic, you're more than welcome. If you want to know the big picture on digital electronics, for instance, read all of Book VI. If you just want to learn about logic gates, just read Chapter 2 in Book VI. On the other hand, if you just want a refresher on what an XOR gate is, look in the index to find the section on XOR gates.

About the Projects

You can find dozens of projects strewn throughout this book's chapters. You'll find a plethora of simple projects you can build to demonstrate the operation of typical circuits. For example, in the chapter on transistors, you'll find several simple projects that demonstrate common uses for transistors, such as driving an LED, creating an oscillator, or inverting an input.

I suggest you build each of the projects as you read the chapters. Reading about electronics circuits is one thing, but to understand how a circuit works, you really need to build it and see it in operation. Most of the projects are simple enough that you can build them in 20–30 minutes, assuming you have the parts on hand.

Because just about every home in America is within 15 minutes of a RadioShack, these projects are designed (whenever possible) to use components that RadioShack stores keep in stock. Thus, if you want to build one of the projects on a Saturday afternoon, you can buzz over to your local RadioShack store, pick up the parts you'll need, take them home, and build the circuit.

Of course, you can also purchase the components you need at any other store that stocks electronic hobbyist components, and you can find many sources for purchasing the parts online.

Staying Safe

Most of the electronic circuits described in this book are perfectly safe: They run from common AAA or 9 V batteries and therefore don't work with voltages large enough to hurt you.

However, you'll occasionally come across circuits that work with higher voltages, which can be dangerous. Any project that involves line voltage (that is, that you plug into an electrical outlet) should be considered potentially dangerous and handled with the utmost care. In addition, even battery-powered circuits that use large capacitors can build up charges that can deliver a potentially painful shock.

When you work with electronics, you'll also encounter dangers other than those posed by electricity. Soldering irons are hot and can burn you. Wire cutters are sharp and can cut you. And there are plenty of small parts that can fall on the floor and find themselves in the mouths of kids or pets.

Safety is an important enough topic that I've devoted a chapter to it in Book I. I strongly urge you to read Book I, Chapter 4 *before* you build anything.

Please be careful! The projects that are presented in Book VIII all work directly with line-level voltage and should be considered dangerous. You must exercise great care if you decide to build any of those projects, as a single mistake could kill you or someone else. In particular, the projects described in Chapters 1 and 3 of Book VIII are offered as educational prototypes that are designed to be operated only within the safe confines of your workbench, where you can control the power connections so that no one is exposed to dangerous voltages.

How This Book Is Organized

Each of the eight minibooks contained in *Electronics All-in-One For Dummies* can stand alone. Here is a brief description of what you find in each minibook.

Book 1: Getting Started in Electronics

This minibook contains the information you need to get started with electronics. Here, you'll learn some of the fascinating concepts of what exactly electricity is, how to set up a workbench so you can build circuits, how to read schematic diagrams, how to solder, and other skills you need.

Book 11: Working with Basic Electronic Components

This minibook covers what you need to know to work with the most common building blocks of electronic circuits: conductors, resistors, capacitors, inductors, diodes, and transistors. You'll learn the concepts behind how these devices work and how to incorporate these little beasties into your own circuits. Along the way, you'll learn about some of the most important concepts of electronics, such as Ohm's law, and simple math formulas for calculating important electrical quantities such as resistance or capacitance.

Book 111: Working with Integrated Circuits

This minibook introduces you to integrated circuits, which are complete electronic circuits all in one small component. First, you'll learn what integrated circuits are and how to properly handle them. Then, you'll learn how to use two of the most commonly encountered integrated circuits: the 555 timer chip and operational amplifiers.

Book 1V: Getting into Alternating Current

This minibook focuses on circuits that use alternating current (also known as AC). You'll learn how to safely work with circuits that use *line current* (current you obtain from an electrical outlet), as well as how to build power-supply circuits that enable you to build circuits that run on household current rather than on batteries.

Book V: Working with Radio and Infrared

In this minibook, you discover the fascinating field of wireless communications. First, you'll learn what radio is and how it works. Then, you'll learn how to build one of the oldest yet still one of the most interesting of all electronic circuits: A crystal radio, which allows you to listen to local radio stations using only a few simple parts. The most interesting thing about crystal radios is that they have no external power source at all — no batteries, no power cord that plugs into an electrical outlet. Instead, they derive all of their power from the radio waves themselves.

After tinkering with radio, you can look at wireless communication using infrared light.

Book VI: Doing Digital Electronics

This minibook introduces you to the field of digital electronics, which deal with zeros and ones in the form of voltages that are either on or off. You'll learn how digital circuits work and how just a few simple circuits can be combined to form modern-day computers, and you'll gain experience building logic circuits using one of the oldest and most widely used families of integrated circuits.

Book VII: Working with BASIC Stamp Processors

In this minibook, you learn how about microcontrollers, which are essentially self-contained computers on a single chip. Microcontrollers let you build circuits that are programmable. You'll learn how to build circuits utilizing one of the most popular types of microcontrollers, called a *BASIC Stamp*. And you'll learn how to write simple programs that you can easily download to the BASIC Stamp.

Book VIII: Special Effects

The chapters in this minibook show you how to build actual projects that put the concepts covered in the other minibooks into actual practice. The chapters present four complete projects. The first is a game show buzzer for question-answer game shows such as *Jeopardy!*. If you're a teacher, you may find this project useful for your classroom.

The second project is a lighting project that lets you fill your yard with lightning on Halloween. If your lightning is scary enough, you'll easily pay for the project with the money you save on candy, since you won't have to pass out candy to kids who are too scared to knock on your door.

The third project is another lighting project, but this time for creating spectacular lighting displays that are sequenced with music. Such displays are often seen on houses around the holidays, but are also used in retail stores or malls, in museums, or for theatrical productions.

The final project is a general-purpose controller board for animatronic props similar to what you'd see in a haunted house or museum exhibit featuring dinosaurs that move. The controller isn't nearly as complicated as the devices that control the animatronic pirates on Disneyland's *Pirates of the Caribbean* ride, but it will let you synchronize simple movements provided by motors, servos, or compressed air with light and sound effects.

Icons Used in This Book

Like any For Dummies book, this one is chock-full of helpful icons that draw your attention to items of particular importance. You find the following icons throughout this book:

Pay special attention to this icon; it lets you know that some particularly useful tidbit is at hand.

Hold it — overly technical stuff is just around the corner. Obviously, because this is an electronics book, almost every paragraph of the entire book could get this icon. So I reserve it for those paragraphs that go into greater depth, down into explaining how something works under the covers — probably deeper than you really need to know to use a feature, but often enlightening.

You also sometimes find this icon when I want to illustrate a point with an example that uses some electronics gadget that hasn't been covered so far in the book, but that is covered later. In those cases, the icon is just a reminder that you shouldn't get bogged down in the details of the illustration and should instead focus on the larger point.

Danger, Will Robinson! This icon highlights information that may help you avert disaster. You should definitely pay attention to the warning icons because they will let you know about potential safety hazards.

Did I tell you about the memory course I took?

Where to Go from Here

You can obtain source code files for the code listings in Book VIII on this book's companion website at www.dummies.com/go/electronicsaiofd.

Yes, you *can* get there from here. With this book in hand, you're ready to plow right into the exciting hobby of electronics. Browse through the table of contents and decide where you want to start. Be bold! Be courageous! Be adventurous! And above all, have fun!

Occasionally we have updates to our technology books. If this book does have any such updates, they will be posted at www.dummies.com/go/electronicsaiofdupdates.

Book I

Getting Started in Electronics

The 5th Wave   By Rich Tennant

"So I guess you forgot to tell me to strip out the components before drilling for blowholes."

Contents at a Glance

Chapter 1: Welcome to Electronics

In This Chapter

✔ Understanding electricity

✔ Defining the difference between electrical and electronic circuits

✔ Perusing the most common uses for electronics

✔ Looking at a typical electronic circuit board

1 thought it would be fun to start this book with a story, so please bear with me. In January of 1880, Thomas Edison filed a patent for a new type of device that created light by passing an electric current through a carbon-coated filament contained in a sealed glass tube. In other words, Edison invented the light bulb. (Students of history will tell you that Edison didn't really invent the light bulb; he just improved on previous ideas. But that's not the point of the story.)

Edison's light bulb patent was approved, but he still had a lot of work to do before he could begin manufacturing a commercially viable light bulb. The biggest problem with his design was that the lamps dimmed the more you used them. This was because when the carbon-coated filament inside the bulb got hot, it shed little particles of carbon, which stuck to the inside of the glass. These particles resulted in a black coating on the inside of the bulb, which obstructed the light.

Edison and his team of engineers tried desperately to discover a way to prevent this shedding of carbon. One day, someone on his team noticed that the black carbon came off of just one end of the filament, not both ends. The team thought that maybe some type of electric charge was coming out of the filament. To test this theory, they introduced a third wire into the lamp to see if it could catch some of this electric charge.

It did. They soon discovered that an electric current flowed from the heated filament to this third wire, and that the hotter the filament got, the more electric current flowed. This discovery, which came to be known as *the Edison Effect*, marks the beginning of technology known as *electronics*. The device, which Edison patented on November 15, 1883, is the world's first electronic device.

When Edison patented his device in 1883, he had no idea what it would lead to. Now, just about 125 years later, it's hard to imagine a world without electronics. Electronic devices are everywhere. There are more television sets in the United States than there are people. No one uses film to take pictures anymore; cameras have become electronic devices. And you rarely see a teenager anymore without headphones in his ears.

Without electronics, life would be very different.

But have you ever wondered what makes these electronic devices tick? In this chapter, I lay some important groundwork that will help the rest of this book make sense. I examine the bits and pieces that make up the most common types of electronic devices, and take a look at the basic concept that underlies all of electronics: electricity.

I promise I won't bore you too much with tedious or complicated physics concepts, but I must warn you from the start: In order to learn how electronics works at a level that will let you begin to design and build your own electronic devices, you need to have at least a basic idea of what electricity is. Not just what it does, but what it actually *is*. So put on your thinking cap and get started.

What Is Electricity?

Before you can understand even the simplest concepts of electronics, you must first understand what electricity is. After all, the whole purpose of electronics is to get electricity to do useful and interesting things.

The concept of electricity is both familiar and mysterious. We all know what electricity is, or at least have a rough idea, based on practical experience. In particular, consider these points:

✦ We are very familiar with the electricity that flows through wires like water flows through a pipe. That electricity comes from power plants that burn coal, catch the wind, or harness nuclear reactions. It travels from the power plants to our houses in big cables hung high in the air or buried in the ground. Once it gets to our houses, it travels through wires through the walls until it gets to electrical outlets. From there, we plug in power cords to get the electricity into the electrical devices we depend on every day, such as ovens and toasters and vacuum cleaners.

✦ We know, because the electric company bills us for it every month, that electricity isn't free. If we don't pay the bill, the electric company turns off our electricity. Thus, we know that electricity is valuable.

✦ We know that electricity can be stored in batteries, which contain a limited amount of electricity that can be used up. When the batteries die, all their electricity is gone.

✦ We know that some kinds of batteries, like the ones in our cellphones, are *rechargeable*, which means that when they've been drained of all their electricity, more electricity can be put back into them by plugging them into a charger, which transfers electricity from an electrical outlet into the battery. Rechargeable batteries can be filled and drained over and over again, but eventually they lose their ability to be recharged — and you have to replace them with new batteries.

✦ We also know that electricity is the stuff that makes lightning strike in a thunderstorm. In grade school, we were taught that Ben Franklin discovered this by conducting an experiment involving a kite and a key, which we should not attempt to repeat at home.

✦ We know that electricity can be measured in *volts*. Household electricity is 120 volts (abbreviated 120 V). Flashlight batteries are 1.5 volts. Car batteries are 12 volts.

✦ We also know that electricity can be measured in *watts*. Traditional incandescent light bulbs are typically 60, 75, or 100 watts (abbreviated 100 W). Modern compact fluorescent lights (CFLs) have somewhat smaller wattage ratings. Microwave ovens and hair dryers are 1,000 or 1,200 watts. The more watts, the brighter the light or the faster your pizza reheats and your hair dries.

✦ We also may know that there's a third way to measure electricity, called *amps*. A typical household electrical outlet is 15 amps (abbreviated 15 A).

✦ The truth is, most of us don't really know the difference between volts, watts, and amps. (Don't worry; by the time you finish Chapter 2 of this minibook, you will!)

✦ We know that there's a special kind of electricity called *static electricity* that just sort of hangs around in the air, but that can be transferred to us by dragging our feet on a carpet, rubbing a balloon against our hairy arms, or forgetting to put an antistatic sheet in the dryer.

✦ And finally, we know that electricity can be very dangerous. In fact, dangerous enough that for almost 100 years electricity was used to administer the death penalty. Every year, hundreds of people die in the United States from accidental electrocutions.

But Really, What Is Electricity?

In the previous section, I list several ideas most of us have about electricity based on everyday experience. But the reality of electricity is something very different. Chapter 2 of this minibook is devoted to a deeper look at the nature of electricity, but for the purposes of this chapter, I want to start by introducing you to three very basic concepts of electricity: namely, *electric charge*, *electric current*, and *electric circuit*.

✦ *Electric charge* refers to a fundamental property of matter that even physicists as smart as Stephen Hawking don't totally understand. Suffice it to say that two of the tiny particles that make up atoms — protons and electrons — are the bearers of electric charge. There are two types of charge: *positive* and *negative*. Protons have positive charge, electrons have negative charge.

Electric charge is one of the basic forces of nature that hold the universe together. Positive and negative charges are irresistibly attracted to each other. Thus, the attraction of negatively-charged electrons to positively-charged protons hold atoms together.

If an atom has the same number of protons as it has electrons, the positive charge of the protons balances out the negative charge of the electrons, and the atom itself has no overall charge.

However, if an atom loses one of its electrons, the atom will have an extra proton, which gives the atom a net positive charge. When an atom has a net positive charge, it goes looking for an electron to restore its balanced charge.

Similarly, if an atom somehow picks up an extra electron, the atom has a net negative charge. When this happens, the atom goes looking for a way to get rid of the extra electron to once again restore balance.

Okay, technically atoms don't really go "looking" for anything. They don't have eyes, and they don't have minds that are troubled when they're short an electron or have a few too many. However, the natural attraction of negative to positive charges causes atoms that are short an electron to be attracted to atoms that are long an electron. When they find each other, something almost magic happens . . . The atom with the extra electron gives its electron to the atom that's missing an electron. Thus, the charge represented by the electron moves from one atom to another, which brings us to the second important concept . . .

✦ *Electric current* refers to the flow of the electric charge carried by electrons as they jump from atom to atom. Electric current is a very familiar concept: When you turn on a light switch, electric current flows from the switch through the wire to the light, and the room is instantly illuminated.

Electric current flows more easily in some types of atoms than in others. Atoms that let current flow easily are called *conductors*, whereas atoms that don't let current flow easily are called *insulators*.

Electrical wires are made of both conductors and insulators, as illustrated in Figure 1-1. Inside the wire is a conductor, such as copper or aluminum. The conductor provides a channel for the electric current to flow through. Surrounding the conductor is an outer layer of insulator, such as plastic or rubber.

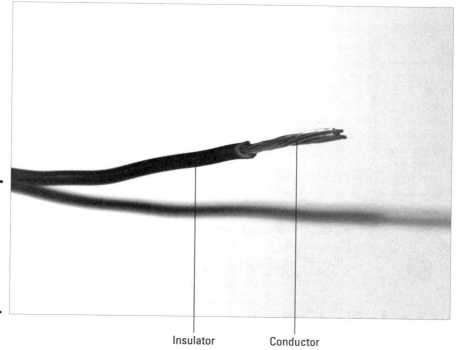

Figure 1-1:
An electric wire consists of a conductor surrounded by an insulator.

Insulator   Conductor

The insulator serves two purposes. First, it prevents you from touching the wire when current is flowing, thus preventing you from being the recipient of a nasty shock. But just as importantly, the insulator prevents the conductor inside the wire from touching the conductor inside a nearby wire. If the conductors were allowed to touch, the result would be a *short circuit*, which brings us to the third important concept . . .

✦ An *electric circuit* is a closed loop made of conductors and other electrical elements through which electric current can flow. For example, Figure 1-2 shows a very simple electrical circuit that consists of three elements: a battery, a lamp, and an electrical wire that connects the two.

The circuit shown in Figure 1-2 is, as I already said, very simple. Circuits can get much more complex, consisting of dozens, hundreds, or even thousands or millions of separate components, all connected with conductors in precisely orchestrated ways so that each component can do its bit to contribute to the overall purpose of the circuit. But all circuits must obey the basic principle of a closed loop.

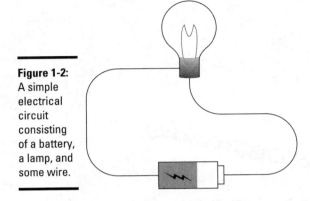

Figure 1-2:
A simple
electrical
circuit
consisting
of a battery,
a lamp, and
some wire.

All circuits must create a closed loop that provides a complete path from the source of voltage (in this case, the battery) through the various components that make up the circuit (in this case, the lamp) and back to the source (again, the battery).

What Is Electronics?

One of the reasons I started this chapter with the history lesson about Thomas Edison was to point out that when the whole field of electronics was invented in 1883, electrical devices had already been around for at least 100 years. For example:

✦ Benjamin Franklin was flying kites in thunderstorms more than 100 years before.

✦ The first electric batteries were invented by a fellow named Alessandro Volta in 1800. Volta's contribution is so important that the common *volt* is named for him. (There is some archeological evidence that the ancient Parthian Empire may have invented the electric battery in the second century BC, but if so we don't know what they used their batteries for, and their invention was forgotten for 2,000 years.)

✦ The electric telegraph was invented in the 1830s and popularized in America by Samuel Morse, who invented the famous Morse code used to encode the alphabet and numerals into a series of short and long clicks that could be transmitted via telegraph. In 1866, a telegraph cable was laid across the Atlantic Ocean allowing instantaneous communication between the United States and Europe.

✦ Contrary to popular belief, Benjamin Franklin wasn't the first to fly a kite in a thunderstorm. In 1850, he published a paper outlining his idea. Then he let a few other people try it first. After they survived, he tried the experiment himself and wound up getting all the credit. Benjamin Franklin was not only very *smart*; he was also very *wise*.

All of these devices, and many other common devices still in use today, such as light bulbs, vacuum cleaners, and toasters, are known as *electrical devices*. So what exactly is the difference between electrical devices and *electronic devices*?

The answer lies in how devices manipulate electricity to do their work. Electrical devices take the energy of electric current and transform it in simple ways into some other form of energy — most likely light, heat, or motion. For example, light bulbs turn electrical energy into light so you can stay up late at night reading this book. The heating elements in a toaster turn electrical energy into heat so you can burn your toast. And the motor in your vacuum cleaner turns electrical energy into motion that drives a pump that sucks the burnt toast crumbs out of your carpet.

In contrast, electronic devices do much more. Instead of just converting electrical energy into heat, light, or motion, electronic devices are designed to manipulate the electrical current itself to coax it into doing interesting and useful things.

That very first electronic device invented in 1883 by Thomas Edison manipulated the electric current passing through a light bulb in a way that let Edison create a device that could monitor the voltage being provided to an electrical circuit and automatically increase or decrease the voltage if it became too low or too high.

Don't worry if you aren't certain what the term *voltage* means at this point. You'll learn about voltage in the next chapter.

One of the most common things that electronic devices do is manipulate electric current in a way that adds meaningful information to the current. For example, audio electronic devices add sound information to an electric current so that you can listen to music or talk on a cellphone. And video devices add images to an electric current so you can watch great movies like *Office Space, Ferris Bueller's Day Off,* or *The Princess Bride* over and over again until you know every line by heart.

Keep in mind that the distinction between electric and electronic devices is a bit blurry. What used to be simple electrical devices now often include some electronic components in them. For example, your toaster may contain an electronic thermostat that attempts to keep the heat at just the right temperature to make perfect toast. (It will probably still burn your toast, but at least it tries not to.) And even the most complicated electronic devices have simple electrical components in them. For example, although your TV set's remote control is a pretty complicated little electronic device, it contains batteries, which are simple electrical devices.

What Can You Do with Electronics?

The amazing thing about electronics is that it's being used today to do things that weren't even imaginable just a few years ago. And of course, that means that in just a few years we'll have electronic devices that haven't even been thought up yet.

That being said, the following sections give a very brief overview of some of the basic things you can do with electronics.

Making noise

One of the most common applications for electronics is making noise. Often in the form of music, though the distinction between *noise* and *music* is often debatable. Electronic devices that make noise are often referred to as *audio devices*. These devices convert sound waves to electrical current, and then store, amplify, and otherwise manipulate the current, and eventually convert the current back to sound waves you can hear.

Most audio devices have these three parts:

✦ A *source*, which is the input into the system. The source can be a microphone, which is a device that converts sound waves into an electrical signal. The subtle fluctuations in the sound waves are translated into subtle fluctuations in the electrical signal. Thus, the electrical signal that comes from the source contains audio information.

   The source may also be a recorded form of the sound, such as sound recorded on a CD or in an MP3.

✦ An *amplifier*, which converts the small electrical signal that comes from the source into a much larger electrical signal that, when sent to the speaker or headphones, can be heard.

   Some amplifiers are small, as they need to boost the signal only enough to be heard by a single listener wearing headphones. Other amplifiers are huge, as they need to boost the signal enough so that 80,000 people can hear, for example, a famous singer forget the words to *The Star Spangled Banner*.

✦ *Speakers*, which convert electrical current into sound you can hear. Speakers may be huge, or they may be small enough to fit in your ear.

Making light

Another common use of electronics is to produce light. The simplest electronic light circuits are LEDs, which are the electronic equivalent of a light bulb.

LED stands for *light-emitting diode*, but that won't be on the test, at least not for this chapter. However, it will definitely be on the test for Book II, Chapter 5, where you learn how to work with LEDs.

Video electronic devices are designed to create not just simple points of light, but complete images that you can look at. The most obvious examples are television sets, which can provide hours and hours of entertainment and ask for so little in return — just a few of your brain cells.

Some types of electronic devices work with light that you can't see. The most common are TV remote controls, which send infrared light to your television set whenever you push a button. (That is, assuming you can find the remote control.) The electronics inside the remote control manipulate the infrared light in a way that sends information from the remote control to the TV, telling it to turn up the volume, change channels, or turn off the power.

Transmitting to the world

Radio refers to the transmission of information without wires. Originally, radio was used as a wireless form of telegraph, broadcasting nothing more than audible clicks. Next, radio was used to transmit sound. In fact, to this day the term *radio* is usually associated with audio-only transmissions, either in the form of music or the ever-popular talk radio. However, the transmission of video information — in other words, television — is also a form of radio, as are wireless networking, cordless phones, and cellphones.

You'll learn much more about radio electronics in Book V.

Computing

One of the most important applications of electronics in the last 50 years has been the development of computer technology. In just a few short decades, computers have gone from simple calculating machines to machines that can beat humans at *Jeopardy!*

Computers are the most advanced form of a whole field of electronics known as *digital electronics*, which is concerned with manipulating data in the binary language of zeros and ones. You'll learn plenty about digital electronics in Books VI and VII.

Looking inside Electronic Devices

Have you ever taken apart an electronic device that no longer works, like an old clock radio or VHS tape player, just to see what it looks like on the inside?

I just took some hefty server computers to an e-waste recycler. You can bet that before I did, I opened them up to see what they looked like on the inside. And I removed a couple of the more interesting pieces to keep on my shelf. (Call me weird if you want. Some people collect teacups, some people collect spoons from around the world, and some people collect shot glasses. I collect old computer CPUs.)

Inside most electronic devices, you'll find a *circuit board* (or circuit *card;* it's all the same), which is a flat, thin board that has electronic gizmos mounted on it. In most cases, one side of the circuit board is populated with tiny devices that look like little buildings. These are the components that make up the electric circuit: the resistors, capacitors, diodes, transistors, and integrated circuits that do the work that the circuit is destined to do. The other side is painted with little lines of silver or copper that look like streets. These are the conductors that connect all the components so that they can work together.

An electronic circuit board looks like a little city! For example, have a look at the typical circuit board pictured in Figure 1-3. The top of the card is shown on the left; it's peopled with a variety of common electronic components. The underbelly of the circuit card is shown on the right; it has the typical silver streaks of conductors that connect the components topside so that they can perform useful work.

Here's the essence of what's going on with these two sides of the circuit card:

✦ The *component* side of the card — the side with the little buildings — holds a collection of electronic components whose sole purpose in life is to bend, turn, and twist electric current to get it to do interesting and useful things. Some of those components restrict the flow of current, like speed bumps on a road. Others make the current stronger. Some work like One-Way street signs that allow current to flow in only one direction. Still others try to smooth out any ripples or variations in the current, resulting in smoother traffic flow.

✦ The *circuit* side of the card — the side with the roads — provides the conductive pathways for the electric current to flow from one component to the other in a certain order. The whole trick of designing and building electronic circuits is to connect all the components together in just the right way so that the current that flows out of one component is passed on to the next component. The circuit side of the board is what lets the components work together in a coordinated way.

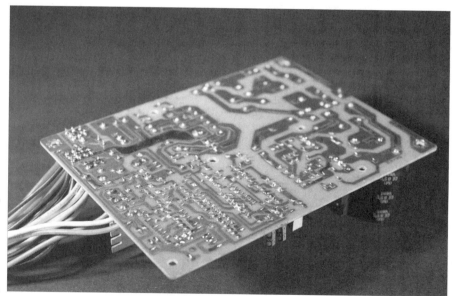

Figure 1-3:
A typical
electronic
circuit
board.

Okay, I couldn't even get through the first chapter of this book without having to give you the first of many warnings about the dangers of working with electronics. So here it goes: Do not under any circumstances plunge carelessly into the disassembly of old electronic circuits until you're certain you know what you're doing.

The little components on a circuit card such as the one shown in Figure 1-3 can be dangerous, even when they are unplugged. In fact, the two tall cylindrical components near the back edge of this circuit card are called *capacitors*. They can contain stored electrical energy that can deliver a powerful — even fatal — shock long after you've unplugged the power cord. Please refer to Chapter 4 of this minibook before you begin disassembling anything!

Chapter 2: Understanding Electricity

In This Chapter

✔ Unraveling the deep mysteries of the matter and energy (well, only a little bit)

✔ Learning about three important aspects of electric circuits: current, voltage, and power

✔ Looking at the difference between direct and alternating current

✔ Learning your first electrical equation (don't worry; it's simple)

*F*rankly, the title of this chapter is a bit ambitious. Before you can do much that's very interesting with electronics, you need to have a basic understanding of what electricity is and how it works, but unfortunately, understanding electricity is a tall order. Don't let this discourage or dissuade you: Even the smartest physicists in the world don't really understand it.

At the start of this chapter, you'll examine the very nature of electricity: what it is and what causes it. This first part of the chapter will remind you of a seventh- or eighth-grade science class, as you delve into the insides of atoms and learn about protons, neutrons, and electrons.

The second part of this chapter introduces you to three things you have to know about electricity if you want to design and build circuits: current, voltage, and power — the Manny, Moe, and Jack of electricity. Or if you prefer, the Huey, Dewey, and Louie, or the Groucho, Chico, and Harpo, or the — you get the idea.

Pondering the Wonder of Electricity

The exact nature of electricity is one of the core mysteries of the universe. Although we don't really know exactly what electricity is, we do know a lot about what it does and how it behaves.

Where did the word electricity come from?

Do you remember the movie *Jurassic Park*, where scientists discovered the DNA of dinosaurs locked inside bits of amber? Amber is fossilized tree resin, and it played a key role in the history of our knowledge about electricity.

Since the days of the ancient Greeks, people have known if you rubbed sticks of amber with fur, the amber could then be used to raise the hair on your head and that lightweight objects like feathers would stick to it. They had no idea why this happened, but they knew that it did happen.

The Greek word for amber is *elektron*. The Latin version of the word was *electricus*.

At the beginning of the seventeenth century, an English scientist named William Gilbert began to study electricity. He used these ancient words to describe the phenomena he was investigating, including the Latin term *electricus*. The influence of Gilbert's book, which was written in Latin, led to the word *electricity* in the English language.

Strange as it may sound, your understanding of electricity will improve right away if you avoid using the term *electricity* to describe it. That's because the term *electricity* isn't very precise. We use the word *electricity* to refer to any of several different but related things. Each has a more precise name, such as *electric charge, electric current, electric energy, electric field*, and so on. All of these things are commonly called *electricity*.

Electricity isn't so much a specific thing, but a phenomenon that has many different faces. So to avoid confusion, I'll try to avoid the word *electricity* in the rest of this book. Instead, I'll use more precise terms such as *charge* or *current*.

I really hate to use the word *phenomenon* here because it sounds so scientific. I feel like I should wear a bow tie whenever I say the word *phenomenon*. I consulted my thesaurus to see if there was a simpler word I could use instead. None of the suggestions really seemed to fit, but the one that came closest was *wonder*.

Wonder isn't a bad substitute. When it comes down to it, the phenomenon we call electricity is pretty amazing. It really does qualify as one of the great wonders of the universe.

Remember the so-called "Seven Wonders of the Ancient World," which included the Great Pyramid of Giza, the Hanging Gardens of Babylon, the Temple of Artemis at Ephesus, the Statue of Zeus at Olympia, the Mausoleum of Halicarnassus, the Colossus of Rhodes, and the Lighthouse of Alexandria? If we were to make a list called "The Seven Wonders of the Universe," I suppose it would have to include Matter, Gravity, Time, Light, Life, Pizza, and Electricity.

Looking for Electricity

One of the most amazing things about electricity is that it is, literally, every-where. By that I don't mean that electricity is commonplace or plentiful, or even that the universe has an abundant supply of electricity. Instead, what I mean is that electricity is a fundamental part of everything.

To get an idea of what I mean, consider a common misconception about electric current. Most of us think that wires carry electricity from place to place. When we plug in a vacuum cleaner and turn on the switch, we believe that electricity enters the vacuum cleaner's power cord at the electrical outlet, travels through the wire to the vacuum cleaner, and then turns the motor to make the vacuum cleaner suck up dirt and grime and dog hair. But that's not the case. The truth is that the electricity was already in the wire. The electricity is always in the wire, even when the vacuum cleaner is turned off or the power cord isn't plugged in. That's because electricity is a fun-damental part of the copper atoms that make up the wire inside the power cord. Electricity is also a fundamental part of the atoms that make up the rubber insulation that protects you from being electrocuted when you touch the power cord. And it's a fundamental part of the atoms that make up the tips of your finger which the rubber keeps from touching the wires.

In short, electricity is a fundamental part of the atoms that make up all matter. So, to understand what electricity is, we must first look at atoms.

Peering Inside of Atoms

As you probably learned in grade school, all matter is made up of unbeliev-ably tiny bits that are called *atoms*. They're so tiny that the period at the end of this sentence contains several trillion of them.

It's hard for us to comprehend numbers as large as trillions. For the sake of comparison, suppose you could enlarge the period at the end of this sen-tence until it was about the size of Texas. Then, each atom would be about the size of — you guessed it — the period at the end of this sentence.

The word *atom* comes from an ancient Greek fellow named Democritus. Contrary to what you might expect, the word *atom* doesn't mean "really small." Rather, it means "undividable." Atoms are the smallest part of matter that can't be divided without changing it to a different kind of matter. In other words, if you divide an atom of a particular element, the resulting pieces are no longer the same thing.

For example, suppose you have a handful of some basic element such as copper and you cut it in half. You now have two pieces of copper. Toss one of them aside, and cut the other one in half. Again, you have two pieces of copper. You can keep doing this, dividing your piece of copper into ever smaller halves. But eventually, you'll get to the point where your piece of copper consists of just a single copper atom.

If you try to cut that single atom of copper in half, the resulting pieces will not be copper. Instead, you'll have a collection of the basic particles that make up atoms. There are three such particles, called neutrons, protons, and electrons.

The neutrons and protons in each atom are clumped together in the middle of the atom, in what is called the *nucleus*. The electrons spin around the outside of the atom.

When I first learned about atoms as a kid, I was taught that the electrons orbit around the nucleus much like planets orbit around the sun in our solar system. Even today, kids are taught this. School children are still being taught to create models of atoms using Styrofoam balls and wires, like the one shown in Figure 2-1.

Figure 2-1:
A common model of an atom.

That turns out to be a really bad analogy. Instead, the electrons whiz around the nucleus in a cloud that's called, appropriately enough, the *electron cloud*. Electron clouds have weird shapes and properties, and strangely enough, it's next to impossible to figure out exactly where in its cloud an electron actually is at any given moment.

Examining the Elements

Several times in this chapter I use the term *element* without explaining it. So here's the deal: An element is a specific type of atom, defined by the number of protons in its nucleus. For example, hydrogen atoms have just one proton in the nucleus, an atom with two protons in the nucleus is helium, atoms with three protons are called lithium, and so on.

The number of protons in the nucleus of an atom is called the *atomic number*. Thus, the atomic number of hydrogen is 1, the atomic number of helium is 2, lithium is 3, and so on. Copper — an element that plays an important role in electronics — is atomic number 29. Thus, it has 29 protons in its nucleus.

What about neutrons, the other particle found in the nucleus of an atom? Neutrons are extremely important to chemists and physicists. But they don't really play that big of a role in the way electric current works, so we can safely ignore them in this chapter. Suffice it to say that in addition to protons, the nucleus of each atom (except hydrogen) contains neutrons — and in most cases, there are a few more neutrons than protons.

The third particle that makes up atoms is the electron. Electrons are what we're most interested in when we work with electricity because they are the source of electric current. They're unbelievably small; a single electron is about 200,000 times smaller than a proton. To gain some perspective on that, if a single electron were the size of the period at the end of this sentence, a proton would be about the size of a football field.

Atoms usually have the same number of electrons as protons, and thus an atom of the element copper has 29 protons in a nucleus that is orbited by 29 electrons. When an atom picks up an extra electron or finds itself short of an electron, things get interesting because of a special property of protons and electrons called *charge*, which I explain in the next section.

Minding Your Charges

Two of the three particles that make up atoms — electrons and protons — have a very interesting characteristic called *electric charge*. Charge can be one of two *polarities*: negative or positive. Electrons have a *negative* polarity, while protons have a *positive* polarity.

The most important thing to know about charge is that opposite charges attract and similar charges repel. Negative attracts positive and positive attracts negative, but negative repels negative and positive repels positive.

As a result, electrons and protons are attracted to each other, but electrons repel other electrons and protons repel other protons.

The attraction between protons and electrons is what holds the electrons and the protons of an atom together. This attraction causes the electrons to stay in their orbits around the protons in the nucleus.

Here are a few more enlightening details about charge:

✦ Charge is a property of one of the fundamental forces of nature known as *electromagnetism*. The other three forces are *gravity*, the *strong force*, and the *weak force*.

✦ As I say in the previous section, an atom normally has the same number of electrons as protons. This is because the electromagnetic force causes each proton to attract exactly one electron. When the number of protons and electrons is equal, the atom itself has no net charge. It is then said to be *neutral*.

   However, it's possible for an atom to pick up an extra electron. When it does, the atom has a net negative charge because of the extra electron. It's also possible for an atom to lose an electron, which causes the atom to have a net positive charge because it has more protons than electrons.

✦ If you've been paying attention, you may have wondered how it can be that the nucleus of an atom can stay together if it consists of two or more protons that have positive charges. After all, don't like charges repel? Yes they do, but the electrical repellent force is overcome by a much more powerful force called, for lack of a better term, the *strong force*. Thus, the strong force holds protons (and neutrons) together in spite of the protons' natural tendency to avoid each other.

✦ The strong force doesn't affect electrons, so you never see electrons clumped together the way protons do in the nucleus of an atom. The electrons in an atom stay well away from each other.

✦ If one were so inclined, one might liken the strong force to the patriotic force that binds the citizens of a nation together in spite of their differences. It's this force that keeps a country together in spite of the fact that its political parties seem to hate each other.

Conductors and Insulators

Some elements don't hold on to their outermost electrons very tightly. These elements frequently lose electrons or pick up extra electrons, and so they frequently get bumped off of neutral and become either negatively or positively charged. Such elements are called *conductors*. The best conductors are the metals silver, copper, and aluminum.

Other elements hold on to their electrons tightly. In these elements, it's hard to pry loose an electron or force another electron in. These elements almost always stay neutral. They're called *insulators*.

In a conductor, electrons are constantly skipping around between nearby atoms. An electron jumps out of one atom — call it Atom A — into a nearby atom, which I'll call Atom B. This creates a net positive charge in Atom A and a net negative charge in Atom B. But almost immediately, an electron will jump out of another nearby atom – call it Atom C — into Atom A. Thus, Atom A again becomes neutral, and now Atom C is negative.

This skipping around of electrons in a conductor happens constantly. Atoms are in perpetual turmoil, giving and receiving electrons and constantly cycling their net charges from positive to neutral to negative and back to positive.

Ordinarily, this movement of electrons is completely random. One electron might jump left, but another one jumps right. One goes up, another goes down. One goes east, the other goes west. The net effect is that although all of the electrons are moving, collectively they aren't going anywhere. They're like Keystone Kops, running around aimlessly in every direction, bumping into each other, falling down, picking themselves back up, and then running around some more. When this randomness stops and the Keystone Kops get organized, the result is electric current, as explained in the next section.

Understanding Current

Electric current is what happens when the random exchange of electrons that occurs constantly in a conductor becomes organized and begins to move in the same direction.

When current flows through a conductor such as a copper wire, all of those electrons that were previously moving about randomly get together and start moving in the same direction. A very interesting effect then happens: The electrons transfer their electromagnetic force through the wire almost instantaneously. The electrons themselves all move relatively slowly — on the order of a few millimeters a second. But as each electron leaves an atom and joins another atom, that second atom immediately loses an electron to a third atom, which immediately loses an electron to the fourth atom, and so on trillions upon trillions of times.

The result is that even though the individual electrons move slowly, the current itself moves at nearly the speed of light. Thus, when you flip a light switch, the light turns on immediately, no matter how much distance separates the light switch from the light.

Here are a few additional points that may help you understand the nature of current:

✦ One way to illustrate this principle is to line up 15 balls on a pool table in a perfectly straight line, as shown in Figure 2-2. If you hit the cue ball on one end of the line, the ball on the opposite end of the line will almost immediately move. The other balls will move a little, but not much (assuming you line them up straight and strike the cue ball straight).

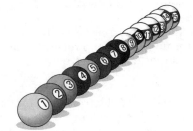

Figure 2-2:
Electrons transfer current through a wire much like a row of pool balls transfer motion.

This is similar to what happens with electric current. Although each electron moves slowly, the ripple effect as each atom loses and gains an electron is lightning fast (literally!).

✦ It's no coincidence that moving water is also called *current*. Many of the early scientists who explored the nature of electricity believed that electricity was a type of fluid, and that it flowed in wires in much the same way that water flows in a river.

✦ The strength of an electric current is measured with a unit called the *ampere*, sometimes used in the short form *amp* or abbreviated *A*. The ampere is nothing more than a measurement of how many charge carriers (in most cases, electrons) flow past a certain point in one second. One ampere is equal to 6,240,000,000,000,000,000 electrons per second. That's 6,240 quadrillion electrons per second.

✦ Most electric incandescent light bulbs have about one amp of current flowing through them when they are turned on. A hair dryer uses about twelve amps.

✦ Current in electronic circuits is usually much smaller than current in electrical devices like light bulbs and hair dryers. The current in an electronic circuit is often measured in thousandths of amps, or *milliamps* (abbreviated *mA*).

✦ Current is often represented by the letter *I* in electrical equations. The *I* stands for *intensity*.

Understanding Voltage

In its natural state, the electrons in a conductor such as copper freely move from atom to atom, but in a completely random way. To get them to move together in one direction, all you have to do is give them a push. The technical term for this push is *electromotive force*, abbreviated *EMF*, or sometimes simply *E*. But you know it more commonly as *voltage*.

A voltage is nothing more than a difference in charge between two places. For example, suppose you have a small clump of metal whose atoms have an abundance of negatively charged atoms and another clump of metal whose atoms have an abundance of positively charged atoms. In other words, the first clump has too many electrons and the second clump has too few. A voltage exists between those two clumps. If you connect those two clumps with a conductor such as a copper wire, you create what is called a *circuit* through which electric current will flow.

This current continues to flow until all the extra negative charges on the negative side of the circuit have moved to the positive side. When that has happened, both sides of the circuit become electrically neutral and the current stops flowing.

Here are some additional points to ponder concerning voltage:

✦ Whenever there's a difference in charge between two locations, there's a possibility that a current will flow between the two locations if those locations are connected by a conductor. Because of this possibility, the term *potential* is often used to describe voltage. Without voltage, there can be no current. Thus, voltage creates the potential for a current to flow.

✦ If current can be compared to the flow of water through a hose, voltage can be compared to water pressure at the faucet. It's water pressure that causes the water to flow in the hose.

✦ Voltage is measured using a unit called, naturally, the *volt*, usually abbreviated *V*. The voltage that's available in a standard electrical outlet in the United States is about 117 V. The voltage available in a flashlight battery is about 1.5 V. A car battery provides about 12 V.

✦ You can find out how much voltage exists between two points by using a device known as a *voltmeter*, which has two wire *test leads* that you can touch to different points in a circuit to measure the voltage between those points. Figure 2-3 shows a typical voltmeter. (Actually, the meter shown in the figure is a *multimeter*, which simply means that it can measure things other than voltage. In the figure, the multimeter is functioning as a voltmeter. For more information about using a voltmeter, refer to Chapter 8 of this minibook.)

✦ Voltages can be considered positive or negative, but only when compared with some reference point. For example, in a flashlight battery, the voltage at the positive terminal is +1.5 V relative to the negative terminal. The voltage at the negative terminal is –1.5 V relative to the positive terminal.

✦ I'd like to tell you the exact definition of a volt, but I can't — at least not yet. The definition of a volt won't make any sense until you learn about the concept of *power*, which is described later in this chapter in the section titled "Understanding Power."

✦ Although current stops flowing when the two sides of the circuit have been neutralized, the electrons in the circuit don't stop moving. Instead, they simply revert to their natural random movement. Electrons are always moving in a conductor. When they get a push from a voltage, they move in the same direction. When there's no voltage to push them along, they move about randomly.

✦ In electrical equations, voltage is usually represented by the letter *E*, which stands for *electromotive force*.

Are you positive about that?

For the first 150 years or so of serious research into the nature of electricity, scientists had electric current backward: They thought that electric current was the flow of positive charges and that electric current flowed from the positive side of a circuit to the negative side.

It wasn't until around 1900 that scientists began to unravel the structure of atoms. They soon figured out that electrons have a negative charge, and current is actually the flow of these negatively-charged electrons. In other words, they discovered that current flows in the opposite direction from what they had long thought.

Old ideas die hard, and to this day most people think of electric current as flowing from positive to negative. This concept of current flow is sometimes called *conventional current*. Modern electronic circuits are almost always described in terms of conventional current, so the assumption is that current flows from positive to negative, even though the reality is that the electrons in the circuit are actually flowing in the opposite direction.

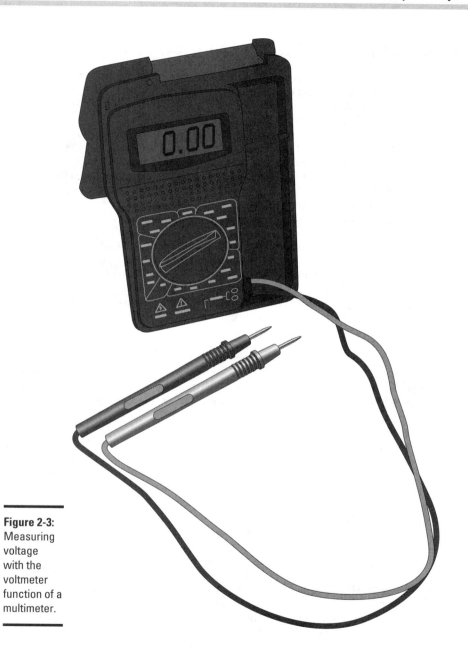

Figure 2-3:
Measuring
voltage
with the
voltmeter
function of a
multimeter.

Comparing Direct and Alternating Current

An electric current that flows continuously in a single direction is called a *direct current*, or *DC*. The electrons in a wire carrying direct current move slowly, but eventually they travel from one end of the wire to the other because they keep plodding along in the same direction.

The voltage in a direct-current circuit must be constant, or at least relatively constant, to keep the current flowing in a single direction. Thus, the voltage provided by a flashlight battery remains steady at about 1.5 V. The positive end of the battery is always positive relative to the negative end, and the negative end of the battery is always negative relative to the positive end. This constancy is what pushes the electrons in a single direction.

Another common type of current is called *alternating current*, abbreviated *AC*. In an alternating-current circuit, voltage periodically reverses itself. When the voltage reverses, so does the direction of the current flow. In the most common form of alternating current, used in most power distribution systems throughout the world, the voltage reverses itself either 50 or 60 times per second, depending on the country. In the United States, the voltage is reversed 60 times per second.

Alternating current is used in nearly all the world's power distribution systems, for the simple reason that AC current is much more efficient when it's transmitted through wires over long distances. All electric currents lose power when they flow for long distances, but AC circuits lose much less power than DC circuits.

The electrons in an AC circuit don't really move along with the current flow. Instead, they sort of sit and wiggle back and forth. They move one direction for 1/60th of a second, and then turn around and go the other direction for 1/60th of a second. The net effect is that they don't really go anywhere.

For your further enlightenment, here are some additional interesting and useful facts concerning alternating current:

✦ A popular toy called *Newton's Cradle* might help you understand how alternating current works. The toy consists of a series of metal balls hung by string from a frame, such that the balls are just touching each other in a straight line, as shown in Figure 2-4. If you pull the ball on one end of the line away from the other balls and then release it, that ball swings back to the line of balls, hits the one on the end, and instantly propels the ball on the other end of the line away from the group. This ball swings up for a bit, and then turns around and swings back down to strike the group from the other end, which then pushes the first ball away from the group. This alternating motion, back and forth, continues for an amazingly long time if the toy is carefully constructed.

Figure 2-4:
Newton's
Cradle
works
much like
alternating
current.

Alternating current works in much the same way. The electrons initially move in one direction, but then reverse themselves and move in the other direction. The back and forth movement of the electrons in the circuit continues as long as the voltage continues to reverse itself.

✦ If you want to see a Newton's Cradle in action, go to YouTube and search for *Newton's Cradle*.

✦ The reversal of voltage in a typical alternating current circuit isn't instantaneous. Instead, the voltage swings smoothly from one polarity to the other. Thus, the voltage in an AC circuit is always changing. It starts out at zero, then increases in the positive direction for a bit until it reaches its maximum positive voltage, and then it decreases until it gets back to zero. At that point, it increases in the negative direction until it reaches its maximum negative voltage, at which time it decreases again until it gets back to zero. Then the whole cycle repeats itself.

✦ The fact that the amount of voltage in an AC circuit is always changing turns out to be incredibly useful. You'll learn why in Book IV, Chapter I when I give you a deeper look at alternating current.

Understanding Power

At the start of this chapter, I mention the three key concepts you need to know about electricity before you can start to work with your own circuits. The first two — current and voltage — are described earlier in this chapter. To recap, *current* is the organized flow of electric charges through a conductor, and *voltage* is the driving force that pushes electric charges to create current.

The third piece of the puzzle is called *power* (abbreviated P in equations). Simply put, power is the work done by an electric circuit. Electric current, in and of itself, isn't all that useful. It becomes useful only when the energy carried by an electric current is converted into some other form of energy, such as heat, light, sound, or radio waves. For example, in an incandescent light bulb, voltage pushes current through a *filament*, which converts the energy carried by the current into heat and light.

Power is measured in units called *watts* (abbreviated *W*). The definition of one watt is simple: One watt is the amount of work done by a circuit in which one ampere of current is driven by one volt.

This relationship lends itself to a simple equation. I promised myself when I started this book that I would use as few equations as possible, but I knew I'd have to include at least some of the basic equations. Fortunately, this one is pretty simple:

$$P = E \times I$$

In other words, *power* (P) equals *voltage* (E) times *current* (I).

To use the equation correctly, you must make sure that you measure power, voltage, and current using their standard units: watts, volts, and amperes. For example, suppose you have a light bulb connected to a 10-volt power supply, and one-tenth of an ampere is flowing through the light bulb. To calculate the wattage of the light bulb, you use the P = E × I formula like this:

$$P = 10 \text{ V} \times 0.1 \text{ A} = 1 \text{ W}$$

Thus, the light bulb is doing 1 watt of work.

Often, you know the voltage and the wattage of the circuit and you want to use those values to determine the amount of current flowing through the circuit. You can do that by turning the equation around, like this:

$$I = \frac{P}{E}$$

For example, if you want to determine how much current flows through a lamp with a 100-watt light bulb when it's plugged into a 117-volt electrical outlet, use the formula like this:

$$I = \frac{100 \ W}{117 \ V} = 0.855 \ A$$

Thus, the current through the circuit is 0.855 amperes.

Here are some final thoughts concerning the concept of power:

+ The term *dissipate* is often used in association with power. As the energy carried by an electric current is converted into another form such as heat or light, the circuit is said to dissipate power.

+ Did you notice that current and voltage are represented by the letters I and E, not the letters C or V as you might expect, but power is represented by the letter P? Sometimes you wonder if the people who make the rules are just trying to confuse everyone.

Maybe the following table will help you keep things sorted out:

Concept	Abbreviation	Unit
Current	I	amp (A)
Voltage	E or EMF	volt (V)
Power	P	watt (W)

+ Earlier in this chapter, in the section "Understanding Voltage," I said that I can't define "one volt" until you know what power is. Now that you know, you can see that the definition of a volt is simple: One volt is the amount of electromotive force (EMF) necessary to do one watt of work at one ampere of current.

+ The P = E × I formula is sometimes called *Joule's Law*, named after the person who discovered it. This little factoid will *not* be on the test.

+ Calculating the power dissipated by a circuit is often a very important part of circuit design. That's because electrical components such as resistors, transistors, capacitors, and integrated circuits all have maximum power ratings. For example, the most common type of resistor can dissipate at most 1/4 watt. If you use a 1/4-watt resistor in a circuit that dissipates more than 1/4 watt of power, you run the risk of burning up the resistor.

Chapter 3: Creating Your Mad-Scientist Lab

In This Chapter

✓ Finding a place where you can build your mad-scientist laboratory

✓ Investing in good tools

✓ Picking out a good assortment of components to get you started

I loved to watch Frankenstein movies as a kid. My favorite scenes were always the ones where Dr. Frankenstein went into his laboratory. Those laboratories were filled with the most amazing and exotic electrical gadgets ever seen. The mad doctor's assistant, Igor, would throw a giant knife switch at just the right moment, and sparks flew, and the music rose to a crescendo, and the creature jerked to life, and the crazy doctor yelled, "It's ALIVE!"

The best Frankenstein movie ever made is still the original 1931 *Frankenstein,* directed by James Whale and starring Boris Karloff. The second-best Frankenstein movie ever made is the 1974 *Young Frankenstein,* directed by Mel Brooks and starring Gene Wilder. Both have great laboratory scenes.

In fact, did you know that the laboratory in *Young Frankenstein* uses the very same props that were used in the original 1931 classic? The genius who created those props was Kenneth Strickfaden, one of the pioneers of Hollywood special effects. Strickfaden kept the original Frankenstein props in his garage for decades. When Mel Brooks asked if he could borrow the props for *Young Frankenstein*, Strickfaden was happy to oblige.

You don't need an elaborate mad-scientist laboratory like the ones in the Frankenstein movies to build basic electronic circuits. However, you will need to build yourself a more modest workplace, and you'll need to equip it with a basic set of tools as well as some basic electronic components to work with.

However, no matter how modest your work area is, you can still call it your mad-scientist lab. After all, most of your friends will think you're a bit crazy and a bit of a genius when you start building your own electronic gadgets.

In this chapter, I introduce you to the stuff you need to acquire before you can start building electronic circuits. You don't have to buy everything all at once, of course. You can get started with just a simple collection of tools and a small space to work in. As you get more advanced in your electronic skills, you can acquire additional tools and equipment as your needs change.

Setting Up Your Mad-Scientist Lab

First, you must create a good place to work. You can build a fancy workbench in your garage or in a spare room, but if you don't have that much space, you can set up an ad-hoc mad-scientist lab just about anywhere. All you need is a place to set up a small workbench and a chair.

I do most of my electronics work in a spare room in my home, which also doubles as a display area for some of the Halloween props I've built over the years for my haunted house. Thus, as Figure 3-1 shows, my Mad-Scientist Lab really is a mad-scientist lab!

Figure 3-1:
My Mad-
Scientist
Lab really
is a mad-
scientist
lab!

Here are the essential ingredients of any good work area for electronic tinkering:

✦ **Adequate space:** You'll need to have adequate space for your work. When you're just getting started, your work area can be small — maybe just two or three feet in the corner of the garage. But as your skills progress, you'll need more space.

It's very important that the location you chose for your work area is secure, especially if you have young children around. Your work area will be filled with perils — things that can cause shocks, burns, and cuts, as well as things that under no circumstances should be ingested. Little hands are incredibly curious, and children are prone to put anything they don't recognize in their mouths. So be sure to keep everything safely out of reach, ideally behind a locked door.

✦ **Good lighting:** The ideal lighting should be overhead instead of from the side or behind you. If possible, purchase an inexpensive fluorescent shop light and hang it directly over your work area. If your chosen spot doesn't allow you to hang lights from overhead, the next best bet is a desk lamp that swings overhead, to bring light directly over your work.

✦ **A solid workbench:** Initially, you can get by with something as simple as a card table or a small folding table. Eventually, though, you'll want something more permanent and substantial. You can make yourself an excellent workbench from an old door set atop a pair of old file cabinets, or you can hit the yard sales on a Saturday morning in search of an inexpensive but sturdy office desk.

If your only option for your workbench is your kitchen table, go to your local big box hardware store and buy a 24-inch square piece of 5⁄8-inch plywood. This will serve as a good solid work surface until you can acquire a real workbench.

✦ **Comfortable seating:** If your workbench is a folding table or desk, the best seating is a good office chair. However, many workbenches stand four to six inches taller than desk height. This allows you to work comfortably while standing. If your workbench is tall, you'll need to get a seat of the correct height. You can purchase a bench stool from a hardware store, or you can shop the yard sales for a cheap bar stool.

✦ **Plenty of electricity:** You will obviously need a source of electricity nearby as you build electronic projects. A standard 15-amp electrical outlet will provide enough current capacity, but it probably won't provide enough electrical outlets for your needs. The easiest way to meet that need is to purchase several multi-outlet power strips and place them in convenient locations behind or on either side of your work area.

✦ **Plenty of storage:** You'll need a place to store your tools, supplies, and components. The ideal storage for hand tools is a small sheet of pegboard mounted on the wall right behind your workbench. Then, you can use hooks to hang your tools within easy reach. For larger tools, such as a drill or saw, built-in cabinets are best.

For small parts, multicompartment storage boxes such as the ones shown in Figure 3-2 are best. I suggest you get one or two of these to store all the little components such as resistors, diodes, capacitors, transistors, and so on. If you get two boxes, get one that has a few larger compartments and another that has a greater number of smaller compartments.

Figure 3-2:
Multicompartment storage boxes are ideal for storing small components.

It's also a good idea to keep a few small, shallow storage bins handy. These are especially useful for storing parts for the project you're working on. It helps to keep your parts together in a shallow bin rather than having them scattered lose all over your work area.

Equipping Your Mad-Scientist Lab

Like any hobby, electronics has its own special tools and supplies. Fortunately, you don't need to run out and buy everything all at once. But the more involved you get with the hobby, the more you will want to invest in a wide variety of quality tools and supplies. The following sections outline some of the essential stuff you'll need at your disposal.

Basic hand tools

For starters, you'll need a basic set of hand tools, similar to the assortment shown in Figure 3-3. Specifically, you'll need these items:

✦ **Screwdrivers:** Most electronic work is relatively small, so you don't need huge, heavy-duty screwdrivers. But you should get yourself a good assortment of small- and medium-sized screwdrivers, both flat-blade and Phillips head.

A set of *jeweler's screwdrivers* is sometimes very useful. The swiveling knob on the top of each one makes it easy to hold the screwdriver in a precise position while turning the blade.

Figure 3-3:
Basic hand
tools you'll
want to
have.

+ **Pliers:** You will occasionally use standard flat-nosed pliers, but for most electronic work, you'll depend on *needle-nose pliers* instead, which are especially adept at working with wires — bending and twisting them, pushing them through holes, and so on. Most needle-nose pliers also have a cutting edge that lets you use them as wire cutters.

  Get a small set of needle-nose pliers with thin jaws for working with small parts, and a larger set for bigger jobs.

+ **Wire cutters:** Although you can use needle-nose pliers to cut wire, you'll want a few different types of wire cutters at your disposal as well. Get something heavy-duty for cutting thick wire, and something smaller for cutting small wire or component leads.

+ **Wire strippers:** Figure 3-4 shows two pieces of wire that I *stripped* (removed the insulation from). I stripped the one on top with a set of wire cutters and the one on the bottom with a set of *wire strippers*. Notice the crimping in the one at the top, at the spot where the insulation ends? That was caused by using just a bit too much pressure on the wire cutters. That crimp has created a weak spot in the wire that may eventually break.

  To avoid damaging your wires when you strip them, I suggest you purchase an inexpensive (under $10) wire stripping tool. You'll thank me later.

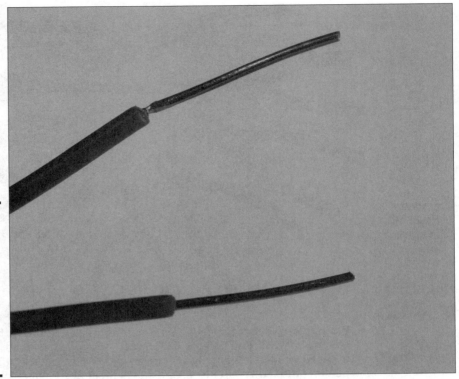

Figure 3-4:
The wire on the top was stripped with wire cutters; the one on the bottom was stripped with a wire stripper.

Magnifying glasses

One of the most helpful items you can have in your tool arsenal is a good magnifying glass. After all, electronic stuff is small. Resistors, diodes, and transistors are downright tiny.

Actually, I suggest you have at least three types of magnifying glasses on hand:

✦ A **handheld magnifying glass** to inspect solder joints, read the labels on small components, and so on.

✦ A **magnifying glass mounted on a base** so that you can hold your work behind the glass. The best mounted glasses combine a light with the magnifying glass, so the object you're magnifying is bright.

✦ **Magnifying goggles,** which provide completely hands-free magnifying for delicate work. Ideally, the goggles should have lights mounted on them. See Figure 3-5.

You get what you pay for

When it comes to tools, the old mantra "you get what you pay for" is generally true. Good tools that are manufactured from the best materials and with the best quality fetch a premium price. Cheap tools are, well, cheap. The price range can be substantial. You can easily spend $20 or $25 on decent wire cutters, or you can buy cheap ones for $3 or $4.

There are two main drawbacks to cheap tools. First, they don't last. The business end of a cheap tool wears out very fast. Each time you cut a wire with a pair of $4 wire cutters, you ding the cutting blade a bit. Pretty soon, the cutters can barely cut through the wire. The second drawback of cheap tools is a consequence of the first: When tools wear out, they tend to damage the materials you use them on. For example, tightening a screw with a badly worn screwdriver can strip the screw. Likewise, attempting to loosen a tight nut with a worn-out wrench can strip the nut.

There are a few situations in which I would endorse spending money on cheap tools. One is as a way of getting started in this fascinating hobby as inexpensively as possible. You can always start with cheap tools, and then replace them one by one with more expensive tools as your experience, confidence, love of the hobby, and budget increases. Another good reason to buy cheap tools is if you're absentminded (like me) and tend to lose things. There's not much point in buying expensive tools if you're going to have to replace them every few months because you keep losing them!

Third hands and hobby vises

A *third hand* is a common tool amongst hobbyists. It's a small stand that has a couple of clips that you use to hold your work, thus freeing up your hands to do delicate work. Most third-hand tools also include a magnifying glass. Figure 3-6 shows an inexpensive third-hand tool holding a circuit card.

The most common use for a third hand in electronics is soldering. You use the clips to hold the parts you want to solder, positioned behind the magnifying glass so you can get a good look.

Although the magnifying glass on the third hand is helpful, it does tend to get in the way of the work. It can be awkward to maneuver your soldering iron and solder behind the magnifying glass. For this reason, I often remove the magnifying glass from the third hand and use my favorite magnifying headgear instead.

The third hand is often helpful for assembling small projects, but it lacks the sturdiness required for larger projects. Eventually you'll want to invest in a small hobby vise such as the one shown in Figure 3-7. This one is made by PanaVise (www.panavise.com).

Figure 3-5:
The author
modeling
his favorite
magnifying
headgear.

Figure 3-6:
A third hand
can hold
your stuff so
both your
hands are
free to do
the work.

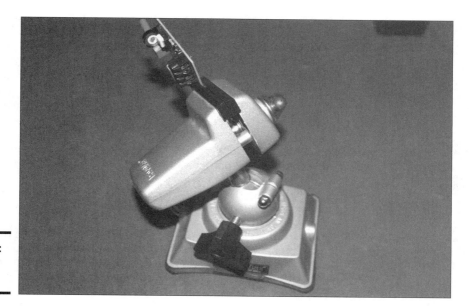

Figure 3-7:
A hobby
vise.

Here are a few things to look for in a hobby vise:

✦ **Mount:** Get a vise that has a base with the proper type of workbench mount. There are three common types of mounts:

- *Bolt mount:* The base has holes through which you can pass bolts or screws to attach the vise to your workbench. This is the most stable type of mount, but it requires that you put holes in your workbench.

- *Clamp mount:* The base has a clamp that you can tighten to fix the base to the top and bottom of your workbench. Clamp mounts are pretty stable but can be placed only near the edge of your workbench.

- *Vacuum mount:* The base has a rubber seal and lever you pull to create a vacuum between the seal and the workbench top. Vacuum mounts are the most portable but work well only when the top of your workbench is smooth.

✦ **Movement:** Get a vise that has plenty of movement so that you can swivel your work into a variety of different working positions. Make sure that when you lock the swivel mount into position, it stays put. You don't want your work sliding around while you are trying to solder on it.

✦ **Protection:** Make sure the vise jaws have a rubber coating to protect your work.

Soldering iron

Soldering is one of the basic techniques used to assemble electronic circuits. The purpose of soldering is to make a permanent connection between two conductors — usually between two wires or between a wire and a conducting surface on a printed circuit board.

The basic technique of soldering is to physically connect the two pieces to be soldered, and then heat them with a soldering iron until they are hot enough to melt *solder* (a special metal made of lead and tin that has a low melting point), then apply the solder to the heated parts so that it melts and flows over the parts.

Once the solder has flowed over the two conductors, you remove the soldering iron. As the solder cools, it hardens and bonds the two conductors together.

You'll learn all about soldering in Chapter 6 of this minibook. For now, suffice it to say that you need three things for successful soldering:

✦ **Soldering iron:** A little hand-held tool that heats up enough to melt solder. An inexpensive soldering iron from RadioShack or another electronics parts store is just fine to get started with. As you get more involved with electronics, you'll want to invest in a better soldering iron that has more precise temperature control and is internally grounded.

✦ **Solder:** The soft metal that melts to form a bond between the conductors.

✦ **Soldering iron stand:** To set your soldering iron on when you aren't soldering. Some soldering irons come with stands, but the cheapest ones don't. Figure 3-8 shows a soldering iron that comes with a stand. You can purchase this type of soldering iron from RadioShack for about $25.

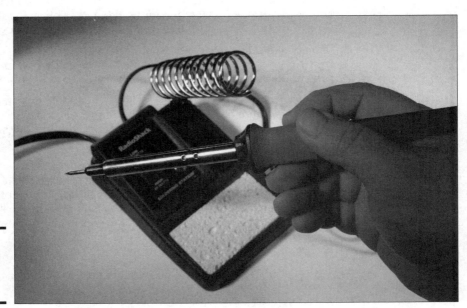

Figure 3-8:
A soldering
iron with a
stand.

Multimeter

In Chapter 2 of this minibook, you learn that you can measure voltage with a voltmeter. You can also use meters to measure many other quantities that are important in electronics. Besides voltage, the two most common measurements you'll need to make are current and resistance.

Rather than use three different meters to take these measurements, it's common to use a single instrument called a *multimeter*. Figure 3-9 shows a typical multimeter purchased from RadioShack for about $20.

Solderless breadboard

A *solderless breadboard* — usually just called a *breadboard* — is a must for experimenting with circuit layouts. A breadboard is a board that has holes in which you can insert either wires or electronic components such as resistors, capacitors, transistors, and so on to create a complete electronic circuit without any soldering. When you're finished with the circuit, you can take it apart, and then reuse the breadboard and the wires and components to create a completely different circuit.

Figure 3-10 shows a typical breadboard, this one purchased from RadioShack for about $20. You can purchase less expensive breadboards that are smaller, but this one (a little bigger than 7 x 4 inches) is large enough for all the circuits presented in this book.

What makes breadboards so useful is that the holes in the board are actually solderless connectors that are internally connected to one another in a specific, well-understood pattern. Once you get the hang of working with a breadboard, you'll have no trouble understanding how it works.

Throughout the course of this book, I show you how to create dozens of different circuits on a breadboard. As a result, you'll want to invest in at least one. I suggest you get one similar to the one shown in Figure 3-10, plus one or two other, smaller breadboards. That way, you won't always have to take one circuit apart to build another.

You can learn more about working with solderless breadboards in Chapter 6 of this minibook.

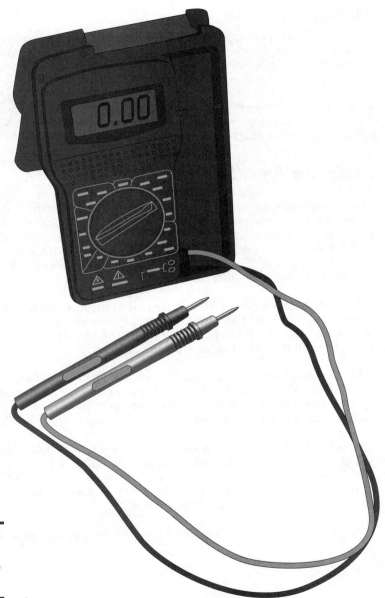

Figure 3-9:
An
inexpensive
multimeter.

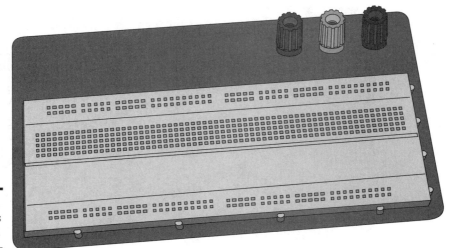

Figure 3-10:
A solderless
breadboard.

Wire

One of the most important items to have on hand in your lab is *wire,* which is simply a length of a conductor, usually made out of copper but sometimes made of aluminum or some other metal. The conductor is usually covered with an outer layer of insulation. In most wire, the insulation is made of polyethylene, which is the same stuff used to make plastic bags.

Wire comes in these two basic types:

✦ **Solid wire:** Made from a single piece of metal

✦ **Stranded wire:** Made of a bunch of smaller wires woven together.

Figure 3-11 shows both types of wire with the insulation stripped back so you can see the difference.

For most purposes in this book, you'll want to work with solid wire because it's easier to insert into breadboard holes and other types of terminal connections. Solid wire is also easier to solder. When you try to solder stranded wire, inevitably one of the tiny strands gets separated from the rest of the strands, which can create the potential for a short circuit.

On the other hand, stranded wire is more flexible than solid wire. If you bend a solid wire enough times, you'll eventually break it. For this reason, wires that are frequently moved are usually stranded.

Figure 3-11:
Solid and
stranded
wire.

Wire comes in a variety of sizes, which are specified by the wire's *gauge,* and is generally coiled in or on the packaging. Strangely, the larger the gauge number, the smaller the wire. For most electronics projects, you'll want 20- or 22-gauge wire. You'll need to use large wires (usually 14 or 16 gauge) when working with household electrical power.

Finally, you may have noticed that the insulation around a wire comes in different colors. The color doesn't have any effect on how the wire performs, but it's common to use different colors to indicate the purpose of the wire. For example, in DC circuits it's common to use red wire for positive voltage connections and black wire for negative connections.

To get started, I suggest you purchase a variety of wires — at least four rolls: 20-gauge solid, 20-gauge stranded, 22-gauge solid, and 22-gauge stranded. If you can find an assortment of colors, all the better.

In addition to wires on rolls, you may also want to pick up *jumper wires,* which are precut, stripped, and bent for use with solderless breadboards. Figure 3-12 shows an assortment I bought at RadioShack for about $6.

Batteries

Don't forget the batteries! Most of the circuits covered in this book use either AA or 9-volt batteries, so you'll want to stock up.

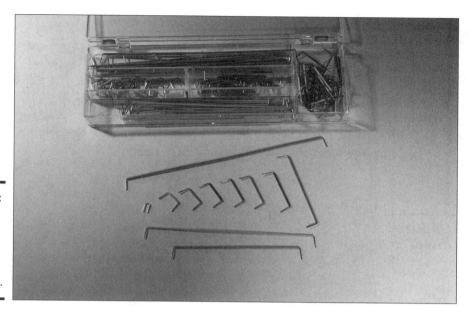

Figure 3-12:
Jumper wires for working with a solderless breadboard.

If you want, you can use rechargeable batteries. They cost more initially, but you don't have to replace them when they lose their charge. If you use rechargeables, you'll also need a battery charger.

To connect the batteries to the circuits, you'll want to get several AA battery holders. Get one that holds two batteries and another that holds four. You should also get a couple of 9-volt battery clips. These holders and clips are pictured in Figure 3-13.

Other things to stock up on

Besides all the stuff I've listed so far, here are a few other items you may need from time to time:

+ **Electrical tape:** Get a roll or two of plain black electrician's tape. You'll use it mostly to wrap around temporary connections to hold them together and keep them from shorting out.

+ **Compressed air:** A small can of compressed air can come in handy to blow dust off an old circuit board or component.

+ **Cable ties:** Also called *zip ties*, these little plastic ties are handy for temporarily (or permanently) holding wires and other things together.

+ **Jumper clips:** These are short (typically 12 or 18 inches) wires that have *alligator clips* attached on either end, as shown in Figure 3-14. You'll use them to make quick connections between components for testing purposes.

9 V battery clip

Figure 3-13:
Battery
holders will
help deliver
power
to your
circuits.

AA battery holder

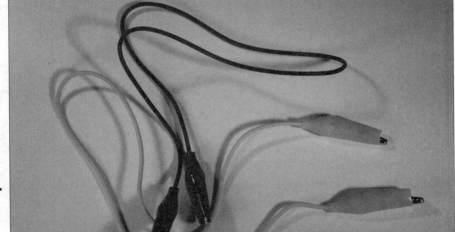

Figure 3-14:
Jumper
clips are
great for
making
quick
connections.

Stocking up on Basic Electronic Components

Besides all of the tools and supplies I've described so far in this chapter, you'll also need to gather a collection of inexpensive electronic components to get you started with your circuits. You don't have to buy everything all at once, but you'll want to gather at least the basic parts before you go much farther in this book.

You can buy many of these components in person at any RadioShack store. If you're lucky enough to have a specialty electronics store in your community, you may be able to purchase the parts there for less than what RadioShack charges. Alternatively, you can buy the parts online from www. radioshack.com or another electronic parts distributor.

Resistors

A *resistor* is a component that resists the flow of current. It's one of the most basic components used in electronic circuits; in fact, you won't find a single circuit anywhere in this book that doesn't have at least one resistor. Figure 3-15 shows three resistors, next to a penny so you can get an idea of how small they are. You'll learn all about resistors in Book II, Chapter 2.

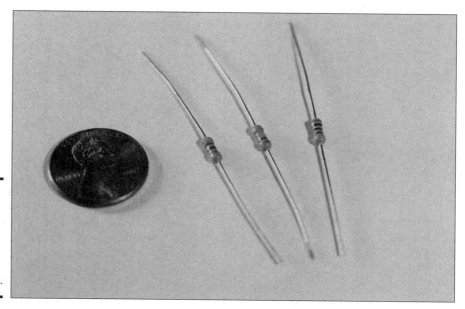

Figure 3-15: Resistors are one of the most commonly used circuit components.

Resistors come in a variety of *resistance values* (how much they resist current, measured in units called ohms and designated by the symbol Ω) and *power ratings* (how much power they can handle without burning up, measured in watts).

All the circuits in this book can use resistors rated for one-half watt. You'll need a wide variety of resistance values. I recommend you buy at least 10 each of the following 12 resistances:

470 Ω	4.7 kΩ	47 kΩ	470 kΩ
1 kΩ	10 kΩ	100 kΩ	1 MΩ
2.2 kΩ	22 kΩ	33 kΩ	220 kΩ

You can save money by purchasing a package that contains a large assortment of resistors. For example, RadioShack sells a package that contains an assortment of 500 resistors — at least 10 of all values listed here, plus a few others, for under $15.

Capacitors

Next to resistors, capacitors are probably the second most commonly used component in electronic circuits. A *capacitor* is a device that can temporarily store an electric charge. You'll learn all about capacitors in Book II, Chapter 3. Figure 3-16 shows some capacitors.

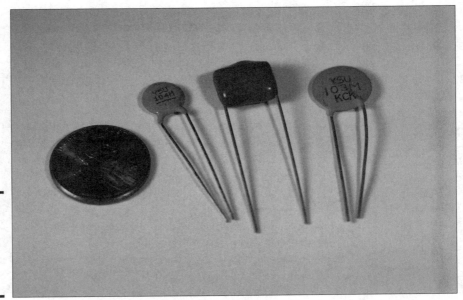

Figure 3-16: Capacitors come in many shapes and sizes.

Capacitors come in several different varieties, the two most common being *ceramic disk* and *electrolytic*. The amount of capacitance of a given capacitor is usually measured in *microfarads*, abbreviated µF. As a starting assortment of capacitors, I suggest you get at least five each of the following capacitors:

+ **Ceramic disk:** 0.01 µF and 0.1 µF.

+ **Electrolytic:** 1 µF, 10 µF, 100 µF, 220 µF, and 470 µF.

Diodes

A *diode* is a device that lets current flow in only one direction. A diode has two terminals, called the *anode* and the *cathode*. Current will flow through the diode only when positive voltage is applied to the anode and negative voltage to the cathode. If these voltages are reversed, current will not flow.

You'll learn all about diodes in Book II, Chapter 5. For now, I suggest you get at least five of the basic diodes known as the 1N4001, as shown in Figure 3-17. You should be able to find these at any RadioShack.

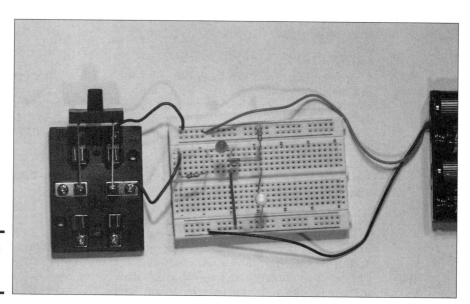

Figure 3-17:
1N4001
diodes.

Light-Emitting Diodes

A *light-emitting diode* (or *LED*) is a special type of diode that emits light when current passes through it. You'll learn about LEDs in Book II, Chapter 5. Although there are many different types of LEDs available, I suggest you get started by purchasing five red diodes. Figure 3-18 shows a typical red LED.

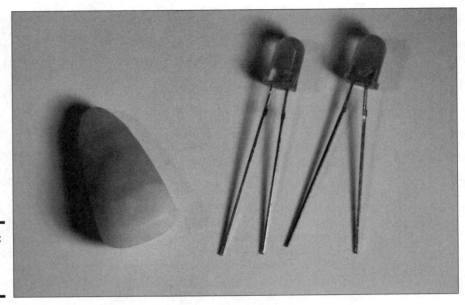

Figure 3-18:
Light-
emitting
diodes.

Transistors

A *transistor* is a three-terminal device in which a voltage applied to one of the terminals (called the *base*) can control current that flows across the other two terminals (called the *collector* and the *emitter*). The transistor is one of the most important devices in electronics, and I cover it in detail in Book II, Chapter 6. For now, you can just get a few simple 2N3904 NPN transistors, shown in Figure 3-19, to have on hand.

Don't worry; by the time you finish Book II, Chapter 6, you'll know what the designation *2N3904 NPN* means.

Integrated Circuits

An *integrated circuit* is a special component that contains an entire electronic circuit, complete with transistors, diodes, and other elements, all photo-graphically etched onto a tiny piece of silicon. Integrated circuits are the building blocks of modern electronic devices such as computers and cellphones.

You'll learn how to work with some basic integrated circuits in Book III. To get started, you'll want to pick up a few each of at least two different types of integrated circuits: a 555 timer and an LM741 op-amp. These chips are depicted in Figure 3-20.

Figure 3-19:
A look at a
2N3904 NPN
transistor.

Figure 3-20:
Two popular
integrated
circuits: A
555 timer
and an
LM741
op-amp.

One Last Thing

There is, of course, one last thing your mad-scientist lab will need to make it complete. That one last thing is a sign that properly warns your friends and family that you are indeed a mad scientist. You have my express permission to photocopy the sign shown in Figure 3-21 and place it in a prominent spot near your workbench.

Figure 3-21: Make sure your friends and family are properly warned.

Chapter 4: Staying Safe

*W*hen I was a kid, I helped a good friend named Barry who built a Tesla coil. By "helped," I mean that I hung out in his garage and watched while he meticulously wrapped thousands of turns of bare copper wire around a huge glass milk bottle, painted it with dozens of coats of lacquer, and polished the brass ball that attached to the very top of the coil. I'm quite certain he couldn't have done it without me.

When it was done, we plugged it in and marveled at what it could do. Sparks flew at random a foot or two into the air from the ball at the top of the coil. If you held a crowbar in one hand, you could draw a spark several feet from the ball to the crowbar. The current coming off the ball flew through the air and into the crowbar, and then passed through our bodies and into the ground. You could also light up a fluorescent light tube simply by holding the tube in your hand within a few feet of the coil.

To this day, I cannot believe Barry's parents let him build it. I know my parents wouldn't have let me build one. My mom was kind of like the mom in *A Christmas Story*, who wouldn't let her son Ralphie have a Red Ryder BB gun (the one with a compass in the stock and this thing that tells time) because "you'll shoot your eye out."

That was my mom. No Tesla coils for me. Too dangerous.

None of the electronic projects described in this book are anywhere near as dangerous as a Tesla coil. In fact, most of them pose no threat at all. That being said, it's important to remember that whenever you're working with electricity, you're working with something that's potentially very dangerous.

The possibility of electric shock is always present whenever you work with electricity, but there are other potential dangers as well. You probably won't shoot your eye out, but if you're not careful, you might start a fire or otherwise injure yourself or someone else.

The purpose of this chapter, then, is to keep you safe while you experiment with electronics. Please read it well, and please heed every bit of advice I give here.

Facing the Realities of Electrical Dangers

There's no escaping the simple fact that an electric shock, if strong enough, can kill you. So whenever you work with electricity, you must be sure to take every precaution you can to avoid being the recipient of a shock strong enough to do damage.

In the United States, somewhere between 500 and 1,000 people die every year from accidental electrocutions. Many of those are industrial or weather-related accidents in which people come into contact with downed power lines. But many of them are completely avoidable accidents that happen in the home. In the sections that follow, I give you specific guidelines for avoiding accidental electrocution.

Household electrical current can kill you!

Too many people are under the false impression that the 120 volts of alternating current running through household electrical wires isn't enough to kill. So let's start by getting one fact straight:

The electricity in your home wiring system is more than strong enough to kill you.

You're exposed to household electrical current primarily in two places: in electrical outlets and in the lamp sockets within light fixtures. As a result, you should be extra careful whenever you plug or unplug something into or from an electrical outlet, and you should be careful whenever you change a light bulb. Specifically, you should follow these precautions:

✦ *Never change a light bulb when the light is turned on*. If the light is controlled by a switch, turn the switch off. If the light isn't controlled by a switch, unplug the light from the wall outlet.

✦ *If an extension cord becomes frayed or damaged in any way, discard it.* When the insulation begins to rub off of an extension cord, the shock hazard is very real.

✦ *Never perform electrical wiring work while the circuit is energized.* If you insist on changing your own light switches or electrical outlets, *always* turn off the power to the circuit by turning off the circuit breaker that controls the circuit before you begin. Many people die every year because they think they can be careful enough to safely work with live power.

Is it true that current, not voltage, kills?

There is an old adage that "it's the current that kills, not the voltage." Although this statement may be technically true, it's also dangerously misleading. In fact, it stems from a fundamental misunderstanding of what current and voltage are. It can cause you to take dangerous risks if you don't understand the relationship between current and voltage.

The danger from electric shock occurs as current passes through vital parts of your body — specifically, your heart. It takes only a few milliamperes of current to stop your heart. At somewhere around 10 mA, muscles seize up, making it impossible to let go if you're holding a live wire. At around 15 mA, the muscles in your chest can seize up, making it impossible to breath. And at around 60 mA, your heart can stop. It takes only a few moments of exposure for these effects to occur.

So yes, it is current passing through your body that can kill you.

But current is inseparable from voltage. Current can't happen without voltage, and all other things being equal, the greater the voltage, the greater the current. As a result, it's very difficult to receive a lethal shock from three volts even if you're dripping wet and standing on bare concrete. But under those conditions, 30 volts may be enough to create a painful and damaging shock.

Saying "it's the current that kills, not the voltage," is kind of like saying "it's lack of oxygen, not water," that causes drowning. Although it may be technically true, isn't it the water that causes the lack of oxygen?

+ *Never work on an AC-powered appliance when it has power applied.* Simply turning the appliance off isn't enough to be safe. If the appliance has a power cord, unplug it before you work on it. If it doesn't have a power cord, turn off the power to the appliance by throwing the circuit breaker on your home's electrical panel.

+ *Take extra precautions when you're working with your own AC circuits.* In Book IV, I tell you more about working with AC circuits, I say more about AC safety then.

Even relatively small voltages can hurt you

Most of the projects in this book work with AA batteries, usually four of them tied together to produce a total of six volts. That's not enough voltage to do serious harm. Even if you do get a shock with six volts, you will probably barely feel it.

However, it's possible to injure yourself with voltages even as low as six volts. If you accidentally create a short circuit between the two poles of a battery, a lot of current will flow very fast. This will very likely cause the wire connecting the two ends of the battery to get very hot, and the battery itself may also heat up. The heat may be enough to inflict a nasty burn.

Staying safe by staying dry

We've all seen murders committed on TV crime dramas by throwing a plugged-in electrical appliance such as a hair dryer into a bathtub while the victim was taking a bath. I've always wondered how often that really happens, and how likely it's fatal. For example, how quickly would the circuit breaker kick in and cut power to the hair dryer? Would the special GFCI-protection devices required in all bathrooms work as designed and cut power to the hair dryer in time?

I've never wanted to conduct an experiment to actually find out — nor should you, under any circumstances. Water and electricity are a very bad combination because water is an excellent conductor of electricity, and it flows everywhere.

Strictly speaking, pure uncontaminated water is actually an insulator. But pure water is very rare. Most water is filled with contaminates, and those contaminates turn the water into an excellent conductor. Thus, it's true that you should avoid water when working with electrical current. Here are a few tips for staying safe by staying dry:

- ✔ **Make sure the floor is dry.** Don't work on electronic or electrical devices in an area where the floor is wet.

- ✔ **Beware of high humidity, especially if it condenses into moisture on your projects.**

- ✔ **Dry your hands before working with electrical current.** Even a small amount of sweat on your hands can lower your body's natural resistance and accentuate the danger of electrical shocks from lower voltages.

If the racing current goes unchecked, there's also the possibility that the battery will explode. Trust me; you don't want to be nearby if that happens. You really don't want to make a trip to the emergency room to have fragments of an exploded battery removed from your eyes.

As a result of this danger, you should take the following precautions when working with the battery-powered circuits described in this book:

- ✦ *Don't connect power to the circuit until the circuit is completely finished and you've reviewed your work to ensure that everything is connected properly.*

- ✦ *Don't leave your circuits unattended when they're connected to power.* Always remove the batteries before you walk away from your workbench.

- ✦ *Periodically touch the batteries with your finger to make sure they aren't hot.* If they're getting warm, remove the batteries and recheck your circuit to make sure you haven't made a wiring mistake.

✦ *If you smell anything burning, remove the batteries and recheck your circuit.*

✦ *Always wear protective eyewear to protect yourself against exploding batteries.* (Under the right circumstances, other components can explode as well!)

Sometimes voltage hides in unexpected places

One of the biggest shock risks in electronics comes from voltages that you didn't expect to be present. It's easy enough to keep your eye on the voltages that you know about, such as in your power supply or batteries, but some electronic circuits are designed to amplify voltages. So even though your circuit runs on six-volt batteries, there may be much larger voltages at specific points within your circuit.

In addition, some electrical devices can actually store electric charge long after the power from your circuit has been disconnected. The most notorious device with this characteristic is the *capacitor*, which alternately builds up and then releases electrical charges. Thus, you should be wary of any circuit that contains capacitors — especially if the capacitors are large. Common ceramic-disk capacitors, which are typically smaller than a tiddlywink, don't store much charge. However, if your circuit has capacitors the size of batteries, you should be very careful when working around them. Such capacitors can hold large charges long after the power has been cut off.

Here are some safety points concerning capacitors:

✦ *One of the most common places to find large capacitors is in the power-supply circuit.* Any electronic device that plugs into a household electrical outlet has a power-supply circuit that may contain a large capacitor. Be very careful around these capacitors. In fact, if the power-supply circuit is inside its own enclosed box, *don't open the box.* Instead, replace the entire power supply if you suspect it's bad.

✦ *Another common place to find high-voltage capacitors is in a flash camera.* Even though the battery may be just 1.5 V, the capacitor that drives the flash unit may well be holding a charge of 300 V or more.

✦ *Before working on a circuit that contains a capacitor, always discharge the capacitor first.* You can discharge small capacitors by shorting out their leads with the blade of a screwdriver. Make sure you touch only the insulated handle of the screwdriver while you short out the leads, and don't touch any other part of the circuit with your free hand.

✦ *Larger capacitors should be discharged by connecting their leads to a lamp or a large resistor.* The easiest way to do this is to wire up a lamp holder to a pair of alligator clips, screw a lamp into the lamp holder, then carefully connect the clips to the capacitor leads. If the capacitor is holding a charge, the lamp will glow for a moment as the capacitor discharges through the lamp.

WARNING!

✦ *If you don't feel completely confident in what you're doing where large capacitors are concerned, walk away from the project.*

Other Ways to Stay Safe

Electric shock isn't the only danger you'll encounter when you work with electronics. The following paragraphs summarize a few of the other risks you may be exposed to and describes the precautions you should take to minimize those risks:

✦ **Soldering poses an obvious fire hazard.** If your soldering iron is hot enough to melt solder, it's also hot enough to ignite combustible materials such as paper, wire insulation, and so on. Therefore:

   • *Always be aware of when your soldering iron is on.* Don't plug it in until you need it, and unplug it when you're finished soldering.

   • *Never set a hot soldering iron down directly on your workbench.* Instead, get a soldering iron holder to safely hold the soldering iron while it's hot. Figure 4-1 shows a soldering iron resting in a simple stand. As you can see, this stand keeps the business end of the soldering iron safely elevated away from the work surface.

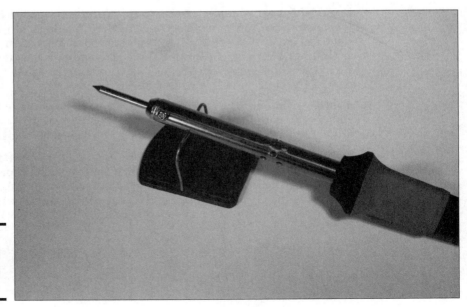

Figure 4-1:
A soldering iron resting on a stand.

- *Give your soldered joints a few minutes to cool down before you handle them.*

- *Watch out for the soldering iron's electrical cord.* Obviously, you want to avoid burning the cord with the soldering iron. As ridiculous as it sounds, I did this myself once when I carelessly set the soldering iron aside, directly on top of its own power cord. Fortunately, I noticed my mistake before the soldering iron melted much of the power cord's insulation.

  Make sure the soldering iron's power cord is placed safely away from your stuff so that you won't bump it as you work, knocking it out of its stand and perhaps causing a burn.

- *Be sure to wear eye protection when you solder.* As solder melts, it occasionally boils and splatters little globules of hot solder through the air. You really don't want molten metal anywhere near your eyes.

+ **Electronics — and especially soldering — can also create a chemical hazard.** When you solder, small amounts of lead are released into the air. Therefore:

  - *Always work in a well-ventilated place.*

  - *Wash your hands after you work with solder or any other electronic components before you touch your face, mouth, nose, or eyes.* Small amounts of lead and potentially other toxic substances are bound to get on your hands. It's best to wash them frequently to keep whatever gunk they pick up from getting into your body.

  - *Keep your soldering tools away from children.* Young children and pets love to stick things in their mouths. If you leave solder or little electronic parts like resistors or diodes sitting loose on top of your workbench, your kids or pets may decide to make a meal of them, so keep such things safely stored in boxes or cabinets and, if possible, keep your entire work area safely off limits and behind closed doors.

  - *Don't get into the habit of sticking parts into your mouth to hold them while you're working.* As crazy as it sounds, I've seen people hold a dozen resistors in their mouth while soldering each one into a printed circuit board. That's definitely a bad idea.

+ **Working with sharp tools such as knives, wire cutters, and power drills creates a risk of cutting injury.** Therefore:

  - *Think before you cut.* Make sure you know exactly where you want to make the cut, and make sure you know exactly where all of your fingers are before you start the cut.

  - *Let the tool do the work.* Don't apply excessive force to coerce a tool into making a bigger, deeper, or wider cut than it's designed to do.

- *Keep your tools sharp.* Working with dull tools causes you to use extra force, which often results in the tool slipping and finding itself lodged in your finger.

- *Remove jewelry such as rings, wristwatches, and long dangling necklaces before you start – especially if you're working with power tools.*

- *Wear safety goggles whenever you're cutting, sawing, or drilling.* Little pieces of the work or blade can easily break off and hit you in the face. Add bits of insulation, copper wire, and broken drill bits to the growing list of things you don't want in your eyes.

Keeping Safety Equipment on Hand

In spite of every precaution you might take, accidents are bound to happen as you work with electronics. Other than preventing an accident from happening in the first place, the best strategy for dealing with an accident is to be prepared for it, so I recommend you keep the following items nearby whenever you're working with electronics:

- **Fire extinguisher:** So you can quickly put out any fire that might start before it gets out of hand.

- **First-aid kit:** For treating small cuts and abrasions as well as small burns. The kit should include bandages, antibacterial creams or sprays, and burn ointments.

- **Phone:** So that you can call for assistance in case something goes really wrong.

- **Friend:** If your project works with household current (120 volts), a friend can help in case you get shocked.

Protecting Your Stuff from Static Discharges

Static electricity — more properly called *electrostatic charge* — results when electric charges (that is, voltage) builds up in the absence of a circuit that allows current to flow. Your own body is frequently the carrier of static charge, which can be created by a variety of causes. The most common is friction that results from simple things such as walking across a carpet. Your clothes can also pick up static charge, and usually do when you toss them around in a clothes dryer.

Static charge accumulated in your body usually discharges itself over time. However, if you touch a conductor — such as a brass doorknob — while you're charged up, the charge will dissipate itself quickly in an annoying shock.

If the conductor happens to be a sensitive electronic component such as a transistor or an integrated circuit rather than a brass doorknob, the discharge can be more than annoying; it can fry the innards of the component, rendering it useless for your projects. For this reason, it's wise to protect your stuff from static discharge when you work on your electronic projects. The easiest way to do that is to make sure you're properly discharged before you start your work. If you have a metal workbench or a large metal tool such as a drill press or grinder near your workbench, simply reach out and touch it after you've settled in to your seat and before you begin your work.

A more reliable way to protect your gear from static discharge is to wear a special *antistatic wristband* on one wrist, as shown in Figure 4-2. Wear the wristband tightly so that it's in good solid contact with your skin all the way around your wrist. Then, plug the alligator clip into a metal surface such as your workbench frame or that nearby drill press.

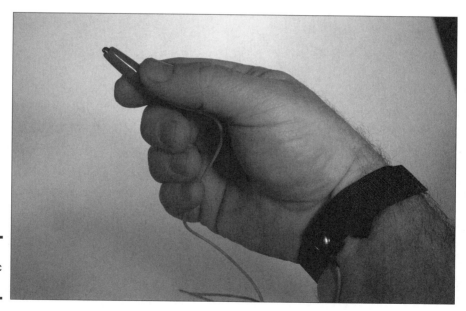

Figure 4-2:
An antistatic
wristband.

For best results, the alligator clip on your antistatic wristband should be connected to a proper *earth ground*. To create a proper earth ground, clamp a long length of wire to a metal water pipe. The wire should be long enough to reach from the pipe to your workbench. Carefully route the wire from the pipe to your workbench, strip off an inch or so of insulation, and staple or clamp the wire to the workbench, leaving the stripped end free so you can attach the alligator clip from your antistatic wristband to it. (Note that this technique works only if the building uses metal pipes throughout. If the building uses plastic pipe, the water-pipe won't provide a proper ground.)

An often-recommended way to connect the wristband to an earth ground is to connect it to the ground receptacle of a properly grounded electrical outlet. I'm definitely not a fan of this method, as the key to its operation lies in the term "properly grounded electrical outlet." All it takes is one stupid wiring mistake, or one wire shaken loose by a sonic boom or a mild earthquake, and suddenly that ground wire might not be a ground wire anymore — it might be energized. Call me paranoid if you wish, but there's no way I can recommend strapping a conductor around your wrist and then plugging it into an electrical outlet.

Chapter 5: Reading Schematic Diagrams

In This Chapter

✔ **Examining how schematic diagrams provide a roadmap for electronic circuits**

✔ **Looking at the most commonly used component symbols**

✔ **Noting how voltage supply and common ground circuits are often drawn**

✔ **Seeing how components are typically labeled**

I love maps. I think I've kept every map I've used on every trip I've taken. I have big maps of entire countries and states, maps of cities, walking maps, maps of parks and museums, and even subway maps. My favorite maps are topographical maps of the areas where I've gone on weeklong backpacking trips. These maps not only show the routes I've hiked, but also have elevation lines that represent every painful uphill step I've carried my 50-pound backpack up.

Without maps, we'd be lost. We'd never get to our destinations because we wouldn't know where the roads are. Think of all the sights we'd miss along the way!

Electronics has its own form of maps. They're called *schematic diagrams*. They show how all the different parts that make up an electronic circuit are connected.

Just as maps use symbols to represent features like cities, bridges, and railroads, schematic diagrams use special symbols to represent the different parts of a circuit, such as batteries, resistors, and diodes, and like maps, schematic diagrams have conventions that almost always are used. For example, positive voltages are almost always shown at the top of a schematic diagram, just as north is almost always shown at the top of a map.

In this chapter, you can learn about the symbols used in schematic diagrams and the conventions used to draw them.

Introducing a Simple Schematic Diagram

I've read a lot of computer programming books in my day, and I've written a few too. In a computer programming book, the first complete computer program usually shown is a program called Hello World, a program that simply displays the text "Hello World!" on a screen, and then quits. It's pretty much the simplest possible computer program that can be written. It doesn't do anything useful, but it's a great starting point for learning how to write computer programs.

Figure 5-1 shows a schematic diagram that is the electronic equivalent of the Hello World program. This diagram is about the simplest schematic diagram possible that actually does something: it lights a lamp, thus announcing to the world that a circuit is indeed working.

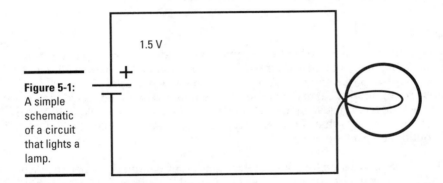

Figure 5-1:
A simple
schematic
of a circuit
that lights a
lamp.

This diagram contains two symbols representing the two components in the circuit: a 1.5 V battery and an incandescent lamp. The lines that connect the two components represent conductors, which could be actual wires or traces of copper in a printed circuit board.

In the circuit depicted in this schematic, the positive side of the battery is connected to one lead from the lamp, and the other lead from the lamp is connected to the negative side of the battery. Once these connections are made, current will flow from the battery to the lamp, through the lamp's filament to produce light, and then back to the battery.

Schematic diagrams always depict *conventional current flow*, which, as you learn in Chapter 2 of this minibook, means that current flows from positive to negative. Thus, the current flows from the positive terminal of the battery through the lamp and then back to the negative terminal of the battery.

In reality, conventional current flow is opposite of the actual flow of electrons through the circuit. The negative side of the battery has an excess of negatively charged particles (extra electrons) whereas the positive side has an excess of positively charged particles (missing electrons). Thus, the electric charge flows through the conductor from the negative side of the battery, through the lamp, and back to the positive side. (For more about the difference between real current flow and conventional current flow, refer to Chapter 2 of this minibook.)

As it passes through the lamp, the resistance of the lamp's filament causes the current to heat the filament, which in turn causes the filament to emit visible light.

Laying Out a Circuit

One of the most important things to realize about a schematic diagram is that the arrangement of components in the diagram doesn't necessarily correspond to the physical arrangement of parts in the circuit when you actually build the circuit.

For example, the circuit depicted in Figure 5-1 shows the battery on the left side of the circuit and the lamp on the right. It also shows the battery oriented so that the positive terminal is at the top and the negative terminal is at the bottom. However, that doesn't mean the circuit would actually have to be built that way. If you want, you could put the lamp on the left and the battery on the right, or you could put the battery at the top and the lamp on the bottom.

The physical arrangement of the circuit doesn't matter as long as the component connections remain the same as shown in the schematic. Thus, in this example, no matter how you physically arrange the components, you must connect the positive terminal of the battery to one lead of the lamp and the negative terminal to the other lead.

Because there are only two components and two conductors in the circuit shown in Figure 5-1, it would be pretty hard to mess up the connections. However, in a more complicated circuit with perhaps dozens of components and dozens of connections, laying out the circuit and making sure that all the connections exactly match the connections indicated in the schematic can be a challenge. Each connection must be checked carefully to make sure it's correct.

To Connect or Not to Connect

One of the goals when laying out a schematic circuit diagram is to keep the diagram as simple as possible. However, the lines in all but the simplest of schematic diagrams will at some places need to cross over each other. When they do, it's vital that you can tell whether the lines that cross represent actual connections (also called *junctions*) between the conductors or the lines cross over each other but don't actually connect.

Unfortunately, there isn't one clear and universally used standard that dictates how to indicate whether crossed lines represent a junction. Figure 5-2 shows some of the ways for showing crossed wires with or without junctions.

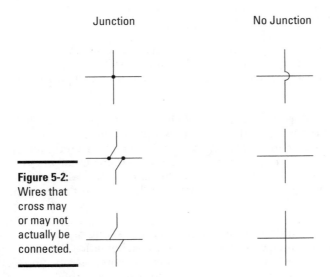

Junction   No Junction

Figure 5-2:
Wires that
cross may
or may not
actually be
connected.

The three examples on the left side of Figure 5-2 show how junctions are indicated. The example at the top left shows the most common way to indicate a junction: by placing a conspicuous dot at the point where the wires cross. Any time you see a dot where two lines intersect, you know that the two lines form a junction.

In the two junction styles shown in the middle-left and bottom-left examples in Figure 5-2, the vertical lines are angled to avoid coming together at the same spot on the horizontal line. With or without the dot, junctions are clearly indicated in both of these examples.

The three examples on the right side of Figure 5-2 show how lines that cross but don't connect to form junctions are most commonly shown. In the top two examples, one line "hops" over the other, and one of the lines is broken at the spot where it crosses the other.

The example in the bottom-right corner of Figure 5-2 is a bit ambiguous. Here, the lines cross each other. However, there's no hop or break to indicate that no junction is present, nor is there a dot to indicate that a junction should be present. So is there a junction here or not? The answer is, in most cases, no. You can usually assume that a junction is *not* present when lines cross but there's no dot. However, you should examine the rest of the diagram to make sure. If you find other places in the diagram where nonjunctions are indicated by a hop or a break, the crossed lines without the hop or break may indeed indicate a junction.

To avoid ambiguity altogether, the schematic diagrams in this book always use a dot to indicate a junction and a hop to indicate a nonjunction. You'll never see lines simply cross without a hop or a dot.

Looking at Commonly Used Symbols

The circuit shown in Figure 5-1 has just two components: a battery and a lamp. Most electronic circuits will have additional components. There are hundreds of different types of electronic components, and each has its own unique schematic diagram symbol. Fortunately, you need to know only a few basic symbols to get you started. These symbols are summarized in Table 5-1. (Note that when used in an actual circuit diagram, the symbols are often rotated.)

Table 5-1	Common Symbols for Schematic Diagrams
Symbol	*Description*
—(A)—	Battery
—|⊦—	Capacitor
—▷|—	Diode
—(V)—	Ground connection
—ᴍᴍ—	Inductor (coil)
(Ⓞ)	Lamp

(continued)

Table 5-1 *(continued)*

Symbol	Description
	Light-emitting diode
	Resistor
	Source voltage connection
	Speaker
	Switch
	Transformer
	Transistor (NPN)
	Transistor (PNP)
	Variable resistor (potentiometer)

Figure 5-3 shows a schematic diagram that includes several of these components. Don't worry — you don't need to understand this diagram right now. I just want you to get an idea of what real-world schematic diagrams look like and how to read them.

As you can see, the circuit depicted in Figure 5-3 contains six components. Working from left to right, they are:

+ 6 V battery
+ NPN transistor
+ Resistor
+ Capacitor
+ PNP transistor (at the top right)
+ Light-emitting diode (at the bottom right)

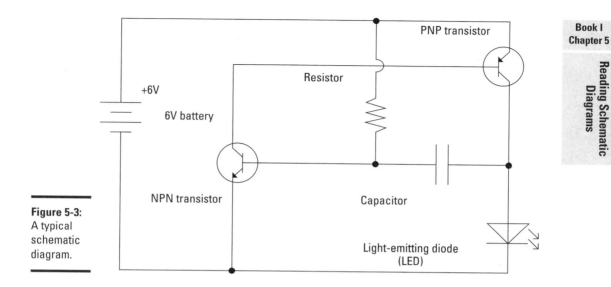

Figure 5-3:
A typical
schematic
diagram.

Throughout the course of this book, I use these and other symbols in the schematic diagrams that describe the circuits. Whenever I use a symbol for the first time, I'll be sure to explain what it is and how it works.

Simplifying Ground and Power Connections

In many electronic circuits, the distribution of voltage connections is one of the most complicated aspects of the circuit. For example, about half of the connections in the schematic diagram shown in Figure 5-3 are used to connect the resistor, transistors, and the LED to either the positive or negative terminal of the battery.

In a more complicated circuit, there can be dozens or even hundreds of power connections. If all the lines representing those connections had to be drawn to the positive or negative side of the battery symbol, schematic diagrams would quickly be overwhelmed by the power connections.

Most circuits have a common path by which current returns to its source. In the case of Figure 5-3, it's the conductor at the very bottom of the diagram that collects current from the LED and the resistor and returns it to the battery. This conductor is necessary to complete the circuit so that current can flow in a complete loop from the battery through the various components and then back to the battery.

This common return path is often called the *ground*, and can be replaced by the ground symbol that was shown in Table 5-1. Figure 5-4 shows a schematic diagram that uses ground symbols instead of a line to show the path by which current returns to the battery. The circuit shown in Figure 5-4 is identical to the circuit shown in Figure 5-3.

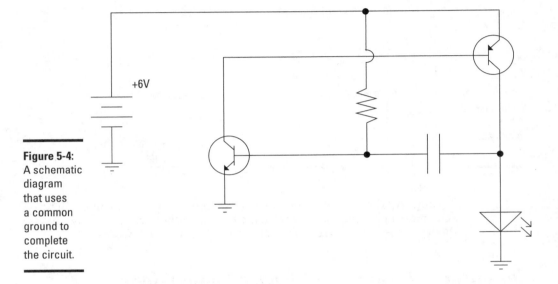

Figure 5-4:
A schematic diagram that uses a common ground to complete the circuit.

In addition to a common ground path, most circuits also have a common voltage path. In the case of the circuit shown in Figures 5-3 and 5-4, the common voltage path goes from the battery to the resistor and on to the second transistor. This conductor can be replaced by symbols representing voltage sources that appear wherever voltage is required in a circuit.

The symbol for a voltage source is either an open circle or an arrow. The quantity of voltage is always indicated next to the circle or arrow. When a voltage source symbol is used in a schematic diagram, the symbol for the battery (or other power source if the circuit isn't powered by a battery) is omitted. Instead, the presence of voltage source symbols implies that voltage is provided by some means, either by a battery or by some other device such as a solar cell or a power supply plugged into an electrical outlet.

Figure 5-5 shows a schematic diagram for the same circuit that was shown in Figures 5-3 and 5-4, but with voltage source symbols instead of a battery symbol. As you can see, +6 V is required in two places in the circuit: at the resistor and at the second transistor. This circuit is functionally identical to the circuits shown in Figures 5-3 and 5-4.

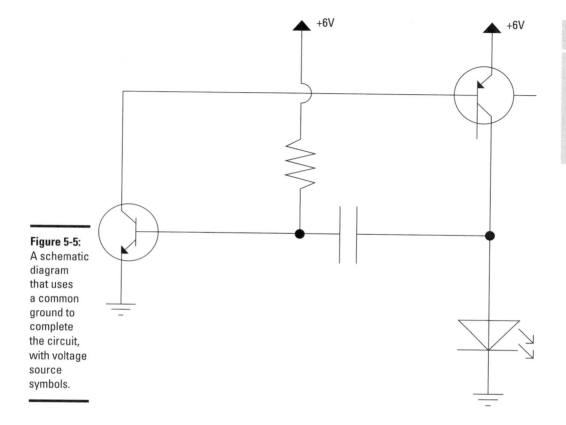

Figure 5-5:
A schematic
diagram
that uses
a common
ground to
complete
the circuit,
with voltage
source
symbols.

Although the circuit shown in Figure 5-5 has a positive voltage source and the ground is negative, this isn't always the case. You can also use the voltage source symbol to refer to negative voltage. In that case, the ground actually carries positive voltage back to the source.

In some cases, a circuit may require both positive and negative voltages at different places within the circuit. Remember from Chapter 2 of this minibook that voltages are always measured with respect to two points in a circuit. Thus, voltages are always relative. For example, the positive pole of a AAA battery is +1.5 V relative to the negative pole. At the same time, the negative pole of the battery is –1.5 V relative to the positive pole.

Now suppose you connect two AAA batteries end to end. Then, the voltage at the positive terminal of the first battery will be +3 V relative to the voltage at the negative terminal of the second battery. But, the voltage at the positive pole of the first battery will be +1.5 V relative to the point between the batteries, and the voltage at the negative pole of the second battery will be –1.5 V relative to the point between the batteries.

Figure 5-6 shows how this arrangement might be drawn in a schematic diagram, with a pair of resistors connected across each battery to the middle point. The diagram on the left shows the batteries and connections to them. The diagram on the right shows the same circuit using ground and voltage source symbols instead.

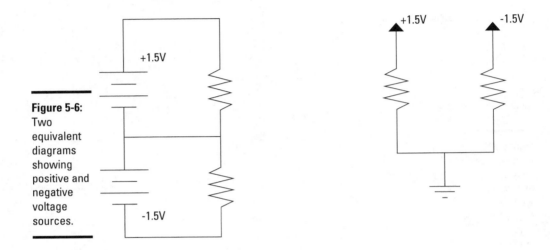

Figure 5-6: Two equivalent diagrams showing positive and negative voltage sources.

Labeling Components in a Schematic Diagram

A symbol alone is not usually enough information to completely identify an electronic component in a schematic diagram. Further information is usually included with text that's placed adjacent to the symbol, as shown in Figure 5-7. This additional information usually includes the following:

+ **Reference identifier:** Each component is usually labeled with a letter that designates the type of component followed by a number that helps identify each component of the same type. For example, if a circuit has four resistors, the resistors are identified as R1, R2, R3, and R4. The most commonly used letters are shown in Table 5-2.

Table 5-2	Commonly Used Reference Identifiers
Letter	*Meaning*
R	Resistor
C	Capacitor
L	Inductor
D	Diode

Letter	Meaning
LED	Light-emitting diode
Q	Transistor
SW	Switch
IC	Integrated circuit

✦ **Value or part number:** For components such as resistors and capacitors, the value is given in ohms (for resistors) and microfarads (for capacitors). Thus, a 470 Ω resistor would have the number 470 next to it, and a 100 μF capacitor would have the number 100 next to it.

The letters K and M are used to denote thousands and millions. For example, a 10,000 Ω resistor is identified as 10K in a schematic.

Components such as diodes, transistors, and integrated circuits don't have values; instead, they have manufacturer's part numbers. Thus, you might find a part number such as 1N4001 (for a diode), 2N2222 (for a transistor), or 555 (for an integrated circuit, IC) next to one of these components.

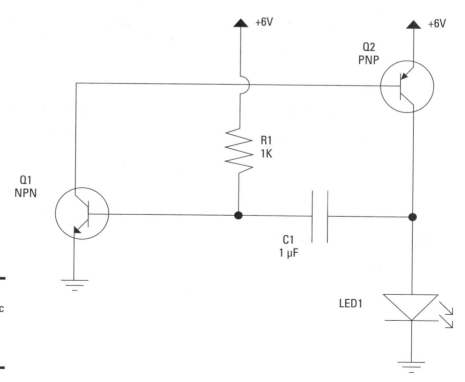

Figure 5-7:
A schematic diagram with parts labeled.

In some cases, the value or part number is omitted from the schematic diagram itself and instead included in a separate *parts list* that identifies the value or part number of each referenced part that appears in the schematic. Then, to find the value or part number of a particular component, you look up the component by its reference identifier in the parts list.

Representing Integrated Circuits in a Schematic Diagram

One important symbol that isn't shown in Table 5-1 is the symbol for an IC (integrated circuit). ICs are small assemblies that usually have multiple leads, called *pins,* which connect to various parts of the circuit contained within the assembly. Some ICs have as few as six or eight pins; others have dozens or even hundreds. These pins are numbered, beginning with pin 1. Each pin in an IC has a distinct purpose, so connecting to the correct pins in your circuit is vital to the circuit's proper operation. If you connect to the wrong pins, your circuit won't work, and you may damage the IC.

The most common way to depict an integrated circuit in a schematic diagram is as a simple rectangle with leads coming out of it to depict the various pins. The arrangement of the pins in the schematic diagram doesn't necessarily correspond to the physical arrangement of pins on the IC itself. Instead, the pins are positioned to provide for the simplest circuit paths in the diagram. The pins in the diagram are numbered to indicate the correct pin to use.

For example, Figure 5-8 shows a schematic diagram that uses a popular IC called a 555 timer IC to make an LED flash. The 555 has eight pins, and you can see that the schematic calls for connections on all eight. However, the pins in the diagram are arranged in a manner that simplifies the connections to be made to the pins. In an actual 555 IC, the pins are arranged in numerical order on either side of the IC, with pins 1 through 4 on one side and pins 5 through 8 on the other side.

Don't worry about any of the details of the operation of this circuit. You learn how it works in Book III, Chapter 2. My only purpose for including it here is so you can see how integrated circuits are depicted in a schematic diagram.

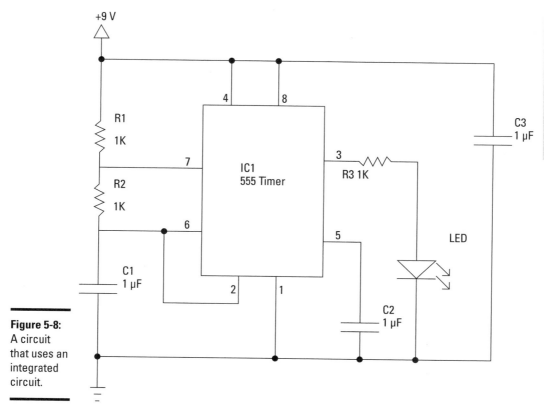

Figure 5-8:
A circuit
that uses an
integrated
circuit.

Chapter 6: Building Projects

In This Chapter

- ✔ Fleshing out an idea for an electronic project
- ✔ Creating a workable circuit design
- ✔ Building a prototype on a solderless breadboard
- ✔ Creating a permanent circuit on a printed circuit board
- ✔ Finishing the project by putting everything into a suitable enclosure

Yogi Berra is alleged to have said, "In theory, there is no difference between theory and practice. But in practice, there is."

Much of this book is theoretical — how electric current works, how individual electronic components like resistors, capacitors, and transistors work, how digital logic works, and so on.

But the heart of electronics is building things. The reason for learning all the theory is so you can practice the art by actually building circuits and putting them to use.

Throughout this book, I back up theoretical explanations about how various types of electronic components work with simple construction projects you can build to demonstrate the theory in actual use. In this chapter, you learn the basic construction techniques needed to build these projects.

Specifically, you learn how to create a prototype of a circuit using a handy device called a *solderless breadboard.* Then, you learn several techniques for creating a more permanent version of the circuit, in which the components and all the circuit's interconnections are soldered together on a circuit board. Finally, you learn how to enclose your circuit board in a project box or other enclosure.

Looking at the Process of Building an Electronic Project

Electronic projects such as the ones you learn about in this book typically follow this predictable sequence of general steps from start to finish:

1. **Decide what you want to build.**

   Before you can design or build an electronic project, you must have a solid idea in mind for what you expect the project to do, what you want it to look like, and how human beings will interact with it.

2. **Design the circuit.**

   Once you've settled on what you want to build, you need to design an electronic circuit that gets the job done. The end result of this step is a schematic diagram.

3. **Build a prototype.**

   Before you invest the time and materials needed to build a permanent circuit, it's a good idea to first build a *prototype,* which lets you quickly test the circuit to make sure it works. Usually, you build the prototype on a solderless breadboard.

4. **Build a permanent circuit.**

   Once your prototype is working, you can build a permanent version of the circuit. Usually, you build the permanent version by soldering components onto a printed circuit board.

5. **Finish the project.**

   To finish the project, you mount the circuit board along with any other necessary components such as batteries, switches, or light-emitting diodes in a suitable enclosure.

The remaining sections in this chapter describe each of these steps in greater detail.

Envisioning Your Project

Before you get lost in the details of designing and building your project, you should step back and look at the big picture. First, you need to make sure you have a solid idea for your project. Why do you want to build it? What will it do, who will use it, and why?

For example, every year I like to build something to scare trick-or-treaters on Halloween. A few years ago, I built a giant jack-in-the-box that pops up and screams when people walk up to it. The box was made of plywood, and the pop-up mechanism that made the door open and the scary clown pop up was driven by compressed air. Figure 6-1 shows the finished contraption. Trust me; I scared a lot of kids and more than a few adults with it.

Figure 6-1:
One of my scarier electronics projects.

I knew right away that I'd need some type of electronic circuit to control the jack-in-the-box. At first, I wasn't sure exactly what type of circuit I'd need, but I knew I needed a circuit of some kind.

Once you have a general idea for a project, you can flesh out the details. You'll need to answer questions like these:

✦ What will its *user interface* be? That is, how will a person work with the device to get it to do what it's supposed to do?

✦ Will it be a stand-alone device, or will it interact with other devices?

✦ Will it be powered by batteries, or will it plug into a wall outlet to get its power? Or will it be solar powered?

✦ How big will it be? Does it need to be small enough to hold in your hand or fit in your pocket? Or will it sit on a shelf?

The jack-in-the-box Halloween prop is a fairly complicated project — too complicated to use as an illustration this early in the book. So, here's a simpler project: an electronic decision maker. Have you ever resorted to tossing a coin to make a difficult decision? For this project, you create an electronic version of a coin toss. Instead of flipping a coin into the air to see if it lands heads or tails, you build an electronic device that does the coin toss. That way, you can make decisions even when you're penniless.

The specifications for the coin-toss project are as follows:

✦ The device will have two LED indicators to indicate heads and tails.

✦ It will also have two small metal contacts, which the user can touch with his or her finger. When the user touches both of the posts, the LEDs start flashing, alternating back and forth, much like a coin flips end over end when you toss it into the air.

✦ When the user removes his finger from the two metal contacts, one of the two lights will stay lit, indicating whether the result of the coin toss is heads or tails. Which light stays lit will be essentially random.

✦ To conserve battery life, the device will have an on/off pushbutton. The user must depress the pushbutton to make the device work; when the button is released, the device is turned off.

✦ The device will be battery powered and contained in an enclosure small enough to hold in your hand.

As you flesh out the details for your project, you may want to start drawing diagrams to show how it will look. Figure 6-2 shows a hand-drawn sketch I created for the electronic coin tosser.

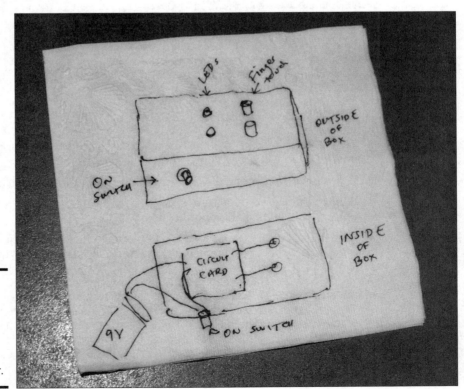

Figure 6-2:
A hand-drawn sketch for an electronic coin tosser.

Designing Your Circuit

Once you have an idea for a project, the next step is to design a circuit that meets the project's needs. At first, you'll find it very difficult to design your own circuits, so you'll turn to books like this one or to the Internet to find other people's circuit designs. With a bit of Google searching, you can probably find a schematic diagram that's very close to what your project needs.

In many cases, you won't be able to find exactly the circuit you're looking for. You may find a circuit that's close, but you may need to make minor modifications to make the circuit fit your project's needs. At first, making modifications to a circuit may seem beyond your abilities. But as you gain experience, you'll find yourself tweaking circuits all the time to fit specific applications.

One helpful strategy for designing circuits is to break complex requirements down into simpler parts. For example, consider the pop-up jack-in-the-box Halloween prop I mention earlier. The complete circuit for this project required several different elements, including these:

✦ A circuit to detect when someone has entered the room to trigger the prop's action

✦ A circuit to open and close the jack-in-the-box

✦ A circuit to time how long the jack-in-the-box should stay open

✦ A circuit that plays a screaming sound

✦ A circuit that provides a 30-second delay before the prop is activated again

The coin-toss project is much simpler than the jack-in-the-box project. In fact, a quick Google search will turn up several possible circuits that do almost exactly what the coin-toss project requires. For example, Figure 6-3 shows the schematic diagram for a typical coin-toss circuit you might find on the Internet. This circuit diagram uses a 555 Timer integrated circuit, four resistors, two LEDs, one capacitor, a switch, and a 9V power supply (most likely a 9V battery).

The schematic diagram shown in Figure 6-3 differs from our project's needs in just two ways. First, it doesn't have an on/off switch. And second, it uses a pushbutton instead of the user's fingers to start and stop the LEDs from flashing.

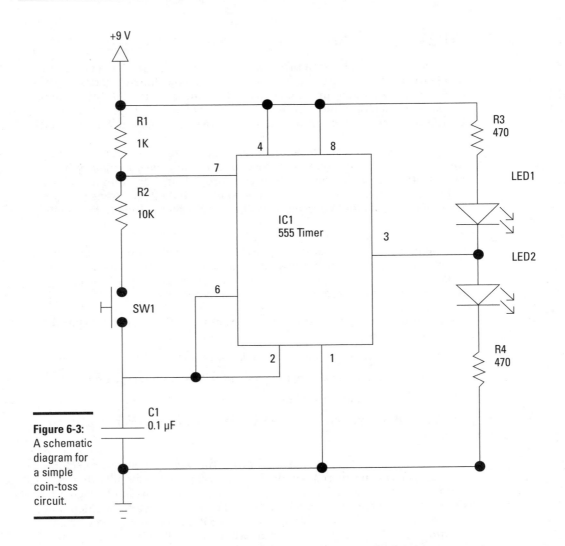

Figure 6-3:
A schematic
diagram for
a simple
coin-toss
circuit.

Figure 6-4 shows the schematic after I made those modifications. As you can see, I added a push-button switch that must be pressed to provide the +9 V voltage needed to run the circuit, and I replaced the pushbutton that was in the original schematic with two open terminals. When the user touches these two terminals, the resistance of his or her finger completes the circuit.

Please don't worry at all if you don't understand how the circuit depicted in Figure 6-4 works. I wouldn't expect you to at this point in the book! Understanding how a circuit works and building that circuit are two entirely different things; you can (and probably will) build plenty of circuits whose operation you don't understand. The only thing you should focus on at this point is how the schematic diagram indicates the various connections between the parts in the circuit. You learn the details of how this circuit works in Book III, Chapter 2.

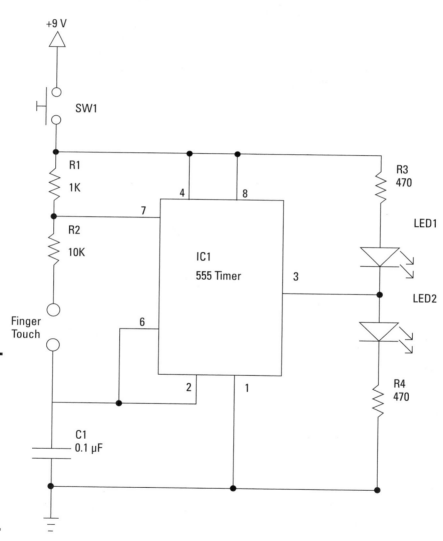

Figure 6-4:
The schematic diagram for the coin-toss circuit after it has been modified a bit for our project.

One final step you might want to consider when designing a circuit is to create a final version of the schematic diagram that indicates what components will be mounted on your final circuit board and what components won't be on the circuit board. This diagram will come in handy later, when you're ready to create the circuit board that will become the permanent home of your circuit.

For example, Figure 6-5 shows a version of the coin-toss circuit that uses a dashed line to delineate the items that won't be mounted on the circuit board: the battery power supply (that is, the +9 V voltage source and the ground), the push-button power switch, the two metal finger contacts, and the two LEDs. Instead, they'll be mounted separately within the project box. Thus, the circuit board will need to hold only six components: the 555 timer integrated circuit, the four resistors, and the capacitor.

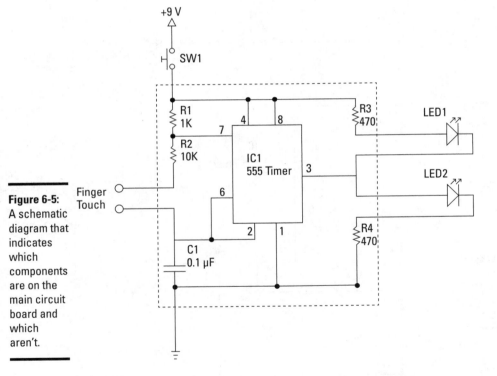

Figure 6-5:
A schematic diagram that indicates which components are on the main circuit board and which aren't.

Once you've completed your circuit design, you'll want to compile a list of all the parts you'll need to build the circuit. Then, you can rummage through your parts bin to figure out what parts you already have at your disposal and what parts you'll need to purchase. Here's a list of the components you'll need to build the coin-toss circuit:

Part ID	Description
R1	1 KΩ, ¼ W resistor
R2	10 KΩ, ¼ W resistor
R3	470 Ω, ¼ W resistor
R4	470 Ω, ¼ W resistor
C1	0.1 µF capacitor
LED1	5 mm red LED
LED2	5 mm green LED
IC1	555 timer IC
SW1	Momentary-contact, normally open pushbutton

Prototyping Your Circuit on a Solderless Breadboard

Before you commit your circuit to a permanent circuit board, you want to make sure it works. The easiest way to do that is to build the circuit on a *solderless breadboard*. The solderless breadboard lets you quickly assemble the components of your circuit without soldering anything. Instead, you just push the bare wire leads of the various components you need into the holes on the breadboard and then use *jumper wires* to connect the components together.

The beauty of working with a solderless breadboard is that if the circuit doesn't work the way you expect it to, you can make changes to the circuit simply by pulling components or jumper wires out and inserting new ones in their place. If you discover that your schematic diagram is missing an important connection, you can add another jumper wire to create the missing connection or, if you want to see how the circuit might work with a different resistor or capacitor, you can pull out the original resistor or capacitor and insert a different one in its place. Figure 6-6 shows a typical solderless breadboard.

Understanding how solderless breadboards work

Although many different manufacturers make solderless breadboards, they all work pretty much the same way. The board consists of several hundred little holes called *contact holes* that are spaced 0.1" apart. This is a convenient spacing because it also happens to be the standard spacing for the pins that come out of the bottom or sides of most integrated circuits. Thus, you can insert all the pins of even a large integrated circuit directly into a solderless breadboard.

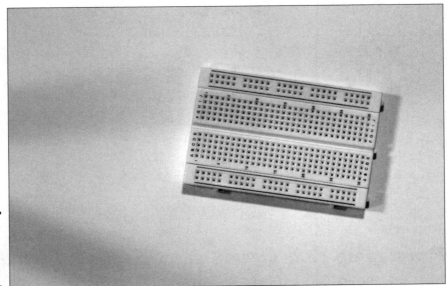

Figure 6-6:
A typical solderless breadboard.

Beneath the plastic surface of the solderless breadboard, the contact holes are connected to one another inside the breadboard. These connections are made according to a specific pattern that's designed to make it easy to construct even complicated circuits. Figure 6-7 shows how this pattern works.

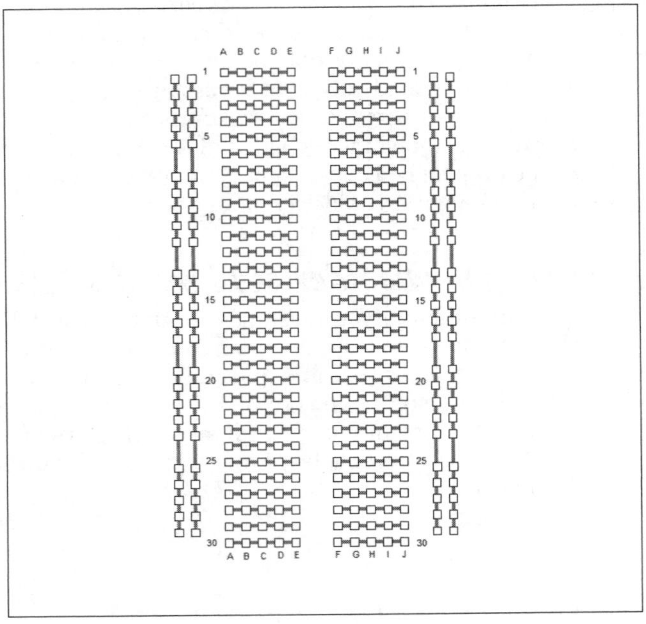

Figure 6-7:
The contact holes in typical solderless breadboards are internally connected following this pattern.

The holes in the middle portion of a solderless breadboard are connected in groups of five that are called *terminal strips*. These terminal strips are arranged in two groups, with a long open slot between the two groups, like a little ditch. It is in these holes that you will connect components such as resistors, capacitors, diodes, and integrated circuits.

It's important to note that the rows of holes are not connected across the ditch. Thus, each row comprises two electrically separate terminal strips: one that connects the holes labeled A through E, the other connecting the holes labeled F through J.

The breadboard is designed so that integrated circuits can be placed over the top of the ditch, with the pins on either side of the integrated circuit pushed into the holes on either side of the ditch.

The holes on the outside edges of the breadboard are called *bus strips*. There are two bus strips on either side of the breadboard. For most circuits, you will use the bus strips on one side of the breadboard for the voltage source and use the bus strips on the other side of the board for the ground circuit.

Most solderless breadboards use numbers and letters to designate the individual connection holes in the terminal strips. In Figure 6-7, the rows are labeled with numbers from 1 through 30, and the columns are identified with the letters A through J. Thus, the connection hole in the top-left corner of the terminal strip area is A1, and the hole in the bottom-right corner is J30. The holes in the bus strips are not typically numbered.

Solderless breadboards come in several different sizes. Small breadboards usually have about 30 rows of terminal strips and about 400 holes altogether. But you can get larger breadboards, with 60 or more rows with 800 or more holes.

Laying out your circuit

The most difficult challenge of creating a circuit on a solderless breadboard is the task of translating a schematic diagram into a layout that can be assembled on the breadboard. Only in rare cases will a circuit assembled on a breadboard look like the circuit's schematic diagram. In most cases, the components are arranged differently and jumper wires are required to connect the components together.

The key when assembling a circuit on a solderless breadboard is to ensure that every connection represented in the schematic diagram is faithfully recreated on the breadboard. For example, the schematic diagram in Figure 6-4 indicates that pin 1 of the 555 timer IC must be connected to ground. Thus, when you build the circuit on a breadboard, you must ensure that this connection is properly made.

One of the first challenges you face when building a circuit on a breadboard is connecting the pins on an integrated circuit. In a schematic diagram, the pin connections on an integrated circuit are rarely drawn in numerical order. For example, in the schematic diagram shown in Figure 6-4, the pin connections on the 555 timer IC are listed in this order, going counterclockwise from the top left: 7, 6, 2, 1, 3, 8, and 4. (Pin 5 is not used.)

But the pins on an actual 555 timer IC chip are arranged in numerical order starting at the top-left corner of the chip, as shown in Figure 6-8. Notice also that there are pins on the left and right side of the chip but none on the top or bottom. (The dot imprinted on the top of the chip is used to identify pin 1.)

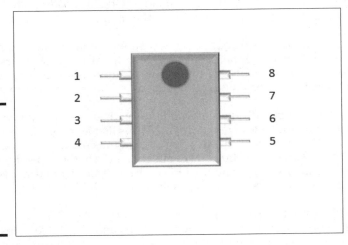

Figure 6-8:
How the pins are numbered on a 555 Timer integrated circuit.

You'll have to use your wits to recreate a circuit represented by a schematic diagram on a solderless breadboard. Here are some pointers to get you started:

✦ Start by designating the top row of bus strips as the positive power supply and the bottom row as the ground. Connect your battery connector to holes in one end of these bus strips, but don't yet connect the battery; it's never a good idea to apply power to your circuit before you've finished assembling it.

✦ Next, insert any ICs required for the circuit. Insert them so that they straddle the ditch in the middle of the terminal rows and, if your circuit has more than one IC, orient them all the same. You'll only confuse yourself if pin 1 is on the bottom-left corner of some of your ICs and on the top-right corner of others.

✦ Each pin of each IC is connected to a terminal strip that has four additional connection holes. Thus, you can connect as many as four additional components or jumper wires to each pin. If your circuit requires more than four component connections to a single pin, use a jumper wire to extend the pin's terminal strip to an unused row anywhere on the breadboard.

✦ Use jumper wires to connect the voltage source and ground pins for each IC to the nearest available connection hole in the voltage and ground buses.

✦ Now work your way around the rest of the pins for each IC, connecting each component as needed. If one end of a component connects to an IC pin and the other end connects to either the voltage source or ground, plug one end of the component into an available connection hole on the terminal strip for the IC pin and plug the other end into the nearest available connection hole on either the voltage supply or ground buses.

✦ If you want, you can trim the leads of the various components so that the parts fit closer to the breadboard. This results in a breadboard circuit that's neater, and with fewer bare leads sticking up high above the breadboard, the likelihood of leads accidentally coming in contact with each other and creating a short-circuit are less likely. I usually don't trim the leads, however, unless the circuit is complex enough that I'm not able to keep the component leads away from each other without cutting them down to size.

Assembling the coin-toss circuit on a solderless breadboard

This section presents a complete procedure for assembling the coin-toss circuit on a small solderless breadboard. Once you get all your materials together, you should be able to complete this project in about an hour.

All the parts required to build this prototype circuit can be purchased from your local RadioShack, or you can order them online from any electronic parts supplier. For your convenience, here is a complete list of the parts you'll need to build this prototype circuit, along with the RadioShack catalog part numbers:

Part Number	Quantity	Description
276-003	1	Small solderless breadboard
276-173	1	Solderless breadboard jumper wire kit
276-1723	1	LM555 timer IC
271-1321	1	1 kΩ, ¼ W resistor (5 per package)
271-1335	1	10 kΩ, ¼ W resistor (5 per package)
271-1317	2	470 Ω, ¼ W resistor (5 per package)
272-1053	1	0.1 µF polyester film capacitor
276-041	1	Red LED 5 mm
276-022	1	Green LED 5 mm
270-325	1	9 V battery snap connector
n/a	1	9 V battery

You can build this circuit using equivalent parts from any supplier. So if you already have equivalent parts on hand, you don't need to run out to RadioShack and purchase them just for the sake of spending money.

You won't need many tools for this project. You can probably assemble it without any tools at all, but you may want to keep your wire cutters, wire strippers, and tweezers handy.

The steps that follow identify specific holes in the terminal strip area of the breadboard using numbers and letters. If you're using a different breadboard than the one listed in the parts list, you might encounter a different numbering system. If so, you can refer to Figure 6-7 to translate the numbers given in the steps for the breadboard you're using.

Once you have everything you need, follow these steps to assemble the circuit:

Follow these three steps to insert the IC and connect it to power.

1. Insert the 555 timer IC.

Take a close look at the 555 timer IC. On the top, notice a small dot in one corner; this dot marks the location of pin 1. Carefully insert the leads of the 555 timer IC into the breadboard near the middle of the board, inserting pin 1 into hole E14 and pin 8 in hole F14. The IC will straddle the groove that runs down the center of the board.

2. **Connect pin 1 of the 555 timer IC to the ground bus.**

   Insert one end of a small jumper wire into hole A14 and the other end into the nearest available hole in the bottommost bus strip.

3. **Connect pin 8 of the 555 timer IC to the +9 V bus.**

   Insert one end of a small jumper wire into hole J14 and the other end into the nearest available hole in the topmost bus strip.

4. **Connect pins 2 and 6 of the 555 timer together.**

   Insert one end of a small jumper wire into hole C15 and the other end in hole H16. The jumper wire will reach over the top of the 555 Timer chip.

   Figure 6-9 shows what the breadboard looks like after these three steps.

Figure 6-9:
The breadboard after the IC has been inserted and connected to the power buses.

The next five steps connect the LEDs and resistors R3 and R4. The LEDs will use the terminal strips in rows 19 and 21.

1. **Connect pin 3 of the IC to row 19.**

   Insert one end of a short jumper wire in hole C16 and the other end into hole C19.

2. **Connect the two segments of row 19.**

   Insert one end of a short jumper wire into hole E19 and the other end into hole F19. This jumper wire bridges the gap between the two terminal strips in row 19, effectively making them a single terminal strip.

3. **Insert the red LED.**

   If you look carefully at the red LED, you'll see that one lead is a bit shorter than the other. This short lead is called the *cathode*. The longer lead is called the *anode*. Insert the cathode (shorter lead) into hole D21. Then, insert the anode (longer lead) into hole D19.

4. **Insert the green LED.**

   The green LED also has a short cathode lead and a longer anode lead. Insert the anode (long) lead in hole G21 and the cathode (short) lead in hole G19.

   Note that the leads of the two LEDs are installed reversed from one another: the red LED's anode and the green LED's cathode are inserted into row 19, while the red LED's cathode and the green LED's anode are inserted into row 21. There's a very good reason for this, but I wouldn't expect you to understand it yet even if I tried to explain it. So for now, take it on faith that you must install the two LED's reversed like this for the circuit to work. (You'll learn more about LEDs, cathodes, and anodes in Book II, Chapter 5.)

5. **Insert resistors R3 and R4.**

   Both of these resistors are 470 Ω. You can identify these resistors by looking at the three color strips painted on the resistors — they're yellow, purple, and brown. Insert one end of the first resistor in hole B21 and the other end in the nearest available hole in the bottommost bus strip (the ground bus). Then, insert one end of the other resistor in hole I21 and the other end in the nearest available hole in the topmost bus strip (the +9 V bus).

   Figure 6-10 shows what the breadboard looks like after these steps.

The next five steps connect the finger-touch circuit that lets the user activate the coin toss by touching the two metal contacts. For the purposes of this prototype, you connect one end of a pair of jumper wires to the circuit and leave the other ends protruding from the end of the breadboard. Touching the bare ends of these wires with your fingers will simulate touching the metal contacts that you use in the final version of the circuit. The two jumper wires will be inserted into holes in row 9.

1. **Insert resistor R1 from pin 7 of the IC to the +9 V bus.**

   Resistor R1 is the 1 kΩ resistor, which should be connected between pin 7 of the IC and the +9 V bus. This resistor has stripes in the following sequence: brown, black, and red. Insert one end of this resistor into hole J15 and the other end into the nearest available hole in the topmost bus strip.

Figure 6-10:
The breadboard after the LEDs have been connected.

2. **Insert capacitor C1 from pin 2 of the IC to the ground bus.**

   Insert one lead of the capacitor (it doesn't matter which) into hole B15, and then insert the other into the nearest available slot in the bottom-most bus strip.

3. **Insert resistor R2 from pin 7 of the IC to one of the metal contacts.**

   This resistor is the 10 kΩ. It should be connected between pin 7 of the IC and one of the metal contacts that the user will touch with his finger to activate the coin-toss action. This resistor has the following sequence of color stripes: brown, black, and orange. Insert one end of it into hole H15 and the other end into hole H9.

4. **Connect a jumper wire from pin 2 of the IC to the other metal contact.**

   Insert one end of a short jumper wire into hole B15 and the other end into hole B9.

5. **Insert the two jumper wires that simulate the metal contacts.**

Pick out a couple of jumper wires long enough to reach from row 9 and dangle an inch or so over the edge of the breadboard. Insert one end of these wires into holes E9 and F9 and leave the other ends free. Separate the ends of the two jumper wires to make sure they're not touching; they should be about ½" apart.

Figure 6-11 shows what the breadboard looks like after these steps.

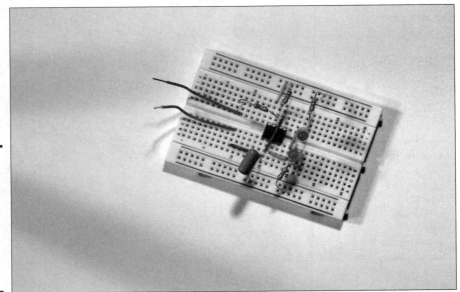

Figure 6-11: The breadboard after the finger contact jumpers have been connected.

The remaining two steps complete the circuit by connecting the power supply.

1. **Connect the battery snap connector.**

The leads on the battery snap connector use stranded rather than solid wire, so you'll need to prepare them a bit before you insert them into the breadboard.

a. Use your wire strippers to strip off about ½" of insulation from the end of both leads.

b. Use your fingers to twist the leads as tightly as you can, so that no individual strands are protruding from the very tip of the wire.

c. Insert the red lead into the last hole of the topmost row and insert the black lead into the last hole of the bottommost row.

2. **Connect the 9 V battery to the snap connector.**

   The red LED should immediately light up. (If not, see the troubleshooting tips in the next section.)

You can now test the circuit by touching both of the two free jumper wires. Pinch them both between your thumb and index finger, but don't let the wires actually touch each other. The resistance in your skin will conduct enough current to complete the circuit, and the LEDs will start alternately flashing, red, green, red, green, and so on. They will continue to flash until you let go of the jumper wires. Then, one or the other will stay lit. When you touch the wires again, the flashing will resume.

Figure 6-12 shows the completed circuit in operation.

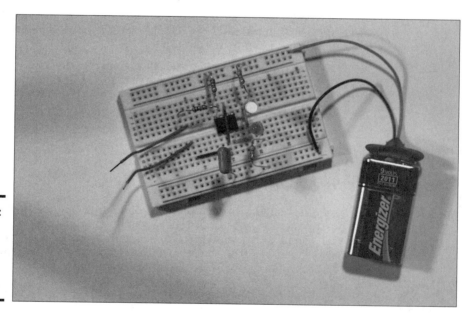

Figure 6-12:
The prototype of the coin tosser in operation.

Notice that if you squeeze the wires tightly, the rate at which the LEDs flash increases. If you squeeze tight enough, the LEDs will flash so fast that both will appear to be solid on. The LEDs are still flashing alternately, but they're flashing faster than your eye's ability to discern the difference, so they appear to be on constantly.

What if it doesn't work?

If your circuit doesn't work, there are a number of troubleshooting steps you can take to find out why and correct the problem. Here are some helpful troubleshooting tips:

✦ **Examine all the component leads to make sure none of them are touching each other.** If any of the leads are touching, gently adjust them so that they're not touching.

✦ **Make sure the circuit is getting power.** Use your multimeter to test the battery voltage (see Chapter 7 of this minibook for information on how to do that), and verify that the leads from the battery snap connector are inserted properly into the solderless breadboard.

✦ **Carefully double-check your wiring to make sure that every jumper and every component has been inserted in the correct spot.**

✦ **Verify the orientation of the 555 timer IC, making sure that pin 1 is in hole E14 and pin 8 is in hole F14.**

✦ **Verify that the LEDs are inserted in the correct direction.** For the red LED, the short lead (the cathode) goes in D21, and the long lead (the anode) goes in D19. For the green LED, the short lead (the cathode) goes in G21, and the long lead (the anode) goes in G19.

Constructing Your Circuit on a Printed Circuit Board (PCB)

Once you're satisfied with the operation of your circuit, the next step is to build a permanent version of the circuit. Although there are several ways to do that, the most common is to construct the circuit on a *printed circuit board*, also called a *PCB*. In the following sections, you'll learn how printed circuit boards work and how to assemble the coin-toss circuit on a PCB.

Note that assembling a circuit on a PCB requires that you know how to solder. To learn that all-important skill, refer to Chapter 7 of this minibook.

Understanding how printed circuit boards work

A printed circuit board is made from a layer of insulating material such as plastic or some similar material. Copper circuit paths are bonded to one side of the board. The circuit paths consist of *traces*, which are like the wires that connect components, and *pads,* which are small circles of copper that the component leads can be soldered to. Figure 6-13 shows a typical printed circuit board.

Figure 6-13:
A printed
circuit
board.

There are two basic styles of printed circuit boards available:

✦ **Through-hole:** A PCB in which the copper circuits are on one side of the board and the components are installed on the opposite side. In a through-hole PCB, small holes (usually $\frac{1}{16}$" in diameter) are drilled through the board at the center of the copper pads. Components are mounted to the blank side of the board by passing their leads through the holes and soldering the leads to the copper pads on the other side of the board. Once the solder joint is completed, any excess wire lead is trimmed away.

✦ **Surface-mount:** A PCB in which the components are installed on the same side of the board as the copper circuits. No holes are drilled.

Surface-mount PCBs are easier for large-scale automated circuit assembly. However, they're much more difficult to work with as a hobbyist because the components tend to be smaller and the leads are closer together. Thus, all of the PCBs used in this book are of the through-hole variety.

Using a preprinted PCB

The easiest way to work with printed circuit boards is to purchase a pre-printed board from RadioShack or another electronics parts supplier. RadioShack carries several different preprinted circuit boards in their stores. You can find an even greater variety of preprinted PCBs if you shop online distributors such as www.hobbyengineering.com, www.jameco.com, or www.allelectronics.com.

As you can see, preprinted PCBs come in a wide variety of shapes and sizes. The most useful, from a hobbyist's point of view, are the ones that mimic the terminal-strip and bus-strip layout of a solderless breadboard. For example, Figure 6-14 shows a PCB that has 550 holes laid out in a standard bread-board arrangement. With a preprinted PCB that has a breadboard layout, you can transfer your breadboard prototype circuit to the PCB without having to come up with an entirely new layout.

Figure 6-14:
A preprinted PCB that uses a standard breadboard layout.

Some preprinted circuit boards have layouts that are similar to standard breadboard layouts, but not identical. So check carefully before you build; you may have to make minor adjustments to your circuit layout to accom-modate the PCB you're using.

If necessary, you can cut a larger PCB to a smaller size. One way to cut a PCB is to score it on both sides with a heavy-duty utility knife, and then snap it at the score. Another way is to cut it with a rotary tool such as a Dremel.

Creating a custom PCB

As you get more advanced in your electronics skills, you may find that you want to create your own custom-printed circuit boards rather than force your circuit designs to fit the limited variety of preprinted circuit boards that are available. Although it isn't a trivial or inexpensive process, it's possible to make your own printed circuit boards customized for your circuit.

Here are the basic steps for creating your own printed circuit boards:

1. **Purchase a blank PCB. The entire surface of one side of this board will be completely coated with copper.**

2. **Create a *mask* on the copper surface that indicates the circuit layout.**

   There are several ways to do this. For simple circuits, you can simply hand-draw the circuit onto the copper using a special pen designed for this purpose. For more complicated circuits, you can purchase special stickers that are shaped like pads and common traces and place the stickers directly on the copper. Or you can design the circuit on your computer using any graphics drawing software, print the design on special paper, and transfer the design to the copper using a hot iron.

3. **Etch the board by dipping it into a special chemical that eats away all the copper that isn't covered by your mask.**

   This is a nasty process that needs to be done outside in a well-ventilated area while wearing gloves, a face mask, and goggles. When the etching is finished, all the copper that wasn't covered by the mask will be gone.

4. **Wash the board to get all that nasty copper etching solution off.**

5. **Scrub away the mask to reveal the beautiful copper circuit pattern left behind.**

6. **Drill holes in the center of each pad, and then assemble your circuit.**

Note that there are several companies on the Internet that will make small printed circuit boards for you. It isn't cheap, but the price per board comes down dramatically if you have them make more than just one. For example, Pad2Pad (www.pad2pad.com) will make 4"-square circuit boards for about $30 per board if you order five.

Building the coin-toss circuit on a PCB

This section presents a complete procedure for building the coin-toss circuit on a small preprinted PCB. Once you get all your materials together, you should be able to complete this project in about an hour.

All the parts required to build this prototype circuit can be purchased from your local RadioShack. Or, you can order them online from any electronic parts supplier. For your convenience, here is a complete list of the parts you'll need to build this prototype circuit, along with the RadioShack catalog part numbers:

Part Number	Quantity	Description
276-159	1	General-purpose, dual-printed circuit board
276-1723	1	LM555 timer IC
271-1321	1	1 kΩ, ¼ W resistor (5 per package)
271-1335	1	10 kΩ, ¼ W resistor (5 per package)
271-1317	2	470 Ω, ¼ W resistor (5 per package)
272-1053	1	0.1 µF polyester film capacitor
276-041	1	Red LED 5 mm
276-022	1	Green LED 5 mm
275-1547	1	Normally open, momentary-contact pushbutton
270-325	1	9 V battery snap connector
n/a	1	9 V battery

You will also need about a foot each of 22-gauge solid insulated wire and 22-gauge stranded insulated wire. The color doesn't matter.

Note: This list is similar to the list I give earlier in this chapter for building the coin-toss circuit on a solderless breadboard. If you have the parts from that project, you can reuse them here.

Figure 6-15 shows the layout of the preprinted circuit board. Before we start building the circuit, study the layout of this board for a moment to familiarize yourself with it. As you can see, this board doesn't contain bus strips like those found on a breadboard. However, the overall layout of the board is similar to the layout of the terminal strips on a breadboard. The center portion of the board contains a total of 20 terminal strips, ten on each side of the ditch. Each strip has three holes but is also connected to a second strip of two holes along the edge of the board. Thus, each strip effectively has five holes.

The strips aren't numbered on the board, but I've numbered them in Figure 6-14. I use the numbers 1 through 10 to number the strips on the left side of the board and the numbers 11 through 20 to number the strips on the right. In the instructions that follow, I use these numbers to indicate which holes to attach components or jumper leads to. **Note:** To keep things simple, I just specify the terminal strip number and leave it up to you to decide which of the five holes in the strip to use.

The numbers used in the PCB layout are relative to the bottom of the board — that is, the surface of the board with the copper traces and pads. Thus, the numbers 1 through 10 are on the left and the numbers 11 through 20 are on the right. When you flip the board over to insert components from the top of the board, you'll have to mentally reverse the numbers: 1 through 10 is on the right and 11 through 20 is on the left.

When installing components onto the PCB, use an alligator clip as a temporary clamp to hold the components flush against the board. This will enable you to turn the board upside down so you can solder the leads to the pads. If you don't clamp the component to the board, the component will fall out when you turn the board upside down, or you'll be tempted to hold the component in place with one finger while you solder the leads. Bad idea: Resistors get really hot when you solder them.

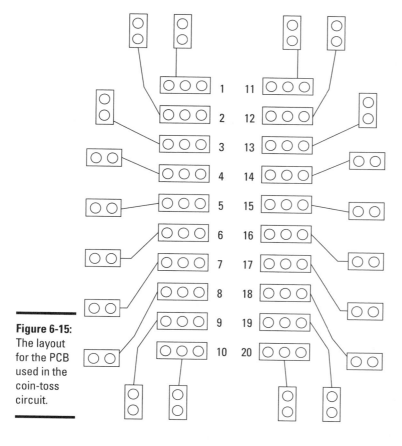

Figure 6-15:
The layout for the PCB used in the coin-toss circuit.

Here are the steps for building the coin-toss circuit on a preprinted PCB.

1. Break the PCB in half.

The preprinted circuit board comes with two identical sections. We need just one of those sections for this project, so you can break the board in half and save the other half for another project. (To break the board, just grab an end in each hand and snap it in two.)

2. Insert the 555 timer IC.

Remember that the dot or notch on the 555 IC marks pin 1. Install the chip so that pin 1 is in strip 4 and pin 8 is in strip 14. Then solder the chip carefully into place. (See Chapter 7 of this minibook for tips on soldering.)

3. Install the jumper wires.

This circuit needs a total of nine jumper wires. Cut the jumper wires from the 22-gauge solid wire and carefully strip the insulation from each end. Use needle-nosed pliers to bend the bare end of each jumper wire down, insert both ends into the appropriate holes, solder the leads to the pads, and then use your wire cutters to snip the excess of the end of each lead.

The following table provides the PCB strip locations for each jumper wire. Use your own judgment to determine which hole in the indicated strip to place the jumper wire in. Whenever possible, use the shortest possible path for each jumper wire.

Jumper #	From strip	To strip
1	9	10
2	19	20
3	5	16
4	4	10
5	7	19
6	2	6
7	1	5
8	2	12
9	14	19

4. Install the resistors.

There are four resistors to be installed. Use the following table to install each resistor into its correct location. Bend the leads down and insert each resistor into the correct holes, solder the resistor in place, and then snip off the excess wire from the ends of the leads.

Resistor#	Value	Colors	From strip	To strip
R1	1 kΩ	Brown, black, red	9	10
R2	10 kΩ	Brown, black, orange	11	15
R3	470 Ω	Yellow, purple, brown	13	19
R4	470 Ω	Yellow, purple, brown	3	10

5. Install the capacitor.

Install the capacitor into strips 5 and 10. Push the capacitor all the way in until it's flush with the board. Then solder the leads to the pad and trim off the excess wire.

6. Connect the two segments of row 19.

Insert one end of a short jumper wire into hole E19 and the other end into hole F19. This jumper wire bridges the gap between the two terminal strips in row 19, effectively making them a single terminal strip.

7. Install the LEDs.

Remember that LEDs are directional and must be installed in the correct direction or they won't work. One lead is shorter than the other to help you tell which lead is which. This short lead is the cathode, and the longer lead is the anode.

The following table shows where to install the LEDs:

LED Color	Cathode (short lead)	Anode (long lead)
Red	12	13
Green	3	2

When you install the LEDs, do *not* push the LED in until it is flush with the circuit board. Instead, push just a little bit of the leads into the holes so that the LED stands up about an inch from the top of the board.

8. Install the jumper wires for the metal contacts.

Cut two, 2" lengths of stranded wire and strip about ⅜" of insulation from each end. Solder one end of each wire into holes in strips 1 and 11 and leave the other ends free. When the circuit is installed in its final enclosure, you connect the ends of these wires to the metal posts that the user will touch to activate the coin-toss circuit.

Feeding the stranded wire through the holes in the circuit board can be tricky. First, carefully twist the loose strands until there are no stragglers protruding from the end of the wire. Then, carefully push the wire through the hole. If any of the strands get caught and refuse to go through the hole, pull the wire out and try again.

9. Connect the pushbutton.

Cut a 2" length of stranded wire and strip about ⅜" of insulation from each end. Solder one end to either terminal on the pushbutton (it doesn't matter which). Push the other end through a hole in strip 10 and solder it in place.

10. Connect the battery snap connector.

Strip off about ⅜" of insulation from the end of both leads. Then, solder the black lead to the free terminal on the pushbutton (the terminal you did not use in Step 9) and solder the red lead to a hole in strip 20 on the PCB.

11. Connect the 9V battery to the snap connector.

The red LED should immediately light up, indicating that the circuit is ready to do its decision making work.

12. Turn off your soldering iron.

You're done!

Test the circuit by pinching both of the free jumper wires between your fingers. The LEDs should alternately flash until you let go, at which time one or the other will remain lit.

Figure 6-16 shows the completed circuit in operation.

Figure 6-16:
The completed coin-tosser PCB.

Finding an Enclosure for Your Circuit

Once your circuit board is finished, the final step to completing your project is to mount it in a nice enclosure such as a plastic, metal, or wooden box. You can purchase plastic or metal boxes specifically designed for electronics projects from most electronic parts suppliers. Most RadioShack stores stock a half dozen or so different sizes in their stores, but you'll find a better assortment of sizes if you shop online. Figure 6-17 shows an assortment of boxes I picked up at my local RadioShack.

Figure 6-17: Project boxes come in a variety of shapes and sizes.

If you don't want to spend the money for a bona fide electronic project box, here are a few alternative ways to find the perfect enclosure for your project:

✦ **Shop discount department stores for small storage boxes.** You might find one that's just the right size and shape for much less money than an official project box of the same size would cost. (If the storage box is an unattractive color, you can always give it a quick coat of paint.)

✦ **In the electrical department of any hardware store, you'll find inexpensive plastic and metal boxes designed for household wiring.** Many of these boxes can be adapted for your electronic projects.

✦ **Before you throw away an old electronic gizmo, take a quick look at the box it's contained in.** If you think it might be useful for a project someday, take it apart and discard all the innards, keeping only the empty carcass.

Be careful whenever you disassemble any electronic device. Make sure you have first completely removed the power source and watch out for large capacitors that may be holding on to their charge.

✦ **If you frequent yard sales, be on the lookout for items that might be useful as containers for your projects.**

Working with a project box

Most project boxes are made of plastic or metal and have a detachable lid that's held on with four screws, one at each corner of the lid. To gain access to the insides of the box, you simply remove the screws to release the lid.

The inside of the box may be completely smooth, or it may contain ridges or mounting studs designed to make it easier to mount components inside the box. If there are no such accoutrements inside the box, you'll have to devise your own method of attaching the various bits and pieces that need to go inside. Here are some tips:

✦ **You'll need a good assortment of small drill bits to drill holes through the box to mount your components.** You'll need to drill holes to mount the circuit board as well such things as battery holders, switches, LEDs, speakers, and whatever else your project may require.

✦ **Make a good sketch of your project box and how its parts will be arranged before you start drilling holes.** When you're sure you have everything laid out the way you want, use a marker to indicate the exact position of the holes you need to drill.

✦ **To mount the circuit board, use *standoffs* to provide some empty space between the board and the case.** A *standoff* is a screw that allows you to mount the board so that it is raised above the bottom of the project box. You can purchase standoffs from any electronic parts supplier, but they are surprisingly expensive, often as much as 45 cents each.

If you have an ample supply of nuts and bolts, you can fashion your own standoffs simply by cutting a short length of plastic tubing — my favorite material is ¼" drip irrigation hose — and feeding a long bolt through it.

✦ **Consider mounting the circuit board on the back of the lid rather than inside the body of the box.** This sometimes frees up more room within the box for larger items such as batteries or a speaker.

+ **Most switches can be mounted to a box by drilling a hole large enough to allow the neck of the switch to pass through.** The switch comes with a nut that you can tighten over the neck of the screw to secure the screw to the box.

+ **Some components don't have mounting nuts to secure them to the box.** For them, a small amount of epoxy or other glue can help set the component in place.

+ **Use stranded wire for connections within the project box.** Stranded wire holds up better to the handling it occasionally gets when you open the box, for example, to change the batteries.

Mounting the coin-toss circuit in a box

In this section, you finish the coin-toss project by mounting its circuit board in a plastic project box along with a 9 V battery, a power button, and the two metal contacts that the user can touch to toss the coin.

All the parts you need for this project can be purchased at most RadioShack stores — with the exception of the standoffs, which I got at a local hardware store. In addition to the circuit board assembled earlier in this chapter, you'll need the following materials:

Part Number	Quantity	Description
270-1803	1	Project enclosure 5" x 2.5" x 2"
270-326	1	9 V battery holder
n/a	8	½" 6-32 standoff male-to-female
n/a	6	6-32 nuts (to fit standoffs)
n/a	4	6-32 ¼" bolts (to fit standoffs)

Once you've collected your parts, here's the procedure for putting the project together:

1. **Drill the required mounting holes in the lid.**

    You need to drill a total of eight holes in the lid. Four are for mounting the circuit board, two are for the LEDs, and two are for the metal finger contacts.

    Figure 6-18 provides a template you can use. The smaller holes are ⅛", the two larger holes are ³⁄₁₆". *Note:* The distance between the four holes at the top of the lid must be exactly 1⅜" so that they line up with the mounting holes in the circuit board. The measurements for the other holes don't need to be as precise.

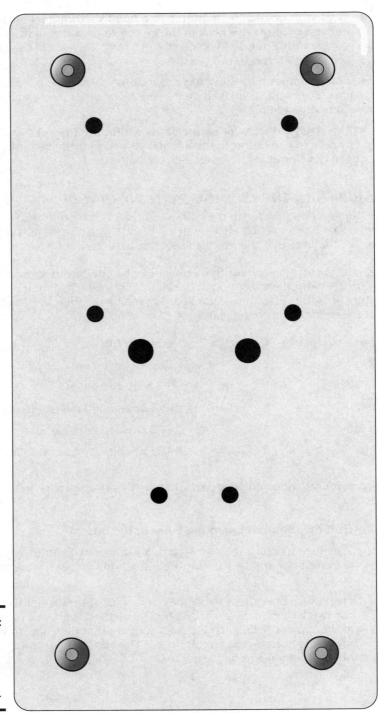

Figure 6-18:
Location
for drilling
holes in the
lid of the
project box.

2. **Drill a ⁵⁄₁₆" hole for the pushbutton near the top of the left side of the box.**

    The exact location isn't that important. I suggest you hold the box in the palm of your left hand and drill the hole at a location where your thumb will be able to easily reach the button.

3. **Mount the pushbutton in the box.**

    Remove the nut from the neck of the pushbutton, pass the neck through the hole you drilled in the preceding step from the inside of the box, and then thread the nut onto the neck from the outside of the box and tighten it down with pliers.

4. **Use Epoxy or hot glue to glue the 9 V battery holder to the base of the box.**

    Position the holder near the top of the box, adjacent to the hole you drilled for the switch.

5. **Mount the circuit board standoffs.**

    Mount four of the ½" standoffs on the underside of the lid by inserting the threaded end of the standoff through the appropriate hole and attaching a 6-32 nut on the other side. See the left side of Figure 6-19 for guidance on how to attach these four standoffs.

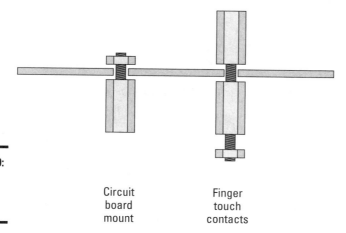

Figure 6-19:
Installing the standoffs.

Circuit board mount

Finger touch contacts

6. **Mount the finger touch contacts on the lid.**

    To make the metal contacts that the user can touch to toss the coin, first insert the threaded end of one of the standoffs through the hole from the top side of the lid, and then screw a second standoff into it from the bottom of the lid. Then attach a 6-32 nut to the threaded end of the standoff that's beneath the lid. See the right side of Figure 6-19 for guidance on installing these standoffs.

Figure 6-20 shows how the box should appear at this point. In this figure, you can see the battery holder and pushbutton already installed in the box, and you can see the standoffs on the underside of the lid.

Figure 6-20:
The box with the pushbutton, battery holder, and standoffs installed.

7. **Mount the circuit board to the standoffs.**

To complete this step, you must first bend the LEDs around so that they wrap over the edge of the circuit card and then face straight down. Be very careful when you bend the LEDs so that you don't break the leads or damage any of the solder joints. Figure 6-21 shows what the circuit board looks like when the LEDs are properly turned around.

Once the LEDs are ready, position the circuit board over the four standoffs. The LEDs should slide right into the two ³⁄₁₆" holes you drilled for them in Step 1. If they don't, just nudge them a bit to make them fit. When everything is in place, secure the circuit board to the standoffs using the 6-32 bolts.

Figure 6-21:
The circuit board with the LEDs turned around.

8. **Connect the finger contacts.**

Attach the free ends of the two jumper wires that are connected to the circuit board to the two finger contacts. To connect each wire, wrap the stripped end of the wire tightly around the threaded part of the standoff. Then, attach a 6-32 nut to the standoff and tighten it with pliers.

Figure 6-22 shows what the project looks like with the circuit board installed into the lid and the jumpers connected to the finger contacts.

9. **Install the battery.**

Insert the battery into the holder, and then connect the snap adapter to the battery.

Figure 6-22:
The circuit
board
attached to
the lid and
the finger
contacts
connected.

10. **Attach the lid to the box.**

Carefully flip the lid over and secure it to the box using the screws that
came with the project box.

11. **Turn it on and toss a coin!**

You're finally ready to use the coin-tosser project to help you make
decisions. Hold the project box in your left hand and depress the pushbutton with your thumb. Then, touch the index finger of your right hand
to the two finger contacts, and watch the LEDs alternate. When you're
ready, let go and see whether the red or green light stays lit.

Figure 6-23 shows the finished project.

Figure 6-23:
The
completed
coin-toss
project.

Chapter 7: The Secrets of Successful Soldering

In This Chapter

✓ Assembling your soldering toolkit

✓ Soldering

✓ Distinguishing a good solder joint from a not-so-good one

✓ Undoing a bad solder joint

*S*oldering is one of the basic skills of building electronic projects. Although you can use solderless breadboards to build test versions of your circuits, sooner or later you'll want to build permanent versions of your circuits. To do that, you need to know how to solder.

In this chapter, you learn the basics of soldering: how soldering works, what tools and equipment you need for soldering, and how to create the perfect solder joint. You also learn how to correct your mistakes when (not if) you make them.

Keep in mind throughout this chapter that soldering is a skill that takes a bit of practice to master. It's not nearly as hard as playing the French horn, but soldering does have a learning curve. When you're first getting started, you'll feel like you're all thumbs, and your solder joints may look less than perfect. But stick with it — with a little practice, you'll get the hang of soldering in no time at all.

Understanding How Solder Works

Before we get into the nuts and bolts of making a solder joint, spend a few minutes thinking about what soldering is and what it isn't.

Soldering refers to the process of joining two or more metal objects by heating them, and then applying *solder* to the joint. Solder is a soft metal made from a combination of tin and lead. When the solder melts, which happens at about 700° Fahrenheit, it flows over the metals to be joined. When the solder cools, it locks the metals together in a connection.

700° Fahrenheit is pretty hot — certainly hot enough to burn your skin instantly on contact — so soldering is an inherently dangerous task. It isn't extremely dangerous, as the amount of solder you actually use and the tools you use to solder are fairly small. So any burns you do receive are likely to be small. But they can be painful, so you should take great care whenever you're soldering.

Soldering is especially useful for electronics because not only does it create a strong physical connection between metals, but it also creates an excellent conductive path for electric current to flow from one conductor to another. This is because the solder itself is an excellent conductor. For example, you can create a reasonably good connection between two wires simply by stripping the insulation off the ends of each wire and twisting them together. However, current can flow through only the areas that are actually physically touching. Even when twisted tightly together, most of the surface area of the two wires won't actually be touching. But when you solder them, the solder flows through and around the twists, filling any gaps while connecting the entire surface area of both wires.

Soldering isn't the same as braising or welding. *Braising* is similar to soldering, but instead of solder, metals with a higher melting point (usually 850° Fahrenheit or more) are used. Braising forms a stronger bond than soldering. *Welding* is an entirely different process. In welding, the metals to be joined are actually melted. In liquid form, the metals intermix and they cool to form an extremely strong bond.

Procuring What You Need to Solder

Before you can start soldering, you need to acquire some stuff, as described in the following sections.

Buying a soldering iron

A *soldering iron* — also called a *soldering pencil* — is the basic tool for soldering. Figure 7-1 shows a typical soldering iron.

Here are some things to look for when purchasing a soldering iron:

✦ **The wattage rating should be between 20 and 50 watts.** Note that the wattage doesn't control how hot the soldering iron gets. Instead, it controls how fast it heats up and how fast it regains its normal operating temperature after completing each solder joint. (Each time you solder, the tip of the soldering iron cools a bit as it transfers its heat to the wires you're joining and to the solder itself. A higher-wattage soldering iron can maintain a stable temperature longer while you're soldering a connection and can reheat itself faster in between.)

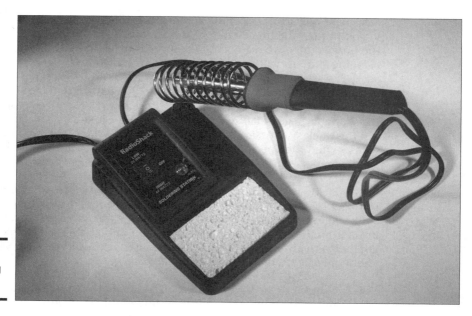

Figure 7-1:
A soldering
iron.

+ **The tip should be replaceable.** When you buy your soldering iron, buy a few extra tips so you'll have replacements handy when you need them.

+ **Although you can buy a soldering iron by itself for under $10, I suggest you spend a few more dollars and buy a *soldering station* that includes an integrated stand.** A good, secure place to rest your soldering iron when not in use is essential. Without a good stand, I guarantee that your workbench will soon be covered with unsightly burn marks. (The soldering iron pictured in Figure 7-1 includes a soldering station.)

+ **A ground three-prong power plug is desirable, but not essential.**

 The grounding helps prevent static discharges that can damage some sensitive electronic components when you solder them.

+ **More expensive models have built-in temperature control.** Although not necessary, temperature control is a nice feature if you'll be doing a lot of soldering.

Stocking up on solder

Solder, the soft metal that's used to create solder joints, is an alloy of tin and lead. Most solders are 60 percent tin and 40 percent lead, but that ratio may vary a bit.

Although solder is wound on spools and looks like wire, it's actually a thin hollow tube that has a thin core of rosin in the center. This rosin, called the *flux,* plays a crucial role in the soldering process. It has a slightly lower melting point than the tin/lead alloy, and so it melts just a few moments before the tin/lead mixture melts. The flux prepares the metals to be joined by cleaning and lubricating the surfaces to be joined.

Solder comes in various thicknesses, and you'll need to have several different thicknesses on hand for different types of work. I suggest you start with three spools: 0.062", 0.032", and 0.020". You'll use the 0.032" for most work, but the thick stuff (0.062") comes in handy for soldering larger stranded wires — and the fine solder (0.020") is useful for delicate soldering jobs on small components.

Solder is about 40 percent lead, and as you probably know, lead poisoning is a very real health hazard. Fortunately, your actual exposure to lead while soldering is pretty small. The smoke that often puffs up when you're soldering is from the rosin flux, not from the lead or tin melting. Even so, it's a good idea to always work in a well-ventilated area, and to wash your hands after soldering to remove any residual lead.

You can purchase lead-free solder, although it's considerably more expensive than regular solder made from lead and tin. However, lead-free solder is more difficult to work with than normal solder. If you're concerned about the long-term effects of working with leaded solder, I suggest you switch to lead-free solder but only after you've become proficient at working with leaded solder.

My attorney advised me to rewrite the previous paragraph to say that under no circumstance should you ever use solder that contains lead. He also insisted that I warn you to never plug your soldering iron in, lest it get hot and create the potential for burns, and that you should never ever connect your electronic circuits to actual electricity because it could create a shock hazard. In fact, you shouldn't even read this book because it could damage your eyes or cause a paper cut. Or you might accidentally drop it on your foot.

Do *not* buy acid-flux solder, which plumbers use to solder pipes. The acid in this type of solder will destroy your electronics projects.

Other goodies you need

Besides a soldering iron and solder, there are a few additional things you'll need to have on hand for successful soldering. To wit:

✦ **Third-hand tool or vise:** It takes at least three hands to solder: one to hold the items you're soldering, one to hold the soldering iron, and one to hold the solder. Unless you actually have three hands, you need to use a third-hand tool, a vise, or some other resourceful device to hold the items you're soldering so you can wield the soldering iron and the solder. (Refer to Chapter 3 of this minibook for photographs of a third-hand tool and a hobby vise.)

✦ **A sponge:** Used to clean the tip of the soldering iron.

✦ **Alligator clips:** They serve two purposes when soldering. First, you can use an alligator clip as a clamp to hold a component in place while you solder it and, second, as a heat sink to avoid damaging a sensitive component when soldering the component's leads. (A *heat sink* is simply a piece of metal attached to a heat-sensitive component that helps dissipate heat radiated by the component.)

✦ **Eye protection:** Always wear eye protection when soldering. Sometimes hot solder pops and flies through the air. Your eye and melted solder aren't a good mix.

✦ **Magnifying glass:** Soldering is much easier if you do it through a magnifying glass so you can get a better look at your work. You can use a table-top magnifying glass, or a magnifying glass attached to a third-hand tool, or special magnifying goggles.

✦ **Desoldering braid and desoldering bulb:** These tools are used to undo soldered joints when necessary to correct mistakes. For more information, see the section "Desoldering" later in this chapter.

Preparing to Solder

Before you start soldering, you should first prepare your soldering iron by cleaning and tinning it. Follow these steps:

1. **Turn on your soldering iron.**

It will take about a minute to heat up.

2. **When the iron is hot, clean the tip.**

The best way to clean your soldering iron is to wipe the tip of the iron on a damp sponge. As you work, you should wipe the iron on the sponge frequently to keep the tip clean.

3. **Tin the soldering iron.**

Tinning refers to the process of applying a light coat of solder to the tip of the soldering iron. Tinning the tip of your soldering iron helps the solder flow more freely once it heats up. To tin your soldering iron, melt a small amount of solder on the end of the tip. Then, wipe the tip dry with your sponge.

You should clean the tip of your soldering iron frequently as you work — ideally, immediately after every joint you solder. You should also tin the soldering iron occasionally. Usually, it's sufficient to tin the iron once at the start of each project. But if the solder disappears from the tip of the iron, you should tin it again.

The Ten Soldering Commandments

Have you heard of the Ten Soldering Commandments? In truth, they're actually more like guidelines than commandments. But if you heed them, it will go well with your solder joints, your children's solder joints, and your children's children's solder joints. So let it be written; so let it be done.

I. Thou shalt wear eye protection whenever thou solderest, lest thy get molten solder in thine eye.

II. Thou shalt not touch the heated end of thy soldering iron, lest thy burn thyself.

III. Thou shalt not fashion molten solder into false globs.

IV. Thou shalt wash thy hands after thou solderest, to remove vile lead contamination from upon thy hands before thou eatest.

V. Thou shalt provide bright illumination upon thine objects which thou solderest, that thou might see clearly the way unto which the solder may be applied.

VI. Thou shalt not spill thy excess solder upon thy neighbor's pad lest thy create unintended pathways through which current may flow.

VII. Thou shalt not leave thine hot soldering iron unattended.

VIII. Thou shalt not covet thy neighbor's professional-grade temperature-controlled soldering station.

IX. Thou shalt not apply solder directly upon thine soldering iron, but shalt instead apply solder to the objects which thou solderest, that their heat may melteth thy solder.

X. Thou shalt always place thine hot soldering iron in a suitable holder.

Soldering a Solid Solder Joint

The most common form of soldering when creating electronic projects is soldering component leads to copper pads on the back of a printed circuit board. If you can do that, you'll have no trouble with other types of soldering, such as soldering two wires together or soldering a wire to a switch terminal.

Just for fun, try saying "Soldering a Solid Solder Joint" real fast ten times.

The following steps outline the procedure for soldering a component lead to a PC board:

1. **Pass the component leads through the correct holes.**

Check the circuit diagram carefully to be sure you have installed the component in the correct location. If the component is polarized (such as a diode or an integrated circuit) verify that the component is oriented correctly. You don't want to solder it in backward.

2. **Secure the component to the circuit board.**

   If the component is near the edge of the board, the easiest way to secure it is with an alligator clip. You can also secure the component with a bit of tape.

3. **Clamp the circuit board in place with your third-hand tool or vise.**

   Orient the board so that the copper-plated side is up. If you're using a magnifying glass, position the board under the glass.

4. **Make sure you have adequate light.**

   If you have a desktop lamp, adjust it now so that it shines directly on the connection to be soldered.

5. **Touch the tip of the soldering iron to both the pad and the lead at the same time.**

   It's important that you touch the tip of the soldering iron to both the copper pad and the wire lead. The idea is to heat them both so that solder will flow and adhere to both.

   The easiest way to achieve the correct contact is to use the tip of the soldering iron to press the lead against the edge of the hole, as shown in Figure 7-2.

Figure 7-2:
Positioning
the
soldering
iron.

6. **Let the lead and the pad heat up for a moment.**

It should take only a few seconds for the lead and the pad to heat up sufficiently.

7. **Apply the solder.**

   Apply the solder to the lead on the opposite side of the tip of the soldering iron, just above the copper pad. The solder should begin to melt almost immediately.

   Figure 7-3 shows the correct way to apply the solder.

Figure 7-3: Applying the solder.

Do *not* touch the solder directly to the soldering iron. If you do, the solder will melt immediately, and you may end up with an unstable connection, often called a *cold joint*, where the solder doesn't properly fuse itself to the copper pad or the wire lead.

8. **When the solder begins to melt, feed just enough solder to cover the pad.**

   As the solder melts, it will flow down the lead and then spread out onto the pad. You want to feed just enough solder to completely cover the pad, but not enough to create a big glob on top of the pad.

   Be stingy when applying solder. It's more common to have too much solder than too little, and it's a lot easier to add a little solder later if you didn't get quite enough coverage than it is to remove solder if you applied too much.

9. **Remove the solder and soldering iron and let the solder cool.**

   Be patient — it will take a few seconds for the solder to cool. Don't move anything while the joint is cooling. If you inadvertently move the lead, you'll create an unstable cold joint that will have to be resoldered.

10. **Trim the excess lead by snipping it with wire cutters right just above the top of the solder joint.**

    Use a small pair of wire cutters so you can trim it close to the joint.

Checking Your Work

After you've completed a solder joint, you should inspect it to make sure the joint is good. Look at it under a magnifying glass, and gently wiggle the component to see if the joint is stable. A good solder joint should be shiny and fill but not overflow the pad, as shown in Figure 7-4.

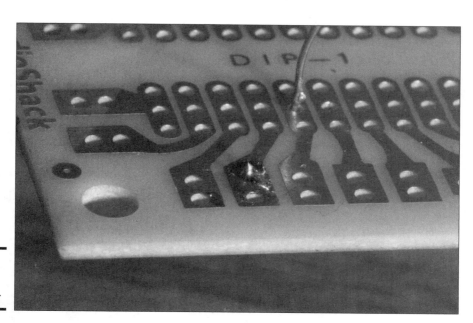

Figure 7-4:
A good
solder joint.

Nearly all bad solder joints are caused by one of three things: not allowing the wire and pad to heat sufficiently, applying too much solder, or melting the solder with the soldering iron instead of with the wire lead. Here are some indications of a bad solder joint:

+ **The pad and lead aren't completely covered with solder, enabling you to see through one side of the hole through which the lead passes.** Either you didn't apply quite enough solder, or the pad wasn't quite hot enough to accept the solder.

+ **The lead is loose in the hole or the solder isn't firmly attached to the pad.** One possible reason for this is that you moved the lead before the solder had completely cooled.

+ **The solder isn't shiny.** Shiny solder indicates solder that heated, flowed, and then cooled properly. If the solder gets just barely hot enough to melt, then flows over a wire or pad that isn't heated sufficiently, it will be dull when it cools. (Unfortunately, the new lead-free solder almost always cools dull, so it looks like a bad solder joint even when the joint is good!)

+ **Solder overflows the pad and touches an adjacent pad.** This can happen if you apply too much solder. It can also happen if the pad didn't get hot enough to accept the solder, which can cause the solder to flow off the pad and onto an adjacent pad. If solder spills over from one pad to an adjacent pad, your circuit may not work right.

Desoldering

Desoldering refers to the process of undoing a soldered joint. You may have to do this if you discover that a solder joint is less than satisfactory, if a component fails, or if you connect your circuit incorrectly.

To desolder a solder connection, you need a desoldering bulb and a desoldering braid. Figure 7-5 shows these tools.

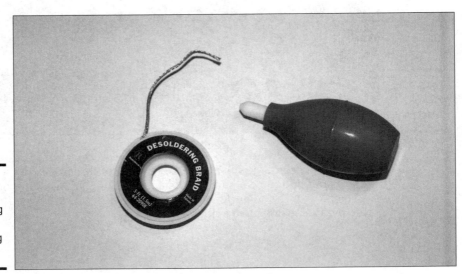

Figure 7-5:
A desoldering bulb and a desoldering braid.

Here are the steps for undoing a solder joint:

1. **Apply the hot soldering iron to the joint you need to remove.**

   Give it a second for the solder to melt.

2. **Squeeze the desoldering bulb to expel the air it contains, and then touch the tip of the desoldering bulb to the molten solder joint and release the bulb.**

   As the bulb expands, it will suck the solder off the joint and into the bulb.

3. **If the desoldering bulb didn't completely free the lead, apply heat again and touch the remaining molten solder with the desoldering braid.**

   The desoldering braid is specially designed to draw solder up much like a candle wick draws up wax.

4. **Use needle-nose pliers or tweezers to remove the lead.**

   Do *not* try to remove the lead with your fingers after you have desoldered the connection. The lead will remain hot for awhile after you have desoldered it.

Chapter 8: Measuring Circuits with a Multimeter

In This Chapter

✔ Familiarizing yourself with a multimeter

✔ Measuring current, voltage, and resistance

✔ Looking at your first electronic equation: Ohm's Law

Wouldn't it be great if every circuit you ever built worked right the first time you built it? You'd quickly develop a reputation as an electronic genius and in no time at all you'd be the president of Intel.

But in the real world, a circuit doesn't always work right the first time. When it doesn't, you can scratch your head, stare at it, put a dead chicken in a bag and swing it over your head, or you can pull out your test equipment and analyze the circuit to find out what went wrong.

In this chapter, you learn how to use one of the electronic guru's favorite tools: the multimeter. Lean how to use it well. It will be your trusty companion throughout your electronic journeys.

Looking at Multimeters

You know those late-night commercials that try to sell you those amazing kitchen gadgets that are like a combination blender, juicer, food processor, mixer, and snow cone maker all in one? A multimeter is kind of like one of those crazy kitchen gadgets, except that, unlike the crazy kitchen gadgets, a multimeter really *can* do all the things it claims it can do. And, unlike the crazy kitchen gadgets, the things a multimeter can do turn out to be genuinely useful.

Along with a good soldering iron, a good multimeter is the most important item in your toolbox. Learn how to use it, and your electronic exploits will be much more fruitful.

Figure 8-1 shows a simple, inexpensive multimeter. This one can be purchased from your local RadioShack store for under $20. If you shop around, however, you can often find a basic multimeter for under $10 and sometimes for as little as $5.

Of course, you can also spend much more, but if you're just getting started, an inexpensive multimeter is fine. Eventually, you'll want to invest a little more money in a better-quality multimeter.

Figure 8-1:
You can
buy a basic
multimeter
like this
one for
under $20.

The multimeter shown in Figure 8-1 is a *digital* multimeter, which displays its values using a digital display that shows the actual numbers for the measurements being taken. The alternative to a digital multimeter is an *analog* multimeter, which shows its readings by moving a needle across a printed scale. To determine the value of a measurement, you simply read the scale behind the needle.

Figure 8-2 shows a typical analog multimeter. This one happens to be one of my favorites. Even though it's old enough to be called vintage, it's still an excellent and accurate meter. One of the benefits of spending a little more to buy quality equipment is that the equipment will give you many years — sometimes decades — of reliable service.

The following paragraphs describe the various parts that make up a typical multimeter:

✦ **Display or meter:** Indicates the value of the measurement being taken. In a digital multimeter, the display is a number that indicates the amperage (current), voltage, or resistance being measured. In an analog meter, the current, voltage, or resistance is indicated by a needle that moves across a printed scale. To read the value, you look straight down at the needle and read the scale printed behind it.

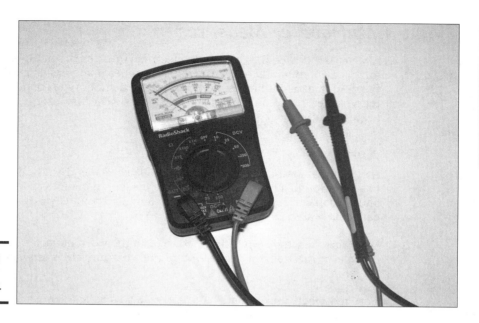

Figure 8-2:
An analog
multimeter.

✦ **Selector:** Most multimeters — digital or analog — have a dial that you can turn to tell the meter what you want to measure. The various settings on this dial indicate not only the type of measurement you want to make (voltage, current, or resistance) but also the range of the expected measurements. The range is indicated by the maximum amount of voltage, current, or resistance that can be measured.

Higher ranges let you measure higher values, but with less precision. For example, the analog multimeter shown in Figure 8-2 has the following ranges for reading DC voltage: 2.5 V, 10 V, 50 V, 250 V, and 500 V. If you use the 2.5 V range, you can easily tell differences of a tenth of a volt, such as the difference between 1.6 and 1.7 V. But when the range is set to the 500 V range, you'll be lucky to pick out differences of 10 volts.

✦ **On/off switch:** Some multimeters don't have an on/off switch. Instead, one of the positions on the selector dial is Off. Other multimeters have a separate on/off switch. If your meter doesn't give you any readings, check to make sure the power switch is turned on.

✦ **Test leads:** The *test leads* are a pair of red and black wires with metal probes on their ends. One end of these wires plugs into the meter. You use the other end to connect to the circuits you want to measure. The red lead is positive; the black lead is negative.

What a Multimeter Measures

A *meter* is a device that measures electrical quantities. A multimeter, therefore, is a combination of several different types of meters all in one box. At the minimum, a multimeter combines three distinct types of meters (ammeter, voltmeter, and ohmmeter) into a single device, as described in the following sections.

Ammeter

As you learn in Chapter 2 of this minibook, *current* is the flow of electric charge through a conductor. Current is measured in units called *amperes*. It should come as no surprise, then, that a meter that measures amperage is called an *ammeter*.

Don't ask me why the *p* is dropped to form the word *ammeter*, rather than *ampmeter*. After all, the short form of the word *ampere* is *amp*, not *am*. Go figure.

Very few electronic circuits have currents so strong they can be measured in actual amperes. So ammeters usually measure current in *milliamperes*, also called a *milliamp* and usually abbreviated *mA*. One mA is one-thousandth of an ampere; in other words, there are 1,000 milliamps in an amp.

The world's first ammeter was invented by a Dutch physicist named Hans Christian Oersted in 1821, when he accidentally left a compass next to a wire that had an electric current flowing through it. Hans noticed that when the current flowed, the needle moved away from its normal northerly orientation and pointed toward the wire. This is because current moving through a wire creates a magnetic field around the wire, and the magnetic field was strong enough to attract the magnetized end of the compass needle.

After fooling around with it a bit, Hans discovered that the more current he ran through the wire, the farther the needle strayed away from north. It didn't take him long to figure out that this discovery could be utilized to measure the amount of current flowing through a circuit. Analog ammeters work by this very same principle even today.

Hans Christian Oersted wasn't the famous writer of children's stories; that was Hans Christian Andersen. In a strange twist of history, though, Hans Christian Oersted was close friends with Hans Christian Andersen. It's entirely possible that they were the founding members of some secret society of Evil Mad-Scientist Children's Book Writers. Perhaps we'll read about them in Dan Brown's next novel.

Voltmeter

In Chapter 2 of this minibook, you learn about a second fundamental quantity of electricity, *voltage*, a term that refers to the difference in electric charge between two points. If those two points are connected to a conductor, a current will flow through the conductor. Thus, voltage is the instigator of current.

The unit of voltage is, naturally, the *volt*, and a device that measures voltage is called a *voltmeter*.

It turns out that, all other things being equal, a change in the amount of voltage between two points results in a corresponding change in current. Thus, if you can keep things equal, you can measure voltage by measuring current, and you already know of a device that can measure current: It's called an ammeter.

The basic difference between an ammeter and a voltmeter is that in an ammeter, you let current run directly through the meter so that you can measure the amount of current. In a voltmeter, the current is first run through a very large resistor and then through the ammeter, and the device makes the necessary calculations as follows.

You haven't learned it yet (unless you're eagerly skipping around the book), but in Book II, Chapter 1, you find out that there's a direct relationship between voltage, resistance, and current in an electrical circuit. In particular, if you know any two of these quantities, the third one is easy to calculate. In a voltmeter, a large fixed resistance is used, and the ammeter measures the current. Because you know the amount of the fixed resistance and the amount of current, you can easily calculate the amount of voltage across the circuit.

Don't worry; you don't have to do any math to calculate this voltage. The voltmeter does the math for you. In an analog voltmeter, the calculation is built in to the scale that's printed on the meter, so all you have to do is look at the position of the needle on the scale to read the voltage. In a digital voltmeter, the voltage is automatically calculated and displayed digitally.

For a brief explanation of the relationship between current, voltage, and resistance, see the sidebar, "A sneak peek at Ohm's law."

Ohmmeter

As you know, a *resistor* is a material that resists the flow of current. How much the current is restricted is a function of the amount of resistance in the resistor, which is measured in units called *ohms*. The symbol for ohms is the Greek letter omega, Ω. A device that measures resistance is called an *ohmmeter*.

A sneak peek at Ohm's law

In Book II, Chapter 2, you take a close look at resistors and also take an in-depth look at *Ohm's law*, which is one of the most important mathematical relationships in electronics. But without going into all the details of Ohm's law here, I want to at least give you a sneak preview.

Ohm's law describes a fundamental relationship between current, voltage, and resistance in an electrical circuit. Remember from Chapter 2 that voltage is a difference in electrical charge between two points, and that if those two points are connected by a conductor, current will flow through the conductor.

Other than exotic superconductors (which exist only in laboratory experiments), no conductor is perfect. All conductors have a certain amount of resistance that inhibits the flow of current. The greater this resistance, the less the current will flow. The less this resistance is, the more the current will flow.

Ohm's law is a mathematical formula that formalizes the relationship between current, voltage, and resistance. The formula is this:

$$I = \frac{V}{R}$$

In other words, the amount of current running through a circuit is equal to the amount of voltage across the circuit divided by the amount of resistance in the circuit. The amount of current in amperes is represented by the letter I (don't ask why; it just is). V represents voltage in volts, and R represents resistance in ohms.

Using basic math, you can use this equation to calculate the voltage if you know the current and the resistance. Then, the formula becomes:

$$V = IR$$

In other words, voltage is equal to current times resistance.

Similarly, you can calculate resistance if you know the current and the voltage. Then, the formula becomes:

$$R = \frac{V}{I}$$

In other words, resistance equals the voltage divided by the current.

Like voltage, resistance can also be measured with an ammeter. Remember how I say in the previous section that there's a direct relationship between voltage, resistance, and current in any circuit, and that if you know any two of these quantities you can easily calculate the third? To recap, to measure voltage, a voltmeter provides a fixed resistance, uses an ammeter to measure the current, and then uses the resistance and current to calculate the voltage.

To measure the resistance of a circuit, an ohmmeter provides a fixed amount of voltage across the circuit, uses an ammeter to measure the current that flows through the circuit, and then uses the amount of voltage provided by the meter and the amount of current read by the meter to calculate the resistance.

As with voltage, you don't have to do this calculation; the calculation is automatically made by digital multimeters and is built in to the meter scale for analog multimeters. Thus, all you have to do is read the display or the needle on the meter to determine the resistance.

Other measurements

All multimeters can measure current, voltage, and resistance. Some multimeters can perform other types of measurements as well. For example, some meters can measure the capacitance of capacitors, and some meters can test diodes or transistors. These features are handy, but not essential.

Schematic symbols for meter functions

Ammeter, voltmeter, and ohmmeters are often included in schematic diagrams. When they are, the following symbols are used:

Symbol	Meaning
—(A)—	Ammeter
—(V)—	Voltmeter
—(Ω)—	Ohmmeter

Using Your Multimeter

In the following sections, I show you how to use your multimeter to measure current, voltage, and resistance in a simple circuit. The circuit being measured consists of just three components: a 9 V battery, a light-emitting diode (LED), and a resistor. The schematic for this circuit is shown in Figure 8-3.

If you want to follow along with the measuring procedures detailed in the following sections, you can build this circuit on a solderless breadboard. You'll need the following parts:

✦ Small solderless breadboard

✦ 470 Ω, ¼ W resistor

✦ Red LED, 5 mm

✦ 9 V battery snap connector

✦ 9 V battery

✦ Short length of jumper wire (1" or less)

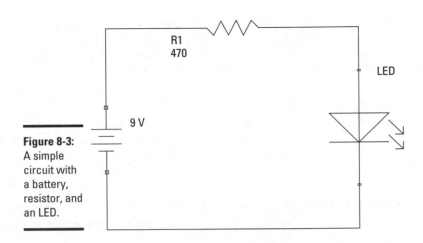

Figure 8-3:
A simple
circuit with
a battery,
resistor, and
an LED.

Figure 8-4 shows the circuit installed on the breadboard. Here are the steps for building this circuit:

1. **Connect the battery snap connector.**

   Insert the red lead in the top bus strip and the black lead in the bottom bus strip. Any hole will do, but it makes sense to connect the battery at the very end of the breadboard.

2. **Connect the resistor.**

   Insert one end of the resistor into any hole in the bottom bus strip. Then, pick a row in the nearby terminal strip and insert the other end into a hole in that terminal strip.

3. **Connect the LED.**

   Notice that the leads of the LED aren't the same length; one lead is shorter than the other. Insert the short lead into a hole in the top bus strip, and then insert the longer lead into a hole in a nearby terminal strip.

   Insert the LED into the same row as the resistor. In the figure, both the LED and the resistor are in row 26.

4. **Use the short jumper wire to connect the terminal strips into which you inserted the LED and the resistor.**

   The jumper wire will hop over the gap that runs down the middle of the breadboard.

5. **Connect the battery to the snap connector.**

The LED will light up. If it doesn't, double-check your connections to make sure the circuit is assembled correctly. If it still doesn't light up, try reversing the leads of the LED (you may have inserted it backwards). If that doesn't work, try a different battery.

Do *not* connect the LED directly to the battery without a resistor. If you do, the LED will flash brightly, and then it will be dead forever.

Figure 8-4:
The LED circuit assembled on a breadboard.

Measuring current

Electric current is measured in amperes, but actually in most electronics work, you'll measure current in milliamps, or mA. To measure current, you must connect the two leads of the ammeter in the circuit so that the current flows through the ammeter. In other words, the ammeter must become a part of the circuit itself.

The only way to measure the current flowing through the LED circuit that's shown in Figure 8-3 is to insert your ammeter into the circuit. Figure 8-5 shows one way to do this. Here, the ammeter is inserted into the circuit between the LED and the resistor.

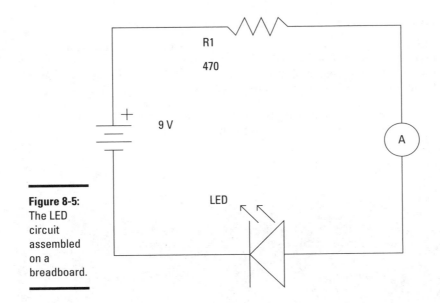

Figure 8-5:
The LED
circuit
assembled
on a
breadboard.

Note that it doesn't matter where in this circuit you insert the amme-ter. You'll get the same current reading whether you insert the ammeter between the LED and the resistor, between the resistor and the battery, or between the LED and the battery.

To measure the current in the LED circuit, follow these steps:

1. **Set your multimeter's range selector to a DC milliamp range of at least 20 mA.**

   This circuit uses direct current (DC), so you need to make sure the multi-meter is set to a DC current range.

2. **Remove the jumper wire that connects the two terminal strips.**

   The LED should go dark, as removing the jumper wire breaks the circuit.

3. **Touch the black lead from the multimeter to the LED lead that con-nects to the terminal strip (not the bus strip).**

4. **Touch the red lead from the multimeter to the resistor lead that con-nects to the terminal strip (not the bus strip).**

   The LED should light up again, as the ammeter is now a part of the cir-cuit, and current can flow.

5. **Read the number on the multimeter display.**

   It should read between 12 and 13 mA. (The exact reading will depend on the exact resistance value of the resistor. Resistor values aren't exact, so even though you're using a 470 Ω resistor in this circuit, the actual resistance of the resistor may be anywhere from 420 to 520 Ω. For more about this effect, refer to Book II, Chapter 2.)

6. **Congratulate yourself!**

   You have made your first official current measurement.

7. **After a suitable celebration, replace the jumper wire you removed in Step 2.**

   If you forget to replace the jumper wire, the procedure described in the next section for measuring voltage won't work.

There are two places in this circuit that you should *not* connect the ammeter. First, don't connect the ammeter directly across the two battery terminals. This effectively shorts out the battery. It will get real hot, real fast. Second, don't connect one lead of the ammeter to the positive battery terminal and the other directly to the LED lead. That will bypass the resistor, which will probably blow out the LED.

If you want to experiment a little more, try measuring the current at other places in the circuit. For example, remove the battery snap connector from the battery, and then reconnect it so that just the negative battery terminal is connected. Then, touch the red meter lead to the positive battery terminal and the black lead to the lead of the resistor that's connected to the bus strip (not the lead that's connected to the terminal strip). This measures the current by inserting the ammeter between the resistor and the battery. You should get the same value that you got when you measured between the LED and the resistor.

You can use a similar method to measure the current between the LED and the negative battery terminal. Again, the result should be the same.

Measuring voltage

Measuring voltage is a little easier than measuring current because to measure voltage, you don't have to insert the meter into the circuit. Instead, all you have to do is touch the leads of the multimeter to any two points in the circuit. When you do, the multimeter displays the voltage that exists between those two points.

For example, Figure 8-6 shows how you can insert a voltmeter into the LED circuit so that you can measure voltage. In this case, the voltage is measured across the battery. It should read in the vicinity of 8.3 V. (9 V batteries rarely provide a full 9 V.)

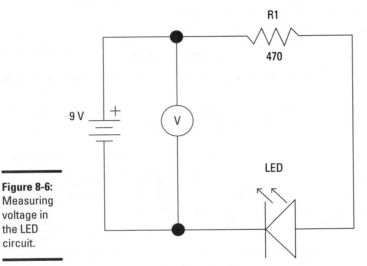

Figure 8-6:
Measuring voltage in the LED circuit.

To measure voltages in the LED circuit, first put the circuit back together (assuming you took it apart to measure currents). Then spin the multimeter dial to a range whose maximum is at least 10 V. Now just touch the leads to different spots in the circuit. To measure the voltage across the entire circuit as shown in Figure 8-6, touch the black lead to the LED lead that's inserted into the negative bus strip, and touch the red lead to the resistor lead that's inserted into the positive bus strip.

Here's an interesting exercise. Write down the following three voltage measurements:

✦ **Across the battery:** Connect the red meter lead to the resistor lead that's inserted into the positive bus strip and the black meter lead to the LED lead that's inserted into the negative bus strip.

✦ **Across the resistor:** Connect the red meter lead to the resistor lead that's inserted into the positive bus and the black meter lead to the other resistor lead.

✦ **Across the LED:** Connect the black meter lead to the LED lead that's inserted into the negative bus and the red meter lead to the other LED lead.

What do you notice about these three measurements? (It's a little bit of a puzzle, so I won't give the answer here. But you'll find it in Book II, Chapter 2.)

Measuring resistance

Measuring resistances is similar to measuring voltages, with a key difference:

You must first disconnect all voltage sources from the circuit whose resistance you want to measure. That's because the multimeter will inject a known voltage into the circuit so that it can measure the current and then calculate the resistance. If there are any outside voltage sources in the circuit, the voltage won't be fixed, so the calculated resistance will be wrong.

Here are the steps for measuring resistance in the LED circuit:

1. Remove the battery.

Just unplug it from the battery snap connector and set the battery aside.

2. Turn the meter selector dial to one of the resistance settings.

If you have an idea of what the resistance is, pick the smallest range that's greater than the value you're expecting. Otherwise, pick the largest range available on your meter.

3. If you're using an analog meter, calibrate it.

Analog meters must first be calibrated before they can give an accurate resistance measurement. To calibrate an analog meter, touch the two meter leads together. Then, adjust the meter's calibration knob until the meter indicates 0 resistance.

4. Touch the meter leads to the two points in the circuit for which you wish to measure resistance.

For example, to measure the resistance of the resistor, touch the meter leads to the two leads of the resistor. The result should be in the vicinity of 470 Ω.

You can learn much more about measuring resistances in Book 2, Chapter 2. But until then, here are a few additional thoughts to tide you over:

✦ When you measure the resistance of an individual resistor or of circuits consisting of nothing other than resistors, it doesn't matter what direction the current flows through the resistor. Thus, you can reverse the multimeter leads and you'll still get the same result.

✦ Some components such as diodes pass current better in one direction than in the other. In that case, the direction of the current does matter. You can learn more about this effect in Book II, Chapter 5.

✦ Resistors aren't perfect. Thus, a 470 Ω resistor rarely provides exactly 470 Ω of resistance. The usual tolerance for resistors is 5%, which means that a 470 Ω resistor should have somewhere between 446.5 and 495.5 Ω of resistance. For most circuits, this amount of imprecision doesn't matter. But in circuits where it does, you can use the ohmmeter function of your multimeter to determine the exact value of a particular resistor. Then, you can adjust the rest of your circuit accordingly. (More on this in Book III, Chapter 2.)

Chapter 9: Catching Waves with an Oscilloscope

In This Chapter

✔ Learning what an oscilloscope is

✔ Getting started with an oscilloscope

✔ Calibrating your scope

✔ Looking at waveforms

lectronics can generate some gnarly waves, dude. You'll encounter all kinds of totally hairy waves in your electronic circuits. You meet sine waves, which just roll along nice and easy, like corduroy on the horizon. Sawtooth waves are epic: They ride up slow, then drop you way fast. And of course, square waves. They're just, you know, *square*.

Your basic multimeter, which you learn about in the preceding chapter, is essential. You can't do electronics without one. In this chapter, I tell you about another incredibly useful tool, called an *oscilloscope*. Although a voltmeter can give you a simple number that represents voltage, an oscilloscope can draw a picture of voltage. And, as they say, a picture is worth a thousand words — er, numbers.

Why is a picture so much more valuable than a number, when it comes to voltage? Because in all but the simplest circuits, voltage is always in motion; it's always changing, and an oscilloscope is the perfect tool for observing voltage in motion.

As I said, an oscilloscope is an incredibly useful tool to have on your workbench. Unfortunately, oscilloscopes are also expensive. A decent oscilloscope will cost at least a few hundred dollars, and really good ones start at $1,000 and go up from there. So although oscilloscopes are useful, most electronic hobbyists get by without one.

The purpose of this chapter is really twofold. First, I want to show you how to use an oscilloscope should you manage to get your hands on one. Second, and perhaps more importantly, I want to convince you to start saving your pocket change so that someday you'll be able to afford one. Once you get an oscilloscope on your workbench, you'll wonder how you ever managed without it.

Understanding Oscilloscopes

Figure 9-1 shows a typical oscilloscope. This one is an older model, but although oscilloscope technology has changed over the years, even older oscilloscopes are useful for basic circuit testing. If you invest in an oscilloscope, you'll have a tool that will last you many, many years.

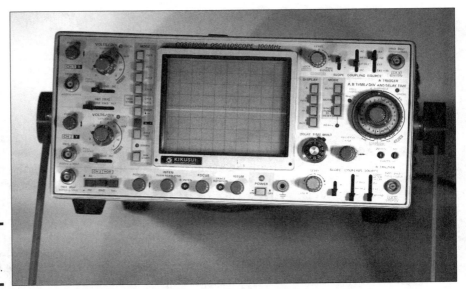

Figure 9-1:
A typical
oscilloscope.

The most obvious feature of any oscilloscope is its screen. On older oscilloscopes, the screen is a cathode-ray tube (CRT) similar to an older television or computer monitor. On newer oscilloscopes, the screen is an LCD display like a flat-screen computer monitor.

Whether CRT or LCD, the purpose of the screen is the same: To display a simple graph of an electric signal. This graph, called a *trace*, shows how voltage changes over time. The horizontal axis of this graph, reading from left to right, represents time. The vertical axis, going up and down, represents voltage.

Figure 9-2 shows a typical oscilloscope display showing a very common type of trace known as a *sine wave*. Before I tell you about the sine wave, though, there are a few things to notice about the display:

✦ Gridlines are printed on the display. On most oscilloscopes, these lines are 1 cm apart, with ten horizontal and eight vertical divisions.

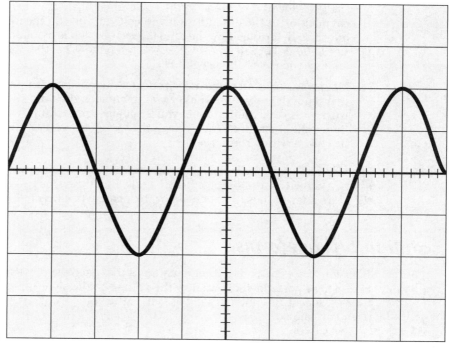

Figure 9-2:
An
oscilloscope
trace
showing a
sine wave.

✦ The vertical and horizontal lines in the middle are thicker than the other lines and include hash marks, usually 2 mm apart. These hash marks help you pinpoint the exact position of the trace between the major intervals.

✦ Various knobs and dials on the oscilloscope let you set the *scale* at which the graph of the waveform is plotted.

✦ The vertical divisions represent voltage. On most oscilloscopes, you can set the voltage scale to as little as 5 mV (millivolts) and as much as 10 V or more. The oscilloscope usually represents 0 V by the horizontal line in the middle — so lines in the top half of the display represent positive voltage, and lines in the bottom half are negative voltage. Thus, if the voltage scale is set to 1 V, the display can show voltages between +4 V and –4 V. If you set the scale to 2 V, the display can show voltages between +8 V and –8 V.

✦ The horizontal divisions represent time. The maximum time per division is typically 0.2 s (seconds), and the minimum time is typically 0.55 μs – that's half a microsecond. There are a million microseconds in a single second.

To draw a waveform, the oscilloscope actually draws a single dot that moves across the screen from left to right. Each passage of the dot from the left edge of the screen to the right edge is called a *sweep*. The vertical position of the dot indicates the voltage, and the speed that the dot moves is determined by the time interval, which is sometimes called the *sweep time*. Thus, if you set the sweep time to 0.2 s, the dot sweeps the display once every two seconds.

Most waveforms in electronics repeat at much smaller intervals than two seconds, so you'll usually want to shorten the time interval. As you work with your oscilloscope, you'll usually need to adjust the sweep time until at least one full cycle of the waveform you're examining can be shown within the screen.

Figure 9-2 shows a typical oscilloscope display showing a very common type of trace known as a *sine wave*. Before I talk about the sine wave, though, there are a few things to notice about the display itself, as follows.

Examining Waveforms

Waveforms are the characteristic patterns that oscilloscope traces usually take. These patterns indicate how the voltage in the signal changes over time — does it rise and fall slow or fast, is the voltage change steady or irregular, and so on.

There are four basic types of waveforms that you'll run into over and over again as you work with electronic circuits. These four waveforms are shown in Figure 9-3. They are:

✦ **Sine wave:** The voltage increases and decreases in a steady curve. If you remember your trigonometry class from high school, you might remember a trig function called *sine*, which has to do with angles measured in right triangles.

Few of us want to go back to high school trigonometry, so that's all I'm going to say about the mathematics behind sine waves. Suffice it to say, however, that sine waves are found everywhere in nature. For example, sine waves can be found in sound waves, light waves, ocean waves — even the bouncing of a slinky is a sine wave.

And, most importantly from the standpoint of electronics, the alternating current voltage that is provided in the public power grid is in the form of a sine wave. In an alternating current sine wave, voltage increases steadily until a peak voltage is reached. Then, the voltage begins to decrease until it reaches zero. At that point, the voltage becomes negative, which causes the current flow to reverse direction. Once it's negative, the voltage continues to change until it reaches its peak negative voltage, and then it begins to increase until it reaches zero again. The voltage then becomes positive, the current reverses, and the sine wave cycle repeats.

The number of times a sine wave (or any other wave, for that matter) repeats itself is called its *frequency*. Frequency is measured in units called *hertz*, abbreviated *Hz*. The alternating current available from a standard electrical outlet changes sixty times a second. Thus, the frequency of household AC is 60 Hz. Most waveforms found in electronic circuits have a much higher frequency than household alternating current, typically in the range of several thousand hertz (*kilohertz*, or *kHz*) or millions of hertz (*megahertz*, or *MHz*).

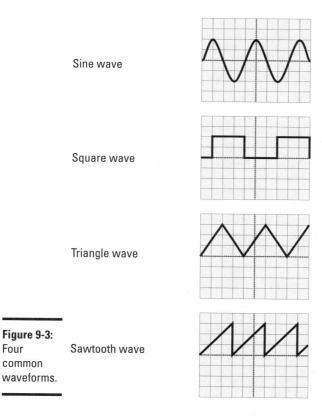

Sine wave

Square wave

Triangle wave

Figure 9-3:
Four
common
waveforms.

Sawtooth wave

✦ **Square wave:** Represents a signal in which a voltage simply turns on, stays on for awhile, turns off, stays off for awhile, and then repeats. The graph of such a wave shows sharp, right-angle turns, which is why it's called a square wave.

In actual practice, most circuits that attempt to create square waves don't do their job perfectly. As a result, the voltage rarely comes on absolutely instantly, and it rarely shuts off absolutely instantly. Thus, the vertical parts of the square wave in Figure 9-3 aren't actually vertical in the real world. In addition, sometimes the initial voltage overshoots the target voltage by a little bit, so the initial vertical uptake goes a little too high for a very brief moment, and then settles down to the right voltage.

Square waves are found in many electronic circuits. For example, the 555 timer IC used in the coin-toss project presented in Chapter 7 of this minibook produces square waves that flash the light-emitting diodes on and off, and digital logic circuits (for example, computer circuits) rely almost exclusively on square waves to represent the ones and zeros of digital electronics. (I tell you more about the 555 timer IC in Book III, Chapter 2, and digital electronics in Book VI.

✦ **Triangle wave:** Voltage increases in a straight line until it reaches a peak value, and then it decreases in a straight line. If the voltage reaches zero and then begins to rise again, the triangle wave is a form of direct current. If the voltage crosses zero and goes negative before it begins to rise again, the triangle wave is a form of alternating current.

✦ **Sawtooth wave:** This one is a hybrid of a triangle wave and a square wave. In most sawtooth waves, the voltage increases in a straight line until it reaches its peak voltage, and then the voltage drops instantly (or as close to instantly as possible) to zero, and immediately repeats.

Sawtooth waves have many interesting applications. One of the most appropriate for the purposes of this chapter is within an oscilloscope that has a CRT display. Here's a very simplified explanation of how the CRT in an oscilloscope works: It shoots a beam of electrons at a specially coated glass surface that glows when electrons hit it and uses electromagnets to steer the beam. Electromagnets above and below the beam steer it vertically; electromagnets to the right and left steer it horizontally.

To create the sweep of the electron beam from left to right, a sawtooth wave is applied to the electromagnets on the left and right of the beam. As the voltage increases, the electromagnet produces an increasingly stronger magnetic field, which pulls the beam toward the right side of display. When the voltage reaches its peak and drops instantly back to zero, the magnetic field collapses, and the electron beam snaps back to the left side of the display.

Changing the sweep rate of the oscilloscope is simply a matter of changing the frequency of the sawtooth wave applied to the horizontal electromagnets in the oscilloscope's CRT.

Calibrating an Oscilloscope

Quick: What were the first words spoken from the surface of the moon?

If you guessed, "That's one small step for a man," you'd be off by a long shot. Neil Armstrong and Buzz Aldrin had been on the moon for several hours by the time Neil said that.

If you guessed, "Houston, Tranquility Base here; the Eagle has landed," you'd be close, but still not quite right.

Contrary to popular belief, the first words spoken from the surface of the moon were not spoken by Buzz Aldrin, not Neil Armstrong. Those first words were: "Engine Stop. ACA out of detent. Auto mode control, both auto. Descent engine command override, off. Engine arm, off. 413 is in."

Before Neil Armstrong could make his historic announcement that the Eagle had landed on the moon, Buzz (being the lunar module pilot) had to quickly verify the settings of some key controls within the lunar module to ensure that everything was working well.

In the same way, before you make your historic first waveform measurement, you must first verify the settings of some key controls on your oscilloscope to ensure that everything is working well. The exact steps you need to follow to set up your oscilloscope vary depending on the exact type and model of your scope, so be sure to read the instruction manual that came with your scope. But the general steps should be as follows:

1. **Examine all the controls on your scope and set them to normal positions.**

   For most scopes, all rotating dials should be centered, all pushbuttons should be out, and all slide switches and paddle switches should be up.

2. **Turn your oscilloscope on.**

   It it's the old-fashioned CRT kind, give it a minute or two to warm up.

3. **Set the VOLTS/DIV control to 1.**

   This sets the scope to display one volt per vertical division. Depending on the signal you're displaying, you may need to increase or decrease this setting, but one volt is a good starting point.

4. **Set the TIME/DIV control to 1 ms.**

   This control determines the time interval represented by each horizontal division on the display. Try turning this dial to its slowest setting. (On my scope, the slowest setting is half a second, so it takes a full five seconds for the dot to travel across the screen.) Then, turn the dial one notch at a time and watch the dot speed up until it becomes a solid line.

5. **Set the Trigger switch to Auto.**

   The Auto position enables the oscilloscope to stabilize the trace on a common trigger point in the waveform. If the trigger mode isn't set to Auto, the waveform may drift across the screen, making it difficult to watch.

6. **Connect a probe to the input connector.**

   If your scope has more than one input connector, connect the probe to the one labeled *A*.

   Oscilloscope probes include a probe point, which you connect to the input signal and a separate ground lead. The ground lead usually has an alligator clip. When testing a circuit, this clip can be connected to any common ground point within the circuit. In some probes, the ground lead is detachable, so you can remove it when it isn't needed.

7. **Touch the end of the probe to the scope's calibration terminal.**

   This terminal provides a sample square wave that you can use to calibrate the scope's display. Some scopes have two calibration terminals, labeled *0.2 V* and *2 V*. If your scope has two terminals, touch the probe to the 2 V terminal.

   For calibrating, it's best to use an alligator clip test probe. If your test probe has a pointy tip instead of an alligator clip, you can usually push the tip through the little hole in the end of the calibration terminal to hold the probe in place.

   It isn't necessary to connect the ground lead of your test probe for calibration.

8. **If necessary, adjust the TIME/DIV and VOLTS/DIV controls until the square wave fits nicely within the display.**

   For example, see Figure 9-4.

9. **If necessary, adjust the Y-POS control to center the trace vertically.**

10. **If necessary, adjust the X-POS control to center the trace horizontally.**

11. **If necessary, adjust the Intensity and Focus settings to get a clear trace.**

12. **Congratulate yourself!**

    You're now ready to begin viewing the waveforms of actual electronic signals.

Remember that the controls of every oscilloscope make and model are unique. Be sure to read the owner's manual that came with your oscilloscope to see if there are any other setup or calibration procedures you need to follow before feeding real signals into your scope.

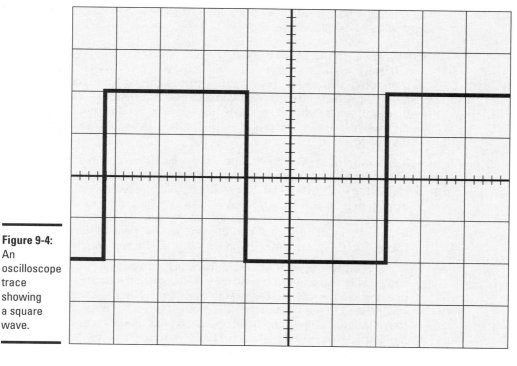

Figure 9-4:
An
oscilloscope
trace
showing
a square
wave.

Displaying Signals

The basic procedure for testing a circuit with an oscilloscope is to attach the ground connector of the scope's test lead to a ground point in the circuit, and then touch the tip of the probe to the point in the circuit that you want to test.

For example, if you want to verify that the output from a pin of an integrated circuit is emitting a square wave, touch the oscilloscope probe to the pin and look at the display on the scope. Note that you may need to adjust the VOLTS/DIV and TIME/DIV settings on the scope to clearly see the waveform. But once you get those settings adjusted correctly, you should be able to visualize the square wave. If the square wave doesn't appear, you likely have a problem with the circuit.

Never connect the oscilloscope probe directly to an electrical outlet. You're likely to kill your scope or yourself. (If you want to measure voltage from an outlet, just use your regular multimeter.)

The following paragraphs give a few ideas for viewing various kinds of waveforms with an oscilloscope:

✦ To view a simple DC waveform, try connecting the oscilloscope to a 1.5 V battery such as a AA or AAA cell. Set the VOLTS/DIV knob to 2 V, and then touch the probe ground connector to the negative battery terminal and the probe tip to the positive terminal. The resulting display should be a simple straight line midway between the second and third vertical division above the centerline. (If the battery is dead or weak, this line may be lower.)

✦ If you want to see the 60 Hz sine wave available from an electrical wall outlet, find a plug-in power supply (commonly called a *wall wart)* that generates low-voltage AC. If you don't have one lying around, you can buy them new at many stores. You can also find them for $1 or so at thrift stores. Plug the wall wart into an electrical outlet, and then connect the oscilloscope probe to the wall wart's low-voltage plug. Adjust the VOLTS/DIV and TIME/DIV settings until you can see the sine wave.

✦ If you want to see what an audio waveform looks like, find a short ⅛" audio cable that's male on both ends. Plug one end into the headphone jack of any audio device, such as a radio or an iPod. Then, connect the probe's ground lead to the shaft of the plug on the free end of the audio cable and touch the probe tip to the tip of the audio plug, as shown in Figure 9-5. After fiddling with the VOLTS/DIV and TIME/DIV settings, you should see a display of the jumbled waveform that's typical of audio signals.

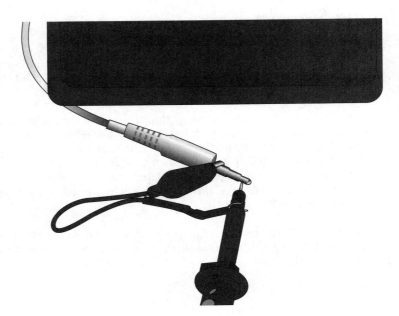

Figure 9-5:
Connecting an oscilloscope probe to an audio plug.

✦ As you build circuits while working your way through this book, keep your oscilloscope handy. Don't hesitate at any time to pick up the oscilloscope probe and check out the signals that are being generated at various points within your circuit. Connect the probe's ground clip to any ground point in the circuit, and then touch the probe tip to every loose wire and exposed pin that you can to see what's going on inside the circuit.

Book II

Working with Basic Electronic Components

"I think I've fixed the intercom. Just remember to speak into the ceiling fan when the doorbell rings."

Contents at a Glance

Chapter 1: Working with Basic Circuits

In This Chapter

✔ Examining the basic elements of circuits

✔ Looking at different batteries that you can use to power your circuits

✔ Categorizing the various switches that you can use to control your circuits

✔ Learning the difference between series and parallel circuits

✔ Building a variety of circuits to demonstrate basic circuits

*B*ook I introduces you to the underlying concepts of electricity and gives you an overview of the tools and skills you need to start working with electronics. Now that you know what electricity is, and you've gone shopping for some basic tools (such as a solderless breadboard, a soldering iron, and a multimeter), it's time to start learning how electronic circuits work.

This chapter explores the basic concepts of a circuit. The circuits I cover in this chapter are very simple, consisting of nothing other than batteries that supply voltage, wires that carry current, and lamps that consume power (because they light up in the process!). I also throw in some switches that let you turn the circuit on and off.

Though simple, the circuits presented here are a great introduction to more complicated circuits. The remaining chapters in this minibook add additional elements that are basic to all circuits, such as resistors, capacitors, coils, diodes, transistors, and integrated circuits. By the time you get through this minibook, you'll be familiar with all the basic components of electronic circuits.

What Is a Circuit?

A *circuit* is a complete course of conductors through which current can travel. Circuits provide a path for current to flow. To be a circuit, this path must start and end at the same point. In other words, a circuit must form a loop.

For example, Figure 1-1 shows a simple circuit that includes two components: a battery and a lamp. The circuit allows current to flow from the battery to the lamp, through the lamp, then back to the battery. Thus, the circuit forms a complete loop.

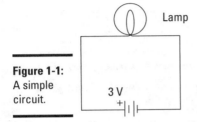

Figure 1-1:
A simple
circuit.

Of course, circuits can be more complex. However, all circuits can be distilled down to three basic elements:

✦ **Voltage source:** A voltage source causes current to flow. In Figure 1-1, the voltage source is the battery.

✦ **Load:** The load consumes power; it represents the actual work done by the circuit. Without the load, there's not much point in having a circuit.

In Figure 1-1, the load is the lamp. In complex circuits, the load is a combination of components, such as resistors, capacitors, transistors, and so on.

✦ **Conductive path:** The conductive path provides a route through which current flows. This route begins at the voltage source, travels through the load, and then returns to the voltage source. This path must form a loop from the negative side of the voltage source to the positive side of the voltage source.

In Figure 1-1, the two lines that travel between the battery and the lamp represent the conductive path. The conductive path can be intricate in a complex circuit, but it must still form a loop from the negative side of the voltage source to the positive side.

The following paragraphs describe a few additional interesting points to keep in mind as you ponder the nature of basic circuits:

✦ When a circuit is complete and forms a loop that allows current to flow, the circuit is called a *closed circuit*. If any part of the circuit is disconnected or disrupted so that a loop is not formed, current cannot flow. In that case, the circuit is called an *open circuit*.

Open circuit is an oxymoron. After all, the components must form a complete path to be considered a circuit. If the path is open, it isn't a circuit. Therefore, *open circuit* is most often used to describe a circuit that has become broken, either on purpose (by the use of a switch, which I discuss later in this chapter) or by some error, such as a loose connection or a damaged component.

✦ *Short circuit* refers to a circuit that does not have a load. For example, Figure 1-2 shows a short circuit; the lamp is connected to the circuit but a direct connection is present between the battery's negative terminal and its positive terminal, too.

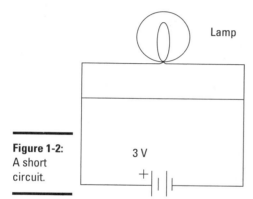

Figure 1-2:
A short
circuit.

Lamp

3 V

Current in a short circuit can flow at dangerously high levels. Short circuits can damage electronic components, cause a battery to explode, or maybe start a fire.

The short circuit shown in Figure 1-2 illustrates an important point about electrical circuits: it is possible — common, even — for a circuit to have multiple pathways for current to flow. In Figure 1-2, the current can flow through the lamp as well as through the path that connects the two battery terminals directly.

Current flows everywhere it can. If your circuit has two pathways through which current can flow, the current doesn't choose one over the other; it chooses both. However, not all paths are equal, so current doesn't flow equally through all paths. For example, in the circuit shown in Figure 1-2, current will flow much more easily through the short circuit than it will through the lamp. Thus, the lamp will not glow because nearly all of the current will bypass the lamp in favor of the easier route through the short circuit. Even so, a small amount of current will flow through the lamp.

To determine how much current flows through a given path, you use the mathematical formula that you read about in Chapter 2 of this minibook. Nevertheless, when one of the available paths is a short circuit, you needn't bother with the formula because nearly all the current will flow through the short circuit.

✦ Imagine electric current flowing through a circuit from the positive side of the voltage source to the negative side because this is usually how you visualize a circuit when you study it. For example, in Figure 1-1, think of the current as flowing in a clockwise direction, starting at the positive terminal of the battery, flowing through the path at the left side of the diagram to the lamp at the top of the diagram, through the lamp and then through the path at the right side of the diagram and finally returning to the negative terminal of the battery.

As you see in Book I, Chapter 2, this way of thinking about current flow is called *conventional current.* In reality, the electric charge — and the electrons — in the circuit flow from the negative side of the voltage source through the circuit to the positive side of the voltage source.

Using Batteries

The easiest way to provide a voltage source for a circuit is to include a battery. There are plenty of other ways to provide voltage, including AC adapters (which you can plug into the wall) and solar cells (which convert sunlight to voltage). However, batteries remain the most practical source of juice for most of the circuits you build in this book.

A *battery* is a device that converts chemical energy into electrical energy in the form of voltage, which in turn can cause current to flow. A battery works by immersing two plates made of different metals into a special chemical solution called an *electrolyte.* The metals react with the electrolyte to produce a flow of charges that accumulate on the negative plate, called the *anode.* The positive plate, called the *cathode,* is sucked dry of charges. As a result, a voltage is formed between the two plates. These plates are connected to external terminals to which you can connect a circuit to cause current to flow.

Figure 1-3 shows a simplified diagram of how a battery works. A bowl filled with the right kind of chemical that includes an anode and cathode made of the right metal gives you a working battery.

Technically, Figure 1-3 shows a *cell,* not a battery. A *battery* is a combination of two or more cells.

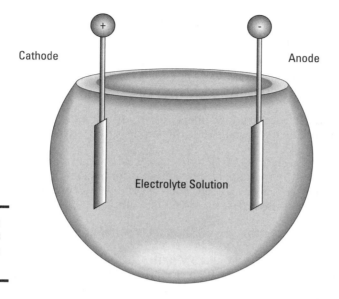

Cathode

Anode

Electrolyte Solution

Figure 1-3:
What goes
on inside a
battery.

Batteries come in many different shapes and sizes, but for the purposes of
this book, you need concern yourself only with a few standard types of batteries, all of which are available at any grocery, drug, or department store.
Figure 1-4 shows the most common sizes available.

Figure 1-4:
Common
batteries.

Cylindrical batteries come in four standard sizes: AAA, AA, C, and D. Regardless of the size, these batteries provide 1.5 V each; the only difference between the smaller and larger sizes is that the larger batteries can provide more current. Technically, AAA, AA, C, and D cells are just that — single cells, not batteries. But if it were a crime to call them batteries, our prisons would be more overloaded than they already are.

The cathode, or positive terminal, in a cylindrical battery is the end with the metal bump. The flat metal end is the anode, or negative terminal.

The rectangular battery in Figure 1-4 is a 9 V battery. It truly is a battery because that little rectangular box actually contains six small cells, each about half the size of a AAA cell. The 1.5 volts produced by each of these small cells combine to create a total of 9 volts.

Here are a few other things you should know about batteries:

+ Besides AAA, AA, C, D, and 9 V batteries, many other battery sizes are available. Most of those batteries are designed for special applications, such as digital cameras, hearing aids, laptop computers, and so on.

+ All batteries contain chemicals that are toxic to you and to the environment. Treat them with care, and dispose of them properly according to your local laws. Don't just throw them in the trash.

+ You can (and should) use your multimeter to measure the voltage produced by your batteries. Set the multimeter to an appropriate DC voltage range (such as 20 V). Then, touch the red test lead to the positive terminal of the battery and the black test lead to the negative terminal. The multimeter will tell you the voltage difference between the negative and positive terminals. For cylindrical batteries (AAA, AA, C, or D) it should be about 1.5 V. For 9 V batteries, it should be about 9 V.

+ *Rechargeable* batteries cost more than non-rechargeable batteries but last longer because you can recharge them when they go dead.

+ The technical term for a cell that cannot be recharged is *primary cell*. A rechargeable cell is properly called a *secondary cell*. However, I would be shocked if you walked into any store that sells batteries asking for secondary cells and the salesperson knew what you were talking about.

+ The easiest way to use batteries in an electronic circuit is to use a battery holder, which is a little plastic gadget designed to hold one or more batteries.

+ Wonder why they sell AAA, AA, C, and D cells but not A or B? Actually, A cell and B cell batteries exist. However, those sizes never really caught on in consumer devices, so they aren't readily available at retail stores.

Building a Lamp Circuit

Project 1-1 shows you how to build a simple circuit that uses a battery to light a lamp. Although this circuit is quite simple, it helps illustrate the basic principles I've explained so far. Figure 1-5 shows the circuit assembled.

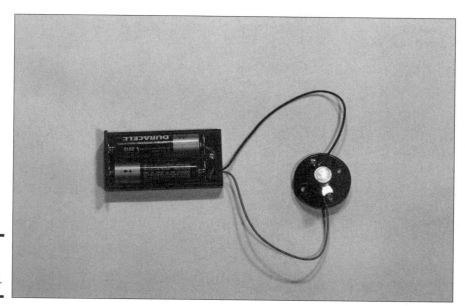

Figure 1-5:
A simple
lamp circuit.

Working with Switches

Switches are an important part of most electronic circuits. In the simplest case, most circuits contain an on/off switch to turn the circuit on and off. In addition to the on/off switch, many circuits contain additional switches that control how the circuit works or activate different features of the circuit.

Switches are mechanical devices with two or more *leads* (or *terminals*) that are internally connected to metal contacts which can be opened or closed by the person operating the switch. When the switch is in the On position, the contacts are brought together to complete the circuit so that current can flow. When the contacts are together, the switch is closed. When the contacts are apart, the switch is open and current cannot flow.

The following sections describe two ways to categorize switches: by the method used to operate the switch and by the connections made by the switch.

Project 1-1: A Simple Lamp Circuit

In this project, you'll build a simple circuit that connects a lamp to a battery. You will also use a multimeter to measure the voltage and current in the circuit. To assemble and test the circuit, you'll need a small Phillips-head screwdriver and a multimeter.

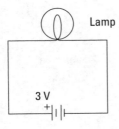

Lamp

3 V

Parts List
2 AA batteries
1 Battery holder (RadioShack 270-408)
1 Lamp holder (RadioShack 272-357)
1 3 V flashlight lamp (RadioShack 272-1175)

Steps

1. **Attach the red lead from the battery holder to one of the screw terminals on the lamp holder.**

   Loosen the screw terminal a bit to create a gap under the screw. Next, bend the stripped end of the red lead into a hook shape (you can do this easily by wrapping the wire around the tip of the screwdriver). Insert the stripped end of the lead beneath the terminal screw and then tighten the screw to secure the wire.

2. **Attach the black lead to the other terminal on the lamp holder.**

3. **Insert the lamp into the lamp holder.**

4. **Insert the batteries into the holder.**

   The lamp should light.

1. Set the multimeter range to the lowest DC Voltage setting that will measure at least 3 volts.

   My meter has a 5VDC range, so I used that setting.

2. Touch the red multimeter lead to the lamp holder terminal that the red battery holder lead is attached to and then touch the black lead to the other terminal.

   The meter should read about 3 volts.

3. Disconnect the red battery lead from the lamp holder.

   This breaks the circuit, so the lamp will go out.

4. Turn the multimeter dial to your highest DC mA setting.

   On mine, the highest setting is 1,000 DC mA.

5. Touch the stripped end of the red battery lead to the tip of the red multimeter lead, then touch the tip of the black multimeter lead to the unconnected terminal on the lamp holder.

   The meter should read approximately 250 mA. If the largest DC mA range on your multimeter is less than 250 mA, you may not get an accurate reading. However, you should get an indication that the current exceeds the maximum for the range.

   Additionally, when you test the current, the lamp comes back on because the meter completes the circuit.

The many ways to throw the switch

One way to categorize switches is by the movement a person uses to open or close the contacts. Figure 1-6 shows many different switch designs. The most common are

+ **Slide switch:** A *slide switch* has a knob that you can slide back and forth to open or close the contacts.

+ **Toggle switch:** A *toggle switch* has a lever that you flip up or down to open or close the contacts. Common household light switches are examples of toggle switches.

+ **Rotary switch:** A *rotary switch* has a knob that you turn to open and close the contacts. The switch in the base of many tabletop lamps is an example of a rotary switch.

+ **Rocker switch:** A *rocker switch* has a seesaw action. You press one side of the switch down to close the contacts, and press the other side down to open the contacts.

+ **Knife switch:** A *knife switch* is the kind of switch Igor throws in a Frankenstein movie to reanimate the creature. In a knife switch, the contacts are exposed for everyone to see.

+ **Pushbutton switch:** A *pushbutton switch* is a switch that has a knob that you push to open or close the contacts. In some pushbutton switches, you push the switch once to open the contacts and then push again to close the contacts. In other words, each time you push the switch, the contacts alternate between opened and closed.

Other pushbutton switches are *momentary contact switches,* where contacts change from their default state only when the button is pressed and held down. The two types of momentary contact switches are

• *Normally open (NO):* In a normally open switch, the default state of the contacts is open. When you push the button, the contacts are closed. When you release the button, the contacts open again. Thus, current flows only when you press and hold the button.

• *Normally closed (NC):* In a normally closed switch, the default state of the contacts is closed. Thus, current flows until you press the button. When you press the button, the contacts are opened and current does not flow. When you release the button, the contacts close again and current resumes.

Figure 1-6:
There are
switches for
every need.

Making connections with poles and throws

Another way to classify switches is by the connections they make. If you were under the impression that switches simply turn circuits on and off, guess again. Two important factors that determine what types of connections a switch makes are

✦ **Poles:** A switch *pole* refers to the number of separate circuits that the switch controls. A *single-pole* switch controls just one circuit. A *double-pole* switch controls two separate circuits.

A double-pole switch is like two separate single-pole switches that are mechanically operated by the same lever, knob, or button.

✦ **Throw:** The number of *throws* indicates how many different output connections each switch pole can connect its input to. Figure 1-7 shows the two most common types: single-throw and double-throw:

• A *single-throw* switch is a simple on/off switch that connects or disconnects two terminals. When the switch is closed, the two terminals are connected and current flows between them. When the switch is opened, the terminals are not connected, so current does not flow.

- A *double-throw* switch connects an input terminal to one of two output terminals. Thus, a double-pole switch has three terminals. One of the terminals is called the *common terminal*. The other two terminals are often referred to as *A* and *B*. When the switch is in one position, the common terminal is connected to the A terminal, so current flows from the common terminal to the A terminal but no current flows to the B terminal. When the switch is moved to its other position, the terminal connections are reversed: current flows from the common terminal to the B terminal, but no current flows though the A terminal.

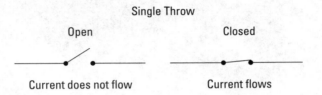

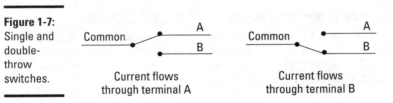

Figure 1-7: Single and double-throw switches.

Switches vary in both the number of poles and the number of throws. In theory, any number of poles and any number of throws is possible. However, most switches have one or two poles and one or two throws. This leads to four common combinations, as described in the following paragraphs. The symbols used in schematic diagrams for each of these switches are shown in the margins.

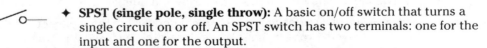

✦ **SPST (single pole, single throw):** A basic on/off switch that turns a single circuit on or off. An SPST switch has two terminals: one for the input and one for the output.

✦ **SPDT (single pole, double throw):** An SPDT switch routes one input circuit to one of two output circuits. This type of switch is sometimes called an A/B switch because it lets you choose between two circuits, called A and B. An SPDT switch has three terminals: one for the input and two for the A and B outputs.

+ **DPST (double pole, single throw):** A DPST switch turns two circuits on or off. A DPST switch has four terminals: two inputs and two outputs.

+ **DPDT (double pole, double throw):** A DPDT switch routes two separate circuits, connecting each of two inputs to one of two outputs. A DPDT switch has six terminals: two for the inputs, two for the A outputs, and two for the B outputs.

Here are a few other points to ponder concerning the arrangement of poles and throws:

+ Switches with more than two poles or more than two throws are not commonplace, but they do exist. Rotary switches lend themselves especially well to having many throws. For example, the rotary switch in a multimeter typically has 16 or more throws, one for each range of measurement the meter can make.

+ A common variation of a double throw switch is to have a middle position that does not connect to either output. Often called *center open,* this type of switch has three positions, but only two throws. For example, an SPDT center open switch can switch one input between either of two outputs, but in its center position, neither output is connected.

+ If you're stocking up on switches just to have them on hand, you're better off buying DPDT switches rather than single pole or single throw switches because a DPDT can be used when a circuit calls for a simpler SPST, SPDT, or DPST switch. You can use a DPDT switch when a simpler type is called for because there's no law that says you have to wire all the contacts on the switch. For example, to use a DPDT switch as an SPST switch, you just use one of the poles and one of the throws and leave the other connections unused.

Building a Switched Lamp Circuit

Project 1-2 presents a simple construction project that lets you explore the use of a simple on/off switch to control a lamp. Figure 1-8 shows the assembled project.

This project and the remaining projects in this chapter use a DPDT knife switch from RadioShack, pictured in Figure 1-9. It's unlikely that you'd use a knife switch in an actual electronic circuit. However, a knife switch like this is a perfect tool for learning the ins and outs of working with switches. For one thing, it is entirely exposed, so you can see how it works. Additionally, because they come on their own base and have screw terminals, connecting them in temporary circuits is simple because you don't have to do any soldering.

Figure 1-8:
The
switched
lamp
project.

As you can see in the figure, the knife switch is a double pole, double throw (DPDT) switch, which means it operates like two SPDT switches that are mechanically linked. I numbered the six terminals on the switch 1X, 1A, 1B, 2X, 2A, and 2B. The 1 and 2 designate which of the two circuits is being switched. The X terminals are the input terminals in the center of the switch, and the A and B terminals are for the two possible outputs. Thus, when the switch is flipped one way, 1X is connected to 1A and 2X is connected to 2A. When the switch is flipped the other way, 1X is connected to 1B and 2X is connected to 2B.

If you want to experiment with different variations of this project, try the following:

✦ Try moving the switch from the positive side of the circuit to the negative side. In other words, connect the red lead from the battery holder to the lamp and connect the black lead to the 1X terminal on the switch.

The circuit will function the same. This shows that the location of a switch in a circuit often doesn't matter. If the circuit is broken anywhere, current cannot flow. Thus, whether the switch is before or after the lamp doesn't matter.

✦ Cut a second 6" piece of wire and strip the insulation from both ends. Then, wire the circuit so that the red battery lead goes to switch terminal 1X, the black lead goes to switch terminal 2X, one of the wires goes from switch terminal 1A to one of the lamp terminals, and the other wire goes from switch terminal 2A to one of the lamp terminals.

Now you've created the circuit shown in Figure 1-10. In this circuit, the knife switch is used as a DPST (double pole, single throw) switch to interrupt the circuit on both the negative and the positive side of the lamp.

Book II
Chapter 1

Working with
Basic Circuits

1A   2A

Figure 1-9:
The DPDT
knife switch.

1X   1B   2B   2X

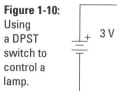

Figure 1-10:
Using
a DPST
switch to
control a
lamp.

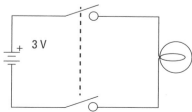

Project 1-2: A Lamp Controlled by a Switch

In this project, you'll build a simple circuit that connects a lamp to a battery and uses a switch to turn the lamp on and off. To assemble and test the circuit, you'll need a small Phillips-head screwdriver, wire cutters, and a wire stripper.

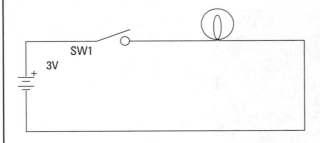

Parts List
2 AA batteries
1 Battery holder (RadioShack 270-408)
1 Lamp holder (RadioShack 272-357)
1 3 V flashlight lamp (RadioShack 272-1175)
1 DPDT knife switch (RadioShack 275-1537)
1 6" length 22-gauge stranded wire

Steps

1. Cut a 6" length of wire and strip ½" of insulation from each end.

2. Cut a 6" length of wire and strip ½" of insulation from each end.

3. Attach the red lead from the battery holder to terminal 1X on the knife switch.

4. Attach the black lead to one of the terminals on the lamp holder.

5. Use the 6" wire to connect terminal 1A of the knife switch to the other terminal of the lamp holder.

6. Insert the batteries into the holder

7. Close the knife switch on the "A" side.

   The lamp should light up.

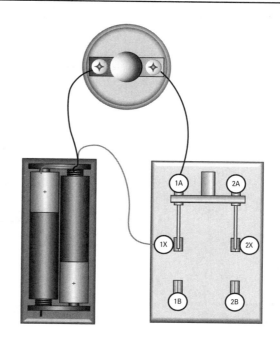

Understand Series and Parallel Circuits

Whenever you have circuits that consist of more than one component, those components must be linked together. The two ways to connect components in a circuit are in series and in parallel. Figure 1-11 illustrates how you might use series and parallel circuits to connect two lamps in a single circuit.

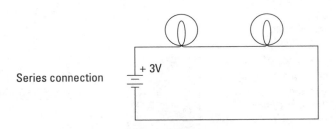

Series connection

Figure 1-11:
Lamps
connected
in series
and in
parallel.

Parallel connection

In a *series* connection, components are connected end to end, so that current flows first through one, then through the other. As you can see in the first circuit in Figure 1-11, the current goes through one lamp and then the other. The lamps are strung together end to end.

One drawback of series connections is that if one component fails in a way that results in an open circuit, the entire circuit is broken and none of the components will work. For example, if either one of the lamps in the series circuit in Figure 1-11 burns out, neither lamp will work. That's because current must flow through both lamps for the circuit to be complete.

In the *parallel* connection shown in Figure 1-12, each lamp has its own direct connection to the battery. This arrangement avoids the if-one-fails-they-all-fail nature of series connections. In a parallel connection, the components do not depend on each other for their connection to the battery. Thus, if one lamp burns out, the other will continue to burn.

An interesting thing happens with voltage when components are connected in series: the voltages present at each component are divided up. For example, in a circuit with a 3 V battery and two identical lamps connected in series, each lamp will see only one and a half volts. If you connected three identical lamps in series, each lamp would see only one volt.

You can measure the voltage seen by any component in a circuit by setting your multimeter to an appropriate voltage range and then touching the leads to both sides of the component. The voltage you measure there is called the component's *voltage drop*.

Building a Series Lamp Circuit

In Project 1-3, you'll build a circuit that connects two lamps in series, a simple circuit. Then, you'll use your multimeter to measure the voltages at various points in the circuit. The completed project is shown in Figure 1-12.

Figure 1-12:
Lamps
connected
in series.

Project 1-3: A Series Lamp Circuit

In this project, you'll connect two lamps together in a series circuit. The lamps are powered by a pair of AA batteries. To build this project, you'll need a small Phillips-head screwdriver, wire cutters, wire strippers, and a multimeter.

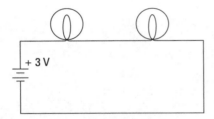

+ 3 V

Parts List

2 AA batteries
1 Battery holder
   (RadioShack 270-408)
2 Lamp holders
   (RadioShack 272-357)
2 3 V flashlight lamps
   (RadioShack 272-1175)
1 6" length 22-gauge
   stranded wire

Steps

1. Cut a 6" length of wire and strip ½" of insulation from each end.

2. Attach the red lead from the battery holder to one of the terminals on one of the lamp holders.

3. Attach the black lead to one of the terminals on the other lamp holder.

4. Use the 6" wire to connect the unused terminal of the first lamp holder to the unused terminal of the second lamp holder.

5. Insert the batteries into the holder.

   Both lamps light.

   Notice that the lamps are dim. That's because in a series circuit made with two identical lamps, each of the two lamps sees only half the total voltage.

6. Remove one of the lamps from its holder.

   The other lamp goes out. This is because in a series circuit, a failure in any one component breaks the circuit so none of the other components will work.

7. Replace the lamp you removed in Step 6.

8. Set your multimeter to a DC voltage range that can read at least 3 volts.

9. Touch the leads to the two terminals on the first lamp holder.

   The multimeter should read approximately 1.5 V. (If you're using an analog meter and the needle moves backwards, just reverse the leads.)

10. Touch the leads to the two terminals on the other lamp holder.

    Again, the multimeter should read approximately 1.5 V.

11. Touch the red lead of the meter to the terminal that the red lead from the battery is connected to and touch the black meter lead to the terminal that the black battery lead is connected to.

    This measures the voltage across both lamps combined. The meter will indicate 3V.

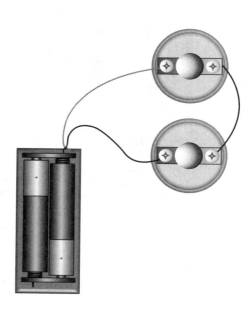

Building a Parallel Lamp Circuit

In Project 1-4, you'll build a circuit that connects two lamps in parallel and you'll use your multimeter measure voltages within various points in the circuit. The completed project is shown in Figure 1-13.

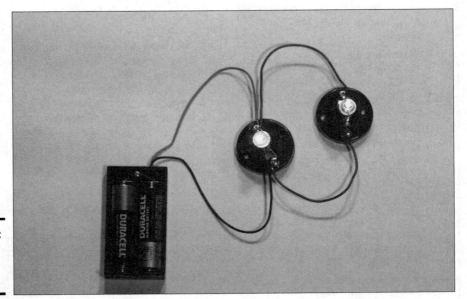

Figure 1-13:
Lamps
connected
in parallel.

Using Switches in Series and Parallel

Just as lamps can be connected in series or parallel, switches can also be connected in series or parallel. For example, Figure 1-14 shows two circuits that each use a pair of SPST switches to turn a lamp on or off. In the first circuit, the switches are wired in series. In the second, the switches are wired in parallel.

The interesting thing to note about wiring switches in series is that both switches must be closed in order to complete the circuit. A great example of switches wired in series is in the typical nuclear-war movie, where two people must flip a switch in order to launch the missiles. Switches wired in series means that Denzel Washington *and* Gene Hackman must both agree to launch the missiles.

When switches are wired in parallel, closing either switch will complete the circuit. Thus, parallel switches are often used when you want the convenience of controlling a circuit from two different locations. If the nuclear missile switches were wired in parallel, either Denzel Washington *or* Gene Hackman could fire the missiles.

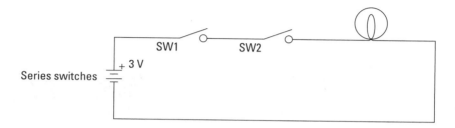

Series switches

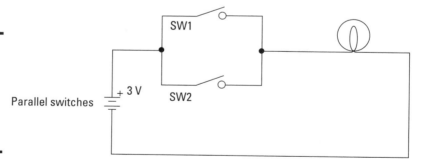

Figure 1-14:
Schematic
diagrams
for series
and parallel
switch
circuits.

Parallel switches

Project 1-4: A Parallel Lamp Circuit

In this project, you'll connect two lamps together in a series circuit. The lamps are powered by a pair of AA batteries. To build this project, you'll need a small Phillips-head screwdriver, wire cutters, wire strippers, and a multimeter.

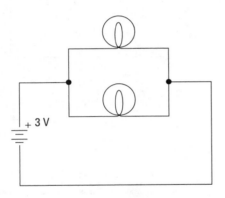

Parts List
2 AA batteries
1 Battery holder (RadioShack 270-408)
2 Lamp holders (RadioShack 272-357)
2 3 V flashlight lamps (RadioShack 272-1175)
2 6" length 22-gauge stranded wires

Steps

1. **Cut two 6" lengths of wire and strip ½" of insulation from each end.**
2. **Attach the red lead from the battery holder to one of the terminals on the first lamp holder.**
3. **Attach the black lead to the other terminals on the first lamp holder.**
4. **Use the two wires to connect each of the terminals on the first lamp holder to the terminals on the second lamp holder.**

   This wiring connects the two lamp holders in parallel.
5. **Insert the batteries.**

   The lamps light, brighter than when you connected them in series in Project 1-3.
6. **Remove one of the lamps.**

   Notice that the other lamp remains lit.
7. **Replace the lamp you removed in Step 6.**
8. **Set your multimeter to a DC voltage range that can read at least 3 volts.**
9. **Touch the leads of your multimeter to the two terminals on the first lamp holder.**

   Make sure you touch the red meter lead to the terminal that the red battery lead is connected to and the black meter lead to the terminal that the black battery lead is attached to.

   Note that the voltage reads a full 3 V.
10. **Touch the meter leads to the**

terminals on the second lamp stand.

Note that the voltage again reads 3 V. When components are connected in parallel, the voltage is not divided among them. Instead, each component sees the same voltage. That is why the lamps light at full intensity in the parallel circuit.

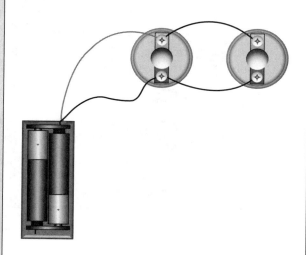

Building a Series Switch Circuit

Project 1-5 presents a simple project that uses two switches to open or close a circuit that lights a lamp. The switches are wired in series, so both switches must be closed to light the lamp. Figure 1-15 shows the completed project.

Figure 1-15: The assembled series switch circuit.

Building a Parallel Switch Circuit

In Project 1-6, you build a simple circuit that uses two switches wired in parallel to control a lamp. Because the switches are wired in parallel, the lamp will light if either of the switches is closed. Figure 1-16 shows the completed project.

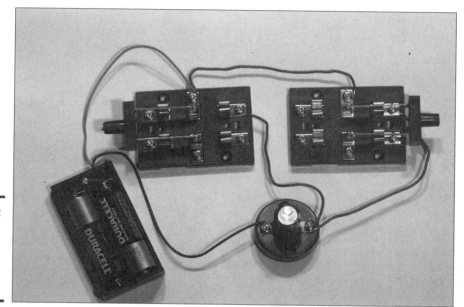

Figure 1-16:
The
assembled
parallel
switch
circuit.

Project 1-5: A Series Switch Circuit

In this project, you'll build a simple circuit that uses two knife switches to control a single lamp. To complete this project, you'll need a small Phillips-head screwdriver, wire cutters, and wire strippers.

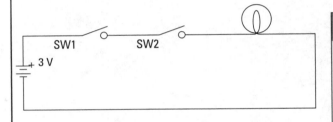

Parts List
2 AA batteries
1 Battery holder (RadioShack 270-408)
1 Lamp holder (RadioShack 272-357)
1 3 V flashlight lamp (RadioShack 272-1175)
2 DPDT knife switches (RadioShack 275-1537)
2 6" length 22-gauge stranded wire

Steps

1. **Cut two 6" lengths of wire and strip ½" of insulation from each end.**

2. **Open both switches.**
   Move the handles to the upright position so the contacts are not connected.

3. **Attach the red lead from the battery holder to terminal 1X of one of the switches.**

4. **Attach the black lead to one of the terminals on the lamp holder.**

5. **Connect one of the 6" wires from Terminal 1A of the first switch to terminal 1X of the second switch.**

6. **Connect the other 6" wire from Terminal 1A of the second switch to on the unused terminal of the lamp holder.**

7. **Insert the batteries into the holder.**

8. **Close the first switch.**
   Notice the lamp does not light.

9. **Close the second switch.**

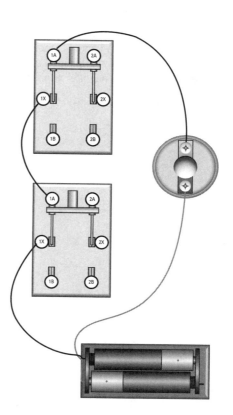

Project 1-6: A Parallel Switch Circuit

This project is a circuit that uses two switches
wired in parallel to control a lamp. To complete
this project, you'll need a small Phillips-head
screwdriver, wire cutters, and wire strippers.

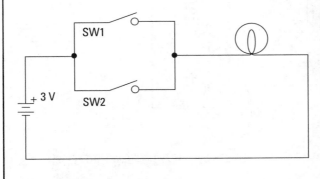

Parts List

2 AA batteries
1 Battery holder
   (RadioShack 270-408)
1 Lamp holder
   (RadioShack 272-357)
1 3 V flashlight lamp
   (RadioShack 272-1175)
2 DPDT knife switches
   (RadioShack 275-1537)
3 6" length 22-gauge
   stranded wire

Steps

1. **Cut three 6" lengths of wire and strip ½" of insulation from each end.**

2. **Open both switches.**
   Move the handles to the upright position so the contacts are not connected.

3. **Attach the red lead from the battery holder to terminal 1X of the first switch.**

4. **Attach the black lead to the first terminal on the lamp holder.**

5. **Use the first 6" wire to connect Terminal 1X of the first switch to Terminal 1X of the second switch.**
   Be sure to leave the red battery wire in place at terminal 1X on the first switch. Terminal 1X on the first switch should have two wires: the red battery terminal and the wire going to terminal 1X of the second switch.

6. **Use the second 6" wire to connect Terminal 1A of the first switch to the second terminal of the lamp holder.**

7. **Use the third 6" wire to connect Terminal 1A of the second switch to the second terminal of the lamp holder.**
   In other words, the unused terminal of the lamp holder should be connected to terminal 1A on both knife switches.

8. **Insert the batteries.**

9. **Close one of the switches.**
   The lamp lights.

10. **Open the switch you closed in**

Step 9 and then close the other switch.

Again, the lamp lights. When switches are connected in parallel, current flows through the circuit when either of the switches is closed.

11. **Close both switches.**
    The lamp remains lit because current continues to flow when both switches are closed.

12. **Open both switches.**
    The lamp goes out. With switches wired in parallel, at least one of the switches must be closed for current to flow.

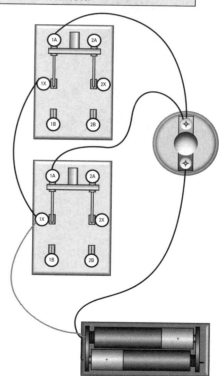

Switching between Two Lamps

In Project 1-7, you build a simple circuit that uses a single pole, double throw (SPDT) to switch a circuit between one of two lamps. In other words, one of two lamps will light depending on the position of the switch. This type of switching is a common requirement in electronic circuits. Figure 1-17 shows the completed circuit.

Figure 1-17:
The switch controls two lamps.

An interesting variant of the circuit in Project 1-7 uses both poles of the DPDT knife switch to switch the circuit on both the negative and positive sides of the lamp. For this circuit, you need four 6" wires. Connect the six terminals of the DPDT switch as follows:

Terminal	Connect To
1X	Red battery lead
1A	Terminal 1 of second lamp
1B	Terminal 1 of first lamp
2X	Black battery lead
2A	Terminal 2 of second lamp
2B	Terminal 2 of first lamp

Figure 1-18 shows this circuit assembled.

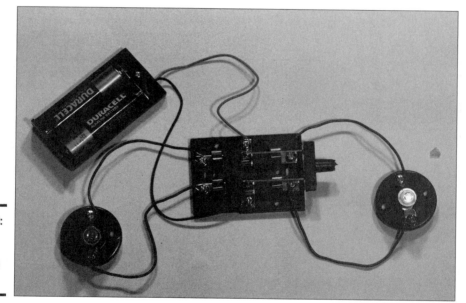

Figure 1-18: Another way to control two lamps.

Project 1-7: Controlling Two Lamps with One Switch

In this project, you'll build a circuit that uses a single switch to control two lamps. When the switch is in the first position, the first lamp lights and the second lamp is dark. When the switch is in the second position, the second lamp lights and the first lamp is dark. You'll need a small Phillips screwdriver, wire cutters, and wire strippers to build this project.

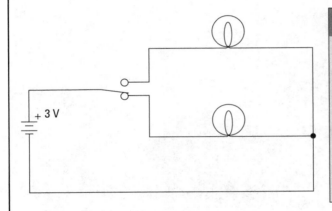

Parts List
2 AA batteries
1 Battery holder (RadioShack 270-408)
1 Lamp holder (RadioShack 272-357)
1 3 V flashlight lamp (RadioShack 272-1175)
2 DPDT knife switches (RadioShack 275-1537)
3 6" length 22-gauge stranded wire

Steps

1. **Cut three 6" lengths of wire and strip ½" of insulation from each end.**

2. **Open the switch.**
   Lift the handle into the upright position so the contacts are not connected.

3. **Attach the red lead from the battery holder to terminal 1X of the switch.**

4. **Attach the black lead to one of the terminals on the first lamp holder.**

5. **Connect one of the 6" wires from Terminal 1A of the switch to one terminal of the second lamp holder.**

6. **Connect another 6" wire from terminal 1B of the switch to the unused terminal of the first lamp holder.**

7. **Use the third 6" wire to connect the unused terminal of the second lamp holder to the terminal of the first lamp holder that the black battery lead is connected to.**

8. **Insert the lamps into the lamp holders.**

9. **Insert the batteries into the holder.**

10. **Flip the switch to the A position.**
    The second lamp lights

11. **Change the switch to the B position.**
    The first lamp lights.

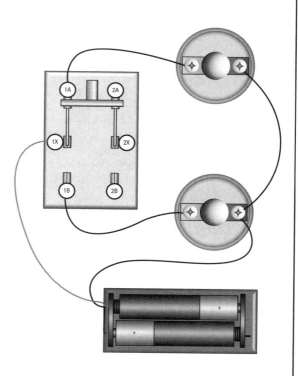

Building a Three-Way Lamp Switch

Many homes and offices have hallways that have a light switch on both ends. You can turn the light on or off by flipping either switch. This kind of switching arrangement is called a *three-way switch*.

Have you ever wondered how these three-way switches work? If you think about it, the switches are puzzling. If the light is on, flipping either switch will turn it off. If the light is off, flipping either switch will turn it on. Say you flip one switch to its On position and the light goes on. Now go to the other switch, flip it to turn the light off, and come back to the first switch. It is still in its On position, but the light is off. To turn the light back on, you can flip the first switch again.

In other words, sometimes the light is on when the switch is up, sometimes it is on when the switch is down. How can this be?

The answer is that both switches are single pole, double throw switches, and they are wired in series such that either both switches must be up or both must be down to complete the circuit. If one switch is up and the other is down, the circuit is open.

Sometimes electricians install one of the switches upside down or wire the three-way switch backwards just to confuse you. Then, the switches work backwards: if both switches are up or if both switches are down, the circuit is open and the lamp lights only when one switch is up and the other is down. But that's not the normal way to wire a three-way switch.

In Project 1-8, you build a simple circuit that uses two single pole, double throw switches to show how a three-way light switch works. Figure 1-19 shows the completed project.

Reversing Polarity

The final project in this chapter — Project 1-9 — shows you a common trick: Using a DPDT switch to reverse the polarity of a circuit. One common use for this trick is powering a DC motor with the circuit. When you reverse the polarity of a DC motor, the motor spins in the opposite direction. Thus, you can use a DPDT switch to control the direction in which a DC motor turns. Figure 1-20 shows the assembled project.

Figure 1-19:
The completed three-way light switch circuit.

Figure 1-20:
The assembled polarity-reversing circuit.

Project 1-8: A Three-Way Light Switch

In this project, you'll create a three-way switch
circuit in which a single lamp is controlled by
either of two switches. You'll need a small
Phillips-head screwdriver, wire cutters, and
wire strippers to complete this project

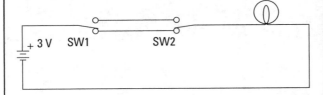

Parts List
2 AA batteries
1 Battery holder (RadioShack 270-408)
1 Lamp holder (RadioShack 272-357)
1 3 V flashlight lamp (RadioShack 272-1175)
2 DPDT knife switches (RadioShack 275-1537)
3 6" length 22-gauge stranded wires

Steps

1. Cut three 6" lengths of wire and strip ½" of insulation from each end.

2. Open both switches.

   Move the handles to their upright positions so the contacts are not connected.

3. Attach the red lead from the battery holder to terminal 1X of the first switch.

4. Attach the black lead to one of the terminals on the lamp holder.

5. Connect one of the 6" wires from Terminal 1A of the first switch to terminal 1A of the second switch.

6. Connect another 6" wire from Terminal 1B of the first switch to Terminal 1B of the second switch.

7. Connect the last 6" wire from Terminal 1X of the second switch to the unused terminal of the lamp socket.

8. Insert the lamps into the lamp holder.

9. Insert the batteries into the holder.

10. Flip the switches to see the operation of the three-way switch.

    The lamp lights only when both switches are in the A position or when both switches are in the B position.

Book II
Chapter 1

Working with
Basic Circuits

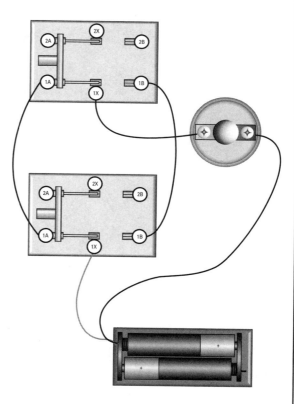

Project 1-9: A Polarity-Reversing Circuit

In this project, you'll build a circuit that uses a
DPDT switch to reverse a circuit's polarity. In
other words, flipping the switch from one
position to the other reverses the direction in
which current flows through the circuit. To
build this circuit, you'll need a small Phillips-
head screwdriver, wire cutters, and wire
strippers.

Parts List
2 AA batteries
1 Battery holder (RadioShack 270-408)
1 Lamp holder (RadioShack 272-357)
1 3 V flashlight lamp (RadioShack 272-1175)
1 DPDT knife switch (RadioShack 275-1537)
4 6" length 22-gauge stranded wires

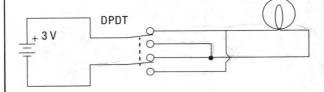

Steps

1. Cut four 6" lengths of wire and strip ½" of insulation from each end.

2. Open the switch.

   Move the handles to their upright positions so the contacts are not connected.

3. Attach the red lead from the battery holder to terminal 1X of the switch.

4. Attach the black lead from the battery to terminal 2X of the switch.

5. Connect the first 6" wire from Terminal 1B of the switch to Terminal 2A of the switch.

6. Connect the second 6" wire from Terminal 2B of the switch to Terminal 1A of the switch.

7. Connect the third 6" wire from Terminal 1A of the switch to one terminal of the lamp socket.

8. Connect the fourth 6" wire from Terminal 2A of the switch to the other terminal of the lamp socket.

9. Insert the lamps into the lamp holder.

10. Insert the batteries into the battery holder.

11. Flip the switch to the A position and measure the voltage at the lamp.

    Touch the red meter probe to the lamp terminal that's connected to switch terminal 1A and touch the black meter probe to the other lamp terminal. You should read approximately +3 V.

12. Flip the switch to the B position

and measure the voltage again.

You should read approximately –3 V. (If you're using an analog meter, you have to reverse the probes to read the negative voltage.)

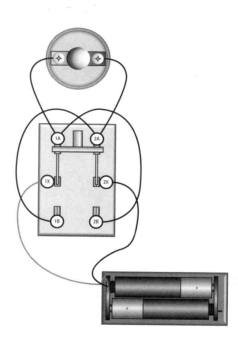

Chapter 2: Working with Resistors

In This Chapter

↗ **Understanding and measuring resistance**

↗ **Calculating resistance with Ohm's law**

↗ **Determining resistor values and tolerance**

↗ **Working with resistors in series, parallel, and combination**

↗ **Creating a voltage-divider circuit**

↗ **Looking at how a potentiometer varies resistance**

My favorite science fiction villains are the Borg from *Star Trek: The Next Generation*. They had a saying: "Resistance is futile."

In the *Star Trek* world, it turned out that resistance against the Borg wasn't futile. Picard and the rest of the crew of the *Enterprise* resisted and eventually triumphed over the Borg.

In the electronics world, resistance is not futile. In fact, resistance can be very useful. Without resistance, electronics would not be possible. Electronics is all about manipulating the flow of current, and one of the most basic ways to manipulate current is to reduce it through resistance. Without resistance, current would flow unregulated and there would be no way to coax it into doing useful work.

In this chapter, you'll learn what resistance is and how to work with *resistors,* which are little devices that let you intentionally introduce resistance into your circuits. You'll also learn about a fundamental relationship in the nature of electricity: the relationship between voltage, current, and resistance. This relationship is expressed in a simple mathematical formula called *Ohm's law.* (Not to worry; the math isn't complicated. If you know how to multiply and divide, you can understand Ohm's law.) And finally, you'll learn the most common ways resistors are used in circuits.

What Is Resistance?

As you already know, a *conductor* is a material that allows current to flow, and an *insulator* is a material that doesn't. Good conductors allow current to flow with abandon, without impediments. Examples of good conductors include the metals copper and aluminum. Carbon is also an excellent conductor. Good insulators, on the other hand, erect solid walls that completely block current. Examples of good insulators include glass, Teflon, and plastic. The key factor that determines whether a material is a conductor or an insulator is how readily its atoms give up electrons to move charge along. Most atoms are very possessive of their electrons, and are therefore good insulators. But some atoms don't have a strong hold on their outermost electrons. Those atoms are good conductors.

If a conductor and an insulator are mixed together, the result is a compound that conducts current, but not very well. Such a compound has *resistance* — that is, it resists the flow of current. The degree to which the compound resists current depends on the exact mix of elements that make up the compound.

For example, a conducting material such as carbon might be mixed with an insulating material such as ceramic. If the mix is mostly carbon, the overall resistance of the mixture will be low. If the mix is mostly ceramic, the overall resistance will be high.

The truth is, *all* materials have some resistance. Even the best conductors have a small but measurable amount of resistance. The only exceptions are certain materials called *superconductors* that, when chilled to unbelievably low temperatures, conduct with 100 percent efficiency. Unfortunately, you can't buy superconductors at RadioShack, and even if you could, you can't buy a freezer at Sears powerful enough to chill the stuff down to absolute zero.

Measuring Resistance

Resistance is measured in units called *ohms,* represented by the Greek letter omega (Ω). The standard definition of *one ohm* is simple: It's the amount of resistance required to allow one ampere of current to flow when one volt of potential is applied to the circuit. In other words, if you connect a one-ohm resistor across the terminals of a one-volt battery, one amp of current will flow through the resistor.

A single ohm ($1\ \Omega$) is actually a very small amount of resistance. Resistances in the hundreds, thousands, or even millions of ohms are usually called for in electronic circuits.

In Book I, Chapter 8, you learn that you can measure resistance of a circuit using an *ohmmeter*, which is a standard feature found in most multimeters. The procedure is simple: First, you disconnect all voltage sources from the circuit; then, you touch the ohmmeter's two probes to the ends of the circuit and read the resistance (in ohms) on the meter.

Here are a few other points to consider about resistance and ohms:

✦ The abbreviations *k* (for *kilo*) and M (for *mega*) are used for thousands and millions of ohms. Thus, a 1,000-ohm resistance is written as 1 kΩ, and a 1,000,000-ohm resistance is written as 1 MΩ.

✦ For the purposes of most electronic circuits, you can assume that the resistance value of ordinary wire is zero ohms (0 Ω). In reality, however, only superconductors have a resistance of 0 Ω. Even copper wire has some resistance. Because of that, the resistance of wire is usually measured in terms of ohms per kilometer or per mile. Electronic circuits usually deal with wires that are at most a few inches or feet long, not kilometers or miles.

✦ Short circuits also have essentially zero resistance.

✦ Just as ordinary wire and short circuits can be considered to have zero resistance, insulators and open circuits can be considered to have infinite resistance, and in reality, there's no such thing as completely infinite resistance. If you connect two wires to the terminals of a battery and hold the wires apart, a voltage difference exists between the ends of those two wires, and a very small current will travel between them — even through the air because air doesn't have infinite resistance. This current is extraordinarily small — too small to even measure — but it's there nonetheless. Electric currents are literally everywhere.

✦ The unit *ohm* is named after the famous German physicist Georg Ohm, who was the first to explain the relationship between voltage, current, and resistance in 1827.

✦ Actually, the discovery was first made by a British scientist named Henry Cavendish more than 45 years earlier, but Cavendish never published his work. If he had, resistance would be measured in cavens, not ohms.

**Book II
Chapter 2**

Working with Resistors

Looking at Ohm's Law

The term *Ohm's law* refers to one of the fundamental relationships found in electric circuits: that, for a given resistance, current is directly proportional to voltage. In other words, if you increase the voltage through a circuit whose resistance is fixed, the current goes up. If you decrease the voltage, the current goes down.

Ohm's law expresses this relationship as a simple mathematical formula:

$$V = I \times R$$

In this formula, *V* stands for voltage (in volts), *I* stands for current (in amperes), and *R* stands for resistance (in ohms). (You may be wondering why *V* stands for voltage here, but in other equations voltage is sometimes represented by the letter *E*. Although scientists sometimes argue over whether V or E should be used in various circumstances, for the most part V and E are interchangeable when referring to voltage.)

Here's an example of how to calculate voltage in a circuit with a lamp powered by the two AA cells. Suppose you already know that the resistance of the lamp is 12 Ω, and the current flowing through the lamp is 250 mA, which is the same as 0.25 A. Then, you can calculate the voltage as follows:

Ohm's law is incredibly useful because it lets you calculate an unknown voltage, current, or resistance. In short, if you know two of these three quantities you can calculate the third.

Go back (if you dare) to your high-school algebra class and remember that you can rearrange the terms in a simple formula such as Ohm's law to create other equivalent formulas. In particular:

$$V = I \times R$$

✦ If you don't know the voltage, you can calculate it by multiplying the current by the resistance.

$$I = \frac{V}{R}$$

✦ If you don't know the current, you can calculate it by dividing the voltage by the resistance.

$$R = \frac{V}{I}$$

✦ If you don't know the resistance, you can calculate it by dividing the voltage by the current.

To convince yourself that these formulas work, look again at the circuit with a lamp that has 12 Ω of resistance connected to two AA batteries for a total voltage of 3 V. Then you can calculate the current flowing through the lamp as follows:

$$I = \frac{V}{R} = \frac{3V}{12\Omega} = 0.25A$$

If you know the battery voltage (3 V) and the current (250 mA, which is 0.25 A), you can calculate the resistance of the lamp like this:

$$R = \frac{V}{I} = \frac{3V}{0.25A} = 12\Omega$$

Wasn't going back to high school algebra fun? Next thing you know, you're going to start looking for a prom date.

The most important thing to remember about Ohm's law is that you must always do the calculations in terms of volts, amperes, and ohms. For example, if you measure the current in milliamps (which you usually will in electronic circuits), you must convert the milliamps to amperes by dividing by 1,000. For example, 250 mA is 0.25 A.

Here are a few other things you should keep in mind concerning Ohm's law:

✦ Remember how I say in the preceding section that the definition of one ohm is the amount of resistance that allows one ampere of current to flow when one volt of potential is applied to it? This definition is based on Ohm's law. If *V* is 1 and *I* is 1, then *R* must also be 1.

$$R = \frac{V}{I} = \frac{1V}{1A} + 1\Omega$$

✦ If you wonder why the symbols for voltage and resistance are *V* and *R*, which make perfect sense, but the symbol for current is *I*, which makes no sense, it has to do with history. The unit of measure for current — the ampere — is named after André-Marie Ampère, a French physicist who was one of the pioneers of early electrical science. The French word he used to describe the strength of an electric current was *intensité* – in English, *intensity.* Thus, amperage is a measure of the intensity of the current. Hence the letter *I.*

✦ In the interest of international cooperation, the term *volt* is named for the Italian scientist Alessandro Volta, who invented the first electric battery in 1800. (Actually, his full name was Count Alessandro Giuseppe Antonio Anastasio Volta. But that won't be on the test.)

Introducing Resistors

A *resistor* is a small component that's designed to provide a specific amount of resistance in a circuit. Because resistance is an essential element of nearly every electronic circuit, you'll use resistors in just about every circuit that you build.

Although resistors come in a variety of sizes and shapes, the most common type of resistor for hobby electronics is the *carbon film resistor,* shown in Figure 2-1. These resistors consist of a layer of carbon laid down on an insulating material and contained in a small cylinder, with wire leads attached to both ends. The resistor itself is about ¼" long, and the leads are about an inch long, making the entire thing about 2-¼" long.

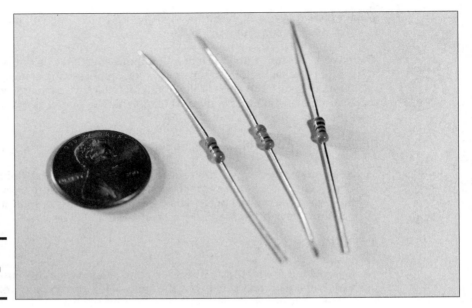

Figure 2-1:
Carbon film
resistors.

Resistors are blind to the polarity in a circuit. Thus, you don't have to worry about installing them backwards. Current can pass equally through a resistor in either direction.

In schematic diagrams, a resistor is represented by a jagged line, like the one shown in the margin. The resistance value is typically written next to the resistor symbol. In addition, an identifier such as R1 or R2 is also sometimes written next to the symbol.

In some schematics, particularly those drawn in Europe, the symbol shown in the margin is used instead of the jagged line.

Resistors are used for many reasons in electronic circuits. The three most popular are

+ **Limiting current:** By introducing resistance into a circuit, resistors can limit the amount of current that flows through the circuit. In accordance with Ohm's law, if the voltage in a circuit remains the same, the current will decrease if you increase the resistance.

Many electronic components have an appetite for current that must be regulated by resistors. One of the best known are light-emitting diodes (LEDs), which are a special type of diode that emits visible light when current runs through it. Unfortunately, LEDs don't know when to step away from the table when it comes to consuming current. That's because they have very little internal resistance. Unfortunately, LEDs don't have much tolerance for current, so too much current will burn them out. As a result, it's always prudent — essential, in fact — to place a resistor in series with an LED to keep the LED from burning itself up. (You can learn more about LEDs in Chapter 5 of this minibook.)

You can use Ohm's law to your advantage when using current-limiting resistors. For example, if you know what the supply voltage is and you know how much current you need, you can use Ohm's law to determine the right resistor to use for the circuit as explained in the preceding section.

✦ **Dividing voltage:** You can also use resistors to reduce voltage to a level that's appropriate for specific parts of your circuit. For example, suppose your circuit is powered by a 3 V battery but a part of your circuit needs 1.5 V. You could use two resistors of equal value to split this voltage in half, yielding 1.5 V. For more information, see the section "Dividing Voltage," later in this chapter.

✦ **Resistor/capacitor networks:** Resistors can be used in combination with capacitors for a variety of interesting purposes. You learn about this use of resistors in Chapter 3 of this minibook.

Book II
Chapter 2

Working with Resistors

Reading Resistor Color Codes

You can determine the resistance provided by a resistor by examining the *color codes* that are painted on the resistor. These little stripes of bright colors indicate two important factoids about the resistor: its resistance in ohms and its *tolerance*, which indicates how close to the indicated resistance value the resistor actually is.

Most resistors have four stripes of color. The first three stripes indicate the resistance value, and the fourth stripe indicates the tolerance. Some resistors have five stripes of color, with four representing the resistance value and the last one the tolerance.

If you're uncertain from which side of the resistor to read the colors, start with the side closest to the color stripe. The first stripe is usually painted very close to the edge of the resistor; the last stripe isn't as close to the edge.

Reading a resistor's value

To read a resistor's color code, refer to Table 2-1. Here's the procedure for determining the value of a resistor with four stripes:

1. Orient the resistor so you can read the stripes properly.

You should read the stripes from left to right. The first stripe is the one that's closest to one end of the resistor. If this stripe is on the right side of the resistor, turn the resistor around so the first stripe is on the left.

2. **Look up the color of the first stripe to determine the value of the first digit.**

   For example, if the first stripe is yellow, the first digit is 4.

3. **Look up the color of the second stripe to determine the value of the second digit.**

   For example, if the first stripe is violet, the second digit is 7.

4. **Look up the color of the third stripe to determine the multiplier.**

   For example, if the third stripe is brown, the multiplier is 10.

5. **Multiply the two-digit value by the multiplier to determine the resistor's value.**

   For example, 47 times 10 is 470. Thus, a yellow-violet-brown resistor is 470 Ω.

If a resistor has five stripes, the first three stripes are the value digits, and the fourth stripe is the multiplier. The fifth stripe is the tolerance, as described in the next section.

Table 2-1	Resistor Color Codes (Resistance Values)	
Color	*Digit*	*Multiplier*
Black	0	1
Brown	1	10
Red	2	100
Orange	3	1 k
Yellow	4	10 k
Green	5	100 k
Blue	6	1 M
Violet	7	10 M
Gray	8	100 M
White	9	1,000 M
Gold		0.1
Silver		0.01

Here are a few examples that should help you understand how to read resistor codes:

Color Stripes	Digit Values	Multiplier (in Ohms)	Resistor Value
Brown - black - brown	10	10	100 Ω
Brown - black - red	10	100	1 kΩ
Red - red - orange	22	1 k	22 kΩ
Red - red – yellow	22	10 k	220 kΩ
Yellow - violet – black	47	0.1	47 Ω

Noticing standard resistor values

In theory, there are 100 different combinations of colors for the first two bands, covering the range of values from 00 through 99. However, in practice there are only a few color combinations that are commonly encountered. These color combinations represent standardized values that allow manufacturers to produce resistors that will be useful in a wide variety of applications.

For example, take the value 47, represented by the color code yellow-violet. 47 happens to be one of the preferred resistor numbers, so you can easily obtain resistors of 4.7W, 47W, 470W, 4.7KW, 47KW, 470KW, and 4.7MW.

But 45 is not one of the preferred values. Thus you won't find 45W or 450W resistors.

Although there are several different systems for standardizing preferred resistor values, the most common system uses 12 different standard values:

First Two Colors	Standard Value
Brown – black	10
Brown – red	12
Brown – green	15
Brown – gray	18
Red – red	22
Red - Violet	27
Orange – Orange	33
Orange – white	39
Yellow – violet	47
Green – blue	56
Blue – gray	68
Gray – red	82

These values are standardized values that are designed to provide a wide range of resistance values. Although not exact, each value is approximately 1.2 times larger than the previous value.

Understanding resistor tolerance

The value indicated by the stripes painted on a resistor provides an estimate of the actual resistance. The exact resistance varies by a percentage that depends on the *tolerance* factor of the resistor.

For example, a 22 kΩ resistor with a 5% tolerance actually has a value somewhere between 5% above and 5% below 22 kΩ, which works out to somewhere between 20.9 and 23.1 kΩ. A 470 Ω resistor with a 10% tolerance has an actual value somewhere between 423 and 517Ω.

Why the approximations? Simply because it costs more money to manufacture resistors to very close tolerances, and for most electronic circuits, a 5% or 10% margin of error is perfectly acceptable. For example, if you're building a circuit to limit the current flowing through a component to 200 mA, it probably won't matter much if the actual current is a little above or below 200 mA. Thus, a 5%- or 10%-tolerance resistor is acceptable.

If your application demands higher precision, you can spend a bit more money to buy higher-tolerance resistors. But 5%- or 10%- tolerance resistors are fine for most work, including all the circuits presented in this book (unless otherwise indicated).

The tolerance of a resistor is indicated in the resistor's last color stripe, as listed in Table 2-2.

Table 2-2	Resistor Color Codes (Tolerance Values)
Color	*Tolerance*
Brown	1%
Red	2%
Orange	3%
Yellow	4%
Gold	5%
Silver	10%
Black	20%

Understanding Resistor Power Ratings

Resistors are like brakes for electric current. Like the brakes in your car, resistors work by applying the electrical equivalent of friction to flowing current. This friction inhibits the flow of current by absorbing some of the current's energy and dissipating it in the form of heat. Whenever you use a resistor in a circuit, you must make sure that the resistor is capable of handling the heat.

The *power rating* of a resistor indicates how much power a resistor can handle before it becomes too hot and burns up. You may remember from Book I, Chapter 2 that power is measured in units called *watts.* The more watts a resistor can handle, the larger and more expensive the resistor is.

Most resistors are designed to handle ⅛ W or ¼ W. You can also find resistors rated for ½ W or 1 W, but they're rarely needed in the types of electronic projects you build in this book. Unless otherwise stated, all the resistors used in this book are rated at ¼ W.

Unfortunately, you can't tell a resistor's power rating just by looking at it. Unlike resistance and tolerance, there are no color codes for wattage. However, the size of the resistor is a good indicator of its power rating. Power ratings are written on the packaging when you buy new resistors. After you work with them for a while, you'll come to quickly recognize the size difference between resistors of different power ratings.

If you want to be safe, you can calculate the power demands required of a particular resistor in your circuits. First, use Ohm's law to calculate the voltage across the resistor and the current that will pass through the resistor. For example, if a 100 Ω resistor will have 3 V across it, you can calculate that 30 mA of current will flow through the resistor by dividing the voltage by the resistance (3 V ÷ 1,000 Ω = 0.03 A, which is 30 mA).

Once you know the voltage and the current, you can calculate the power that will be dissipated by the resistor by using the power formula you learn about back in Book I, Chapter 2:

$$P = I \times V$$

Thus, the power dissipated by the resistor will be just 0.09 W, well under the maximum that can be handled by a ¼ W (0.25 W) resistor. (A ⅛ W resistor should be able to handle this amount of power too, but it's always better to err on the large side when it comes to power ratings.)

Limiting Current with a Resistor

One of the most common uses for resistors is to limit the current flowing through a component. Some components, such as light-emitting diodes, are very sensitive to current. A few milliamps of current is enough to make an LED glow; a few hundred milliamps is enough to destroy the LED.

**Book II
Chapter 2**

**Working with
Resistors**

Project 2-1 shows you how to build a simple circuit that demonstrates how a resistor can be used to limit current to an LED. The finished circuit, which you will assemble on a small solderless breadboard, is shown in Figure 2-2.

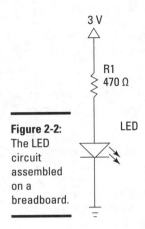

Figure 2-2: The LED circuit assembled on a breadboard.

Before we get into the construction of the circuit, here's a simple question: Why a 120 Ω resistor? Why not a larger or a smaller value? In other words, how do you determine what size resistor to use in a circuit like this?

The answer is simple: Ohm's law, which can easily tell you what size resistor to use, but you must first know the voltage and current. In this case, the voltage is easy to figure out: You know that two AA batteries provide 3 V. To figure out the current, you just need to decide how much current is acceptable for your circuit. The technical specifications of the LED tell you how much current the LED can handle. In the case of a standard 5 mm red LED (the kind you can buy at RadioShack for about $1.50), the maximum allowable current is 28 mA. To be safe and make sure that you don't destroy the LED with too much current, round the maximum current down to 25 mA.

To calculate the desired resistance, you divide the voltage (3 V) by the current (0.025 A). The result is 120 Ω.

Do *not* connect the LED directly to the battery without a resistor. If you do, the LED will flash brightly, and then it will be dead forever.

Combining Resistors

Suppose you've designed the perfect circuit, and it calls for a 1,100 Ω resistor in a critical spot. You run down to your local RadioShack and are miffed because you can't find any 1,100 Ω resistors. So you go online and search your favorite online suppliers and are even more miffed to discover that they don't have 1,100 Ω resistors either. You can buy 1 kΩ resistors and 100 Ω resistors, but no one seems to have any 1,100 Ω resistors.

Is it time to give up? Or must you settle for a 1 kΩ resistor and hope it will be close enough?

Certainly not!

All you have to do is use two or more resistors in combination to create the resistance that you need. Such a combination of resistors is sometimes called a *resistor network*. You can freely substitute a resistor network for a single resistor whenever you want.

There are two basic ways to combine resistors: in series (strung end to end) and in parallel (side by side). The following sections explain how you calculate the total resistance of a network of resistors in series and in parallel.

You'll need to put your thinking cap on when you read through the following sections. Ohm's law is simple enough, but the math calculations required to calculate parallel resistors can get a little complex. The math isn't horribly complicated, but it isn't trivial, either.

Book II
Chapter 2

Working with
Resistors

Combining Resistors in Series

Calculating the total resistance for two or more resistors strung end to end — that is, in series — is simple: You simply add the resistance values to get the total resistance.

For example, if you need 1,100 ohms of resistance and can't find an 1,100 Ω resistor, you can combine a 1,000 Ω resistor and a 100 Ω resistor in series. Adding these two resistances together gives you a total resistance of 1,100 Ω.

You can place more than two resistors in series if you want. You just keep adding up all the resistances to get the total resistance value. For example, if you need 1,800 Ω of resistance, you could use a 1 kΩ resistor and eight 100 Ω resistors in series.

Project 2-1: Using a Current-Limiting Resistor

In this project, you'll build a simple circuit that uses a resistor to limit the current to an LED. Without this current-limiting resistor, too much current would flow through the LED and the LED would be destroyed.

The only tools you need for this project are perhaps a small Phillips-head screwdriver to open the battery holder, and wire cutters and strippers to cut and strip the jumper wire.

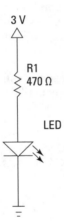

3 V

R1
470 Ω

LED

Parts List
2 AA batteries
1 Battery holder (RadioShack 270-408)
1 Lamp holder (RadioShack 272-357)
1 3 V flashlight lamp (RadioShack 272-1175)

Steps

1. Connect the battery holder.

Orient the breadboard as shown in the diagram, so that hole A30 is near the top left. Then, insert the black lead in the top bus strip and the red lead in the bottom bus strip. Any location on the correct bus will be fine, but I recommend you insert the leads into the holes at the very end of the breadboard.

2. Connect the resistor.

Insert one end of the resistor into the breadboard in hole G24, and insert the other lead into any nearby hole in the positive voltage bus strip (the strip that's connected to the red battery wire).

3. Connect the LED.

Notice that the leads of the LED aren't the same length; one lead is shorter than the other. Insert the longer lead (called the *anode*) into hole A24 and the shorter lead (called the *cathode*) into any nearby hole in the negative voltage bus strip (the one that the black battery lead is inserted into).

Note that the circuit won't work if you insert the LED backward. The short lead must be in the negative voltage bus strip.

4. Use the short jumper wire to connect the terminal strips into which you inserted the LED and the resistor.

Insert the jumper in holes E24 and F24, so that the jumper hops over the gap that runs down the middle of the breadboard. This connects the resistor to the LED.

5. Insert the batteries.

The LED lights up. If it doesn't, double-check your connections to make sure the circuit is assembled correctly. If it still doesn't light up, try a different battery.

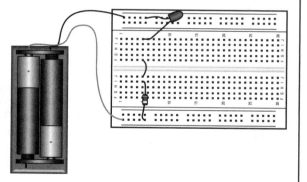

**Book II
Chapter 2**

**Working with
Resistors**

Figure 2-3 shows how serial resistors work. Here, the two circuits have identical resistances. The circuit on the left accomplishes the job with one resistor; the circuit on the right does it with three. Thus, the circuits are equivalent.

Figure 2-3:
Combining resistors in series.

 Any time you see two or more resistors in series in a circuit, you can substitute a single resistor whose value is the sum of the individual resistors. Similarly, any time you see a single resistor in a circuit, you can substitute two or more resistors in series as long as their values add up to the desired value.

 The total resistance of resistors in series is always greater than the resistance of any of the individual resistors. That's because each resistor adds its own resistance to the total.

Combining Resistors in Parallel

You can also combine resistors in parallel to create equivalent resistances. However, calculating the total resistance for resistors in parallel is a bit more complicated than calculating the resistance for resistors in series.

As the circuit shown in Figure 2-4 illustrates, when you combine two resistors in parallel, current can flow through both resistors at the same time. Although each resistor does its job to hold back the current, the total resistance of two resistors in parallel is always less than the resistance of either of the resistors because the current has two pathways through which to go.

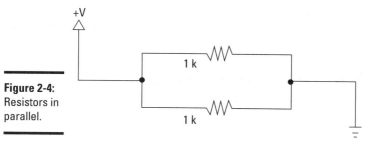

Figure 2-4:
Resistors in
parallel.

So how do you calculate the total resistance for resistors in parallel? Very carefully. Here are the rules:

✦ First, the simplest case: Resistors of equal value in parallel. In this case, you can calculate the total resistance by dividing the value of one of the individual resistors by the number of resistors in parallel. For example, the total resistance of two, 1 kΩ resistors in parallel is 500 Ω and the total resistance of four, 1 kΩ resistors is 250 Ω.

Unfortunately, this is the only case that's simple. The math when resistors in parallel have unequal values is more complicated.

✦ If only two resistors of different values are involved, the calculation isn't too bad:

$$R_{total} = \frac{R1 \times R2}{R1 + R2}$$

In this formula, R1 and R2 are the values of the two resistors.

Here's an example, based on a 2 kΩ and a 3 kΩ resistor in parallel:

$$R_{total} = \frac{R1 \times R2}{R1 + R2} = \frac{2,000\Omega \times 3,000\Omega}{2,000\Omega + 3,000\Omega} = 1,200\Omega$$

✦ For three or more resistors in parallel, the calculation begins to look like rocket science:

$$R_{total} = \frac{1}{\frac{1}{R1} + \frac{1}{R2} + \frac{1}{R2} \cdots}$$

The dots at the end of the expression indicate that you keep adding up the reciprocals of the resistances for as many resistors as you have.

In case you're crazy enough to actually want to do this kind of math, here's an example for three resistors whose values are 2 kΩ, 4 kΩ, and 8 kΩ:

$$R_{total} = \frac{1}{\frac{1}{2,000\ \Omega} + \frac{1}{4,000\ \Omega} + \frac{1}{8,000\ \Omega}} = \frac{1}{0.000875\ \Omega} = 1,142.857\ \Omega$$

As you can see, the final result is 1,142.857 Ω. That's more precision than you could possibly want, so you can probably safely round it off to 1,142 Ω, or maybe even 1,150 Ω.

Mixing Series and Parallel Resistors

Resistors can be combined to form complex networks in which some of the resistors are in series and others are in parallel. For example, Figure 2-5 shows a network of three, 1 kΩ resistors and one 2 kΩ resistor. These resistors are arranged in a mixture of serial and parallel connections.

Conducting your way through parallel resistors

The parallel resistance formula makes more sense if you think about it in terms of the opposite of resistance, which is called *conductance*. Resistance is the ability of a conductor to block current; conductance is the ability of a conductor to pass current. Conductance has an inverse relationship with resistance: When you increase resistance, you decrease conductance, and vice versa.

Because the pioneers of electrical theory had a nerdy sense of humor, they named the unit of measure for conductance the *mho*, which is *ohm* spelled backward. The mho is the reciprocal (also known as inverse) of the ohm. To calculate the conductance of any circuit or component (including a single resistor), you just divide the resistance of the circuit or component (in ohms) into 1. Thus, a 100 Ω resistor has $\frac{1}{100}$ mho of conductance.

When circuits are connected in parallel, current has multiple pathways it can travel through. It turns out that the total conductance of a parallel network of resistors is simple to calculate: You just add up the conductances of each individual resistor. For example,

suppose you have three resistors in parallel whose conductances are 0.1 mho, 0.02 mho, and 0.005 mho. (These are the conductances of 10 Ω, 50 Ω, and 200 Ω resistors, respectively.) The total conductance of this circuit is 0.125 mho (0.1 + 0.02 + 0.005 = 0.125).

One of the basic rules of doing math with reciprocals is that if one number is the reciprocal of a second number, the second number is also the reciprocal of the first number. Thus, since mhos are the reciprocal of ohms, ohms are the reciprocal of mhos. To convert conductance to resistance, you just divide the conductance into 1. Thus, the resistance equivalent to 0.125 mho is 8 Ω (1 ÷ 0.125 = 8).

It may help you remember how the parallel resistance formula works when you realize that what you're really doing is converting each individual resistance to conductance, adding them up, and then converting the result back to resistance. In other words, convert the ohms to mhos, add them up, and then convert them back to ohms. That's how — and why — the resistance formula actually works.

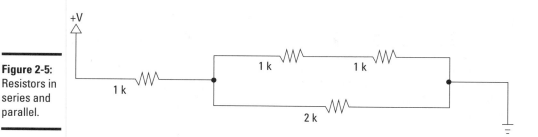

Figure 2-5:
Resistors in
series and
parallel.

The way to calculate the total resistance of a network like this is to divide
and conquer. Look for simple series or parallel resistors, calculate their
total resistance, and then substitute a single resistor with an equivalent
value. For example, you can replace the two 1 kΩ resistors that are in series
with a single 2 kΩ resistor. Now, you have two, 2 kΩ resistors in parallel.
Remembering that the total resistance of two resistors with the same value
is half the resistance value, you can replace these two, 2 kΩ resistors with a
single 1 kΩ resistor. You're now left with two, 1 kΩ resistors in series. Thus,
the total resistance of this circuit is 2 kΩ.

Deceptively simple, eh?

Combining Resistors in Series and Parallel

Project 2-2 lets you do a little hands-on work with some simple series and
parallel resistor connections so you can see firsthand how the calculations
described in the previous three sections actually work in the real world.
You'll probably find that due to the individual variations of actual resistors
(due to their manufacturing tolerances) the calculated resistances don't
always match the resistance of the actual circuits. But in most cases, the
variations aren't significant enough to affect the operation of your circuits.

In this project, you will assemble five resistors into three different configurations. The first has all five resistors in series. The second has them all in parallel. And the third creates a network of two sets of parallel resistors that are
connected in series. Figure 2-6 shows how these three configurations appear
when assembled.

Figure 2-6:
The
assembled
resistors for
Project 2-2.

Resistors in series   ...in parallel   ...and in series and parallel

Project 2-2: Resistors in Series and Parallel

In this project, you'll experiment with resistors in series and in parallel. You'll need a multimeter with an ohmmeter function to measure the resistances.

Series

Parallel

Steps

1. **Set your multimeter to its ohmmeter function with a range large enough to measure at least 5 k Ω of resistance.**

   On my ohmmeter, the closest range is 20 k Ω.

2. **Insert the five, 1 k Ω resistors in series.**

   Use the following holes on the breadboard:

Resistor	First Lead	Second Lead
1	A5	A10
2	B10	B15
3	C15	C20
4	D20	D25
5	E25	E30

   Refer to the Series Layout diagram for the layout of these resistors.

3. **Using your multimeter, measure the resistance of each resistor individually, and then measure the total resistance of two, three, four, and five resistors in series.**

   To do these measurements, place the meter probes on the leads at the breadboard holes indicated in the following Series Resistance Measurements table. Write your actual measurements in the column on the far right

4. **Rearrange the resistors into a parallel circuit.**

   Remove the resistors and reinsert them into the following holes:

Resistor	First Lead	Second Lead
1	A5	A10
2	B5	B10
3	C5	C10
4	D5	D10
5	E5	E10

   Refer to the Parallel Layout diagram for the layout of these resistors.

5. **Use your ohmmeter to measure the resistance of the parallel resistor circuit.**

   Because there are five 1 k Ω resistors, your measurement should be approximately 200 Ω.

6. **Rearrange the resistors into a series/parallel network.**

   Remove resistors 3, 4, and 5 and insert them as follows:

Resistor	First Lead	Second Lead
1	A5	A10
2	B5	B10
3	C10	C15
4	D10	D15
5	E10	E15

   Refer to the Series/Parallel Layout diagram for the layout of these resistors.

   This configuration has the first two resistors in one parallel circuit and the other three in a second parallel circuit. The two parallel circuits are connected together to form a series circuit.

7. **Use your ohmmeter to measure the resistance of two parallel circuits and the entire circuit.**

   Record your measurements in the Series/Parallel Resistance Measurements table.

8. **You're done!**

   Unless you're a glutton for punishment. In that case, feel free to experiment with other combinations of series and parallel resistor circuits. Grab some resistors with other values and throw them into the mix. Each time, do the math to predict what the resulting resistance should be. With some practice, you'll get good at calculating resistor networks.

Series Resistance Measurements

Red Lead	Black Lead	Number Of Resisters	Expected Measurement	Your Your Measurement
A5	A10	1	1 k Ω	
B10	B15	1	2 k Ω	
C15	C20	1	3k Ω	
D20	D25	1	1 k Ω	
E25	E30	1	1k Ω	
A5	B15	2	2 k Ω	
A5	C20	3	3 k Ω	
A5	D25	4	4 k Ω	
A5	E30	5	5 k Ω	

Series/Parallel Resistance Measurements

A5	A10	2	500 k Ω
C10	C15	3	333 k Ω
A5	C15	5	833 k Ω

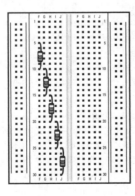

Series Layout

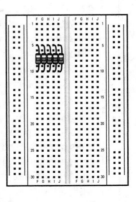

Parallel Layout

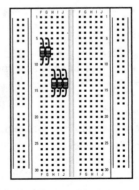

Series/Parallel Layout

Dividing Voltage

One interesting and useful property of resistors is that if you connect two resistors together in series, you can tap into the voltage at the point between the two resistors to get a voltage that is a fraction of the total voltage across both resistors. This type of circuit is called a *voltage divider,* and is a common way to reduce voltage in a circuit. Figure 2-7 shows a typical voltage-divider circuit.

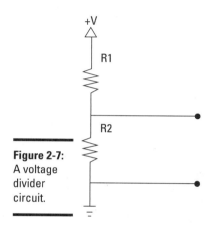

Figure 2-7:
A voltage
divider
circuit.

When the two resistors in the voltage divider are of the same value, the voltage is cut in half. For example, suppose your circuit is powered by a 9 V battery, but your circuit really only needs 4.5 V. You could use a pair of resistors of equal value across the battery leads to provide the necessary 4.5 V.

When the resistors are of different values, you must do a little math to calculate the voltage at the center of the divider. The formula is as follows:

$$V_{out} = \frac{V_{in} \times R2}{R1 + R2}$$

For example, suppose you're using a 9 V battery, but your circuit requires 6 V. In that case, you could create a voltage divider using a 1 kΩ resistor for R1 and a 2 kΩ resistor for R2. Here's the math:

$$V_{out} = \frac{9V \times 2,000\ \Omega}{1,000\ \Omega + 2,000\ \Omega} = \frac{18,000V}{3,000} = 6V$$

As you can see, these resistor values cut the voltage down to 6 V.

Dividing Voltage with Resistors

In Project 2-3, you build a simple voltage divider circuit on a solderless breadboard to provide either 3 V or 6 V from a 9 V battery. The assembled circuit is shown in Figure 2-8.

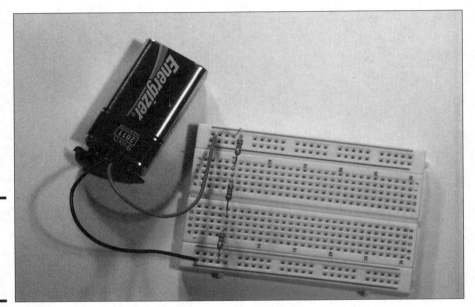

Figure 2-8:
The assembled voltage divider circuit.

Varying Resistance with a Potentiometer

Many circuits call for a resistance that can be varied by the user. For example, most audio amplifiers include a volume control that lets the user turn the volume up or down, and you can create a simple light dimmer by varying the resistance in series with a lamp.

A variable resistor is properly called a *potentiometer,* or just *pot* for short. A potentiometer is simply a resistor with three terminals. Two of the terminals are permanently fixed on each end of the resistor, but the middle terminal is connected to a wiper that slides in contact with the entire surface of the resistor. Thus, the amount of resistance between this center terminal and either of the two side terminals varies as the wiper moves.

Figure 2-9 shows how a typical potentiometer looks from the outside. The resistive track and slider (properly called the *wiper*) are enclosed within the metal can, and the three terminals are beneath it. The rod that protrudes from the top of the metal can is connected to the wiper so that when the user turns the rod, the wiper moves across the resistor to vary the resistance.

Figure 2-9:
A potenti-
ometer.

Figure 2-10 shows how a potentiometer works on the inside. Here, you can see that the resistor is made of a semicircular piece of resistive material such as carbon. The two outer terminals are connected to either end of the resistor. The wiper, to which the third terminal is connected, is mounted so that it can rotate across the resistor. When the wiper moves, the resistance between the center terminal and the other two terminals changes.

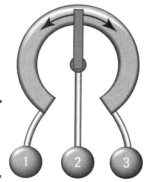

Figure 2-10:
How a
potentiom-
eter works.

Project 2-3: A Voltage Divider Circuit

In this project, you'll build a simple voltage divider circuit using three 1 KΩ resistors and you'll use the ohmmeter function of your multimeter to measure the effect of the divider.

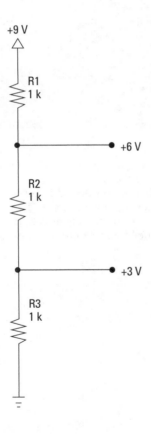

Parts List		
1	Solderless breadboard (RadioShack 276-003)	
1	9 V battery	
1	9 V battery snap holder (RadioShack 270-325)	
3	1 k Ω, ¼" W resistor (brown-black-red)	
1	Multimeter with a voltmeter function	

Steps

1. Connect the battery holder.

Insert the black lead into the first hole in the ground bus (on the side nearest column A). Then, insert the red lead in the first hole in the positive voltage bus on the side nearest column J).

2. Insert the resistors.

Place the resistor leads into the holes indicated in the following table:

Resistor	First Lead	Second Lead
1	Ground bus near row 5	A5
2	D5	G5
3	I5	Positive bus near row 5

3. Connect the battery.

4. Use the voltmeter to observe how the voltage is divided.

Set the voltmeter to a range that will measure at least 10 V DC. Then take the following measurements by touching the meter leads to the resistor leads that are plugged into the holes indicated in the table:

5. You're done!

Be sure to unplug the battery from the circuit. If you leave it plugged in, current will continue to flow through the series resistor circuit, and the battery will soon go dead.

Series Resistance Measurements

Black Lead	Red Lead	Expected Measurement	Your Measurement
Ground bus	Positive bus	9 V	
Ground bus	H5	6 V	
Ground bus	A5	3 V	
A5	H5	3 V	

9 V
- +

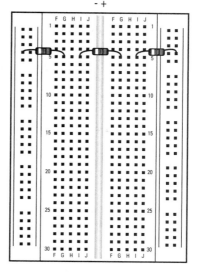

The symbol used for a potentiometer in schematic diagrams is shown in the margin. As you can see, the center tap of the resistor is indicated by an arrow that is meant to reflect that the value of the resistance at this terminal varies when the wiper moves.

Potentiometers are rated by their total resistance. The resistance between the center terminal and the two other terminals always adds up to the total resistance rating of the potentiometer. For example, the two resistances split by a 100 kΩ potentiometer always add up to 100 Ω. When the dial is exactly in the center, both resistances are 50 Ω. As you move the wiper one way or the other, one resistance increases while the other decreases; but in all cases, the total of the two resistances always adds up to 100 Ω.

Here are a few other rambling thoughts to keep in mind about potentiometers:

✦ Potentiometers come in a wide variety of shapes and sizes. With a little hunting around in stores or on the Internet, you should be able to find the perfect potentiometer for every need.

✦ Some potentiometers are very small and can be adjusted only by the use of a tiny screwdriver. This type of pot is called a *trim pot*, designed to make occasional fine-tuning adjustments to your circuits.

✦ Some potentiometers have switches incorporated into them so that when you turn the knob all the way to one side or pull the knob out, the switch operates to either open or close the circuit.

✦ When the wiper reaches one end of the resistor or the other, the resistance between the center terminal and the terminal on that end is essentially zero. Keep this point in mind when you're designing circuits. To avoid circuit paths with no resistance, it's common to put a small resistor in series with a potentiometer.

✦ In some potentiometers, the resistance varies evenly as you turn the dial. For example, if the total resistance is 10 kΩ, the resistance at the halfway mark is 5 kΩ, and the resistance at the one-quarter mark is 2.5 kΩ. This type of potentiometer is called a *linear taper* potentiometer because the resistance change is linear.

Many potentiometers, however, aren't linear. For example, potentiometers designed for audio applications usually have a *logarithmic taper*, which means that the resistance doesn't vary evenly as you move the dial.

✦ Some variable resistors have only two terminals: one on an end of the resistor itself, the other attached to the wiper. This type of variable resistor is properly called a *rheostat*, but most people use the term *potentiometer* or *pot* to refer to both two- and three-terminal variable resistors.

Chapter 3: Working with Capacitors

In This Chapter

✔ Digging in to the mysteries of capacitors

✔ Learning about the measurements used with capacitors

✔ Calculating important capacitor numbers

✔ Building some basic capacitor circuits

*M*y favorite science fiction gadget is the device invented by Doc Brown in *Back to the Future,* which enabled him send his friend Marty McFly back to 1955: the flux capacitor. Doc Brown's famous flux capacitor was able to convert 1.21 gigawatts of power into a distortion in the space-time continuum to alter the course of history.

In this chapter, you can examine ordinary capacitors, the little brothers of the fictitious flux capacitor. Although a time-travelling flux capacitor would be pretty useful if such a thing existed, real capacitors do exist, and are incredibly useful. In fact, capacitors are among the most useful of all electronic components. You'll find one or more capacitors in almost every circuit you build.

What Is a Capacitor?

To begin to understand what a capacitor is, consider this question: What makes current flow within a conductor? You may remember from Book I, Chapter 2 that opposite charges attract one another and like charges repel one another. The attraction or repulsion is caused by a force known as the *electromagnetic force*. All charged particles exhibit an electric field that is associated with this force. This field is what draws *electrons* (negative charges) toward *protons* (positive charges), and it's also this field that pushes electrons away from one another.

Within a conductor such as copper, the electric fields produced by individual electrons create a constant movement of electrons within the conductor. However, this movement is completely random. When voltage is applied across the two ends of a conductor, an electric field is created, which causes the movement of electrons to become organized. The electric field pushes electrons through the conductor in an orderly fashion from the negative side of the voltage to the positive side. The result: current.

Electric fields are magical. Unlike current, an electric field doesn't need a conductor to travel. An electric field can reach right across an insulator and push away or tug on charges on the other side of the insulator.

This might seem like magic, but you've encountered it before. Blow up a balloon, rub it fast against your shirt, and then hold it up to your hair. Rubbing the balloon creates a static charge in the balloon. When this charge comes close to your hair, its electric field tugs the charges in your hair, which is magically drawn upward toward the balloon.

A *capacitor* is an electronic component that takes advantage of the apparently magical ability of electric fields to reach out across an insulator. It consists of two flat plates made from a conducting material such as silver or aluminum, separated by a thin insulating material such as Mylar or ceramic, as shown in Figure 3-1. The two conducting plates are connected to terminals so that a voltage can be applied across the plates.

Figure 3-1:
A capacitor creates an electric field between two charged plates separated by an insulator.

Note that because the two plates are separated by an insulator, a closed circuit *isn't* formed. Nevertheless, current flows — for a moment, anyway.

How can this be? When the voltage from a source such as a battery is connected, the negative side of the battery voltage immediately begins to push negative charges toward one of the plates. Simultaneously, the positive side of the battery voltage begins to pull electrons (negative charges) away from the second plate.

What permits current to flow is the electric field that quickly builds up between the two plates. As the plate on the negative side of the circuit fills with electrons, the electric field created by those electrons begins to push electrons away from the plate on the other side of the insulator, toward the positive side of the battery voltage.

As this current flows, the negative plate of the capacitor builds up an excess of electrons, whereas the positive side develops a corresponding deficiency of electrons. Thus, voltage is developed between the two plates of the capacitor. (Remember, the definition of *voltage* is the difference of charge between two points.)

But there's a catch: This current flows only for a brief time. As the electrons build up on the negative plate and are depleted from the positive plate, the voltage between the two plates increases because the difference in charge between the two plates increases. The voltage continues to increase until the capacitor voltage equals the battery voltage. Once the voltages are the same, current stops flowing through the circuit, and the capacitor is said to be *charged*.

At this point the magic gets even better. Once a capacitor has been charged, you can disconnect the battery from the capacitor, and the voltage will remain in the capacitor. In other words, although the voltage in the capacitor is created by the battery, this voltage isn't dependent on the battery for its continued existence. Disconnect the battery, and the voltage remains across the two plates of the capacitor.

Thus, capacitors have the ability to store charge — an ability known as *capacitance*.

Note that from a pure physics standpoint, saying that a capacitor has the ability to store charge isn't quite accurate. What a capacitor actually stores is energy in the form of an electric field that creates a voltage across the two plates. Strictly speaking, the capacitor doesn't have any more charge in it when it's fully "charged" than it does when it isn't. The difference is that when a capacitor isn't charged, the total negative and positive charges of the capacitor are divided evenly among the plates, so there's no voltage across the plates. In contrast, when a capacitor is charged, the negative charges are concentrated on one plate, and the positive charges are on the other plate. The total amount of charge in the capacitor is the same either way.

Here comes some more magic: If a charged capacitor is connected to a circuit, the voltage across the plates will drive current through the circuit. This is called *discharging* the capacitor. Just as the current that charges a capacitor lasts only for a short time, the current that results when a capacitor discharges also lasts only for a short time. As the capacitor discharges, the charge difference between the two plates decreases, and the electric field collapses. Once the two plates have reached equilibrium, the voltage across the plates reaches zero, and no current flows.

Here are a few additional things you should know about capacitors before moving on:

✦ The most common symbol used for capacitors in schematic diagrams is simply two parallel lines separated by a gap, as shown in the margin.

✦ An alternative symbol uses a straight line and a curved line to represent the plates. The curved line is generally used on the negative side of the circuit.

✦ Although some capacitors aren't sensitive to polarity, many others are. This sensitivity has to do with the choice of materials used to create the capacitors: With some materials, connecting the voltage in the wrong direction can damage the capacitor. Capacitors that have distinct positive and negative terminals are called *polarized capacitors*. A plus sign is used in the schematic diagram to indicate the polarity, as shown in the margin.

✦ Sometimes capacitors are called simply *caps*.

✦ The insulating material between the two conducting plates is properly called *dielectric,* a term that refers to the ability of the insulating layer to become polarized by the electric field that exists between the two plates when they become charged.

Counting Capacitance

Capacitance is the term that refers to the ability of a capacitor to store charge. It's also the measurement used to indicate how much energy a particular capacitor can store. The more capacitance a capacitor has, the more charge it can store.

Capacitance is measured in units called *farads* (abbreviated F). The definition of one farad is deceptively simple. A one-farad capacitor holds a voltage across the plates of exactly one volt when it's charged with exactly one ampere per second of current.

Note that in this definition, the "one ampere per second of current" part is really referring to the amount of charge present in the capacitor. There's no rule that says the current has to flow for a full second. It could be one ampere for one second, or two amperes for half a second, or half an ampere for two seconds. Or it could be 100 mA for 10 seconds or 10 mA for 100 seconds.

One ampere per second corresponds to the standard unit for measuring electric charge, called the *coulomb*. So another way of stating the value of one farad is to say that it's the amount of capacitance that can store one coulomb with a voltage of one volt across the plates.

It turns out that one farad is a huge amount of capacitance, simply because one coulomb is a very large amount of charge. To put it into perspective, the total charge contained in an average lightning bolt is about five coulombs, and you need only five, one-farad capacitors to store the charge contained in a lightning strike. (Some lightning strikes are much more powerful, as much as 350 coulombs.)

It's a given that Doc Brown's flux capacitor was in the farad range because Doc charged it with a lightning strike. But the capacitors used in electronics are charged from much more modest sources. *Much* more modest. In fact, the largest capacitors you're likely to use have capacitance that is measured in millionths of a farad, called *microfarads* and abbreviated μ*F*. And the smaller ones are measured in millionths of a microfarad, also called a *picofarad* and abbreviated *pF*.

Here are a few other things you should know about capacitor measurements:

✦ Like resistors, capacitors aren't manufactured to perfection. Instead, most capacitors have a margin of error, also called *tolerance*. In some cases, the margin of error may be as much as 80%. Fortunately, that degree of impression rarely has a noticeable effect on most circuits.

✦ The μ in μ*F* isn't an italic letter *u*; it's the Greek letter *mu*, which is a common abbreviation for *micro*.

✦ It's common to represent values of 1,000 pF or more in μF rather than pF. For example, 1,000 pF is written as 0.001 μF, and 22,000 pF is written as 0.022 μF.

✦ Historical note: Like many units of measure in electronics, the farad was named after one of the all-time great pioneers of electricity, an Englishman named Michael Faraday, who did the groundbreaking research into magnetism and its relationship with electric current.

✦ As an example of how much Faraday and physicist James Maxwell are respected for their work, consider that Albert Einstein kept portraits of three people on the wall of his study at Princeton. They were Sir Isaac Newton, considered by many to be the greatest physicist of all time; James Maxwell, who developed the theory of electromagnetism, which is considered by many to be as great an accomplishment as Newton's work with gravity; and Michael Faraday.

Reading Capacitor Values

If there's enough room on the capacitor, most manufacturers print the capacitance directly on the capacitor along with other information such as the working voltage and perhaps the tolerance. However, small capacitors don't have enough room for all that. Many capacitor manufacturers use a shorthand notation to indicate capacitance on small caps.

If you have a capacitor that has nothing other than a three-digit number printed on it, the third digit represents the number of zeros to add to the end of the first two digits. The resulting number is the capacitance in pF. For example, *101* represents 100 pF: the digits 10 followed by one additional zero.

If there are only two digits listed, the number is simply the capacitance in pF. Thus, the digits *22* indicate a 22 pF capacitor.

Table 3-1 lists how some common capacitor values are represented using this notation.

Table 3-1	Capacitance Markings	
Marking	*Capacitance (pF)*	*Capacitance (μF)*
101	100 pF	0.0001 μF
221	220 pF	0.00022 μF
471	470 pF	0.00047 μF
102	1,000 pF	0.001 μF
222	2,200 pF	0.0022 μF
472	4,700 pF	0.0047 μF
103	10,000 pF	0.01 μF
223	22,000 pF	0.022 μF

Marking	Capacitance (pF)	Capacitance (µF)
473	47,000 pF	0.047 µF
104	100,000 pF	0.1 µF
224	220,000 pF	0.22 µF
474	470,000 pF	0.47 µF
105	1,000,000 pF	1 µF
225	2,200,000 pF	2.2 µF
475	4,700,000 pF	4.7 µF

You may also see a letter printed on the capacitor to indicate the tolerance. You can interpret the tolerance letter according to Table 3-2.

Table 3-2	Capacitor Tolerance Markings
Letter	Tolerance
A	±0.05 pF
B	±0.1 pF
C	±0.25 pF
D	±0.5 pF
E	±0.5%
F	±1%
G	±2%
H	±3%
J	±5 %
K	±10%
L	±15%
M	±20%
N	±30%
P	−0%, + 100%
S	−20%, + 50%
W	−0%, + 200%
X	−20%, + 40%
Z	−20%, + 80%

Notice that the tolerances for codes P through Z are a little odd. For codes P and W, the manufacturer promises that the capacitance will be no less than the stated value but may be as much as 100% or 200% over the stated value. For codes S, X, and Z, the actual capacitance may be as much as 20% below the stated value or as much as 50%, 40%, or 80% over the stated value. For example, if the marking is 101P, the actual capacitance is no less than 100 pF but may be as much as 200 pF. If the marking is 101Z, the capacitance is between 80 pF and 180 pF.

The Many Sizes and Shapes of Capacitors

Capacitors come in all sorts of shapes and sizes, influenced mostly by three things: the type of material used to create the plates, the type of material used for the dielectric, and the capacitance. Figure 3-2 shows some of the most common varieties of capacitors.

Figure 3-2: Capacitors are made in many different shapes and sizes.

The most common types of capacitors are

✦ **Ceramic disk:** The plates are made by coating both sides of a small ceramic or porcelain disk with silver solder. The ceramic or porcelain disk is the dielectric, and the silver solder forms the plates. Leads are soldered to the plates, and the entire thing is dipped in resin.

Ceramic disk capacitors are small and usually have low capacitance values, ranging from 1 pF to a few microfarads. Because they're small, their values are usually printed using the three-digit shorthand notation described earlier in this chapter, in the section "Reading Capacitor Values."

Ceramic disk capacitors aren't polarized, so you don't have to worry about polarity when you use them.

✦ **Silver mica:** The dielectric is made from mica, and this capacitor is sometimes referred to simply as a *mica capacitor*. As with ceramic capacitors, the plates in a silver mica capacitor are made from silver. Electrodes are joined to the plates, and then the capacitor is dipped in epoxy.

Silver mica capacitors come in about the same capacitance range as ceramic disk capacitors. However, they can be made to much higher tolerances — as close as 1% in some cases. Like ceramic disk capacitors, silver mica capacitors aren't polarized.

Although ceramic disk and mica capacitors are constructed in a similar way, they're easy to tell apart. Ceramic disk capacitors are thin, flat disks and are nearly always a dull, light-brown color. Silver mica capacitors are thicker, bulge at the ends where the leads are attached, and are shiny and sometimes colorful — red, blue, yellow, and green are common colors for silver mica capacitors. (Interestingly enough, I've never seen a silver mica capacitor.)

✦ **Film:** The dielectric is made from a thin film-like sheet of insulating material, and the plates are made from film-like sheets of metal foil. In some cases, the plates and the dielectric are then tightly rolled together and enclosed in a metal or plastic can. In other cases, the layers are stacked and then dipped in epoxy.

Depending on the materials used, capacitance for film capacitors can be as small as 1,000 pF or as large as 100 μF. Film capacitors aren't polarized.

✦ **Electrolytic:** One of the plates is made by coating a foil film with a highly conductive, semiliquid solution called *electrolyte*. The other plate is another foil film on which an extremely thin layer of oxide has been deposited; this thin layer serves as the dielectric. The two layers are then rolled up and enclosed in a metal can.

Electrolytic capacitors are polarized, so you must be sure to connect voltage to it in the proper direction. If you apply voltage in the wrong direction, the capacitor may be damaged and might even explode.

You find these two common types of electrolytic capacitors:

• *Aluminum:* Can be quite large, with as much as a tenth of a farad or more (100,000 μF).

• *Tantalum:* Are smaller, ranging up to about 1,000 μF.

✦ **Variable:** A capacitor whose capacitance can be adjusted by turning a knob. One common use for a variable capacitor is to tune a radio circuit to a specific frequency.

In the most common type of variable capacitor, air is used as the dielectric, and the plates are made of rigid metal. As shown in Figure 3-3, several pairs of plates are typically used in an intermeshed arrangement. One set of plates is fixed (not moveable), but the other set is attached to a rotating knob. When you turn the knob, you change the amount of surface area on the plates that overlap. This, in turn, changes the capacitance of the device.

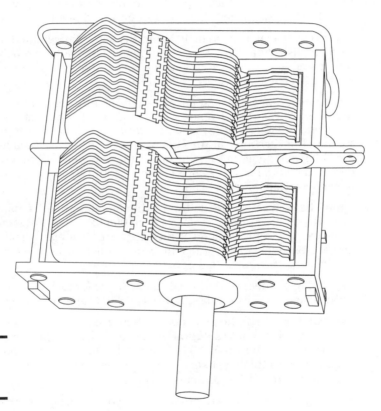

Figure 3-3:
A variable
capacitor.

 The schematic symbol for a variable capacitor is shown in the margin.

Calculating Time Constants for Resistor/Capacitor Networks

As I mention earlier in the "What Is a Capacitor" section, when you put a voltage across a capacitor, it takes a bit of time for the capacitor to fully charge. During this time, current flows through the capacitor. Similarly, when you discharge a capacitor by placing a load across it, it takes a bit of time for the capacitor to fully discharge. Knowing exactly how much time it takes to charge a capacitor is one of the keys to using capacitors correctly in your circuits, and you can get that information by calculating the RC time constant.

Calculating the exact amount of time required to charge or discharge a capacitor requires more math than I'm willing to throw at you in this book. I don't have any qualms about tossing up a little simple algebra, but I draw the line at calculus. Fortunately, you can skip the calculus and use just simple algebra if you're willing to settle for close approximations rather than exactitudes. Given that most resistors and capacitors have a tolerance of plus or minus five or ten percent anyway, using approximate calculations won't have any significant effect on your circuits.

But before I tell you about making these approximate calculations, it's important that you understand the concepts behind the calculus, even if you don't actually do the calculus. So bear with me.

When a capacitor is charging, current flows from a voltage source through the capacitor. In most circuits, a resistor is working in series with the capacitor as well, as shown in Figure 3-4.

Actually, a circuit always has resistance, even if you don't use a resistor. That's because there's no such thing as a perfect conductor, so even solid wire has some resistance. For the purposes of this discussion, assume that there is a resistor in the circuit, but its resistance might be 0 Ω.

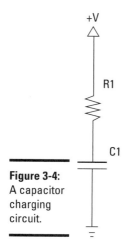

Figure 3-4:
A capacitor
charging
circuit.

The rate at which the capacitor charges through a resistor is called the *RC time constant* (the *RC* stands for *resistor-capacitor*), which can be calculated simply by multiplying the resistance in ohms by the capacitance in farads. Here's the formula:

T = R × C

For example, suppose the resistance is 10 kΩ and the capacitance is 100 μF. Before you do the multiplication, you must first convert the μF to farads. Since one μF is one-millionth of a farad, you can convert μF to farads by dividing the μF by one million. Therefore, 100 μF is equivalent to 0.0001 F. Multiplying 10 kΩ by 0.0001 F gives you a time constant of 1 second.

Note that if you want to increase the RC time constant, you can increase either the resistance or the capacitance, or both. Also note that you can use an infinite number of combinations of resistance and capacitance values to reach a desired RC time constant. For example, all the following combinations of resistance and capacitance yield a time constant of one second:

Resistance	*Capacitance*	*RC Time Constant*
1 kΩ	1,000 μF	1 s
10 kΩ	100 μF	1 s
100 kΩ	10 μF	1 s
1 MΩ	1 μF	1 s

So what is the significance of the RC time constant? It turns out that in each interval of the RC time constant, the capacitor moves 63.2% closer to a full charge. For example, after the first interval, the capacitor voltage equals 63.2% of the battery voltage. So if the battery voltage is 9 V, the capacitor voltage is just under 6 V after the first interval, leaving it just over 3 V away from being fully charged.

In the second time interval, the capacitor picks up 63.2%, not of the full 9 V of battery voltage, but 63.2% of the of the difference between the starting charge (just under 6 V) and the battery voltage (9 V). Thus, the capacitor charge picks up just over two additional volts, bringing it up to about 8 V.

This process keeps repeating: In each time interval, the capacitor picks up 63.2% of the difference between its starting voltage and the total voltage. In theory, the capacitor will never be fully charged because with the passing of each RC time constant the capacitor picks up only a percentage of the remaining available charge. But within just a few time constants, the capacity becomes very close to fully charged.

The following table gives you a helpful approximation of the percentage of charge that a capacitor reaches after the first five time constants. For all practical purposes, you can consider the capacitor fully charged after five time constants have elapsed.

RC Time Constant Interval	Percentage of Total Charge
1	63.2%
2	86.5%
3	95.0%
4	98.2%
5	99.3%

As I mentioned a moment ago, in theory a capacitor will never become fully charged. But in reality, the capacitor does eventually become fully charged. The difference between theory and reality here is that in mathematics, we can use as many digits beyond the decimal point as we want. In other words, there's no limit to how small a number can be. But there is a limit to how small a charge can be: A charge can't be smaller than a single electron. When enough time constants have elapsed, the difference between the battery voltage and the capacitor charge is a single electron. Once that electron joins the charge, the capacitor is full.

Combining Capacitors

In the preceding chapter, you learn that you can combine resistors in series or parallel networks to create any arbitrary resistance value you need. You can do the same with capacitors. But as you'll learn in the following sections, the formulas for calculating the total capacitance of a capacitor network are the reverse of the rules you follow to calculate resistor networks. In other words, the formula you use for resistors in series applies to capacitors in parallel, and the formula you use for resistors in parallel applies to capacitors in series. Isn't it funny how science sometimes likes to mess with your mind?

Combining capacitors in parallel

Calculating the total capacitance of two or more capacitors in parallel is simple: Just add up the individual capacitor values to get the total capacitance. For example, if you combine three 100 μF capacitors in parallel, the total capacitance of the circuit is 300 μF.

This rule makes sense if you think about it for a moment. When you connect capacitors in parallel, you're essentially connecting the plates of the individual capacitors. So connecting two identical capacitors in parallel essentially doubles the size of the plates, which effectively doubles the capacitance.

Figure 3-5 shows how parallel capacitors work. Here, the two circuits have identical capacitances. The first circuit accomplishes the job with one capacitors, the second does it with three. Thus, the circuits are equivalent.

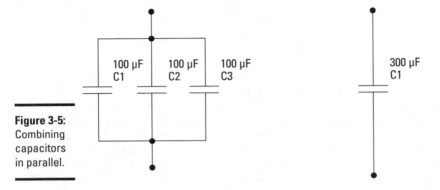

Figure 3-5:
Combining capacitors in parallel.

Whenever you see two or more capacitors in parallel in a circuit, you can substitute a single capacitor whose value is the sum of the individual capacitors. Similarly, any time you see a single capacitor in a circuit, you can substitute two or more capacitors in parallel as long as their values add up to the original value.

The total capacitance of capacitors in parallel is always greater than the capacitance of any of the individual capacitors. That's because each capacitor adds its own capacitance to the total.

Connecting capacitors in series

You can also combine capacitors in series to create equivalent capacitances, as shown in Figure 3-6. When you do, however, the math is a little more complicated. It turns out that the calculations required for capacitors in series are the same as calculating resistors in parallel.

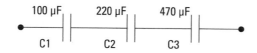

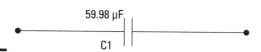

Figure 3-6:
Combing
capacitors
in series.

Here are the rules for calculating capacitances in series:

✦ If the capacitors are of equal value, you're in luck. All you must do is divide the value of one of the individual capacitors by the number of capacitors. For example, the total capacitance of two, 100 μF capacitors is 50 μF.

✦ If only two capacitors are involved, use this calculation:

$$C_{total} = \frac{C1 \times C2}{C1 + C2}$$

In this formula, C1 and C2 are the values of the two capacitors.

Here's an example, based on a 220 μF and 470 μF capacitor in series:

$$C_{total} = \frac{C1 \times C2}{C1 + C2} = \frac{220\ \mu F \times 470\ \mu F}{220\ \mu F + 470\ \mu F} = 149.855\ \mu F$$

✦ For three or more capacitors in series, the formula is this:

$$C_{total} = \frac{1}{\frac{1}{C1} + \frac{1}{C2} + \frac{1}{C3} \cdots}$$

Note that the ellipsis at the end of the expression indicates that you keep adding up the reciprocals of the capacitances for as many capacitors as you have.

Here is an example for three capacitors whose values are 100 μF, 220 μF, and 470 μF:

$$C_{total} = \frac{1}{\frac{1}{100\ \mu F} + \frac{1}{220\ \mu F} + \frac{1}{470\ \mu F}} = \frac{1}{0.016673\ \mu F} = 59.977\ \mu F$$

As you can see, the final result is 59.9768 µF. Unless your name happens to be Spock, you probably don't care about the answer being so precise, so you can safely round it to an even 60 µF.

Putting Capacitors to Work

Now that you know all about how capacitors work and you understand about charging, discharging, time constants, and series and parallel capacitors, you're probably wondering what capacitors are actually used for in real-life circuits. It turns out that capacitors are incredibly useful. The following paragraphs describe the most common ways that capacitors are used:

✦ **Storing energy:** Once charged, a capacitor has the ability to store a lot of energy and discharge it when needed. One of the most familiar uses of this ability is in a camera flash. Other examples are in devices such as alarm clocks that need to stay powered up even when household power is disrupted for a few moments.

✦ **Timing circuits:** Many circuits depend on capacitors along with resistors to provide timing intervals. For example, in Chapter 6 of this minibook you learn how to use a resistor and capacitor together along with a transistor to create a circuit that flashes an LED on and off.

✦ **Stabilizing shaky direct current:** Many electronic devices run on direct current but derive their power from household AC. These devices use a power-supply circuit that converts the AC to DC. As you learn in Book IV, Chapter 3, an important part of this circuit is a capacitor that creates steady DC voltage from the constantly changing voltage of an AC circuit.

✦ **Blocking DC while passing AC:** Sometimes you need to prevent direct current from flowing while allowing alternating current to pass. A capacitor can do this. The basic trick is to choose a capacitor that will fully charge and discharge within the alternating current cycle. Remember that as the capacitor is charging and discharging, it allows current to flow. So as long as you can keep the capacitor charging and discharging, it will pass the current.

✦ **Filtering certain frequencies:** Capacitors are often used in audio or radio circuits to select certain frequencies. For example, a variable capacitor is often used in a radio circuit to tune the circuit to a certain frequency. You see how that works in Book V, Chapter 2.

Charging and Discharging a Capacitor

One of the most common uses for capacitors is to store a charge that you can discharge when it's needed. Project 3-1 presents a simple construction project on a solderless breadboard that demonstrates how you can use a capacitor to do this. You connect an LED to a 3 V battery power supply and use a capacitor so that when you disconnect the battery from the circuit, the LED doesn't immediately go out. Instead, it continues to glow for a moment as the capacitor discharges.

Figure 3-7 shows the completed project.

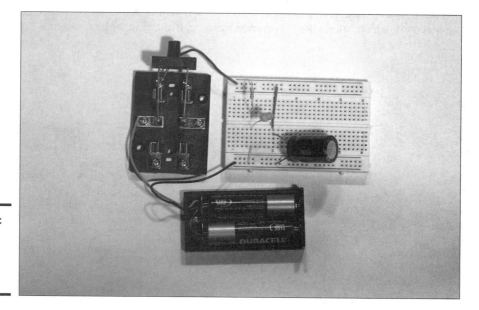

Figure 3-7:
The capacitor discharge circuit.

Here are some additional points you might want to ponder or things you might want to try after you complete Project 3-1:

✦ Pull the LED out of the breadboard, and then close the switch to charge the capacitor. After a few seconds, open the switch. Then, use the volt-meter function of your multimeter to measure the voltage across the two leads of the capacitor. Notice that even though the battery is dis-connected from the circuit, the capacitor reads 3 V.

✦ Try varying the size of the capacitor to see what happens. If you use a smaller capacitor, the LED extinguishes more quickly when you remove the power. If you use a larger capacitor, the LED fades more slowly.

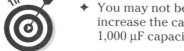

✦ You may not be able to find a capacitor larger than 1,000 μF, but you can increase the capacitance by adding more capacitors in parallel with the 1,000 μF capacitor.

✦ Try adding a resistor in series with the capacitor. To do that, simply replace the jumper wire you connected in Step 6 with a resistor.

Project 3-1: Charging and Discharging a Capacitor

In this project, you build a circuit that places a capacitor in parallel with an LED. When power is applied to the LED, the capacitor charges. When power is disconnected, the capacitor discharges, which causes the LED to remain lit for a short while after the power has been disconnected. You need a Phillips-head screwdriver to secure the connections to the knife switch and wire cutters and strippers to cut and strip the jumper wires.

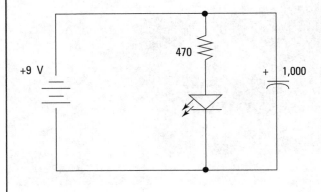

Parts List

1 Solderless breadboard
    (RadioShack 276-003)
2 AA batteries
1 Battery holder
    (RadioShack 270-408)
1 470 Ω, ¼ W resistor
    (yellow-violet-brown)
1 1,000 µ F electrolytic capacitor
    (RadioShack 272-1032)
1 Red LED, 5mm
    (RadioShack 276-209)
1 DPDT knife switch
    (RadioShack 275-1537)
1 ½", 22-gauge solid jumper wire
1 1", 22-gauge solid jumper wire
1 3" length 22-gauge solid
    jumper wire

Steps

1. **Connect the battery holder.**

   Insert the black lead in the first hole on the ground bus strip, located at the bottom of the solderless breadboard near hole A1.

   Connect the red lead of the battery holder to terminal 1X on the switch.

2. **Connect the switch to the breadboard.**

   Use the 3" solid jumper wire to connect terminal 1A on the switch to the first hole in the positive bus strip on the top of the solderless breadboard (near hole J1).

3. **Connect the resistor.**

   Insert the resistor into the breadboard so that one lead is on the positive voltage bus strip (the strip that's connected to the jumper leading to the switch) and the other lead is in hole 5G.

4. **Connect the LED.**

   Notice that one lead of the LED is shorter than the other. Insert the short lead into hole 5E and the long lead into hole 5F.

   Note that the circuit won't work if you insert the LED backward. The short lead must be on the negative side of the circuit.

5. **Use the ½" jumper wire to connect the ground bus to the LED.**

   Insert one end of the jumper wire into a hole in the ground bus near row 5. Then, insert the other end into hole 5A.

6. **Use the 1" jumper wire to connect the capacitor to the positive bus.**

   Insert one end of the jumper wire in hole 10E and the other end in a hole in the positive bus strip near row 10.

7. **Insert the capacitor.**

   The negative terminal of the capacitor, marked by a series of dashes on the capacitor, should go into a hole in the ground strip near row 10. The other lead should go in hole 10D.

8. **Insert the batteries.**

9. **Close the switch.**

   Push the lever into the A position to close the circuit. The LED lights up. You won't be able to see it, but the capacitor will charge up in just a few seconds.

10. **After a few seconds, open the switch.**

    Watch how the LED goes out. Instead of turning off instantly, if fades over the course of a few seconds. That's because when the battery is disconnected from the circuit, the capacitor discharges through the LED. As the capacitor discharges, the LED goes gradually dim and then goes out altogether.

Book II
Chapter 3

Working with
Capacitors

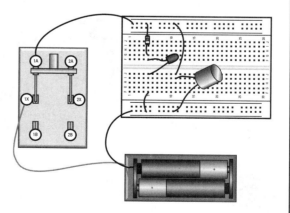

Blocking DC while Passing AC

One of the important features of capacitors is their ability to block direct current while allowing alternating current to pass. This works because of the way a capacitor allows current to flow while the plates are charging or discharging, but halts current in its tracks once the capacitor is fully charged. Thus, as long as the source voltage is constantly changing, the capacitor allows current to flow. When the source voltage is steady, the capacitor blocks current from flowing.

In Project 3-2, you build a simple circuit that demonstrates this principle in action. The circuit places a resistor and a light-emitting diode in series with a capacitor. Then, it uses a knife switch so that you can switch the voltage source for the current between a 9 V battery (direct current) and a 9 VAC power adapter (alternating current). Figure 3-8 shows the finished project.

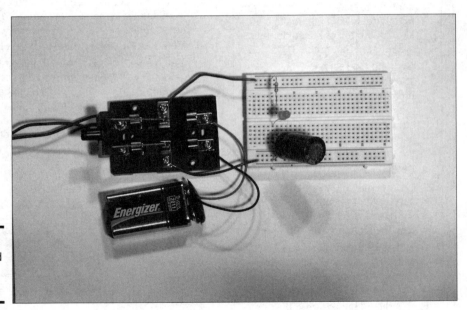

Figure 3-8:
The finished
AC/DC
circuit.

The trick to successfully building this circuit is finding a 9 VAC power adapter. If you go looking for one at your local RadioShack or at a department store, you might find one, but it will probably cost $20 or more. If you search the Internet, you can probably find one for more like $10 or $15.

Figure 3-9 shows the one I used for this project; I purchased it at a thrift store for $1. Notice the markings on the back of this power adapter:

INPUT: 120VAC 60Hz 15W

OUTPUT: 9VAC 1000mA

These are the specifications you want to look for when you search for an AC adapter to use for this project. In particular, the output voltage must be listed at 9 VAC.

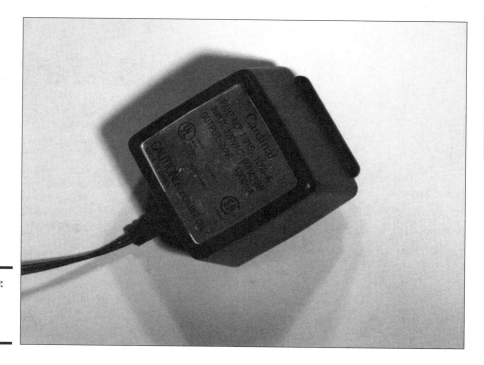

Figure 3-9:
A 9 VAC
power
adapter.

Project 3-2: Blocking Direct Current

In this project, you build a circuit that uses a capacitor to allow alternating current to pass but blocks direct current. The circuit simply places a capacitor in series with an LED, then uses a DPDT switch to connect the LED/capacitor circuit to either 9 VDC or 9 VAC.

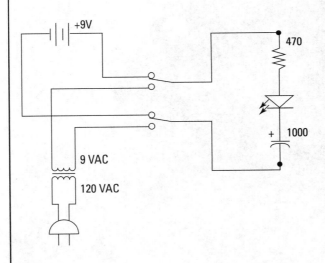

Steps

1. **Connect the battery snap holder to the switch.**

   Connect the red lead to terminal 1A on the switch, and then connect the black lead to terminal 2A.

2. **Connect the 9 VAC adapter to the switch.**

   The 9 VAC adapter should be connected to terminals 1B and 2B on the switch. The easiest way to connect the power adapter is to cut the plug off the end of the power adapter's cable, and then strip about ⅜" of insulation from the ends of the two wires that make up the cable.

3. **Connect the switch to the breadboard.**

   Use one of the 3"dp solid jumper wire to connect terminal 1X on the switch to the first hole in the positive bus strip on the top of the solderless breadboard. Then, use the other 3" jumper wire to connect terminal 2X on the switch with the first hole in the ground bus strip at the bottom of the breadboard.

4. **Connect the resistor.**

   Insert the resistor into the breadboard so that one lead is on the positive voltage bus strip (the strip that's connected to the jumper leading to switch terminal 1X) and the other lead in hole 5I.

5. **Connect the LED.**

   Insert the short LED lead into hole 5E and the long lead into hole 5F.

Note that the circuit won't work if you insert the LED backwards. The short lead must be on the negative side of the circuit.

6. **Insert the capacitor.**

   The negative terminal of the capacitor, marked by a series of dashes on the capacitor, should go into a hole in the ground strip near row 5. The other lead should go in hole 5C.

7. **Move the switch to the vertical position so that neither side of the switch is closed.**

8. **Insert the 9 V battery.**

9. **Plug the 9 VAC power adapter into a wall outlet.**

   The circuit is now done and ready for testing.

10. **Close the switch toward the A contacts.**

    This connects the 9 V battery to the circuit. The LED initially turns on, but after a few moments it begins to fade. When it goes completely out, the capacitor is fully charged. Once charged, the capacitor won't allow direct current to pass, so the LED remains dark.

11. **Flip the switch to the B contacts.**

    This action disconnects the battery from the circuit and instead connects the AC power adapter. Now, the LED comes on and stays on. As long as the AC power source is applied across the circuit, the capacitor allows the current to flow, and the LED lights up.

12. **Flip the switch back to the A contacts.**

    Because the capacitor is now fully charged, the LED goes out immediately.

13. **Open the switch, and then short out the capacitor leads.**

    With the switch in the neutral position so that neither power source is connected to the circuit, touch both leads of the capacitor with the blade of your screwdriver. This shorts out the leads, which causes the capacitor to quickly discharge itself.

14. **You're done!**

Chapter 4: Working with Inductors

In This Chapter

✔ Examining the mystery of induction

✔ Learning what reactance is

✔ Learning how to combine inductors in series and parallel

✔ Seeing how inductors are used in electronic circuits

I have a lot of books about electronics on my bookshelf, covering a wide range of topics. Some of them are new; some are a few decades old. My favorite of them all is a called *A Course in Electrical Engineering,* by Chester L. Dawes. It was written in 1920, when electronics was in its infancy. Radio was brand new. Vacuum tubes were just catching on. The transistor wouldn't be invented for another quarter of a century.

You'd think that an electrical engineering book written in this era would be completely obsolete. But it's amazing how much of this book is still accurate. For example, Ohm's law hasn't changed since 1920. Dawes's explanation of Ohm's law is as good as any I've ever read.

What fascinates me most about this old book is the way it starts. Chapter 1 is titled "Magnetism and Magnets," and it begins like this:

> Magnets and magnetism are involved in the operation of practically all electrical apparatus. Therefore an understanding of their underlying principles is essential to a clear conception of the operation of all such apparatus.

This makes perfect sense when you consider the state of electrical technology in those days. We've moved beyond simple magnetism in our ability to exploit the amazing properties of electricity. Yet the basic relationship that exists between electric current and magnetism is still at the heart of many different types of essential circuits.

For example, a power adapter that converts 120 VAC from a wall outlet to a safer level, say 9 V, uses a transformer that relies on magnetism to step down the voltage. An analog multimeter (the kind with a needle that moves with voltage, current, or resistance) relies on magnetism to deflect the needle. And electric motors convert electric current into magnetism, which is then converted into motion.

In this chapter, I turn your attention to a special class of components called *inductors* that exploit the nature of magnetism and its relationship to electric current. Inductors are sort of like cousins to capacitors, in that they can be used to do similar things and they play by similar rules. For example, an important characteristic of a capacitor is its ability to oppose changes in voltage. Inductors have a similar ability to oppose changes, not in voltage but rather in current.

This chapter is a bit of a departure from the other chapters in this minibook, in that it contains no construction projects. The practical applications for inductors are in circuits that you might not be ready to build yet, such as radio tuners or power supplies. You learn the important concepts of how inductors work in this chapter so that later, when it's time to use an inductor in an actual circuit, you'll know what it does. You can always return to this chapter at that time to refresh your memory on how inductors work. In fact, I won't hold it against you if you just skim through this chapter on your first reading through this book, or even skip it altogether for now. You can always come back to this chapter when the need arises.

What Is Magnetism?

When Albert Einstein was five years old and sick in bed, his father gave him a compass to play with. Young Albert saw that no matter how he spun the compass around the needle always swung back to point north, and he was amazed. Years later, he wrote that this compass was what launched his life-long interest in physics. He realized then that there was "something behind things, something deeply hidden."

In the case of the compass, that deeply hidden thing that Einstein marveled at was *magnetism*. We're all familiar with magnetism, though exactly what it is remains mysterious. A *magnetic field* extends through space and either attracts or repels certain materials. Materials that are strongly affected by magnetic fields are called *magnetic*. Materials that create magnetic fields are called *magnets*.

The north and south of magnetism

Magnetic fields and electric fields are distinct things, but they're closely related and have much in common. For example, magnetic fields are polarized in much the same way that electric fields are polarized. Electric fields exist between electric charges of opposite polarity (negative and positive). Magnetic fields exist between opposite magnetic poles, called *north* and *south*.

Just as opposite electric charges attract and like charges repel, opposite magnetic poles attract and like magnetic poles repel. That's why if you take two strong magnets and try to push the two north poles together or the two south poles together, the magnet fights back, and you won't be able to get the poles together. But if you turn one of the magnets around and try to hook up the north pole with the south pole, the magnets attract each other, and you'll have trouble keeping them apart.

A magnetic field has a distinct shape that you can visualize by using a simple experiment that's commonly done in grade school. All you need is a bar magnet, some iron filings, and a piece of paper. Put the paper over the magnet and sprinkle the filings on the paper. The filings magically line up to reveal the shape of the magnetic field, as shown in Figure 4-1. As you can see, the filings align themselves in lines that are aligned from one pole of the magnet to the other.

Figure 4-1:
The shape of a magnetic field revealed by iron filings.

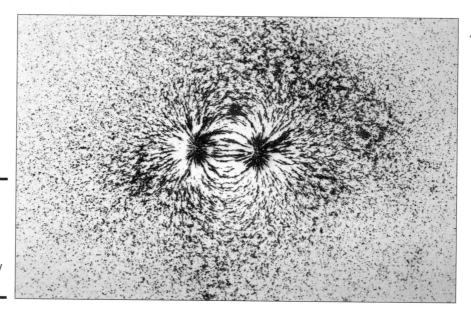

Pondering permanent magnets

A *permanent magnet* is a material that creates its own magnetic field. Some naturally occurring materials, such as loadstone, are inherently magnetic and produce magnetic fields all on their own. But most permanent magnets are made from materials that aren't inherently magnetic but become magnetic when they're exposed to a powerful magnetic field.

You probably have several permanent magnets around your home. In fact, you probably have a few on your refrigerator, holding up your kid's kindergarten picture or your shopping list.

Examining Electromagnets

Permanent magnets create a magnetic field all by themselves. An *electromagnet* relies on the key relationship between electricity and magnetism to create a magnetic field. Specifically, whenever electric current flows, a magnetic field is created. This magnetic field is created by the moving charges. In short, an electron in motion becomes a magnet.

As I say several times in this book, electrons are always in motion, so aren't there little magnets everywhere? The answer is yes, in the same sense that electric current is everywhere. But when the motion of electrons within a material is random, the magnetic fields created by the electrons are oriented randomly, and so they end up just canceling each other out. But when you give the electrons a nudge with voltage, they all start moving in the same direction. This strengthens and organizes the magnetic fields so they can combine to form one large magnetic field.

The magnetic field created by current flowing through a single wire is measurable but small. However, if you coil the wire tightly as shown in Figure 4-2, the magnetic fields are strengthened because of their proximity to one another. For example, you can create a simple electromagnet by wrapping several feet of small, insulated wire around a pencil, a ballpoint pen, or any other rigid tube or cylinder.

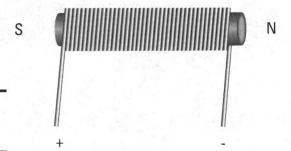

Figure 4-2:
An electro-
magnet.

The strength of the magnetic field of such a coil depends on several factors, the most important being these:

+ **The number of turns in the coil.**

+ **The amount of current flowing through the electromagnet.** Increasing the current increases the strength of the electromagnet exponentially. For example, if you double the current, the electromagnet is four times as strong.

✦ **The material used for the core that the electromagnet is wrapped around.** Any coil wrapped around an iron core will be about 10 times as strong as the same coil wrapped around an inert core such as air, plastic, or wood.

The polarity of an electromagnet is easy to determine: The negative side of the coil is the magnet's north pole, while the positive side is the south pole.

Inducing Current

Electromagnets are possible because a moving current creates a magnetic field. Interestingly enough, the reverse is also true: A moving magnetic field creates an electric current. In other words, if you wave a magnet over a wire, you create a current in the wire. This effect is called *electromagnetic induction,* or sometimes just *magnetic induction.*

Remember our old friend Michael Faraday? In Chapter 3 of this minibook, I mention that even though he didn't invent capacitors, his name is the basis of the *farad,* which is the unit of measure we use to measure capacitors. In 1831, Faraday discovered electromagnetic induction. It's one of the most important discoveries in the history of electrical science, as it's at the heart of nearly all forms of electrical power generation. Coal-burning, hydroelectric, and even nuclear power plants all use moving water to spin turbines that are connected to generators. Those generators use the principle of electromagnetic induction to turn the spinning motion of magnetic fields into electric current.

You can increase the strength of the current induced in a wire by coiling the wire into turns so that you can expose a greater length of wire to the magnetic field. Also, it doesn't matter whether it's the magnet or the coil that moves. Either way, a current is induced in the coil if the coil is moving relative to the magnet, provided of course that the coil passes through the magnet's magnetic field.

Note that whenever current is flowing, there must be voltage. Thus, in addition to causing current to flow through the coil, electromagnetic induction creates a voltage across the coil.

Inductance and the art of resisting change

An *inductor* is a coil that's designed for use in electronic circuits. Inductors take advantage of an important characteristic of coils called *self inductance,* also called just *inductance.* Inductors are simple devices, consisting of nothing more than a coil of wire, often wrapped around an iron core. But their ability to exploit the idea of self-inductance is a stroke of genius.

Self-inductance is similar to electromagnetic induction as described in the previous section, but with a subtle twist. Whereas *electromagnetic induction* refers to the ability of a coil to generate a current when it moves across a magnetic field, *self inductance* refers to the ability of a coil to create the very magnetic field that then induces the voltage. In other words, with self-inductance, the coil feeds back upon itself. A voltage applied across the coil causes current to flow, which creates a magnetic field, which in turn creates more voltage.

Inductance happens only when the current running through the coil changes. That's because only a *moving* magnetic field induces voltage in a coil. Whenever the current changes in a coil, the magnetic field created by the current grows or shrinks, depending on whether the current increases or decreases. When the magnetic field grows or shrinks, it's effectively moving, so a voltage is inducted in the coil as a result of this movement. When the current stays steady, no inductance occurs.

Self-inductance is a tricky concept to get your mind around, so don't panic if this doesn't make sense to you the first time through. It took me awhile to get this concept when I first learned about it. Let's go over the idea in more detail, point by point:

✦ **When voltage is applied across a coil, the voltage causes current to flow through the coil.** Remember, current always requires voltage, and voltage always results in a current when applied across a conductor.

✦ **The current flowing through the coil creates a magnetic field around the coil.** Keep in mind that the coil that creates the magnetic field is itself within the field and can therefore be influenced by it.

✦ **If the current flowing through the coil changes, the magnetic field created by the current also changes.** The magnetic field grows or shrinks, depending on whether the current increases or decreases. Either way, the changing magnetic field is in effect moving.

✦ **Because the magnetic field is moving, voltage is induced in the coil.** This is an additional voltage, on top of the voltage that's driving the main current through the coil.

✦ **The amount of voltage induced by the changing magnetic field depends on the speed in which the current changes.** The faster the current changes, the more the magnetic field moves and therefore the more voltage is induced.

✦ **The polarity of the induced voltage depends on whether the current is increasing or decreasing.** This is because the direction of movement in the magnetic field depends on whether the field is growing or shrinking, and the voltage induced by a moving magnetic field depends on which direction the field is moving, according to the following rules:

- When the current increases, the polarity of the induced voltage is opposite to the polarity of the voltage driving the coil. This inducted voltage is often called *back voltage* because it has the opposite polarity as the supply voltage.

- When the current decreases, the resulting self-induced voltage has the same polarity as the supply voltage.

✦ **The induced voltage creates a current in the coil that flows either with or against the main coil current, depending on whether the coil current is decreasing or increasing, according to the following rules:**

- When the coil current is increasing, the additional current flows against the main coil current. This has the effect of pushing back against the increasing main current, which effectively slows down the rate at which the current can change.

- When the coil current is decreasing, the additional current flows with the main coil current, thus counteracting the decrease in coil current.

✦ **When the coil current stops changing, self-inductance stops.** Thus, when current is steady, an inductor is simply a straight conductor. (It is also an electromagnet, as the current traveling through it produces a magnetic field.)

Because of self-inductance, an inductor is said to oppose changes in current. If the current increases, an opposite voltage is induced across the coil, which slows the rate at which the current can increase. If the current decreases, a forward voltage is induced across the coil, which slows the rate at which the current decreases. An inductor applies equal opposition to both increases and decreases in current.

It turns out that this ability to oppose changes in current is quite useful in electronic circuits, as I explain later in this chapter, in the section "Putting Inductors to Work."

Here are some additional important details concerning inductors:

✦ Inductors can't stop changes in current; they can only slow them down. Measuring how much an inductor can slow down a change in current is the topic of the next section.

✦ The magnetic field from one inductor can spill over into a nearby inductor and induce voltage in it too. To prevent this from happening, many inductors have special shielding around them to keep them magnetically isolated. If you use unshielded inductors in your circuits, be sure to space them as far apart as possible – unless, of course, you *want* them to interact.

Regarding henry

Inductance is only a momentary thing. Exactly how much of a momentary thing depends on the amount of inductance an inductor has. Inductance is measured in units called *henrys*.

The definition of one henry is simple: One henry is the amount of inductance necessary to induce one volt when the current in coil changes at a rate of one ampere per second.

As you might guess, one henry is a fairly large inductor. Inductors in the single-digit henry range are often used when dealing with household current (120 VAC at 60 Hz). But for most electronics work, you'll use inductances measured in thousandths of a henry (*millihenrys*, abbreviated *mH)* or in millionths of a henry (*microhenrys*, abbreviated μ*H).*

Here are a few additional things to know about inductors and henrys:

✦ The plural of *henry* is *henrys*, not *henries.*

✦ The letter *L* is often used to represent inductance in formulas. Inductors in schematic diagrams are usually referenced by the letter L. For example, if a circuit calls for three inductors, they will be identified L1, L2, and L3.

✦ Speaking of schematics, the symbol used to represent inductors in a schematic diagram is shown in the margin.

✦ When I first learned about self-inductance and henrys, I hoped that since the henry is a measure of a coil's ability to resist change in current, the henry should be named after someone who was famous for resisting change, such as Professor Henry Higgins from *My Fair Lady.*

Imagine my disappointment to discover that the henry is named after Joseph Henry. All he ever did was discover the self-inductance and invent the inductor. Well, that plus he was the first Secretary of the Smithsonian Institution. He must have known someone really important to get the henry named after him.

Calculating RL Time Constants

In the preceding chapter, you learn how to calculate the RC time constant for a resistor-capacitor circuit. A similar calculation can be done for inductors, except that instead of calling it the RC time constant, we call it the *RL time constant.* (Remember, *L* is the symbol used to represent inductance.)

The RL time constant indicates the amount of time that it takes to conduct 63.2% of the current that results from a voltage applied across an inductor. (Hmm. Where have we seen 63.2% before? Right! It's the same percentage used to calculate time constants in resistor-capacitor networks. The value 63.2% derives from the calculus equations used to determine the exact time constants for both resistor-capacitor and resistor-inductor networks.)

Here's the formula for calculating an RL time constant:

$$T = \frac{L}{R}$$

In other words, the RL time constant in seconds is equal to the inductance in henrys divided by the resistance of the circuit in ohms.

For example, suppose the resistance is 100 Ω, and the capacitance is 100 mH. Before you do the multiplication, you first convert the 100 mH to henrys. Because one millihenry (mH) is one, one-thousandth of a henry, you can convert millihenrys to farads by dividing the millihenrys by 1,000. Therefore, 100 mH is equivalent to 0.1 H. Dividing 100Ω by 0.1 F gives you a time constant of 0.001 second (s), or one millisecond (ms).

The following table gives you a helpful approximation of the percentage of current that an inductor passes after the first five time constants. For all practical purposes, you can consider the current fully flowing after five time constants have elapsed.

RL Time Constant Interval	Percentage of Total Current Passed
1	62.3%
2	86.5%
3	95.0%
4	98.2%
5	99.3%

Thus, in my example circuit in which the resistance is 100 Ω and the inductance is 0.1 H, you can expect that current will be flowing at full capacity within 5 ms of when the voltage is applied.

Five milliseconds is a very short amount of time. But electronic circuits are often designed to respond within very short time intervals. For example, the sine wave of standard household alternating current swings from its peak positive voltage to its peak negative voltage in about 8 ms. Sound waves at the upper end of the human ear's ability to hear cycle in about 25 µs (microseconds), and the time interval for radio waves can be in small fractions of microseconds. Thus, very small RL time constants can be very useful in certain types of electronic circuits.

Calculating Inductive Reactance

Although inductors oppose changes in current, they don't oppose all changes equally. All inductors present more opposition to fast current changes than they do to slower changes, or put another way, inductors oppose current changes in higher-frequency signals more than they do in lower-frequency signals.

The degree to which an inductor opposes current change at a particular frequency is called the inductor's *reactance*. Inductive reactance is measured in ohms, just like resistance, and can be calculated with the following formula:

$$X_L = 2\pi \times f \times L$$

Here, the symbol X_L represents the inductive reactance in ohms, f represents the frequency of the signal in hertz (cycles per second), and L equals the inductance in henrys. Oh, and π is that magic mathematical constant you learned about in high school, whose value is approximately 3.14, except here in Fresno where we round it down to 3 because it makes the math so much easier.

For example, suppose you want to know the reactance of a 1 mH inductor to a 60 Hz sine wave (not coincidentally, the frequency of household power). The math looks like this:

$$X_L = 2\pi \times f \times L = 2 \times 3.14 \times 60 \times .001 = 0.3768 \ \Omega$$

Thus, a 1 mH inductor has a reactance of about a third of an ohm at 60 Hz.

How much reactance does the same inductor have at 20 kHz? Much more:

$$X_L = 2 \times 3.14 \times 20,000 \times .001 = 125.6 \ \Omega$$

Increase the frequency to 100 MHz and see how much resistance the inductor has:

$$X_L = 2 \times 3.14 \times 100,000 \times .001 = 628,000 \ \Omega$$

At low frequencies, inductors are much more likely to let current pass than at high frequencies. As the next section explains, this characteristic can be exploited to create circuits that block frequencies above or below certain values.

Combining Inductors

Just like resistors or capacitors, you can combine inductors in series or parallel and use simple equations to calculate the total inductance of the circuit. Note, however, that for the calculations to be valid, the inductors must be shielded. If the inductors aren't shielded, they'll not only be affected by their own magnetic fields but also by the magnetic fields of other inductors around them. In that case, all bets are off.

You calculate inductor combinations just like resistor combinations, using exactly the same formulas except substituting henrys for ohms. Here are the formulas:

**Book II
Chapter 4**

Working with Inductors

+ **Series inductors:** Just add up the value of each individual inductor.

+ **Two or more identical parallel inductors:** Add them up and divide by the number of inductors.

+ **Two parallel and unequal inductors:** Use this formula:

$$L_{total} = \frac{L1 \times L2}{L1 + L2}$$

+ **Three or more parallel and unequal inductors:** Use this formula:

$$L_{total} = \frac{1}{\frac{1}{L1} + \frac{1}{L2} + \frac{1}{L3} \cdots}$$

Here's an example in which three inductors valued at 20 mH, 100 mH, and 50 mH are connected in parallel:

$$L_{total} = \frac{1}{\frac{1}{L1} + \frac{1}{L2} + \frac{1}{L3} \cdots} = \frac{1}{\frac{1}{20} + \frac{1}{100} + \frac{1}{50}}$$

$$= \frac{1}{0.05 + 0.01 + 0.02} = \frac{1}{0.08} = 12.5 \text{ mH}$$

In this example, the total inductance of the circuit is 12.5 mH.

Putting Inductors to Work

Now that you know all about how inductors work and you understand inductive reactance, time constants, and all that other good stuff, you may be wondering what inductors are actually used for in real-life circuits. Here are a few of the more common uses for inductors:

+ **Smoothing voltage in a power supply:** The final stage of a typical power supply circuit that converts 120 VAC household current to a useable direct current is often a filter circuit that removes any residual irregularities in the voltage due to the fact that it was derived from a 60 Hz AC input.

+ **Filter:** Selects frequencies that should be allowed to pass, or it selects frequencies that should be blocked. You're probably familiar with the tone settings on a stereo system, which let you bump up the bass, tone down the midrange, and may ease up the upper range just for brightness. There are three different types of filters: *high-pass filters* which allow only frequencies above a certain value to pass, *low-pass filters* which allow only frequencies below a certain value to pass, and *band-pass filters* which allow only frequencies that fall between an upper and a lower value to pass.

+ **Radio tuning circuits:** Coils can be used to help a radio tuning circuit tune to a particular frequency signal and hold it.

+ **Transformers:** One of the most common uses of inductors is in transformers. The simplest transformer consists of a pair of inductors placed next to each other. The transformer serves two functions. First, it can increase or decrease voltage, and second, it can electrically isolate one part of a circuit from another. You'll learn more about transformers in Book IV, Chapter 1.

Chapter 5: Working with Diodes and LEDs

In This Chapter

✓ Digging deep into how semiconductors work

✓ Looking at the simplest of semiconductors, the lowly diode

✓ Understanding how diodes can change AC to DC

✓ Seeing how some diodes can make light

So far in this book, all the components I tell you about were invented in the first 100 years or so since Alessandro Volta invented the first electric battery in 1800. The next 50 years of electronics, from roughly 1900 to 1950, were dominated by a technology that's now all but obsolete: vacuum tubes, which were large, expensive, fragile (they were made of glass), and required a lot of current.

In the 1940s, researchers developed a new technology that lead to new components that were, literally, a quantum-leap improvement over vacuum tubes. These new components were called *semiconductors*.

This chapter introduces you to the most basic kind of semiconductor, called a *diode*. Diodes come in many shapes and sizes, but most of them resemble the ones shown in Figure 5-1. These diodes are about the size of resistors and are available from any electronics supply store, such as RadioShack.

Although diodes look a little like resistors, they behave very differently. Diodes have one ability that sets them apart: They let current flow freely in one direction, but block current if it tries to flow in the other direction. In other words, a diode is like a turnstile gate that you can walk through in one direction but not the other. This characteristic turns out to be incredibly useful in electronic circuits.

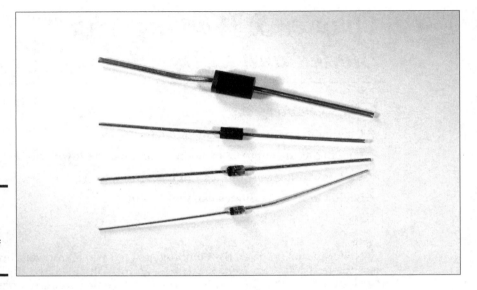

Figure 5-1:
Several
common
varieties of
diodes.

I want to warn you from the outset that the first part of this chapter is fairly difficult, as it delves just a little into the physics of how semiconductors work at the atomic level. You don't really have to understand the physics of diodes to use them in your circuits, so you can skip over the entire first section if you want, and start with the section "Introducing Diodes," later in this chapter. However, I recommend plowing through this first section no matter what. Don't worry much if you have difficulty understanding how semiconductors work. Just take note of the key terms, such as *p-type, n-type,* and *p-n junction,* and move on.

What Is a Semiconductor?

As its name implies, a *semiconductor* is a material that conducts current, but only partly. The conductivity of a semiconductor is somewhere between that of an insulator, which has almost no conductivity, and a conductor, which has almost full conductivity. Most semiconductors are crystals made of certain materials, most commonly silicon.

To understand how semiconductors work, you must first understand a little about how electrons are organized in an atom. The electrons in an atom are organized in layers, kind of like the layers of an onion. These layers are called *shells.* The outermost shell is called the *valence* shell. The electrons in this shell are the ones that form bonds with neighboring atoms. Such bonds are called *covalent bonds* because they share valence electrons, which are also the electrons that sometimes go wandering off in search of other atoms.

Most conductors, including copper and silver, have just one electron in the valence shell. Atoms with just one valence electron have a hard time keeping that electron. That's what makes copper and silver such good conductors: When valence electrons travel, they create moving electric fields that push other electrons out of their way. That's what causes current to flow.

Semiconductors, on the other hand, typically have four electrons in their valence shell. The best known elements with four valence electrons are carbon, silicon, and germanium. Atoms with four valence electrons rarely lose one of them. However, they do like to share them with neighboring atoms.

If all the neighboring atoms are of the same type, it's possible for all the valence electrons to bind with valence electrons from other atoms. When that happens, the atoms arrange themselves into neat and orderly structures called *crystals*. Semiconductors are made out of such crystals.

The most plentiful element with four valence electrons is carbon. However, carbon crystals are rarely used as semiconductors because they have other uses, such as in wedding rings. So instead, semiconductors are usually made from crystals of silicon and occasionally germanium.

Figure 5-2 shows the covalent bonds formed in a silicon crystal. Here, each circle represents a silicon atom, and the lines between the atoms represent the shared electrons. Each of the four valence electrons in each silicon atom is shared with one neighboring silicon atom. Thus, each silicon atom is bonded with four other silicon atoms.

**Book II
Chapter 5**

**Working with
Diodes and LEDs**

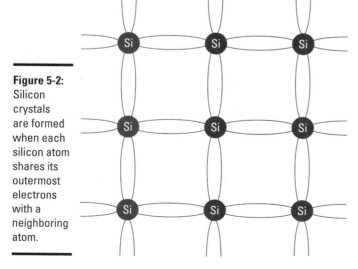

Figure 5-2:
Silicon crystals are formed when each silicon atom shares its outermost electrons with a neighboring atom.

Doping: It's not just for athletes

By themselves, pure silicon crystals are pretty, but not all that useful electronically. With all four valence electrons interlocked with neighboring atoms, it isn't easy to get current to flow through a pure silicon crystal. But things get interesting if you introduce small amounts of other elements into a crystal. Then, the crystal starts to conduct in an interesting way.

The process of deliberately introducing other elements into a crystal is called *doping*. The element introduced by doping is called a *dopant*. By carefully controlling the doping process and the dopants that are used, silicon crystals can transform into one of two distinct types of conductors:

✦ **N-type semiconductor:** Created when the dopant is an element that has five electrons in its valence layer. Phosphorus is commonly used for this purpose. The phosphorus atoms join right in the crystal structure of the silicon, each one bonding with four adjacent silicon atoms just like a silicon atom would. Because the phosphorus atom has five electrons in its valence shell, but only four of them are bonded to adjacent atoms, the fifth valence electron is left hanging out with nothing to bond to.

The extra valence electrons in the phosphorous atoms start to behave like the single valence electrons in a regular conductor such as copper. They are free to move about, as shown in Figure 5-3. Because this type of semiconductor has extra electrons, it's called an *N-type semiconductor*.

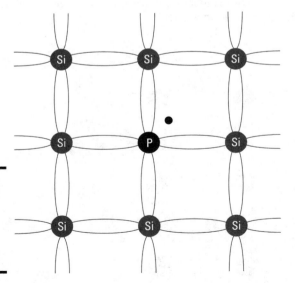

Figure 5-3:
An N-type semiconductor has extra electrons.

✦ **P-type semiconductor:** Happens when the dopant is an element (such as boron) that has only three electrons in the valence shell. When a small amount of boron is incorporated into the crystal, the boron atom is able to bond with four silicon atoms, but since it has only three electrons to offer, a gap is created where there would normally be an electron, as shown in Figure 5-4. This electron gap is called a *hole*. The hole behaves like a positive charge, so semiconductors doped in this way are called *P-type semiconductors.*

Like a positive charge, holes attract electrons. But when an electron moves into a hole, the electron leaves a new hole at its previous location. Thus, in a P-type semiconductor, holes are constantly moving around within the crystal as electrons constantly try to fill them up. A P-type semiconductor is like crazed pieces of Swiss cheese in which you can't quite get a fix on the holes because they're always moving.

**Book II
Chapter 5**

**Working with
Diodes and LEDs**

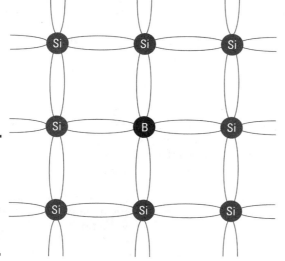

Figure 5-4:
A P-type semicon-ductor has holes where electrons should be.

When voltage is applied to either an N-type or a P-type semiconductor, current flows, for the same reason that it flows in a regular conductor: The negative side of the voltage pushes electrons, and the positive side pulls them. The result is that the random electron and hole movement that's always present in a semiconductor becomes organized in one direction, creating measurable electric current.

Understanding p-n junctions

By themselves, P-type and N-type semiconductors are just conductors. But if you put them together, an interesting and very useful thing happens: Current can flow through the resulting semiconductor, but only in one direction.

The *p-n junction*, as it's called, is like a one-way gate. If you put positive voltage on the p side of the junction and negative voltage on the n side, current flows through the junction. But if you reverse the voltage, putting negative voltage on the p side and positive voltage on the n side, current doesn't flow.

Picture a turnstile gate such as the gates you must go through to get into a baseball stadium or a subway station: You can walk through the gate in one direction but not the other. That's essentially what a p-n junction does. It allows current to flow one way but not the other.

To understand why p-n junctions allow current to flow in only one direction, you must first understand what happens right at the boundary between the p-type material and the n-type material. Because opposite charges attract, the extra electrons on the n-type side of the junction are attracted to the holes on the p-type side. So they start to drift across to the other side.

When an electron leaves the n-type side to fill a hole in the p-type side, a hole is left in the n-type side where the electron was. Thus, it's as if the electron and the hole trade places. The boundary of a p-n junction ends up being populated by defectors: Electrons and holes have crossed the boundary and are now on the wrong side of the junction.

This region that is occupied by electrons and holes that have crossed over is called the *depletion zone*. Because one side of the depletion zone has electrons (negative charges) and the other side has holes (positive charges), a voltage exists between the two edges of the depletion zone. This voltage has an interesting effect on the defectors: It beckons them to turn around and come home. In other words, the holes that have jumped to the negative side of the junction attract the electrons that have jumped to the positive side.

Imagine what it's like to be an electron that has jumped over the boundary and into the p-type side of the junction. Being negatively charged, you're attracted to move further into the p side by the positively charged holes you see ahead of you. But you're also attracted by the positively charged holes that now lie behind you — the very same hole you traded places with now exerts a pull on you that discourages you from going any further.

Unable to make up your mind, you decide to just stay put. That's exactly what happens to the electrons and holes that have crossed over to the other side. The depletion zone becomes stable — a state that's called *equilibrium*.

Now consider what happens when the equilibrium is disturbed by a voltage placed across the p-n junction. The effect depends on which direction the voltage is applied, as follows:

✦ If you apply positive voltage to the p-type side and negative voltage to the n-type side, the depletion zone is pushed from both sides toward the center, making it smaller. Electrons in to the n-type side of the junction are pushed by the voltage toward the depletion zone and eventually collapse it completely. When that happens, the p-n junction becomes a conductor, and current flows.

✦ When voltage is applied in the reverse direction, the depletion zone is pulled from both sides of the junction, and thus it expands. The larger it gets, the more of an insulator the p-n junction becomes. Thus, when voltage is applied in the reverse direction, current doesn't flow through the junction.

Introducing Diodes

A *diode* is a device made from a single p-n junction. As Figure 5-5 shows, leads are attached to the two ends of the p-n junction. These leads allow you to easily incorporate the diode into your circuits.

The lead attached to the n-type semiconductor is called the *cathode*. Thus, the cathode is the negative side of the diode. The positive side of the diode — that is, the lead attached to the p-type semiconductor — is called the *anode*.

Figure 5-5:
A diode has a single p-n junction.

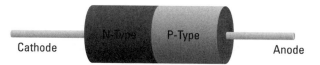

When a voltage source is connected to a diode such that the positive side of the voltage source is on the anode and the negative side is on the cathode, the diode becomes a conductor and allows current to flow. Voltage connected to the diode in this direction is called *forward bias*.

But if you reverse the voltage direction, applying the positive side to the cathode and the negative side to the anode, current doesn't flow. In effect, the diode becomes an insulator. Voltage connected to the diode in this direction is called *reverse bias*.

Forward bias allows current to flow through the diode. Reverse bias doesn't allow current to flow. (Up to a point, anyway. As you'll discover in just a few moments, there are limits to how much reverse bias voltage a diode can hold at bay.)

The schematic symbol for a diode is shown in the margin. The anode is on the left, and the cathode is on the right. Here are two useful tricks for remembering which side of the symbol is the anode and which is the cathode:

+ Think of the anode side of the symbol as an arrow that indicates the direction of conventional current flow — from positive to negative. Thus, the diode allows current to flow in the direction of the arrow.

+ Think of the vertical line on the cathode side as a giant minus sign, indicating which side of the diode is negative for forward bias.

Figure 5-6 illustrates forward and reverse bias with two very simple circuits that connect a lamp to a battery with diodes. In the circuit on the left, the diode is forward biased, so current flows through the circuit and the lamp lights up. In the circuit on the right, the diode is reverse biased, so current doesn't flow and the lamp remains dark.

Figure 5-6:
Forward
and reverse
bias on a
rectifier
diode.

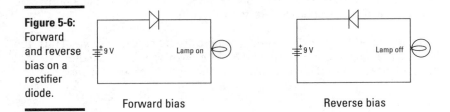

Forward bias   Reverse bias

Note that in a typical diode, a certain amount of forward voltage is required before any current will flow. This amount is usually very small. In most diodes, this voltage is around half a volt. Up to this voltage, current doesn't flow. Once the forward voltage is reached, however, current flows easily through the diode.

This minimum threshold of voltage in the forward direction is called the diode's *forward voltage drop*. That's because the circuit loses this voltage at the diode. For example, if you were to place a voltmeter across the leads of the diode in the forward-biased circuit in Figure 5-6 (left), you would read the forward voltage drop of the diode. Then, if you were to place the voltmeter across the lamp terminals, the voltage would be the difference between the battery voltage (9 V) and the forward voltage drop of the diode.

For example, if the forward voltage drop of the diode was 0.7 V and the battery voltage was exactly 9 V, the voltage across the lamp would be 8.3 V.

Diodes also have a maximum reverse voltage they can withstand before they break down and allow current to flow backward through the diode. This reverse voltage (sometimes called *PIV*, for *peak inverse voltage*, or *PRV* for *peak reverse voltage*) is an important specification for diodes you use in your circuits, as you need to ensure that your diodes won't be exposed to more than their PIV rating.

Besides the forward-voltage drop and peak inverse voltage, diodes are also rated for a maximum current rating. Exceed this current, and the diode will be damaged beyond repair.

The Many Types of Diodes

Diodes come in many different flavors. Some of them are exotic and rarely used by hobbyists, but many are commonly used. The following sections describe three types of diodes you're likely to encounter in your electronic pursuits: rectifier, signal, and Zener diodes. Later in this chapter, you also learn about light-emitting diodes, or LEDs.

Rectifier diodes

A *rectifier diode* is designed specifically for circuits that need to convert alternating current to direct current. The most common rectifier diodes are identified by the model numbers 1N4001 through 1N4007. These diodes can pass currents of up to 1 A, and they have peak inverse voltage (PIV) ratings that range from 50 to 1,000 V.

Table 5-1 lists the peak inverse voltages for each of these common diodes. When choosing one of these diodes for your circuit, pick one that has a PIV that's at least double the voltage you expect it to be exposed to. For most battery-power circuits, the 50 V PIV of the 1N4001 is more than sufficient.

Table 5-1	Peak Inverse Voltages for 1N400x Diodes		
Model Number	*Diode Type*	*Peak Inverse Voltage*	*Current*
1N4001	Rectifier	50 V	1 A
1N4002	Rectifier	100 V	1 A
1N4003	Rectifier	200 V	1 A
1N4004	Rectifier	400 V	1 A
1N4005	Rectifier	600 V	1 A
1N4006	Rectifier	800 V	1 A
1N4007	Rectifier	1,000 V	1 A

Most rectifier diodes have a *forward voltage drop* of about 0.7 V. Thus, a minimum of 0.7 V is required for current to flow through the diode.

Signal diodes

A *signal diode* is designed for much smaller current loads than a rectifier diode and can typically handle about 100 mA or 200 mA of current.

Freewheeling signal diodes

Signal diodes such as the 1N4148 variety often find themselves used with inductors — most commonly relays — to protect surrounding circuits from voltage spikes that can occur when the inductor suddenly shuts off. You learn more about relays in Book IV, Chapter 4, but for now just realize that a relay is a mechanical switch that's operated not by a human finger, but by a magnetic field created by a coil. Relays are incredibly useful because they allow you to use low-voltage circuits to control high-voltage circuits. For example, you can use a circuit powered by a 9 V battery to turn a 120 VAC lamp on or off.

The problem with relays is that they have coils, and as Joseph Henry discovered in the 19th century, coils have an interesting property called *self-inductance*. (If you need a review of what self-inductance is and how it works, refer to the Chapter 4 in this minibook.)

Self-inductance can be a problem in circuits that include relays. When a current is applied to the relay, a magnetic field is formed around the coil. When the current is suddenly shut off, the magnetic field quickly collapses. This collapse induces a voltage in the coil that is opposite in polarity to the voltage that turned the coil on. If you don't do something to contain that voltage, it will send current traveling around your circuit like a drunk driver going down a freeway in the wrong direction.

Signal diodes are just the ticket for stopping this wrong-way voltage in its tracks. Simply place a signal diode across the relay, backward from the voltage applied to the relay coil, as shown in the following schematic:

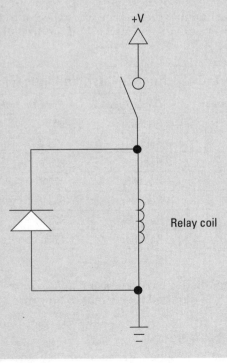

+V

Relay coil

Consider now how this diode functions alongside the relay. When you close the switch, voltage is applied, and current flows. All of this current flows through the relay because the diode is reverse biased: The positive voltage is at the cathode end of the diode.

When you open the switch, the voltage across the relays is cut. Current stops flowing, and the magnetic field around the coil suddenly collapses. This creates a voltage across the coil that's opposite to the voltage shown in the circuit. In other words, the induced voltage is negative at the top of the coil, positive at the bottom. Now the diode is forward biased, so it becomes a conductor. The diode then creates a circuit through which current can flow, thus safely containing the self-induced voltage within the coil and the diode.

Diodes used in this way go by several names, including *freewheeling diodes, flyback diodes, catch diodes,* and *snubber diodes.*

The most commonly used signal diode is the 1N4148. This diode has a close brother called 1N914 that can be used in its place if you can't find a 1N4148. This diode has a forward-voltage drop of 0.7 and a peak inverse voltage of 100 V, and can carry a maximum of 200 mA of current.

Here are a few other interesting points to ponder about signal diodes:

✦ They're noticeably smaller than rectifier diodes and are often made of glass. You have to look at them closely to see it, but the cathode end of a signal diode is marked by a small black band.

✦ They're better than rectifier diodes when dealing with high-frequency signals, so they're often used in circuits that process audio or radio frequency signals. Because of its ability to respond quickly at high frequencies, signal diodes are sometimes called *high-speed diodes.* They're also sometimes called *switching diodes* because digital circuits (which you learn about in Book VI) often use them as high-speed switches.

✦ Some signal diodes are made of germanium rather than silicon. (Germanium the crystal, not to be confused with geranium the flower.) Germanium diodes have a much smaller forward-voltage drop than silicon diodes — as low as 0.15 V. This makes them useful for radio applications, which often deal with very weak signals.

Zener diodes

In a normal diode, the peak inverse voltage is usually pretty high — 50, 100, even 1,000 V. If the reverse voltage across the diode exceeds this number, current floods across the diode in the reverse direction in an avalanche, which usually results in the diode's demise. Normal diodes aren't designed to withstand a reverse avalanche of current. *Zener diodes* are. They're specially designed to withstand current that flows when the peak inverse voltage is reached or exceeded. And more than that, Zener diodes are designed so that as the reverse voltage applied to them exceeds the threshold voltage, current flows more and more in a way that holds the voltage drop across the diode at a fixed level. In other words, Zener diodes can be used to regulate the voltage across a circuit.

In a Zener diode, the peak inverse voltage is called the *Zener voltage*. This voltage can be quite low — in the range of a few volts — or it can be hundreds of volts.

Zener diodes are often used in circuits where a predictable voltage is required. For example, suppose you have a circuit that will be damaged if you feed it with more than 5 V. In that case, you could place a 5 V Zener diode across the circuit, effectively limiting the circuit to 5 V. If more than 5 V is applied to the circuit, the Zener diode conducts the excess voltage away from the sensitive circuit.

Zener diodes have their own variation of the standard diode schematic symbol, as shown in the margin. You can learn more about working with Zener diodes in Book IV, Chapter 3.

Using a Diode to Block Reverse Polarity

Project 5-1 presents a simple little construction project that demonstrates a diode's ability to conduct current in only one direction. For this project, you wire up a rectifier diode in series with a 3 V flashlight lamp and a pair of AA batteries. I have you use a DPDT knife switch between the battery and the diode/lamp circuit so that when the switch is changed from one position to the other, the polarity of the voltage across the diode and lamp is reversed. Thus, the lamp lights in only one of the two switch positions. Figure 5-7 shows the completed project.

Figure 5-7:
The completed circuit for Project 5-1.

Voltage Measurements

Black Lead	Red Lead	Voltage
Switch 2X	Switch 1X	
Lamp 1	Lamp 2	
Diode cathode	Diode anode	

Putting Rectifiers to Work

**Book II
Chapter 5**

Working with
Diodes and LEDs

One of the most common uses for rectifier diodes is to convert household alternating current into direct current that can be used as an alternative to batteries. You learn all about creating complete power supplies for this purpose in Book IV, Chapter 3. For now, just concentrate on one part of a complete DC power supply: the rectifier circuit, which is typically made from a set of cleverly interlocked diodes. (A *rectifier* is a circuit that converts alternating current to direct current.)

In household current, the voltage swings from positive to negative in cycles that repeat sixty times per second. If you place a diode in series with an alternating current voltage, you eliminate the negative side of the voltage cycle, so you end up with just positive voltage, as shown in Figure 5-8.

Figure 5-8:
Using a diode to rectify alternating current.

If you look at the waveform of the voltage coming out of this rectifier diode, you'll see that it consists of intervals that alternate between a short increase of voltage and periods of no voltage at all. This is a form of direct current because it consists entirely of positive voltage. However, it pulsates: first it's on, then it's off, then it's on again, and so on.

Overall, voltage rectified by a single diode is off half of the time. So although the positive voltage reaches the same peak level as the input voltage, the average level of the rectified voltage is only half the level of the input voltage. This type of rectifier circuit is sometimes called a *half-wave rectifier* because it passes along only half of the incoming alternating current waveform.

Project 5-1: Blocking Reverse Polarity

In this project, you'll build a simple circuit that uses a diode to allow current to flow through a lamp circuit in only one direction. The circuit uses a DPDT knife switch to reverse the polarity of the battery. When the polarity is reversed, the diode blocks the current so the LED does not light up.

You also need a small Philips-head screwdriver, wire cutters wire strippers, and a multimeter to complete this project.

Parts List
2 AA batteries
1 Battery holder (Radio Shack 270-408)
1 Lamp holder (Radio Shack 272-357)
1 3 V flashlight lamp (Radio Shack 272-1175)
1 DPDT knife switch (Radio Shack 275-1537)
1 N4001 rectifier diode (Radio Shack 276-1101)
3 5" 22-guage stranded wire

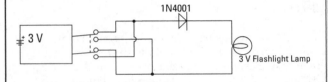

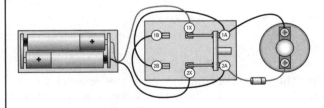

Steps

1. **Cut three, 5" lengths of wire and strip ½" of insulation from each end.**

2. **Open the switch.**

   Move the handles to their upright positions so none of the contacts are connected.

3. **Attach the red lead from the battery holder to terminal 1X of the switch.**

4. **Attach the black lead from the battery to terminal 2X of the switch.**

5. **Use one of the three, 5" wires to connect terminal 1B of the switch to terminal 2A.**

6. **Use a second 5" wire to connect terminal 2B of the switch to terminal 1A.**

7. **Use the third 5" wire to connect terminal 1A of the switch to one terminal of the lamp socket.**

8. **Connect the diode to termnal 2A of the switch and the empty terminal of the lamp socket.**

   The cathode (the side with the stripe) should be connected to the lamp.

9. **Insert the lamp into the lamp holder.**

10. **Insert the batteries into the battery holder.**

    The circuit is now ready to test.

11. **Flip the switch to the A position.**

    The lamp lights because positive voltage is connected to the diode's anode, and negative voltage is applied to the cathode side of the circuit. This allows current to flow through the diode, and the lamp turns on.

12. **Flip switch to the B position.**

    This time, the lamp goes out because voltage is reversed across the diode.

13. **Use your multimeter to measure the voltage across the battery, lamp, and diode.**

    Set the multimeter to a DC volts range that will accomodate the 3 V battery voltage. Then, take the measurements indicated in the table "Voltage Measurements." The third measurement (the voltage across the diode) should be very close to 0.7 V. The second measurements (the voltage across the lamp) plus the voltage across the diode should add up to the first measurement (the voltage across the battery). Don't worry if they're slightly off, as the difference may be accounted for by inaccuracies in your voltmeter — especially if you have an analog meter or a very inexpensive digital meter.

Voltage Measurements		
Black Lead	**Red Lead**	**Voltage**
Switch 2X	Switch 1X	
Lamp 1	Lamp 2	
Diode cathode	Diode anode	

A better type of rectifier circuit uses four rectifier diodes, in a special circuit called a *bridge rectifier*. Figure 5-9 shows a bridge rectifier circuit.

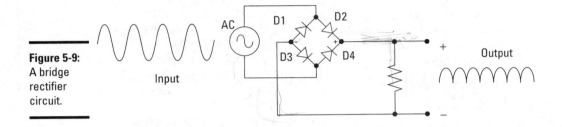

Figure 5-9:
A bridge rectifier circuit.

Look at how this rectifier works on both sides of the alternating current input signal:

✦ In the first half of the AC cycle, D2 and D4 conduct because they're forward biased. Positive voltage is on the anode of D2 and negative voltage is on the cathode of D4. Thus, these two diodes work together to pass the first half of the signal through.

✦ In the second half of the AC cycle, D1 and D3 conduct because they're forward biased: Positive voltage is on the anode of D1, and negative voltage is on the cathode of D3.

The net effect of the bridge rectifier is that both halves of the AC sine wave are allowed to pass through, but the negative half of the wave is inverted so that it becomes positive.

The resulting DC signal doesn't drop to zero for half of the cycle. However, it still doesn't provide a steady DC voltage level. Think about what you learn in Chapters 3 and 4 of this minibook, though. Both capacitors and inductors can be used to slow down changes in current and voltage. Thus, capacitors and inductors are often used in power supply circuits to improve the quality of DC voltage coming out of a rectifier circuit. You can learn more about how that's done in Book IV, Chapter 3. But first, let's look at rectifiers in action.

Building Rectifier Circuits

In Project 5-2, you build simple half-wave and bridge rectifier circuits to see how diodes can convert alternating current to direct current. You don't light any lamps or anything with this circuit; instead, just use a voltmeter to verify that the diodes are doing their job. And to keep things safe, use a 9 VAC power adapter as the source for the alternating current. You should be able to find a 9 VAC power adapter for about $15 online, or you may find one for a dollar or two at a thrift store.

To make the voltage of the 9 VAC power adapter easier to work with, I suggest you cut off the power plug, separate the two wire leads, strip a bit of insulation from the ends, and attach alligator clips to the leads. The alligator clips make it much easier to connect the supply to your experimental circuits.

Before you build the project, use the multimeter to measure the AC voltage actually created by your power adapter. Although it may be marked as 9 VAC, the actual voltage you measure will probably be slightly more than 9 VAC.

You can find more information about rectifiers and how they're used in power supply circuits in Book IV, Chapter 3. For the remainder of this chapter, turn your attention to another useful type of diode: the light-emitting kind.

Introducing Light Emitting Diodes

A *light-emitting diode* (also called an *LED*) is a special type of diode that emits visible light when current passes through it. The most common type of LED emits red light, but LEDs that emit blue, green, yellow, or white light are also available.

The word *LED* is pronounced by spelling the letters out (*el-ee-dee*), not like the word *lead*.

The schematic diagram symbol for an LED is shown in the margin, and Figure 5-10 shows typical LEDs. The two leads protruding from the bottom of an LED aren't the same length: The shorter lead is the cathode, while the anode is the longer lead.

Figure 5-10: LEDs. The shorter lead is the cathode.

Project 5-2: Rectifier Circuits

In this project, you'll build a simple half-wace rectifier and a bridge rectifier. You'll connect the rectifiers to a 9 VAC power source, then use you multimeter to measure the DC output to verify that the AC voltage has been converted to DC.

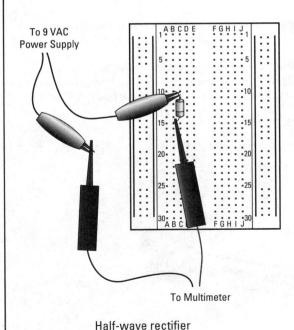

To 9 VAC Power Supply

To Multimeter

Half-wave rectifier

Steps

1. **Start by building a simple half-wave rectifier: Insert one of the diodes in the solderless breadboard.**

   Insert the anode in hole B10 and the cathode in hole B15. (The cathode is the lead that has the stripe next to it.)

2. **Set the multimeter to a suitable VDC range.**

   10 VDC or 20 VDC should be fine.

3. **Connect one lead of the power adapter to the anode.**

   The anode is the lead inserted into hole B10.

4. **Connect the other lead of the power adapter to the black multimeter probe.**

   Use an alligator clip.

5. **Plug the power adapter into a wall outlet.**

   Once the power adapter is plugged in, be careful to prevent the two leads of the power adapter from coming into contact with one another. The resulting short circuit could damage the power adapter.

6. **Touch the red multimeter probe to the diode cathode to measure the DC voltage rectified by the diode.**

   Note that the DC voltage is about half of the input AC voltage. That's because the single diode is allowing only half of the AC sine wave to pass.

7. **Unplug the power adapter.**

   That's it for the half-wave rectifier. Now build the bridge rectifier:

8. **Insert the three remaining diodes to complete the bridge rectifier.**

   When you're done, the four diodes should be inserted into the following holes.

Cathode (Striped End)	Anode
B15	B10
A5	A10
E9	E5
D9	D15

   It's important to note that the first two diodes share a common lead in row 10, while the second two diodes share a common lead in row 9. Make sure you don't insert all four of these diodes into the same row. If you do, the bridge rectifier circuit won't work.

9. **Connect one lead of the AC power adapter to the diode lead in hole A5.**

10. **Connect the other lead of the AC power adapter to the diode lead in hole B15.**

11. **Plug in the 9 VAC power adapter.**

12. **Measure the voltage across the bridge rectifier.**

    Touch the black multimeter probe to the diode lead in hole A10 and the red probe to the diode lead in hole E9.

Book II
Chapter 5

Working with
Diodes and LEDs

Be careful not to accidentally short out any of the leads when you touch them with the voltmete probes.

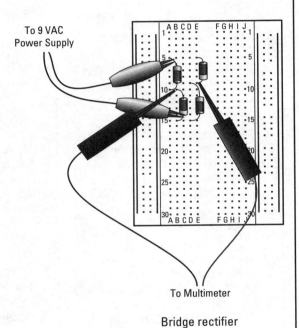

To 9 VAC
Power Supply

To Multimeter

Bridge rectifier

Whenever you use an LED in a circuit, you must provide some resistance in series with the LED, as shown in the schematic diagram in Figure 5-11. Otherwise, the LED will light brightly for an instant, and then burn itself out. In this example, the LED is connected to a 9 V DC supply through a 470 Ω resistor.

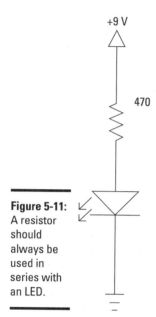

Figure 5-11: A resistor should always be used in series with an LED.

To determine the value of the resistor you should use, you need to know these three things:

+ **The supply voltage:** For example, 9 V.

+ **The LED forward-voltage drop:** For most red LEDs, the forward-voltage drop is 2 V. For other LED types, the voltage drop may be different. Check the specifications on the package if you use other types of LEDs.

+ **The desired current through the LED:** Usually, the current flowing through the LED should be kept under 20 mA.

Once you know these three things, you can use Ohm's law to calculate the desired resistance. The calculation requires just four steps, as follows:

1. **Calculate the resistor voltage drop.**

You do that by subtracting the voltage drop of the LED (typically 2 V) from the total supply voltage. For example, if the total supply voltage is 9 V and the LED drops 2 V, the voltage drop for the resistor is 7 V.

2. Convert the desired current to amperes.

In Ohm's law, the current must be expressed in amperes. You can convert milliamperes to amperes by dividing the milliamperes by 1,000. Thus, if your desired current through the LED is 20 mA, you must use 0.02 in your Ohm's law calculation.

3. Divide the resistor voltage drop by the current in amperes.

This gives you the desired resistance in ohms. For example, if the resistor voltage drop is 7 V and the desired current is 20 mA, you need a 350 Ω resistor.

4. Round up to the nearest standard resistor value.

The next higher resistor value from 350 Ω is 390 Ω. If you can't find a 390 Ω resistor, a 470 Ω will do the trick.

Note that the minor increases in resistances mean that slightly less current will flow through the resistor, but the difference won't be noticeable. However, you should avoid going to a lower resistor value. Lowering the resistance increases the current, which can damage the LED.

Here are a few additional things to remember when you work with LEDs:

✦ If you're going to place more than one LED together in series, just add up the voltage drops to calculate the size of the resistor you need. For example, if you have a 9 V battery that will supply voltage to three LEDs, each with a 2 V drop, the total voltage drop is 6 V, so the voltage drop across the resistor is 3 V. Using Ohm's law, you can then calculate that the resistor needed to restrict the current flow to 20 mA is 150 Ω.

✦ For 9 V and less, ¼ W resistors are more than adequate. If you're applying larger voltages to the LED circuit, you may need to use resistors that can handle more power. To calculate how much power in watts the resistor should be rated for, just multiply the voltage dropped across the resistor by the current in amps. For example, if the voltage drop on the resistor is 3 V and the current is 20 mA, the power dissipated by the resistor will be 0.06 W, which is well under the limits of a ⅛ W resistor.

✦ You can get two LEDs, usually of different colors, combined into a single package. Green and red are a common combination. When two LEDs are bundled in a single package, they're usually wired opposite of one another. That way, you can control which LED lights by changing the polarity of the voltage applied across the LED. In some cases, a third lead is used. This lead is connected to the cathode of both LEDs. The third lead enables you to light both LEDs, which yields a third color. For example, when both LEDs are lit in a green and red combination LED, the resulting color is yellow.

Using LEDs to Detect Polarity

In this section, you build a project that uses two LEDs to indicate the polarity of an input voltage. The voltage is provided by a 9 V battery connected to the circuit via a DPDT knife switch that's wired to reverse the battery polarity. The two LEDs and their corresponding resistors are mounted on a small solderless breadboard. Figure 5-12 shows the completed circuit.

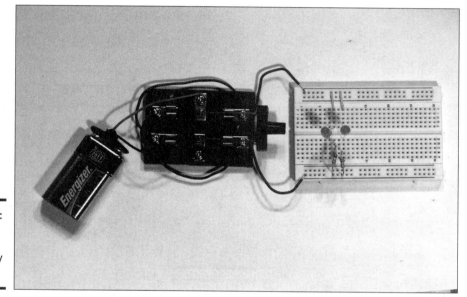

Figure 5-12:
The completed LED polarity detector.

Book II
Chapter 5

Working with Diodes and LEDs

Project 5-3: An LED Polarity Detector

In this project, you build a polarity detector that uses LEDs to indicate the polarity of a power supply. You will need a small Phillips-head screwdriver, wire cutters, and wire strippers to assemble the circuit.

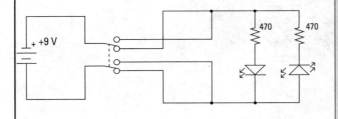

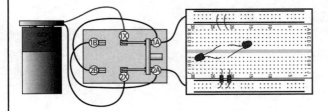

Steps

1. Cut four 5" lengths of wire and strip —½"— of insulation from each end.

2. Open the switch.

   Move the handles to their upright positions so none of the contacts are connected.

3. Attach the red lead from the battery holder to terminal 1X of the switch.

4. Attach the black lead from the battery to terminal 2X of the switch.

5. Use a 5" wire to connect terminal 1B of the switch to terminal 2A.

6. Use a 5" wire to connect terminal 2B of the switch to terminal 1A.

7. Use a 5" wire to connect terminal 1A of the switch to any hole on the positive breadboard bus.

8. Use a 5" wire to connect terminal 2A of the switch to any hole on the negative breadboard bus.

9. Connect a two short jumper wire on the breadboard to connect hole J8 to any nearby hole on the positive bus strip.

10. Use a short jumper wire to connect hole 10A to any nearby hole on the positive bus strip.

11. Insert one of the resistors in hole B8 and any nearby hole on the negative bus strip.

12. Insert the other resistor in hole B10 and any nearby hole in the negative bus strip.

13. Connect the two LEDs on the breadboard as follows:

Cathode (short lead)	Anode
Hole 8E	Hole 8F
Hole 10F	Hole 10E

   Note that the cathode is the short lead.

14. Insert the 9V battery into the battery holder.

15. Flip the switch to the A position.

   The LED in row 8 lights up because the voltage from the battery is forward biased on this LED.

16. Flip the switch to the B position.

Chapter 6: Working with Transistors

In This Chapter

✔ Learning about the miracle device of electronics

✔ Comprehending the basic operation of a transistor

✔ Surveying the many different types of transistors

✔ Using a transistor as an amplifier

✔ Using a transistor as a switch

✔ Building some simple transistor circuits

I was born just a few years after the beginning of the transistor age. As a young boy, I sometimes took the back off of our old black-and-white television set (when my parents weren't home) and marveled at the neat electronic stuff inside. Besides the huge picture tube, the most interesting gadgets in there were the little glass tubes the size of my thumb. As I remember it, there were hundreds of them, each one glowing with a comfortable warm orangish glow.

By the time I was old enough to start studying electronics, that old TV set was replaced by a new color set that wasn't nearly as interesting to look at inside. The big picture tube was still there, but all the little glowing thumb-sized tubes had been replaced by little silver cans smaller than a thimble.

Transistors had taken over, and I hated them. Tubes were much more interesting than transistors. They got hot and glowed. You could see little wire structures inside them — little towers with meshes and grids and who knows what else. Transistors just looked like little thimbles. It was as if someone had built a television set out of stuff they found in my mom's sewing drawer.

I was quite certain that the spaceships on my favorite TV shows like *Star Trek*, *Lost in Space*, and *Land of the Giants* were all run on tubes, not transistors. It wasn't until I found out that real spaceships like Gemini and Apollo were filled with transistors and had nary a tube that I decided maybe transistors were okay. If they were good enough for NASA, they were good enough for me.

In the 40 years since my dad bought that first color television set, the little thimble-sized transistors have given way to transistors that are literally millions of times smaller. Nowadays, we can put 100 million transistors on a single piece of silicon crystal about the size of your fingernail.

In this chapter, you take a look at what transistors are and how you can put them to use in your own circuits. Along the way, you build a few simple transistor circuits to learn how they work.

What's the Big Deal About Transistors?

Here's an interesting thing about the transistor: When it was invented in 1947, it didn't really do anything that hadn't already been done before. It just did it in a radically different way.

The basic idea behind a transistor is that it lets you control the flow of current through one channel by varying the intensity of a much smaller current that's flowing through a second channel.

Think of a transistor as an electronic lever. A lever is a device that lets you lift a large load by exerting a small amount of effort. In essence, a lever amplifies your effort. That's what a transistor does: It lets you use a small current to control a much larger current.

Figure 6-1 shows several of the many different kinds of transistors that are available today. As you can see, transistors come in a variety of different sizes and shapes. One thing all of these transistors have in common is that they each have three leads.

Why were transistors invented?

Devices that performed the function of transistors had been around for 30–40 years prior to the invention of the transistor. They were called *vacuum tubes*. A vacuum tube consisted of a vacuum chamber made from glass or metal, a heating element that heated the space inside the chamber, and electrodes that protruded into the chamber. (I use past tense here because although vacuum tubes still exist, they really aren't used all that often.)

One specific type of vacuum tube was called a *triode*; it had three electrodes. In a triode, a large current flowing through two of the electrodes (called the *anode* and the *cathode*) could be regulated by placing a wire grid (called the *control grid*) between the cathode and the anode. Applying a small current to this grid slowed down the flow of electrons between the cathode and the anode.

Figure 6-1:
Transistors
come
in many
shapes and
sizes.

It didn't take long to figure out that you could use a fluctuating signal such as a radio or audio wave on the control grid. When you did that, the current on the anode followed the fluctuations of the control grid current, but with much larger variations. Thus, the triode was an electronic lever: Small variations in current at the control grid were amplified to create large variations in current at the anode.

The vacuum tube triode was patented in 1907 and was the key invention that enabled the development of radio, television, and computers. But vacuum tubes had many serious limitations: They were expensive to manufacture, big (the small ones were about the size of your thumb), required a lot of power to operate, generated a lot of heat, and lasted only a few years before they burned themselves out.

The transistor changed all that. A transistor performs the same function as a vacuum tube triode, but using semiconductor junctions instead of heated electrodes in a vacuum chamber. Although the transistor didn't do anything that the vacuum tube triode didn't already do, it did it in a radically different way that had huge advantages over the vacuum tube. The earliest transistors were small, required very little power to operate, generated much less heat, and lasted much longer than vacuum tubes.

Looking inside a transistor

There are many different kinds of transistors. The most basic kind is called a *bipolar transistor*. Bipolar transistors are the easiest to understand, and they're the ones you're most likely to work with as a hobbyist. As a result, most of this chapter focuses on bipolar transistors. I describe some of the other types of transistors later in this chapter. But for now — indeed throughout this entire book — you can assume that whenever I use the term *transistor* by itself, I'm referring to a bipolar transistor.

Now let's peer inside a transistor to see how it works.

In the previous chapter, you learn that a *diode* is the simplest kind of semiconductor, made from a single p-n junction, which is simply a junction of two different types of semiconductors, one that's missing a few electrons and thus has a positive charge (*p-type* semiconductor) and the other with a few extra electrons, thus having a negative charge (*n-type* semiconductor).

By itself, a p-n junction works as a one-way gate for current. In other words, a p-n junction allows current to flow in one direction but not the other. A diode is simply a p-n junction with a lead attached to both ends.

A transistor is like a diode with a third layer of either p-type or n-type semiconductors on one end. Thus, a transistor has three regions rather than two. The interface between each of the regions forms a p-n junction. So another way to think of a transistor is as a semiconductor with two p-n junctions.

Figure 6-2 shows the structure of two common types of transistors along with their schematic diagram symbols. The details shown in this figure are explained in the following paragraphs.

One way to make a transistor is with a p-type semiconductor sandwiched between two n-type semiconductors. This type of transistor is called an *NPN transistor* because it has three regions: n-type, p-type, and n-type. It's shown in the top part of Figure 6-2.

The other way to make a transistor is just the opposite, with an n-type semiconductor sandwiched between two p-type semiconductors. This type is called a PNP transistor because its three regions are p-type, n-type, and p-type. It's shown in the bottom part of Figure 6-2.

Each of the three regions of semiconductor material in a transistor has a lead attached to it, and each of these leads is given a name:

✦ **Collector:** This lead is attached to the largest of the semiconductor regions. Current flows through the collector to the emitter as controlled by the base.

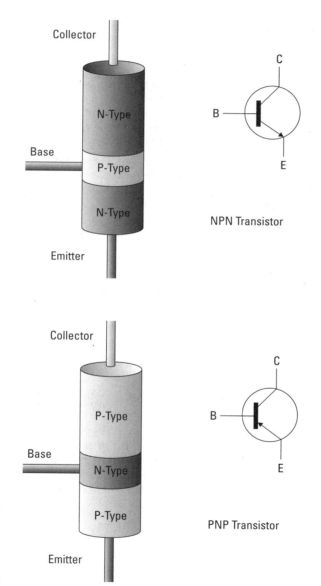

Collector

N-Type

Base

P-Type

N-Type

NPN Transistor

Emitter

Collector

P-Type

Base

N-Type

P-Type

PNP Transistor

Emitter

Figure 6-2: NPN and PNP transistors.

✦ **Emitter:** Attached to the second largest of the semiconductor regions. When the base voltage allows, current flows through the collector to the emitter.

✦ **Base:** Attached to the middle semiconductor region. This region serves as the gatekeeper that determines how much current is allowed to flow through the collector-emitter circuit. When voltage is applied to the base, current is allowed to flow.

These two current paths are important in a transistor:

✦ **Collector-emitter:** The main current that flows through the transistor. Voltage placed across the collector and emitter is often referred to as V_{ce}, and current flowing through the collector-emitter path is called I_{ce}.

✦ **Base-emitter:** The current path that controls the flow of current through the collector-emitter path. Voltage across the base-emitter path is referred to as V_{BE} and is also sometimes called *bias voltage*. Current through the base-emitter path is called I_{BE}.

Here are a few additional points to ponder concerning transistors before we move on to more details:

✦ In an NPN transistor, the emitter is the negative side of the transistor. The collector and base are the positive sides.

✦ In a PNP transistor, the emitter is the positive side of the transistor. The collector and base are the negative sides.

✦ Most circuits that you can build with an NPN transistor can also be built with a PNP transistor. But if you do, you must remember to flip the power connections.

✦ In a schematic diagram, transistors are usually represented by the letter Q.

✦ Yes, there are reasons why the terms *collector, emitter,* and *base* were chosen for three leads of a transistor. Unfortunately, those reasons have to do with the internal operation of the transistor at a level that's deeper than you really need to go for this book. So please take my word for it: The guys who invented the transistor didn't choose the terms *collector, emitter,* and *base* just to confuse you.

Considering field-effect transistors

Bipolar transistors aren't the only kids on the semiconductor block. Another variety of transistor, called a *field-effect transistor (FET)*, has become extremely popular in recent years, especially as the building blocks for *integrated circuits* (also known as *ICs*). Field-effect transistors can be made much smaller than bipolar transistors, and they use much less current.

Field-effect transistors behave much like bipolar transistors, but they have a nomenclature all their own: Instead of *base, emitter,* and *collector,* the terminals in a field-effect transistor are called the *gate, drain,* and *source.* Internally, a field-effect transistor is very different from a bipolar transistor. Instead of using a pair of p-n junctions, a field-effect transistor consists of a single piece of n- or p-type semiconductor with a special substance placed on it that can control the current flow through the semiconductor.

There are a dozen or so different types of field-effect transistors in existence, but the most commonly used are called *MOSFET* (for metal-oxide-semiconductor field-effect transistor) and *JFET* (for junction field-effect transistor).

Be warned that field-effect transistors are quite susceptible to accidental static discharge. If you touch one and hear a little pop as static in your skin discharges through the FET, you may as well throw it away. You should always take precautions against static discharge whenever you handle a field-effect transistor or an integrated circuit that contains field-effect transistors.

Examining transistor specifications

Transistors are more complicated devices than resistors, capacitors, inductors, and diodes. Whereas those components have just a few specifications to wrangle with, such as ohms of resistance and maximum watts of power dissipation, transistors have a bevy of specifications.

You can find the complete specifications for any transistor by looking up its *data sheet* on the Internet; just plug the part number into your favorite search engine. The data sheet gives you dozens of interesting facts about the transistor you're interested in, with charts and graphs only a rocket scientist could love.

If you happen to be a rocket scientist and you're thinking about using the transistor in a missile, by all means please pay attention to every detail in the data sheet. But if you're just trying to do a little on-the-side circuit design, you need to pay attention to only the most important specifications — these in particular:

✦ **Current gain (H_{FE}):** This is a measure of the amplifying ability of the transistor. It refers to the ratio of the base current to the collector current. Typical values range from 50 to 200. The higher this number, the more the transistor is able to amplify an incoming signal.

✦ **Collector-emitter voltage (V_{CEO}):** The maximum voltage across the collector and the emitter. This is usually 30 V or more, which is well above the voltage levels you work with in most hobby circuits.

✦ **Emitter-base voltage (V_{EBO}):** The maximum voltage across the emitter and the base. This is usually a relatively small number, such as 6 V. Most circuits are designed to apply only small voltages to the base, so this limit isn't usually a concern.

✦ **Collector-base voltage (V_{CBO}):** The maximum voltage across the collector and the base. This is usually 50 V or more.

✦ **Collector current (I_C):** The maximum current that can flow through the collector-emitter path. Most circuits use a resister to limit this current flow; the value of the resistor must be calculated using Ohm's law to keep the collector current below the limit. If you exceed this limit for long, the transistor may be damaged.

✦ **Total power dissipation (P_D):** This is the total power that can be dissipated by the device. For most small transistors, the power rating is on the order of a few hundred milliwatts (mW).

You need to worry about these specifications only if you're designing your own circuits. If you're building a circuit you found in a book or on the Internet, all you need to know is the transistor part number specified in the circuit's schematic.

Amplifying with a Transistor

The most common way to use a transistor as an amplifier is shown in Figure 6-3. This type of circuit is sometimes called a *common-emitter* circuit because, as you can see, the emitter is connected to ground, which means that both the input signal and the output signal share the emitter connection.

There are two other ways to use a transistor as an amplifier, called *common-base* and *common-collector*. As you might guess, they involve connecting the base and the collector to ground, respectively. Common-emitter circuits are used more often than common-base or common-collector, so that's what I show you in this chapter.

The circuit in Figure 6-3 uses a pair of resistors as a voltage divider to control exactly how much voltage is placed across the base and emitter of the transistor. The AC signal from the input is then superimposed on this bias voltage to vary the bias current. Then, the amplified output is taken from the collector and emitter. Variations in the bias current are amplified in the output current.

You might remember from Chapter 2 of this minibook that a *voltage divider* is simply a pair of resistors. The voltage across both resistors equals the sum of the voltages across each resistor individually. You can divide the voltage any way you want by picking the correct values for the resistors. If the resistors are identical, the voltage divider cuts the voltage in half. Otherwise, you can use a simple formula to determine the ratio at which the voltage is divided. (If you want to review this formula, feel free to refer to Chapter 2 of this minibook.)

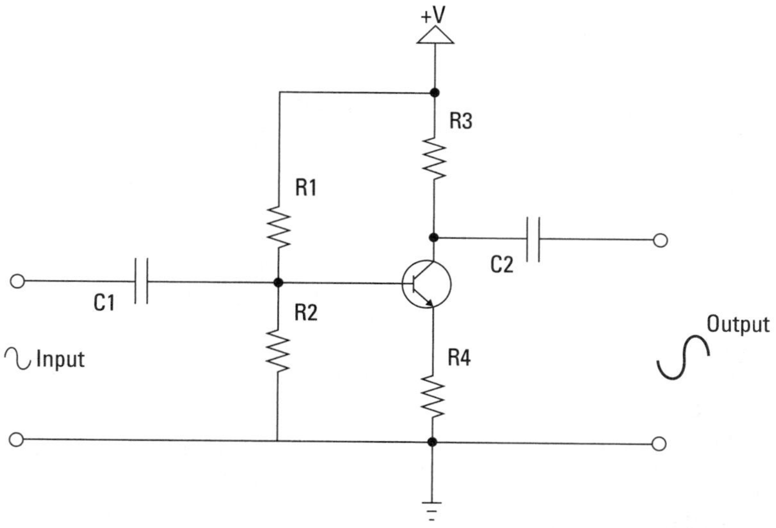

Figure 6-3:
A basic
transistor
amplifier
circuit.

If you look at the schematic diagram in Figure 6-3 and squint your eyes just a bit, you might see that there are actually two voltage dividers in the circuit. The first is the combination of resistors R1 and R2, which provide the bias voltage to the transistor's base. The second is the combination of resistors R3 and R4, which provide the voltage for the output.

The magic pot

My high-school electronics instructor explained that a transistor works like a "magic potentiometer," or as we kids liked to call it, the "magic pot." We used to joke that Puff the Magic Dragon would have enjoyed electronics class because of the magic pot. (Hey, it was the seventies!)

The basic idea is this: A transistor works like a combination of a diode and a variable resistor, also called a *potentiometer* or *pot*. But this isn't just an ordinary pot; it's a magic pot whose knob is mysteriously connected to the diode by invisible rays, kind of like this:

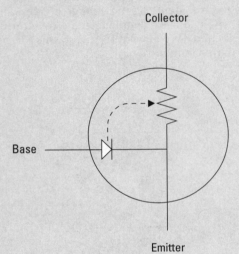

When forward voltage is applied to the diode, the knob of the magic pot turns much like the needle on a voltmeter. This changes the resistance of the potentiometer, which in turn changes the amount of current that can flow through the collector-emitter path.

Note that a magic potentiometer is wired so that when bias voltage increases, resistance decreases. When bias voltage decreases, resistance increases.

Besides being connected to the diode by invisible rays, the magic pot is magic in one more way: Its maximum resistance is infinite. Real-world potentiometers have a finite maximum

resistance, such as 10 kΩ or 1 MΩ, but the magic pot has infinite maximum resistance.

With this knowledge of the magic pot's properties in mind, you can visualize how a transistor works. There are three positions that the magic knob can be in, which correspond to these three operating modes for a transistor.

✔ **Infinite resistance:** When there's no bias voltage, the magic pot's knob is spun all the way in one direction, providing infinite resistance. Thus no current flows through the transistor. (Actually, remember that the base of the transistor is like a diode, which means that a certain amount of forward voltage is required before current begins to flow through the base. The magic pot stays at its infinite setting until that voltage — usually about 0.7 V — is reached.) This state is called *cut-off* because current is cut off. No amps for you!

✔ **Some resistance:** As the bias voltage moves past 0.7 V, the diode begins to conduct, and the invisible rays start turning the knob on the magic pot. Thus current begins to flow. How much current flows depends on how far the bias voltage has caused the knob to turn.

✔ **No resistance:** Eventually, the bias voltage turns the knob to its stopping point, and there's no resistance at all. Current flows unrestricted through the collector-emitter circuit. You can continue to increase the bias voltage, but you can't lower the resistance below zero! This state is called saturation.

In reality, there's a third resistor in the output voltage divider: the collector-emitter path in the transistor itself. In fact, one common way to explain how a transistor works is to think of the collector-emitter path as a potentiometer (a variable resistor), whose knob is turned by the bias voltage. For a more detailed explanation of this, see the sidebar titled "The magic pot."

This second voltage divider is a variable voltage divider: The ratio of the resistances changes based on the bias voltage, which means the voltage at the collector varies as well. The amplification occurs because very small variations in an input signal are reflected in much larger variations in the output signal.

I used the term *reflected* in the preceding paragraph on purpose. In a common-emitter amplifier circuit, the amplified output is a reflection of the input signal. In other words, positive voltage variations in the input appear as negative variations in the output. Another way to put this is to say that the output signal is *inverted* — which is just a fancy way of saying it's upside down.

Because this is tricky stuff, look at this circuit more closely:

✦ The input arrives at the left side of the circuit in the form of a signal, which usually has both a DC and an AC component. In other words, the voltage fluctuates but never goes negative.

✦ One side of the input is connected to ground, to which the battery's negative terminal is also connected. The transistor's emitter is also connected to ground (through a resistor), as is one side of the output.

✦ The purpose of C1 is to block the DC component of the input signal. Only pure AC gets past the capacitor. Without this capacitor, any DC voltage in the input signal would be added to the bias voltage applied to the transistor, which could spoil the transistor's ability to faithfully amplify the AC part of the input signal.

✦ R1 and R2 form a voltage divider that determines how much DC voltage is applied to the transistor base. The AC portion of the signal that gets past C1 is combined with this DC voltage, which causes the transistor's base current to vary with the voltage.

✦ R3, R4, and the variable resistance of the collector-emitter circuit form a voltage divider on the output side of the amplifier. Amplification occurs because the full power supply voltage is applied across the output circuit. The varying resistance of the collector-emitter path reflects the small AC input signal on the much larger output signal.

✦ C2 blocks the DC component of the output signal so that only pure AC is passed on to the next stage of the amplifier circuit.

The trick in designing transistor amplifiers is picking the right values for all the resistors and capacitors. That involves more than a little bit of math and engineering knowledge, and is just beyond the scope of this book. Most hobbyists can get along with published circuits that you can find in kits or on the Internet. But if you really want to know how to calculate these values for yourself, you can find excellent tutorials on the subject on the Internet. Just search for *common emitter* and you'll find what you're looking for.

Using a Transistor as a Switch

One of the most common uses for transistors is as simple switches. In short, a transistor conducts current across the collector-emitter path only when a voltage is applied to the base. When no base voltage is present, the switch is off. When base voltage is present, the switch is on.

In an ideal switch, the transistor should be in only one of two states: off or on. The transistor is off when there's no bias voltage or when the bias voltage is less than 0.7 V. The switch is on when the base is saturated so that collector current can flow without restriction.

Figure 6-4 shows a schematic diagram for a circuit that uses an NPN transistor as a switch that turns an LED on or off.

Look at this circuit component by component:

+ **LED:** This is a standard 5 mm red LED. This type of LED has a voltage drop of 1.8 V and is rated at a maximum current of 20 mA.

+ **R1:** This 330 Ω resistor limits the current through the LED to prevent the LED from burning out. You can use Ohm's law to calculate the amount of current that the resistor will allow to flow. Because the supply voltage is +6 V, and the LED drops 1.8 V, the voltage across R1 will be 4.2 V (6 − 1.8). Dividing the voltage by the resistance gives you the current in amperes, approximately 0.127 A. Multiply by 1,000 to get the current in mA: 12.7 mA, well below the 20 mA limit.

+ **Q1:** This is a common NPN transistor. I used a 2N2222A transistor, but just about any NPN transistor will work. R1 and the LED are connected to the collector, and the emitter is connected to ground. When the transistor is turned on, current flows through the collector and emitter, thus lighting the LED. When the transistor is turned off, the transistor acts as an insulator, and the LED doesn't light.

+ **R2:** This 1 kΩ resistor limits the current flowing into the base of the transistor. You can use Ohm's law to calculate the current at the base. Because the base-emitter junction drops about 0.7 V (the same as a diode), the voltage across R2 is 5.3 V. Dividing 5.3 by 1,000 gives the current at 0.0053 A, or 5.3 mA. Thus, the 12.7 mA collector current (I_{CE}) is controlled by a 5.3 mA base current (I_{BE}).

+ **SW1:** This switch controls whether current is allowed to flow to the base. Closing this switch turns on the transistor, which causes current to flow through the LED. Thus, closing this switch turns on the LED even though the switch isn't placed directly within the LED circuit.

You might be wondering why you'd need or want to bother with a transistor in this circuit. After all, couldn't you just put the switch in the LED circuit and do away with the transistor and the second resistor? Of course you could, but that would defeat the principle that this circuit illustrates: that a transistor allows you to use a small current to control a much larger one. If the entire purpose of the circuit is to turn an LED on or off, by all means omit the transistor and the extra resistor. But as you work with more advanced circuits, you'll find plenty of cases when the output from one stage of a circuit is very small and you need that tiny amount of current to switch on a much larger current. In that case, the transistor circuit shown here is just what you need.

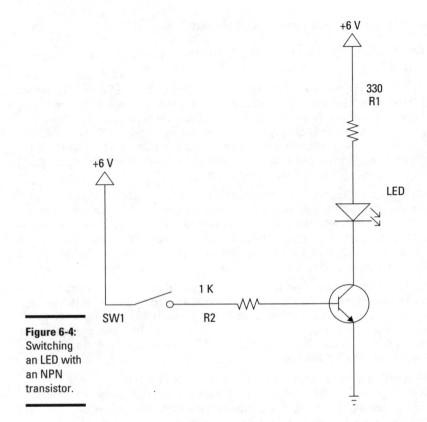

Figure 6-4:
Switching
an LED with
an NPN
transistor.

An LED Driver Circuit

In Project 6-1, you build a circuit similar to the one shown in the preceding section. The circuit uses a transistor to switch on an LED using a current that's much smaller than the LED current.

The only difference between this circuit and the one in Figure 6-4 is that the circuit in Project 6-1 adds an LED to the base circuit. When you close the switch, both LEDs light up. However, LED1 is brighter than LED2 because the collector current is larger than the base current.

Figure 6-5 shows the completed project.

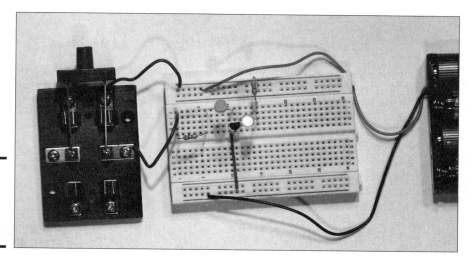

Figure 6-5:
The circuit
for the
LED driver
circuit.

If you feel like experimenting a bit after you complete this project, here are couple of suggestions:

✦ Try replacing R2 with larger resistors, such as 10 kΩ and 22 kΩ, or insert a 100 kΩ potentiometer in series with R2. What effect does the increased resistance have on the brightness of the two LEDs?

✦ Measure the base current and the collector current, as follows:

  • *To measure the base current,* remove the R2 lead in hole F1 and touch your multimeter probes to the lead you removed and the 2X terminal on the switch.

  • *To measure the collector current,* remove the R1 lead at J15 and touch the multimeter probes to the lead you removed and the LED1 anode.

Looking at a Simple NOT Gate Circuit

A *gate* is a basic component of digital electronics, which you learn all about in Book VI. Gate circuits are built from transistor switches that are either ON or OFF. There are a total of 16 different kinds of gates, and you learn about all 16 of them in Book VI, Chapter 2. For now, though, I want to introduce you to one of the simplest of all gate circuits, called a *NOT gate,* which simply takes an input that can be either ON or OFF and converts it to an output that's the opposite of the input. In other words, if the input is ON, the output is OFF. If the input is OFF, the output is ON.

Project 6-1: A Transistor LED Driver

In this project, you build a circuit that uses a transistor to drive an LED. The transistor allows a small amount of current to control a larger amount of current that passes through an LED. The circuit includes two LEDs: one on the base and one on the emitter. Both LEDs light up when the knife switch is closed, but the LED on the emitter circuit glows brighter than the one on the base circuit, demonstrating the transistor's current gain.

To complete this project, you'll need wire cutters, wire strippers, and a multimeter.

Parts List

4 AA batteries
1 Four AA battery holder (RadioShack 270-391)
1 Small solderless breadboard (RadioShack 276-003)
1 DPDT knife switch RadioShack 275-1537
1 NPN switching transistor, 2N2222A or similiar (RadioShack 276-1617)
2 5mm red LED (RadioShack 276-209)
1 330 Ω resistor (orange-orange-brown)
1 1 k Ω resistor (brown-black-tred)
1 1¼" jumper wire
2 2" jumper wires

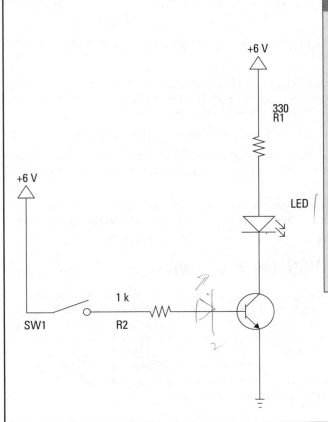

Steps

1. **Attach the red lead from the battery holder to any hole in the positive bus strip of the solderless breadboard.**

2. **Attach the black lead from the battery holder to any hole in the negative bus strip of the breadboard.**

3. **Insert the 1 KΩ resistor.**

   Insert the resistor in holes F1 and F6.

4. **Insert the 330 Ω resistor.**

   Insert the resistor in holes J15 and any hole on the positive bus.

5. **Insert LED1.**

   Insert the cathode (the shorter of the two leads) in hole H12 and the anode (the longer lead) in hole H15.

6. **Insert LED2.**

   Insert the cathode (the short lead) in hole I11 and the anode (the long lead) in hole I6.

7. **Insert the transistor.**

   The following table shows the connections for each of the three transistor leads:

Lead	Hole
Emitter	G10
Base	G11
Collector	G12

8. **Connect the transistor emitter to ground.**

   Use the 1¼" jumper wire. One end goes in hole F10; the other goes in any nearby hole in the negative bus on the left side of the breadboard.

9. **Use the two 2" jumper wires to connect the switch.**

10. **Insert the four AA batteries into the battery holder.**

11. **Flip the switch to the A position.**

    The two LEDs light up. When you open the switch, the LEDs go out.

    Notice that LED1 is brighter than LED2. This is because more current is flowing through the collector than the base.

    Note: If you have trouble seeing the difference in brightness, look down at them from directly overhead

12. **Congratulate yourself!**

    You've built your first transistor circuit!

Cathode

Anode

LED

Collector

Base

Emittor

Transistor

Figure 6-6 shows the schematic diagram for a circuit that uses a single transistor to implement a NOT gate. Here's how the circuit works:

✦ The input is controlled by a single-pole switch. When the switch is closed, the input is ON. When the switch is open, the input is OFF.

✦ The input is sent through the R1and LED1 to bias the transistor. Thus, when the input is ON, LED1 lights up, and the transistor is turned on, which enables the collector-emitter path to conduct. When the input is OFF, LED2 is dark, the transistor turns off, and no current flows through the collector.

✦ LED2 is connected directly between the +6 V power supply and ground, through a current-limiting resistor, of course, to keep the LED from burning itself out.

✦ The anode of LED2 is connected to the transistor's collector.

✦ When the transistor is off, current flows through R2 and LED2 and the LED lights up, indicating an ON output. But when the transistor turns on, a short circuit is created through the transistor's collector and emitter. This causes the current to bypass LED2, so the LED goes dark to indicate an OFF output condition.

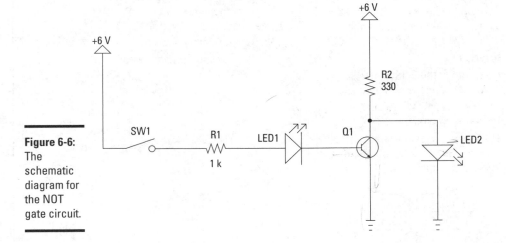

Figure 6-6:
The schematic diagram for the NOT gate circuit.

Building a NOT Gate

Project 6-2 shows you how to build the one-transistor NOT gate circuit on a solderless breadboard. The completed project is shown in Figure 6-7.

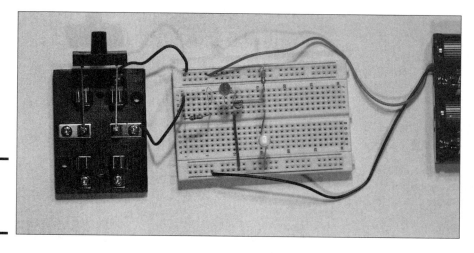

Figure 6-7:
The
transistor
NOT gate.

When you complete this project, you should celebrate with great joy that you've built your first digital logic circuit.

If you feel like experimenting a bit, here are some suggestions:

✦ Try replacing R2 with larger resistors, such as 10 kΩ and 22 kΩ, or insert a 100 kΩ potentiometer in series with R2. What effect does the increased resistance have on the brightness of the two LEDs?

✦ Measure the base current and the collector current.

• *To measure the base current,* remove the R2 lead in hole F1 and touch your multimeter probes to the lead you removed and the 2X terminal on the switch.

• *To measure the collector current,* remove the R1 lead at J15 and touch the multimeter probes to the lead you removed and the LED1 anode.

Oscillating with a Transistor

An *oscillator* is an electronic circuit that generates repeated waveforms. The exact waveform generated depends on the type of circuit used to create the oscillator. Some circuits generate sine waves, some generate square waves, and others generate other types of waves. Oscillators are essential ingredients in many different types of electronic devices, including radios and computers.

Project 6-2: A NOT Gate

In this project, you build a simple NOT gate circuit that uses a transistor to invert an input signal. That is, if the input signal is on, the output if off; if the input signal is off, the output is on. One LED is used to indicate the status of the input, and a second LED is used to indicate the status of the output.

You will need just a few tools to complete this project: a small Phillips head screwdriver, wire cutters, and wire strippers.

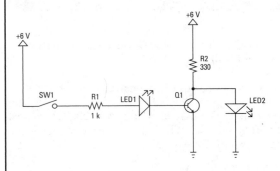

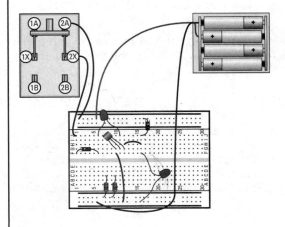

Parts List

4 AA batteries

1 Four AA battery holder (RadioShack 270-391)

1 Small solderless breadboard (RadioShack 276-003)

1 DPDT knife switch (RadioShack 275-1537)

1 NPN switching transistor, 2N2222A or similar (RadioShack 276-1617)

2 5 mm red LED (RadioShack 276-209)

1 1 kΩ resistor (brown-black-red)

1 330 Ω resistor (orange-orange-brown)

1 1¼" jumper wire

1 ½" jumper wire

2 2" jumper wires

Steps

1. **Attach the red lead from the battery holder to any hole in the positive bus strip of the solderless breadboard.**

2. **Attach the black lead from the battery holder to any hole in the negative bus strip of the breadboard.**

3. **Insert the resistors.**

   Insert the two resistors as follows:

Resistor	From	To
1 kΩ	F1	F6
330 Ω	J15	Any hole on the positive bus

4. **Insert LED1.**

   Insert the cathode (the shorter of the two leads) in hole H12 and the anode (the longer lead) in hole H15.

5. **Insert LED2.**

   Insert the anode (the long lead) in hole F16 and the cathode (the short lead) in any convenient hole in the negative bus on the left side of the board.

6. **Insert the transistor.**

   The following table shows the connections for each of the three transistor leads:

Lead	Hole
Emitter	G10
Base	G11
Collector	G12

7. **Connect the transistor emitter to ground.**

   Use the 1¼" jumper wire. One end goes in hole F10; the other goes in any nearby hole in the negative bus on the left side of the breadboard.

   When you choose the Run command, the program is downloaded to the BASIC Stamp. Once the program is downloaded, it automatically starts to run on the Stamp.

8. **Connect the collector to LED2's anode.**

   Use the 1/2" jumper wire. Insert one end in hole H12, the other in hole H16.

9. **Use the two 2" jumper wires to connect the switch.**

   Here are the connections:

From	To
Switch terminal 2X	Hole JI
Switch terminal 2A	Any hole on the positive bus on the right side of the breadboard

10. **Insert the four AAA batteries into the battery holder.**

    Once the batteries are connected, LED2 lights up (assuming the switch is open).

11. **Flip the switch to the A position.**

    LED1 lights up to show that the input is ON, and LED2 goes dark to show that the input is OFF.

LED — Cathode, Anode

Transistor — Collector, Base, Emittor

One of the most commonly used oscillator circuits is made from a pair of transistors that are rigged up to alternately turn on and off. This type of circuit is called a *multivibrator*. If the circuit is designed to continuously cycle between the two transistors, it's called an *astable multivibrator* because the circuit never reaches a point of stability — that is, it never decides which of the two transistors should be on, so it just keeps flipping back and forth between the two. Astable multivibrators are great for producing square waves.

Figure 6-8 is a generalized schematic diagram for an astable multivibrator made from a pair of NPN transistors.

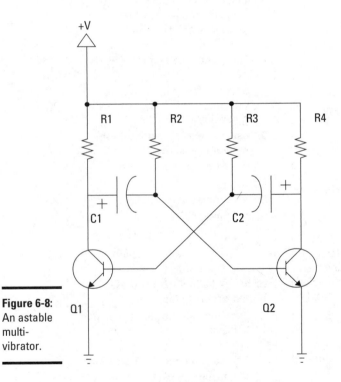

Figure 6-8: An astable multi-vibrator.

Examine this circuit to see how it works. When you first power it on, only one of the transistors turns on. You might think that they would both turn on, because the bases of both transistors are connected to +V, but it doesn't happen that way: One of them goes first. For the sake of discussion, assume that Q1 is the lucky one.

When Q1 comes on, current flows through R1 into the collector and on through the transistor to ground. Meanwhile, C1 starts to charge through R2, developing a positive voltage on its right plate. Because this right plate is connected to the base of Q2, positive voltage also develops on the base of Q2.

When C1 is charged sufficiently, the voltage at the base of Q2 causes Q2 to start conducting. Now the current flows through the collector of Q2 via R4, and C2 starts charging through R3. Because the right-hand plate of C2 is bombarded with positive charge, the voltage on the left plate of C2 goes negative, which drops the voltage on the base of Q1. This causes Q1 to turn off.

**Book II
Chapter 6**

**Working with
Transistors**

C1 discharges while C2 charges. Eventually, the voltage on the left plate of C2 reaches the point where Q1 turns back on, and the whole cycle repeats.

Don't worry if this all seems confusing. It is. If the details seem baffling, just think of the big picture: The dueling capacitors alternately charge and discharge, turning the two transistors on and off, which in turn allows current to flow through their collector circuits. Back and forth it goes, like an amazing rally at Wimbledon that no one ever wins . . . the players just keep lobbing the ball back and forth forever, until their batteries run out.

Here are a few other interesting things to know about astable multivibrators:

✦ The time that each half of the multivibrator is on is determined by the RC time constant formed by the capacitor charging circuits. Thus, you can vary the speed at which the circuit oscillates by adjusting the capacitor and resistor values. For more information about calculating resistor and capacitor time constants, refer to Chapter 3 of this minibook.

✦ You can, if you want, create an astable multivibrator from PNP transistors simply by switching the ground with the +V voltage source.

✦ Output from the multivibrator circuit can be taken directly from the collector of either transistor. For example, you could place an LED or a speaker in series with R1 or R4 to see or hear the oscillator in action. You can see an example of this in the next section.

✦ Alternatively, you can use a third transistor to couple the multivibrator with an output load, as shown in Figure 6-9. Just connect the emitter of one of the multivibrator transistors to the base of the third transistor and connect the load to the collector, as shown in the figure.

This arrangement has two advantages. First, the load itself interferes with the multivibrator circuit if you take it directly from the collector of Q1 or Q2. By using a third transistor, you isolate the load from the multivibrator circuit. And second, the output is much closer to a true square wave when the coupling transistor is used; without it, the output isn't a clean square wave because of the effects of the capacitor charging.

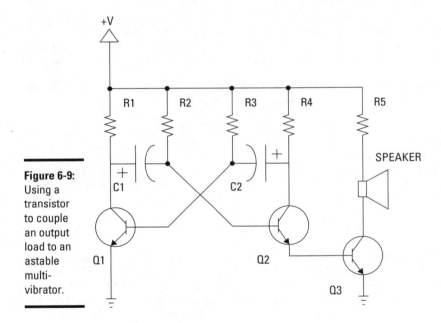

Figure 6-9:
Using a
transistor
to couple
an output
load to an
astable
multi-
vibrator.

Building An LED Flasher

In Project 6-3, you build a circuit that uses an astable multivibrator to alternately flash two LEDs. LED flasher circuits are a favorite of electronic hobbyists because flashing LEDs have all sorts of fun uses. For example, you can add creepy blinking eyes to a jack-o'-lantern for Halloween, or you can add blinking warning lights to your model railroad layout.

The LED flasher circuit is simply an astable multivibrator similar to the one shown in Figure 6-8. The only differences are that I've added LEDs to the collector circuit of each transistor and filled in the resistor and capacitor values. With the values I selected for this project, the lights alternate quickly, a bit faster than once per second.

If you feel like experimenting a bit after you complete this project, here are a couple of suggestions:

✦ Try replacing R2 and R3 with larger resistors, such as 220 kΩ and 470 kΩ. What effect does this have on the flasher?

✦ Try adding a potentiometer in series with R2 or R3. This allows you to vary the flash rate by turning the potentiometer knob.

✦ Try changing the capacitors.

Wrapping Up Our Exploration of Discrete Components

That about does it for our foray into transistors — and in fact, that wraps up our survey of the most common types of *discrete* electronic components (that is, components that contain just a single part in a single case). If you've worked your way through all the chapters in this minibook, you now know the basics of resistors, capacitors, inductors, diodes, and transistors.

In a way, that's all there is. What remains are clever ways to combine these components into different kinds of circuits that you can put to different kinds of uses.

In Book III, you learn how to use *integrated circuits*, which enable you to replace a whole circuitful of discrete components with a single tiny chip. Integrated circuits are simply a way to combine a bunch of semiconductors, resistors, and capacitors into a single tiny package.

On you go!

Project 6-3: An LED Flasher

In this project, you build a circuit that uses a two-transistor astable multivibrator to flash two LEDs in sequence. You'll need a small Phillips-head screwdriver, wire cutters, and wire strippers to build this project.

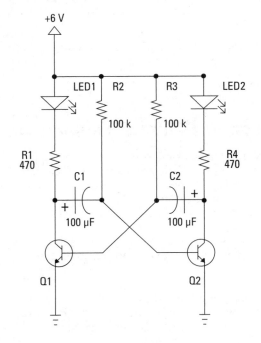

Parts List

4 AAA batteries

1 Four AA battery holder (RadioShack 270-391)

1 Small solderless breadboard (RadioShack 276-003)

2 NPN switching transistors, 2N2222A or similar (RadioShack 276-1617)

2 5 mm red LED (RadioShack 276-209)

2 100 µF electrolytic capacitors (RadioShack 272-1044)

2 100 kΩ resistor (brown-black-orange)

2 470 Ω resistor (yellow-violet-brown)

2 1¼" jumper wire

1 1" jumper wire

1 ¾" jumper wires

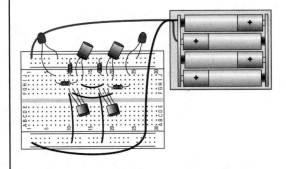

Steps

1. Insert the jumper wires.

Length	From	To
1¼"	F10	Any hole in the ground bus on the left side of the breadboard
1¼"	F17	Any hole in the ground bus on the left side of the breadboard
1"	F11	F20
¾"	H13	H18

2. Insert the capacitors.

Insert the two electrolytic capacitors as follows:

Capacitor	–	+
C1	I13	J12
C2	I20	J19

Be sure to observe the correct polarity of the capacitors. The negative lead is marked with minus signs.

3. Insert the transistors.

The following table shows the connections for each of the three transistor leads of the two transistors:

Lead	Q1	Q2
Emitter	G10	G17
Base	G11	G18
Collector	G12	G19

4. Insert the LEDs.

LED	Cathode (Short)	Anode (Long)
LED1	J7	Any hole in the positive bus
LED2	J24	Any hole in the positive bus

5. Insert the resistors.

Resistor	From	To
R1 (470)	I7	I12
R2 (100K)	J11	Any hole in the positive bus
R3 (100K)	J18	Any hole in the positive bus
R4 (470)	H19	H24

6. Connect the battery holder to the breadboard.

Attach the red lead from the battery holder to any hole in the positive bus strip on the right side of the solderless breadboard. Attach the black lead from the battery holder to any hole in the negative bus strip on the left side of the breadboard.

7. Insert the four AAA batteries into the battery holder.

8. Enjoy the show!

The LEDs alternately flash as long as you leave the batteries in.

LED

Cathode

Anode

Transistor

Collector

Base

Emittor

Book III

Working with Integrated Circuits

Contents at a Glance

Chapter 1: Introducing Integrated Circuits

In This Chapter

✔ Examining the workings of integrated circuits

✔ Looking at IC packages

✔ Learning how to use ICs in your circuits

✔ Looking at the most popular types of integrated circuits

On April 25, 1961, an engineer named Robert Noyce in Palo Alto, CA, got word that a patent application he had submitted a year before was finally approved. The patent was for a new type of device which would soon come to be known as an *integrated circuit.*

Exactly one month later, on May 25, President John F. Kennedy announced to the world that the United States was going to the moon.

What do these two events have in common? A lot. Without the integrated circuit, it's doubtful NASA could have pulled off Kennedy's challenge.

In May 1961, NASA had no idea how to get to the moon. But one thing they did know was that their moonship would need a computer like none the world had ever seen before. They would have to figure out a way to shrink a computer that at the time filled an entire room into a box about the size of a picnic basket.

And so the very first contract NASA awarded for the Apollo program was for the computer. Many more contracts would follow — for the command module, the lunar module, the launch vehicle, the space suit, and thousands of other vital elements that would all have to be built to make the moon landing possible. But NASA's first priority was to build the computer.

The computer contract went to MIT, and the engineers there quickly decided that the only way to build the computer would be to take advantage of the brand-new technology of integrated circuits. NASA became the first large-scale user of integrated circuits.

In this chapter, you learn about the device that helped get us to the moon and then went on to change the world of electronics forever. You can learn how these devices are built, what they can do, and why they keep getting smaller and cheaper. You can also learn how to incorporate them into your own electronic projects.

What Exactly Is an Integrated Circuit?

An *integrated circuit* (also called an *IC* or just a *chip*) is an entire electronic circuit consisting of multiple individual components such as transistors, diodes, resistors, capacitors, and the conductive pathways that connect all the components, all made from a single piece of silicon crystal.

To be clear, an integrated circuit isn't a really small circuit board that has components mounted on it. In an integrated circuit, the individual components are embedded directly into the silicon crystal. Previous circuit fabrication techniques relied on mounting smaller and smaller parts on smaller and smaller circuit boards, but an integrated circuit is all one piece. Instead of just two or three p-n junctions (as in a diode or a triode), an integrated circuit has thousands of individual p-n junctions. In fact, many modern integrated circuits have millions or even billions of them, all fashioned from a single piece of silicon.

The miracle of Moore's law

You've probably heard of *Moore's law*, which in a nutshell predicts that the number of transistors that can be placed on a single integrated circuit doubles about every two years. Gordon Moore, one of the founders of Intel, first stated his prediction in 1965. Back then, the prediction was even more ambitious. Originally, Moore's law said that the transistor count would double every year, not every two years. In the mid-1970s, the pace slowed a bit, so the prediction was scaled back.

The staggering reality of Moore's law is that the increase in complexity of electronic technology is exponential, not incremental as most technologies are. For example, consider the automotive industry, where gas mileage gets incrementally better every year. Gordon Moore once said that if Moore's law applied to automobiles, a Rolls-Royce would get half a million miles per gallon, and it would be cheaper to buy a new one than pay to park the one you have.

Several times over the years, scientists have feared that the end of Moore's law was on the horizon, as the chip manufacturing technology was approaching some physical limit that could not be exceeded, such as the wavelength of the light used to etch the circuits. But each time, some new technological breakthrough has enabled manufacturers to simply bypass the old limit. Moore's law has held true now for nearly fifty years and is expected to continue into the foreseeable future.

One possible explanation for the uncanny accuracy of Moore's law is that it has become a self-fulfilling prophesy. Integrated circuit manufacturers rely on Moore's law to set their own engineering goals, and they then work feverishly to achieve those goals. Thus, Moore's law has become the objective of the semiconductor industry.

The earliest integrated circuits were simple transistor amplifier circuits with just a few transistors, resistors, and capacitors. In fact, they weren't much more complicated than the circuits you breadboarded in Book II, Chapter 6.

Now, integrated circuits are unbelievably complex. At the time I wrote this, the most advanced Intel computer chip had 2.6 *billion* transistors.

Most of the integrated circuits you'll work with for hobby projects will be much more modest, having something on the order of a few dozen transistors. For example, the 555 timer IC, which you learn about in Chapter 3 of this minibook, has 20 transistors, 2 diodes, and 15 resistors and costs about a dollar.

Looking at How Integrated Circuits Are Made

Every time I see an episode of *Modern Marvels* on The History Channel and they go on for an hour about topics like shovels or trucks or cold cuts or grease or dog food, I get mad and yell at my TV. Why don't they do one about integrated circuits?

You don't have to know how integrated circuits are made to use them, so you can skip this section if you want. However, the process is pretty interesting — certainly worthy of an episode of *Modern Marvels.*

The process for manufacturing an IC is complex, and varies depending on the type of chip being made. But the following process is typical:

1. A large, cylindrical piece of silicon crystal is shaved into thin wafers about one-hundredth of an inch thick. Each of these wafers will be used to create several hundred or thousand finished integrated circuits.

2. A special photoresist solution is deposited on top of the wafer.

3. A mask is applied to over the photoresist. The mask is an image of the actual circuit, with some areas transparent to allow light through and others opaque to block the light.

4. The wafer is exposed to intense ultraviolet light, which etches the wafer under the transparent portions of the mask but leaves the areas under the opaque parts of the mask untouched.

5. The mask is removed and any remaining photoresist is cleaned off.

6. The wafer is then exposed to a doping material, which creates n-type and p-type regions in the etched areas of the wafer. (For a review on doping and n-type and p-type semiconductors, please refer to Book II, Chapter 5.)

7. If the circuit design calls for multiple layers stacked on top of one another, the process is repeated for each layer until all of the layers have been created.

8. The individual integrated circuits are then cut apart and mounted in their final packaging.

Here are a few other tidbits worth knowng about how integrated circuits are made:

✦ As you've probably seen on Intel commercials, the manufacturing process for integrated circuits is done in a clean room, where workers wear special suits and masks. This is necessary because at the scale of the integrated circuits, even the smallest speck of dust is enormous.

✦ Each integrated circuit goes through a variety of complicated quality tests after the circuit is finished. The manufacturing process is by no means perfect, so many are discarded.

Integrated Circuit Packages

Integrated circuits come in a variety of different package types, but nearly all of the ICs you'll work with in hobby electronics come in a type of package called *dual inline package*, or *DIP*. Figure 1-1 shows several ICs in DIP packages.

Yes, I know that the phrase "DIP package" is redundant because the *P* in *DIP* already stands for *package*, but the phrase "DIP package" is commonly used. Some justify this usage by claiming that the *P* actually stands for *pin*, so there's no redundancy. But that's just easy rationalization. Like it or not, *DIP* stands for *dual inline package*, and the phrase "DIP package" is commonly used and considered correct. Get used to it.

The phrase "DIP chip" is also sometimes used to describe ICs in DIP packages. It has a nice ring to it and sounds like something you would serve at a Super Bowl party.

A DIP package consists of a rectangular plastic or resin case that encloses the IC itself, with two rows of pins on the long sides of the rectangle. The pins on each side jut out a bit from the case, then turn straight down. This arrangement makes the package look like an insect. (In fact, one common way of wiring circuits that use DIP chips is to glue them to a board upside down and solder wires directly to the pins; this technique is called *dead-bug wiring*.)

Figure 1-1:
Most
integrated
circuits
come in DIP
packaging.

The pins on each side of a DIP package are spaced exactly 0.1" apart, and the two rows of pins are usually spaced 0.3" apart, though some larger DIP packages may have wider spacing. In any event, the standard tenth-of-an-inch spacing is perfect for use with solderless breadboards, which have holes spaced at 0.1" intervals. In fact, the gap that runs down the center of a solderless breadboard happens to be 0.3", which makes it easy to mount DIP chips such that they straddle the gap, as shown in Figure 1-2.

**Book III
Chapter 1**

Introducing
Integrated Circuits

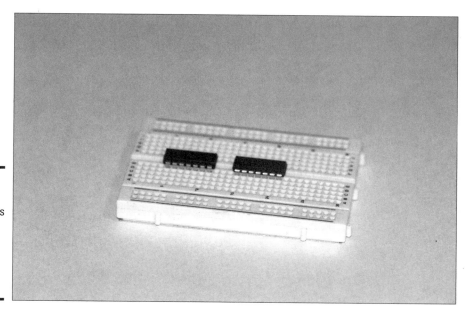

Figure 1-2:
Solderless
breadboards
are
designed
with DIP
chips in
mind.

Each pin in a DIP package is numbered. Looking down on the package from above, and you'll see an orientation mark, usually a notch, groove, or dot. Orient the package so that this mark is on the top, and pin 1 is immediately to the left of the mark. The pins are numbered counter-clockwise, working down the left side and then back up the right side until you get to the last pin, which is immediately to the right of the orientation mark.

In case my explanation of DIP pin numbering doesn't make sense, have a look at Figure 1-3.

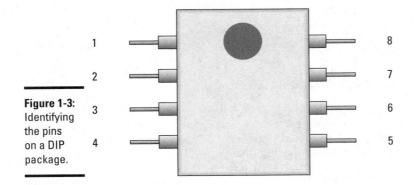

Figure 1-3: Identifying the pins on a DIP package.

The DIP package in Figure 1-3 is for an eight-pin DIP. Larger DIPs have more pins, but the numbering scheme is always the same: Pin 1 is to the left of the orientation mark, and the remaining pins are numbered counter-clockwise from pin 1.

Using ICs in Schematic Diagrams

In a schematic diagram, an IC is usually represented simply as a rectangle with circuit connections placed conveniently around the rectangle without regard for the physical positioning of the pins. Each pin connection is labeled, as shown in Figure 1-4.

Notice that the pins in this schematic diagram aren't in the same order as they are in the actual DIP package. Thus, when you build this circuit, you have to adjust the wiring layout to accommodate the pin arrangement of the DIP package.

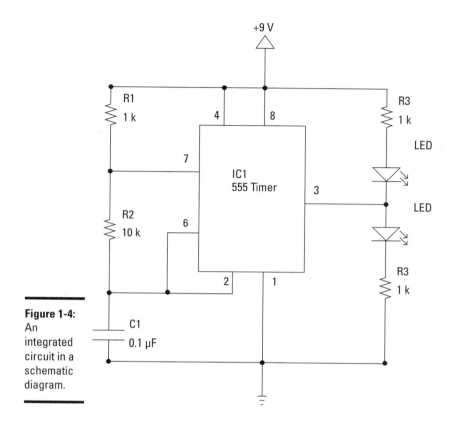

Figure 1-4:
An integrated circuit in a schematic diagram.

Notice also that not all of the pins on an Integrated Circuit are always used. Unused pins are usually omitted from the schematic diagram. For example, pin 5 is not used in the circuit shown in Figure 1-4, so it is omitted from the schematic.

Some integrated circuits contain two or more independent circuits that share a common power supply, sort of like conjoined twins. For example, the 556 dual timer chip contains two complete 555 timer circuits within a single 14-pin package. When chips like this are used in a circuit, the schematic diagram may show them separately. For example, Figure 1-5 shows a circuit that calls for a 556 dual timer chip, but each section of the timer is listed separately in the schematic.

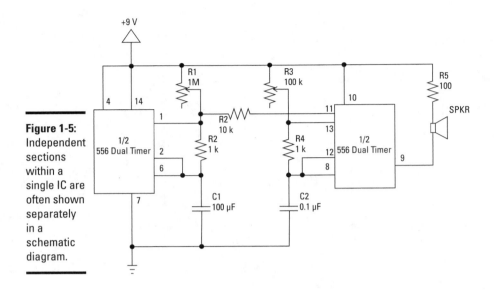

Figure 1-5:
Independent
sections
within a
single IC are
often shown
separately
in a
schematic
diagram.

Please don't worry about the details of this circuit (or of the circuit shown in Figure 1-4). My intent here isn't to explain how these circuits work, but only to show you how the integrated circuits are depicted in the schematic diagrams. You can learn more about how these circuits work in the next chapter.

Powering ICs

In most DIP integrated circuits, two of the pins are used to provide power to the circuit. One of these is designated for positive voltage, typically identified with the symbol V_{CC}. The other is the ground pin. For example, the 555 timer chip (which you learn about in Chapter 3 of this minibook) requires a positive supply voltage between 4.5–15 V at pin 8, and pin 1 is connected to ground.

In the case of ICs that contain two or more separate circuits, the circuits usually share a common power supply. Thus, even though a 556 dual timer chip contains two separate 555 timer circuits, the chip has just one positive voltage pin and one ground pin.

Also, you should be aware that some integrated circuits call for separate positive and negative supply voltages, not just a positive and a ground connection. You can create a power supply like that using the circuit shown in Figure 1-6.

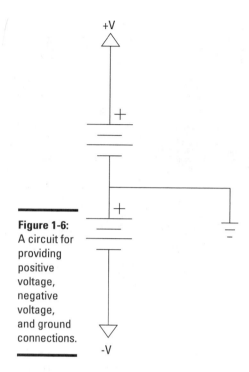

Figure 1-6:
A circuit for providing positive voltage, negative voltage, and ground connections.

Avoiding Static and Heat Damage

When you build a circuit board that contains one or more integrated circuits, be careful that you don't damage the IC when you build your circuit. In particular, you should watch out for these two possible problems:

✦ **Static discharge:** Many integrated circuits can be damaged by static electricity discharged though your fingers when you handle the chips. As a result, make sure you discharge yourself by touching a grounded metal surface before handling an integrated circuit. You may also want to use an antistatic wristband when handling ICs.

✦ **Heat damage:** Some integrated circuits are sensitive to heat, so you should take precautions whenever you solder an IC to a circuit board. If possible, attach an alligator clip or other type of heat sink to the pin to help dissipate some of the heat away from the IC itself.

You can avoid soldering integrated circuits altogether by using DIP sockets such as the ones shown in Figure 1-7. When you use these sockets, you solder the socket to your circuit board. Then, once the socket is safely soldered in place, you simply insert the IC into the socket.

Figure 1-7:
DIP sockets
let you avoid
soldering
delicate
integrated
circuits.

Reading IC Data Sheets

Before you work with a specific type of IC, you should download a copy of the data sheet for the IC. An IC datasheet contains loads of useful information. In addition to basic information such as the manufacturer's name and the IC part number, you'll find information such as

✦ A description of what the circuit does.

✦ Detailed pinout descriptions that tell you the purpose of each pin.

✦ A diagram of the internal circuitry of the chip. For simple circuits, you may get the entire detailed diagram. For more complicated chips, you'll get a conceptual diagram instead of a detailed schematic.

✦ Detailed electrical specifications, such as maximum voltage you can feed the circuit via the V_{CC} pin or the maximum current loads for output pins.

✦ Operating conditions such as maximum and minimum temperatures.

✦ Charts and graphs that illustrate the circuit's behavior for different operating conditions.

✦ Formulas for calculating operating characteristics of the circuit. For example, if the operation of the circuit depends on an external RC (resistor/capacitor) circuit, you'll get formulas for calculating how these external components will affect the operation of the circuit.

✦ Sample circuit diagrams.

✦ Mechanical descriptions including dimensions.

IC datasheets are available from many sources on the Internet. Use a search engine such as Google to search for the IC part number and the word *datasheet*. For example, to find a datasheet for a 555 timer chip, search for *555 datasheet*.

Popular Integrated Circuits

Literally thousands of different types of integrated circuits are available. Most of these were designed for very specific applications. However, many integrated circuits have been designed for general-purpose use and so are used in a wide variety of circuits.

In the sections that follow, I briefly describe some of the most popular general-purpose integrated circuits. All of these ICs have been around for decades, but their multipurpose design, wide availability, and low cost have given them enduring popularity.

555 Timer

The 555 timer chip was invented in 1971 but remains today one of the most popular integrated circuits in use. By some estimates, more than a billion of them are made and sold every year.

As its name implies, the 555 is a timer circuit. The timing interval is controlled by an external resistor/capacitor (RC) network. In other words, by carefully choosing the values for the resistors and capacitors, you can vary the timing duration.

The 555 can be configured in several different ways. In one configuration (called *monostable*), it works like an egg timer: You set it, and then it goes off after a certain period of time has elapsed. In a different configuration (called *astable*), it works like a metronome, triggering pulses at regular intervals.

Besides the basic 555 chip, which comes in an 8-pin DIP package, you can also get a 556 dual timer, which contains two independent 555 timers in a single 14-pin DIP package. Because many common circuits call for two 555 timers working together, the 556 package is very popular.

You can learn more about the 555 and 556 timers in the next chapter.

741 and LM324 Op-Amp

An *op-amp* is a special type of amplifier circuit that has many applications throughout electronics. Although there are many different types of op-amp circuits, the 741 and LM324 are the most common.

The 741 is a single op-amp circuit in an eight-pin DIP package. It was first introduced in 1968 and is still one of the most widely used integrated circuits ever made. The 741 is one of those ICs that require both positive and negative voltage, as described earlier in the section, "Powering ICs."

The LM324 was introduced in 1972. It consists of four separate op-amp circuits in a single 14-pin DIP package. Unlike the 741, the LM324 doesn't require separate negative and positive voltage supplies.

You can learn more about op-amps in Chapter 3 of this minibook.

78xx Voltage Regulator

The 78xx is a family of simple voltage regulator integrated circuits. A *voltage regulator* is a circuit that accepts an input voltage that can vary within a certain range and produces an output voltage that is a constant value, regardless of fluctuations in the input voltage.

The *xx* in *78xx* represents the actual voltage regulated by the chip. For example, a 7805 produces a 5 V output. The input voltage must be at least a couple of volts over the output voltage, and can be as high as 35 V.

You can learn more about this useful integrated circuit in Chapter 4 of this minibook.

74xx Logic Family

One of the primary uses for integrated circuits is for digital electronics, and the 74xx is one of the oldest and still most widely used families of digital integrated circuits. The 74xx family includes a wide variety of chips that provide basic building blocks for digital circuits. Thus, you won't find complete microprocessors in the 74xx family. But you will find circuits such as logic gates, flip-flops, counters, buffers, and so on.

You can learn about this useful family of ICs in Book IV, Chapter 3.

Chapter 2: The Fabulous 555 Timer Chip

In This Chapter

✓ Learning about the 555 timer chip

✓ Exploring the different ways to configure a 555 timer chip

✓ Using the 556 dual-timer chip

✓ Building basic 555 timer chip circuits

The 555 timer chip, developed in 1970, is probably the most popular integrated circuit ever made. By some estimates, more than a billion of them are manufactured every year. Its popularity is well deserved. The 555 is a single-chip version of a commonly used circuit called a *multivibrator*, which is useful in a wide variety of electronic circuits. You can use the 555 chips for basic timing functions, such as turning a light on for a certain length of time, or you can use it to create a warning light that flashes on and off. You can use it to produce musical notes of a particular frequency, or you can use it to control positioning of a servo device. (You learn about servos in Book VII, Chapter 4.) The list goes on and on.

In this chapter, you learn how to use this versatile chip in a variety of circuits. First, I explain how the 555 works and what each of its pins do. Then, you see and build a variety of common 555 circuits.

Looking at How the 555 Works

The 555 can be considered a hybrid of an analog and digital circuit. The output produced by a 555 is purely digital: It's either off (0 V) or it's on (with positive voltage of at least 2.5 V). The timing mechanism within the 555 determines how long the output is on and how long it's off.

The analog part of the circuit lies in how you control how long the output signal is on and how long it's off. You do that by creating an RC network using a resistor and a capacitor. The values you choose for the resistor and capacitor determine the timing interval. I give you all the information you need to choose the correct values for the resistor and capacitor later in this chapter, but if you feel you need a refresher on how RC networks work, please refer to Book II, Chapter 3.

You can also control whether the 555 does its timing job just once, like an egg timer, or whether it cycles the output on and off repeatedly, like a metronome that keeps ticking over and over again. You do that by connecting the pins of the 555 chip in various ways.

Figure 2-1 shows the arrangement of the eight pins in a standard 555 IC. As you can see, the 555 comes in an 8-pin DIP package.

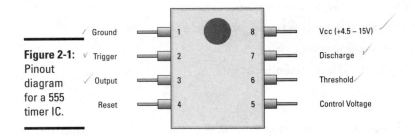

Figure 2-1: Pinout diagram for a 555 timer IC.

The following paragraphs describe the function of each of the eight pins (not in numerical order):

✦ **Ground:** Pin 1 is connected to ground.

✦ **V$_{CC}$:** Pin 8 is connected to the positive supply voltage. This voltage must be at least 4.5 V and no greater than 15 V. It's common to run 555 circuits using four AA or AAA batteries, providing 6 V, or a single 9 V battery.

✦ **Output:** Pin 3 is the output pin. The output is either low, which is very close to 0 V, or high, which is close to the supply voltage that's placed on pin 8. (On some models of the 555, the output voltage may be as much as 2 V below the supply voltage.) The exact shape of the output — that is, how long it's high and how long it's low, depends on the connections to the remaining five pins.

For more information about using the output from pin 3, see the section "Using the 555 Timer Output" later in this chapter.

✦ **Trigger:** Pin 2 is the *trigger*, which works like a starter's pistol to start the 555 timer running. The trigger is an *active low* trigger, which means that the timer starts when voltage on pin 2 drops to below one-third of the supply voltage. When the 555 is triggered via pin 2, the output on pin 3 goes high.

For example, if you want to start the timer by pushing a button, you would connect pin 2 to the supply voltage. Then, to trigger the timer, you simply interrupt this supply voltage. There are several ways to do that, but the most common is to connect a normally open pushbutton between pin 2 and ground. Then, the button is pushed, and the supply voltage is short-circuited to ground; so the voltage at pin 2 drops to zero, and the timer is triggered. You see an example of this type of triggering later in this chapter, in the section "Using the 555 in Monostable (One-Shot) Mode."

✦ **Discharge:** Pin 7 is called the *discharge*. This pin is used to discharge an external capacitor that works in conjunction with a resistor to control the timing interval. In most circuits, pin 7 is connected to the supply voltage through a resistor and to ground through a capacitor.

✦ **Threshold:** Pin 6 is called the *threshold*. The purpose of this pin is to monitor the voltage across the capacitor that's discharged by pin 7. When this voltage reaches two thirds of the supply voltage (Vcc), the timing cycle ends, and the output on pin 3 goes low.

✦ **Control:** Pin 5 is the *control* pin. In most 555 circuits, this pin is simply connected to ground, usually through a small 0.01 μF capacitor. (The purpose of the capacitor is to level out any fluctuations in the supply voltage that might affect the operation of the timer.)

In some circuits, a resistor is used between the control pin and Vcc to apply a small voltage to pin 5. This voltage alters the threshold voltage, which in turn changes the timing interval. Most circuits don't use this capability, however. In this chapter, all the 555 circuits simply connect pin 5 to ground through a 0.01 μF capacitor.

✦ **Reset:** Pin 4 is the reset pin, which can be used to restart the 555's timing operation. Like the trigger input, reset is an active low input. Thus, pin 4 must be connected to the supply voltage for the 555 timer to operate. If pin 4 is momentarily grounded, the 555 timer's operation is interrupted and won't start again until it's triggered again via pin 2.

When used in a schematic diagram, the pins of a 555 timer chip are almost always shown in the arrangement depicted in Figure 2-2. As you can see, this arrangement follows the customary order of elements in a schematic diagram: Supply voltage is at the top, ground is at the bottom, inputs are at the left, and outputs are at the right. This arrangement makes it easy to determine the operation of the circuit from the schematic diagram.

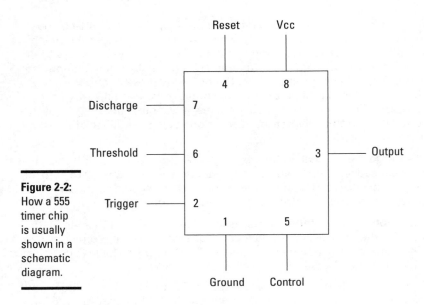

Reset   Vcc

Discharge — 7

Threshold — 6   3 — Output

Trigger — 2

Ground   Control

Figure 2-2:
How a 555
timer chip
is usually
shown in a
schematic
diagram.

Understanding 555 Modes

There are three basic ways — called *modes* — that you can wire up a 555 timer IC:

✦ **Monostable mode:** Works like an egg timer. When you start it, the timer turns on the output, waits for the time interval to elapse, and then turns the output off and stops.

✦ **Astable mode:** Works like a metronome: It keeps running until you turn it off.

✦ **Bistable mode:** Isn't really a timer mode. Instead, it uses the trigger input to alternately turn the output on and off. This type of circuit is often called a *flip-flop* and is very commonly used in digital electronics.

The following sections explain how each of these three operating modes work.

Using the 555 in Monostable (One-Shot) Mode

Monostable mode lets you use the 555 timer chip as a single-event timer. This mode is called *monostable* because when wired this way, the 555 has just one stable mode, with the output at pin 3 off. When the 555 is sent a trigger pulse, this stable state is temporarily interrupted for an interval that's determined by the value of a resistor and a capacitor. During this interval, the output at pin 3 goes high, but once the time interval has passed, the 555 returns to its stable state, with pin 3 going low.

Monostable mode is sometimes called *one-shot mode*, which seems a little more descriptive to me. *One-shot mode* conveys the idea that when triggered, the 555 gives one and only one output pulse. When the time interval is reached, the output pulse stops, and the circuit goes quiet until another trigger pulse is detected. Each trigger pulse results in a single output pulse.

Looking at a typical 555 monostable circuit

Figure 2-3 shows the typical wiring for a 555 timer used in monostable mode.

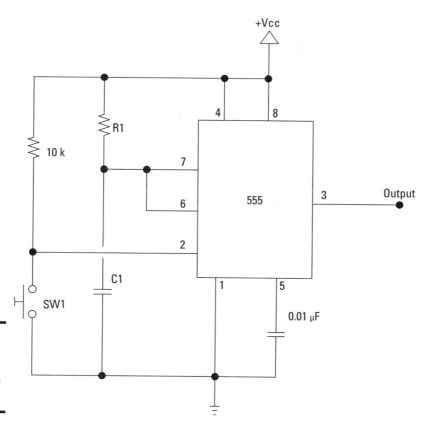

Figure 2-3:
A 555 timer chip in monostable mode.

To understand how this circuit works, first look at the way the 10 kΩ resistor and the switch are wired to pin 2, the trigger input. The switch is a normally open pushbutton. When the button isn't depressed, the 10 kΩ resistor provides a voltage input to pin 2, which keeps the trigger input high. With the trigger input high, the output voltage at pin 3 is near zero.

When the pushbutton switch is depressed, the supply voltage is short-circuited to ground. This causes the voltage at pin 2 to drop to zero, and the timer is triggered. Once the timer is triggered, the output voltage at pin 3 goes high and the timing interval begins.

Looking at the resistor-capacitor circuit in a monostable timer

Now that you understand how the trigger circuit works, look at how the RC circuit (R1 and C1) works. The resistor and capacitor work together to determine how long the output will remain high. In a nutshell, once the circuit is triggered, C1 begins to charge.

Pins 6 and 7 — the threshold and discharge pins — are tied together in a monostable 555 circuit. Pin 6 watches the voltage across the capacitor. As the capacitor charges, this voltage increases. When the capacitor voltage reaches two thirds of the Vcc supply voltage, the timing cycle ends, and the output at pin 3 goes low.

The discharge pin (pin 7), charges and discharges the capacitor. To understand how pin 7 works, it may be helpful to visualize the internal workings of pin 7 using the model shown in Figure 2-4. Here, pin 7 is connected to a switch that's controlled by the status of the output at pin 3. When the output is high, the switch is open; when the output is low, the switch is closed. When the switch is closed, a small 10 Ω resistor within the 555 connects pin 7 to ground. (This isn't really a switch inside the 555, but this model may help you understand how pin 7 works.)

When the output on pin 3 is low, the imaginary switch inside the 555 is closed, and pin 7 is connected to ground through the 10 Ω resistor. This allows the voltage on C1 to discharge through the 555.

But when the output on pin 3 goes high, the imaginary switch inside the 555 is opened. This forces the current flowing through R1 to go through C1, which in turn causes the capacitor to charge at a rate that depends on the values of R1 and the capacitor.

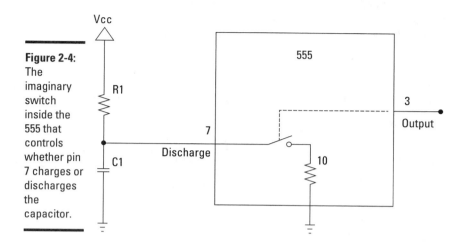

Figure 2-4:
The imaginary switch inside the 555 that controls whether pin 7 charges or discharges the capacitor.

While the capacitor is charging, pin 6 monitors the voltage that builds up across the capacitor. Once this voltage reaches two-thirds of the supply voltage, pin 6 signals the 555 that the timing interval is ended, and the output goes low. This, in turn, closes the imaginary switch inside the 555, which allows the capacitor to discharge.

Calculating the time interval for a monostable circuit

The time interval for a 555 monostable circuit is a measure of how long the output stays high when it's triggered. To calculate the time interval, just use this formula:

$$T = 1.1 \times R \times C$$

Here, T is the time interval in seconds, R is the resistance of R1 in ohms, and C is the capacitance of C1 in farads.

For example, suppose R1 is 500 kΩ, and C1 is 10 µF. Then, you would calculate the time interval like this:

$$T = 1.1 \times 500,000 \ \Omega \times 0.00001 \ F$$

$$T = 5.5 \ s$$

Thus, the circuit will stay on for 5.5 seconds after it's triggered.

When you make this calculation, it's imperative that you use the correct number of zeros for both the resistance and the capacitance. Use Table 2-1 as a guide:

Table 2-1   Conversion of Resistance and Capacitance Values

Resistance	*Capacitance*
1 kΩ = 1,000 Ω	0.01 µF = 0.00000001F
10 kΩ = 10,000 Ω	0.1 µF = 0.0000001F
100 kΩ = 100,000 Ω	1 µF = 0.000001F
1M Ω = 1,000,000 Ω	10 µF = 0.00001F
10M Ω = 10,000,000 Ω	100 µF = 0.0001F
100M Ω = 100,000,000 Ω	1,000 µF = 0.001F

If you don't want to do these calculations yourself, you can find 555 timer calculators on the web. Just go to Google or any other search engine and search *555 timer calculator*.

Using the 555 in Astable (Oscillator) Mode

Another common way to use a 555 timer is in *astable mode*. The term *astable* simply means that the 555 has no stable state: Just as it gets settled into one state (say, the output at pin 3 high), it switches to the opposite state (output low). Then it switches back to the first state, and so on, ad infinitum. This mode is also called *oscillator mode*, because it uses the 555 as an oscillator, which creates a square wave signal.

Looking at a typical astable circuit

Figure 2-5 shows the basic circuit for a 555 in astable mode.

To understand how this circuit works, first notice that the trigger pin (pin 2) is connected directly to C1. In the monostable circuit, the timer was triggered by a switch that short-circuited the voltage applied to pin 2. In the astable circuit, the timer is triggered when the capacitor discharges — once the voltage across the capacitor drops to one-third of the supply voltage, pin 2 triggers the timer to start another cycle.

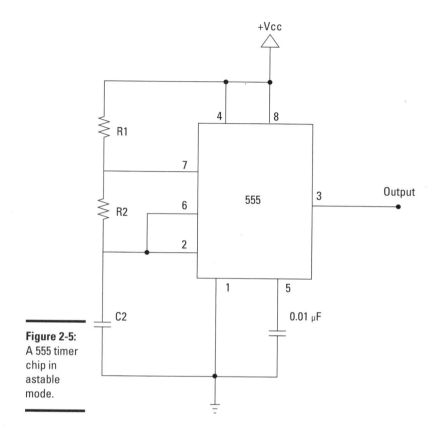

Figure 2-5:
A 555 timer chip in astable mode.

Here I examine how this timing cycle works, step by step, starting with the output at pin 3 in the high condition.

1. With the output high, the discharge pin (pin 7) is open, forcing current through resistors R2 and C1. This causes the capacitor to charge at a rate that depends on the combined value of R1 and R2 and the value of C1.

2. As the capacitor charges, the voltage at pins 2 and 6 increases.

3. When the voltage at pin 6 (the threshold pin) reaches two-thirds of the supply voltage, the threshold circuitry within the 555 causes the output voltage at pin 3 to go low.

4. When the output at pin 3 goes low, the discharge pin (pin 7) is connected to ground within the 555. This allows C1 to discharge. This discharge occurs through R2, so the value of R2 as well as the value of the capacitor determines the rate at which the capacitor discharges.

5. As the capacitor discharges, the voltage at pins 2 and 6 decreases.

6. When the voltage at pin 2 (the trigger pin) drops to one third of the supply voltage, the trigger circuitry inside the 555 causes the output at pin 3 to go high.

7. When the output at pin 3 goes high, the discharge pin (pin 7) is opened, and the cycle starts over again.

Controlling the time intervals in an astable 555 circuit

The output of a 555 circuit in astable mode is a square wave, as depicted in Figure 2-6. There are three important time measurements for a square wave:

✦ **T:** The total duration of the wave, measure from the start of one high pulse to the start of the next high pulse.

✦ T_{high}: The length of the high portion of the cycle.

✦ T_{low}: The length of the low portion of the cycle.

Naturally, the total time T is the sum of T_{high} and T_{low}.

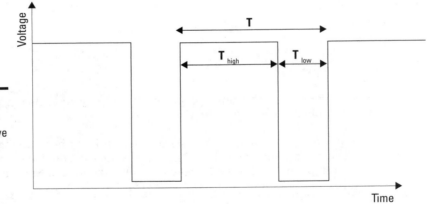

Figure 2-6:
Timing the output wave created by an astable 555 timer circuit.

The values of these time constants depend on the values for the two resistors (R1 and R2) and the C1.

Here are the formulas for calculating each of these time constants:

$$T = 0.7 \times (R1 + 2R2) \times C1$$

$$T_{high} = 0.7 \times (R1 + R2) \times C1$$

$$T_{low} = 0.7 \times R2 \times C1$$

There's an interesting fact buried in these calculations that you need to be aware of: C1 charges through both R1 and R2, but it discharges only through R2. That's why you must add the two resistor values for the T_{high} calculation, but you use only R2 for the T_{low} calculation. It's also why you must double R2 but not R1 for the total time (T) calculation.

Now, plug in some real numbers to see how the equations work out. Suppose both resistors are 100 kΩ and the capacitor is 10 μF. Then, the total length of the cycle is calculated like this:

$$T = 0.7 \times (100,000 \ \Omega + 2 \times 100,000 \ \Omega) \times 0.00001 \ F$$
$$T = 2.1 \ s$$

$$T_{high} = 0.7 \times (100,000 \ \Omega + 100,000 \ \Omega) \times 0.00001 \ F$$
$$T_{high} = 1.4 \ s$$

$$T_{low} = 0.7 \times 100,000 \ \Omega \times 0.00001 \ \Omega$$
$$T_{low} = 0.7 \ s$$

Thus, the total cycle time will be 2.1 s, with the output high for 1.4 s and low for 0.7 s.

If you want, you can also calculate the frequency of the output signal by dividing the total cycle time into 1. So, for the above calculations, the frequency is 0.47619 Hz.

If you use smaller resistor and capacitor values, you'll get shorter pulses and higher output frequencies. For example, if you use 1 kΩ resistors and a 0.1 μF capacitor, the output signal will be 48 kHz, and each cycle will last just a few millionths of a second.

You may have also noticed that if the two resistors are the same value, the signal will be high for twice as long as it's off. By using different resistor values, you can vary the difference between the high and low portions of the signal.

Calculating the duty cycle

The *duty cycle* in a 555 circuit is the percentage of time that the output is high for each cycle of the square wave. For example, if the total cycle time is 1 s and the output is high for the first 0.4 s of each cycle, the duty cycle is 40%.

With an astable circuit such as the one shown in Figure 2-5, the duty cycle must always be greater than 50%. In other words, the duration for which the output is high must always be more than the duration during which the output is low.

The explanation for this is pretty simple: For the duty cycle to be 50%, the capacitor would have to charge and discharge through the same resistance. The only way to accomplish that would be to omit R1 altogether, so that the capacitor charged and discharged through R2 only. But the problem with that is that you would end up connecting pin 7 directly to Vcc. With no resistance between pin 7 and the voltage source, the current flowing through pin 7 would exceed the maximum that can be handled by the circuitry inside the 555, and the chip would be damaged.

There's a clever way around this limitation: Place a diode across R2, as shown in Figure 2-7. This diode bypasses R2 when the capacitor is charged. That way, the capacitor charges through R1 and discharges through R2.

When a diode is used in this way, you have complete control over the duration of both the charge and discharge time. If R1 and R2 have the same value, the capacitor takes the same amount of time to charge as it does to discharge, so the duty cycle will be 50%. If R2 is smaller than R1, the duty cycle is less than 50% because the capacitor discharges faster than it charges.

If you use the diode as depicted in Figure 2-7, you must adjust the formulas for calculating the time intervals as follows:

$$T = 0.7 \times (R_1 + R_2) \times C_1$$

$$T_{high} = 0.7 \times R_1 \times C_1$$

$$T_{low} = 0.7 \times R_2 \times C_1$$

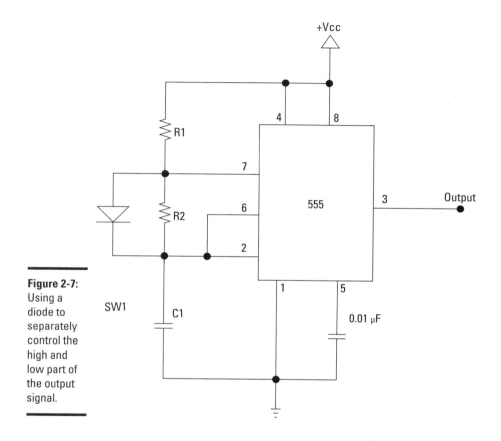

Figure 2-7:
Using a diode to separately control the high and low part of the output signal.

Using the 555 in Bistable (Flip-Flop) Mode

A *flip-flop* is a circuit that alternates between two output states. In a flip-flop, a short pulse on the trigger causes the output to go high and stay high, even after the trigger pulse ends. The output stays high until a reset pulse is received, at which time the output goes low.

This type of circuit is called *bistable* because the circuit has two stable states: high and low. The circuit stays low until it's triggered. Then, it stays high until it's reset. This type of circuit is used extensively in computers and other digital circuits.

For computer applications, the 555 is a poor choice for use as a flip-flop. That's because its output doesn't change fast enough in response to trigger or reset pulses in computer circuits that are driven by high-speed clock pulses. For computer applications, better flip-flop chips are readily available, as you learn in Book VI, Chapter 5.

That being said, the 555 is often used in bistable mode for noncomputer applications where high-speed response isn't necessary. For example, imagine a simple robot that drives itself forward until it bumps into something in front of it, and then drives backward until it bumps into something behind it. The robot would be equipped with contact switches on the front and rear connected to the trigger and reset inputs of a 555 in bistable mode. The robot's drive motor would be connected to the output such that when the output is low, the motor runs forward, and when the output is high, the motor runs backward. Then, the bistable 555 would cause the robot to drive back and forth between two obstacles.

Figure 2-8 shows the schematic for a 555 used in bistable mode. As you can see, this circuit doesn't require a capacitor. That's because in bistable mode, the 555 isn't used as a timer. The highs and lows of the output signal are controlled by the trigger and reset inputs, not by the charging and discharging of a capacitor.

Both the trigger (pin 2) and the reset (pin 4) inputs are connected to Vcc through a 10 kΩ resistor. When the set switch is depressed, pin 2 is shorted to ground. This causes the voltage to bypass pin 2, resulting in a momentary low pulse, which triggers the 555. Once triggered, the output pin goes high.

In astable or monostable mode, the output pin would remain high until the voltage at the threshold pin (pin 6) reaches two-thirds of the supply voltage. However, because pin 6 isn't connected to anything in this circuit, no voltage is ever present on pin 6. Thus, the threshold is never reached, so the output remains high indefinitely until the 555 is reset by a low pulse on the reset pin (pin 4).

The reset input (pin 4) is connected to Vcc in the same manner as the trigger input. As a result, when the reset switch is pressed, pin 4 is short circuited to ground, creating a low pulse, which resets the 555 and brings the output back to high.

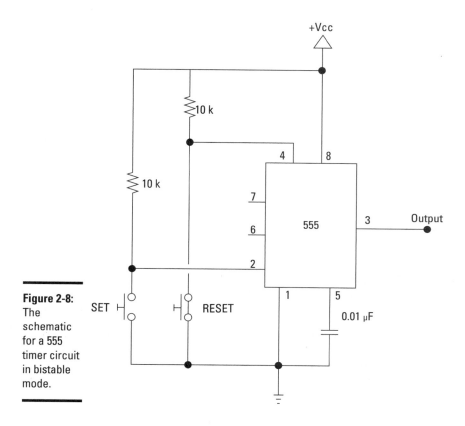

Figure 2-8:
The schematic for a 555 timer circuit in bistable mode.

Using the 555 Timer Output

The output pin (pin 3) of a 555 can be in one of two states: high and low. In the high state, the voltage at the pin is close to the supply voltage. The low state is 0 V.

There are two ways you can connect output components to the output pin. Figure 2-9 illustrates these two configurations, using an LED as the output device. A resistor is also included in the circuit to limit the current flow. Without the resistor, current will flow through the circuit unimpeded, which will quickly burn out the LED and probably ruin the 555 as well.

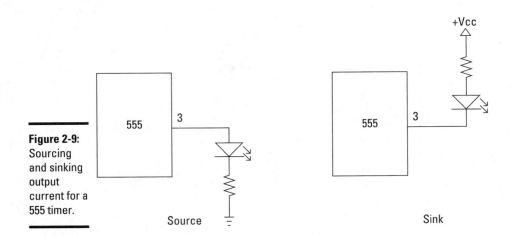

Figure 2-9:
Sourcing
and sinking
output
current for a
555 timer.

In the circuit on the left, current flows through the LED circuit when the output is high. The current flows from the output pin through the LED and resistor to ground. This output configuration is called *sourcing* because the 555 is the source of the current that drives the output.

In the circuit on the right, current flows through the LED circuit when the output is low. The current flows from the Vcc supply, through the LED and resistor, and into the 555 where it's internally routed to ground through pin 1. This output configuration is called *sinking* because the current is sent into the 555.

Whether you source or sink your output circuit depends on whether you want your output circuit to turn on when the output is high or low.

Figure 2-10 shows that you can combine both sourcing and sinking in a single circuit. Here, two LEDs are connected to the output pin. One is sourced; the other is sunk. In this circuit, the LEDs alternately flash as the output switches from high to low. LED1 lights when the output is low, LED 2 when the output is high.

The output circuit of a 555 timer can handle as much as 200 mA of current, which is actually much more current than most integrated circuits can source or sink. If you need to drive a device that requires more than 200 mA, you can isolate the output device from the 555 by using a transistor, as shown in Figure 2-11. For more information about working with transistors, please refer to Book II, Chapter 6.

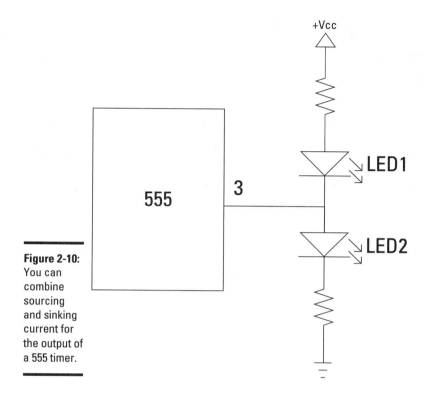

Figure 2-10:
You can
combine
sourcing
and sinking
current for
the output of
a 555 timer.

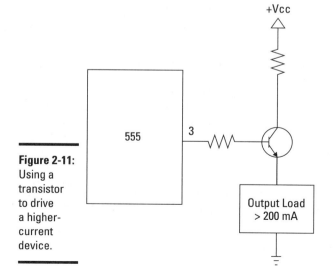

Figure 2-11:
Using a
transistor
to drive
a higher-
current
device.

**Book III
Chapter 2**

**The Fabulous
555 Timer Chip**

Doubling Up with the 556 Dual Timer

If one 555 timer chip is good, two are even better! In fact, it turns out that there are many uses for two (or more) 555 timers in a single circuit — useful enough that you can get two 555 timers in a single chip, called the 556 dual-timer chip.

The 556 dual-timer chip comes in a 14-pin DIP package. The two 555 timers share a common supply and ground pin. The remaining 12 pins are allocated to the inputs and outputs of the individual 555 timers. Table 2-2 lists the pin connections for each of the 555 timers in a 556 dual-timer chip. As an added bonus (no charge!), I also list the pinouts for a standard 555 timer chip in Table 2-2.

Table 2-2   Pinouts for the 555 Timer and 556 Dual-Timer Chips

Function	*555 Timer*	*556 First Timer*	*556 Second Timer*
Ground	1	7	7
Trigger	2	6	8
Output	3	5	9
Reset	4	4	10
Control	5	3	11
Threshold	6	2	12
Discharge	7	1	13
Vcc	8	14	14

One common way to use a 556 dual timer is to connect both 555 circuits in monostable (one-shot) mode, with the output pin from the first 555 timer connected to the trigger pin of the second 555 timer. Then, when the output of the first timer goes low, the second timer is triggered. You can connect as many 555 timers as you want in this way, with each timer's output connected to the next timer's trigger so that the timers work in sequence, one after the other.

For example, Figure 2-12 shows a cascaded timer circuit that uses two separate 555 timer chips. In this circuit, both of the 555 timer chips are configured in monostable mode, much like the circuit in Figure 2-3. The time interval for the first 555 is controlled by R1 and C1. For the second 555, the interval is controlled by R2 and C2. You can choose whatever values you want for these components to achieve whatever time intervals suit your fancy.

The first 555 chip is triggered when SW1 is depressed, taking pin 2 to ground. This takes the output on pin 3 high, which lights LED1. Notice, however, that pin 3 of the first 555 is connected through a small capacitor to the trigger input of the second 555. As soon as the time interval expires on the first 555, its output goes low, which turns off LED1 and at the same time triggers the second 555, which in turn lights up LED2. LED2 stays lit until C2 charges, and then it goes out. The circuit then waits to be triggered again by a press of the switch.

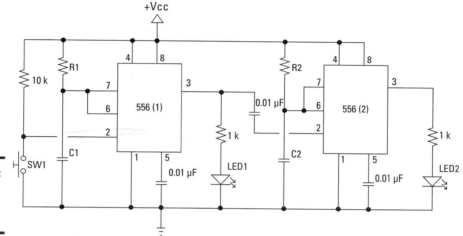

Figure 2-12: 555 timers can be cascaded.

Figure 2-13 shows how this same circuit can be implemented using a single 556 dual-timer chip. This schematic is nearly identical to the schematic shown in Figure 2-13, but with a few important differences:

✦ The two 555 timer circuits are designated as *556 (1)* and *556 (2)* to indicate that these timer circuits are part of a single 556 dual-timer chip.

✦ The pin numbers indicate the pin assignments for the two timer circuits of a 556 instead of the pin assignments for a 555.

✦ The second timer circuit doesn't show a supply or ground connection. That's because the two timer circuits share a common supply and ground connection, which is shown connected to the first timer.

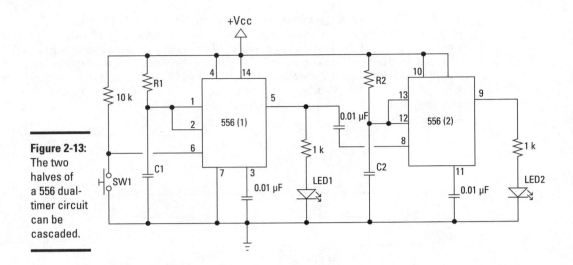

Figure 2-13:
The two
halves of
a 556 dual-
timer circuit
can be
cascaded.

Although it's convenient to show the two halves of a 556 dual timer as separate components in a schematic diagram, you can show the 556 as a single component if you wish. Figure 2-14 shows how the schematic for the cascaded timer circuit can be drawn using a single component for the 556 dual timer. This circuit is nearly identical to the circuit shown in Figure 2-13; the only difference is the way the schematic depicts the two sections of the 556 dual-timer chip.

When you draw a 556 as a single component, it's helpful to draw the connections for the two timers on opposite sides of the component. In Figure 2-14, I drew the connections for the first timer on the left and the second timer on the right, with the exception of the trigger input for the second timer (pin 8). Drawing that connection on the right side of the component would complicate the diagram, so I placed it on the left side just beneath the output pin for the first timer.

Because it's difficult to keep track of which pin is which in a 556, it's helpful to label the function of each pin as I did in the schematic shown in Figure 2-14.

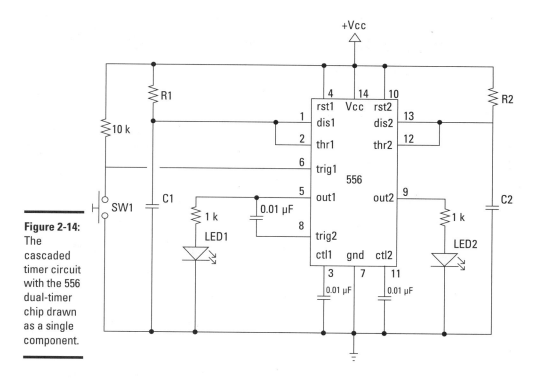

Figure 2-14:
The cascaded timer circuit with the 556 dual-timer chip drawn as a single component.

Making a One-Shot Timer

In this section, you build a circuit that uses a 555 timer chip in monostable mode. When a trigger switch is pressed, an LED lights and stays lit for approximately five seconds. Then, the LED goes dark until the button is pressed again.

Project 2-1 provides all the information you need to assemble this circuit. The circuit is based on the schematic for the monostable circuit that was shown in Figure 2-3. The only differences are that an LED is added to the output pin (pin 3) and the resistor, and capacitor values are included in the capacitor charging circuit.

This project uses a pushbutton switch that isn't designed to be inserted in a solderless breadboard. To make the pushbutton usable with the solderless breadboard, solder a small length of 18- or 20-guage solid wire to the end of each lead of the test strip. You can then insert these leads directly into holes on the breadboard.

Project 2-1: A One-Shot 555 Timer Circuit

In this project, you'll build a circuit that uses a 555 Timer chip in monostable mode to create a one-shot timer. The circuit includes a pushbutton and an LED. When you press the pushbutton, the LED will turn on and stay on for about five seconds. Then, the LED will turn off and stay off until you press the pushbutton again.

The only tools you'll need to complete this project are wire cutters and wire strippers.

Parts List

1 9 V battery
1 9 V battery snap holder (RadioShack 270-325)
1 Small solderless breadboard (RadioShack 276-003)
1 Normally open pushbutton switch
1 555 timer chip (RadioShack 276-1718)
1 5 mm red LED (RadioShack 276-209)
1 10 kΩ resistor (brown-black-orange)
1 470 Ω resistor (yellow-violet-brown)
1 470 kΩ resistor (yellow-violet-yellow)
1 10 µF electrolytic capacitor
1 0.01 µF ceramic disk capacitor
1 Normally open momentary contact pushbutton (RadioShack 275-1547)
8 Jumper wires (various lengths)

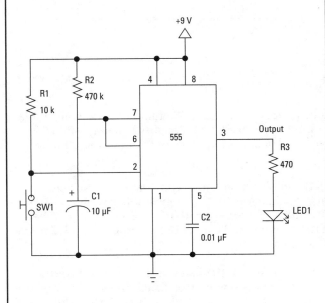

Steps

Throughout these steps, use the bus strip on the bottom of the board for the ground bus and the bus strip on the top of the board as the positive voltage bus.

1. **Insert the 555 timer chip.**

   Insert the chip so that it straddles the gap in the center of the solderless breadboard, with pin 1 in hole E5 and pin 8 in hole F5.

2. **Insert the jumper wires.**

   If you're using precut jumper wires, choose an appropriate length for each segment. Otherwise, cut your own jumper wires as needed. You need a total of eight jumper wires, inserted into the solderless breadboard as follows:

From	To
A5	Any hole in the ground bus
D3	D6
D8	G5 (Note that this wire crosses over the top of the 555.)
F9	Any hole in the ground bus
B11	Any hole in the ground bus
G7	G11
H6	H11
J5	Any hole in the positive voltage bus

3. **Insert the resistors.**

Resistor	From	To
10 kΩ	E3	Any hole in the positive bus
470 kΩ	J11	Any hole on the positive bus

470 Ω	C7	C13

4. **Insert the capacitors.**

   Insert the two capacitors as follows:

Capacitor	From	To
0.01 µF	J8	Any hole in the positive bus
10 µF	E11	F11

   Make sure the negative lead of the electrolytic capacitor is in hole E11.

5. **Insert the LED.**

   The anode (the long lead) should be inserted in hole A13, and the cathode (the short lead) in any hole in the ground bus.

6. **Insert the pushbutton.**

   Insert one of the leads you soldered to the pushbutton in hole A3 and the other one in the close available hole in the ground bus.

7. **Connect the battery.**

   Plug the 9 V battery into the battery snap connector, and then connect the red lead to the positive voltage bus and the black lead to the ground bus.

8. **You're finished!**

When you are finished, the circuit will look like Figure 2-15. To test the circuit, push the pushbutton. The LED should light, stay lit for just over five seconds, and then go back off. It should light again only when you press the pushbutton again.

If the circuit doesn't work, here are a few things to check:

✦ Make sure the battery is good. (Use a voltmeter to test it.)

✦ Carefully double-check all the jumper wires and other components to make sure they're connected properly.

✦ Make sure the LED isn't inserted backwards. As a test, pull it out and insert it with the leads reversed.

✦ Make sure the electrolytic capacitor is inserted with the negative end on ground side of the circuit.

✦ Make sure the solder connections to the pushbutton are solid.

Figure 2-15:
The finished one-shot timer project.

Making an LED Flasher

In this section, you build a circuit that uses a 555 timer chip to alternately flash two LEDs on and off. For this circuit, the 555 is configured in 555 in astable mode, and the resistor values are chosen so that they cause the high and low timings to be very close to one another, about one-tenth of a second each.

Project 2-2 shows you how to build the LED flasher circuit. The schematic for this project is similar to the one that was shown in Figure 2-5, but adds a pair of LEDs to the output circuit. LED1 lights when the output is high, and LED2 lights when the output is low. Note that this circuit uses both sinking and sourcing of the output current.

Figure 2-16 shows the completed project.

Figure 2-16: The finished LED flasher.

If the circuit doesn't work, here are a few things to check:

✦ Check the battery voltage.

✦ Double-check all the jumper wires and resistors to make sure they're inserted in the proper holes.

✦ Make sure the diodes and the electrolytic capacitor are inserted correctly. For the LEDs, the anodes must be on the positive side of the circuit and the cathodes on the negative side. For the capacitor, the negative side must be in the ground bus.

After you finish playing with the completed project, you may want to leave it assembled. The next section (Project 2-2) adds a second 555 timer to this circuit to include additional functionality.

Project 2-2: An LED Flasher

In this project, you'll build a circuit that alternately flashes a pair of LEDs. The circuit uses a 555 Timer configured in astable mode to flash the LEDs. The resistor and capacitor values are chosen so that the duty cycle is close to 50% and each LED stays on for about 1/10th of a second.

The only tools you'll need to complete this project are wire cutters and wire strippers.

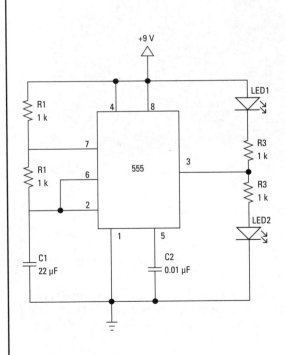

Parts List
1 9 V battery
1 9 V battery snap holder (RadioShack 270-325)
1 Small solderless breadboard (RadioShack 276-003)
1 555 timer chip (RadioShack 276-1718)
2 5 mm red LED (RadioShack 276-209)
1 10 k Ω resistor (brown-black-orange)
3 1 k Ω resistor (brown-black-red)
1 10 µF electrolytic capacitor
1 0.01 µF ceramic disk capacitor
8 Jumper wires (various lengths)

Steps

1. **Insert the 555 timer chip.**

   Insert the chip so that it straddles the gap in the center of the solderless breadboard, with pin 1 in hole E5 and pin 8 in hole F5.

2. **Insert the jumper wires.**

   If you're using precut jumper wires, choose an appropriate length for each segment.

   Otherwise, cut your own jumper wires as needed. You need a total of eight jumper wires, inserted the solderless breadboard as follows:

From	To
A5	Any hole in the ground bus
C3	C6
C7	C10
D6	G7 (This wire crosses over the top of the 555.)
D8	G5 (This wire also crosses over the top of the 555.)
H3	H6
F9	Any hole in the ground bus
J5	Any hole in the positive voltage bus

3. **Insert the resistors.**

   Insert the four resistors as follows:

Resistor	From	To
10 kΩ	D3	G3
1 kΩ	J3	Any hole on the positive bus
1 kΩ	B10	B13
1 kΩ	D10	F13

4. **Insert the capacitors.**

   Insert the two capacitors as follows:

Capacitor	From	To
0.01 µF	G8	G9
10 µF	A3	Any nearby hole in the ground bus

5. **Insert the LEDs.**

   Insert the two LEDs as follows:

LED	Cathode (Short Lead)	Anode (Long Lead)
LED1	J13	Any hole in the positive bus
LED2	Any hole in the ground bus	A13

6. **Connect the battery.**

   Plug the 9 V battery into the battery snap connector, and then connect the red lead to the positive voltage bus and the black lead to the ground bus. The LEDs should begin flashing immediately when the battery is connected.

7. You're finished!

After you finish this project, leave it assembled if you intend to build Project 2-3.

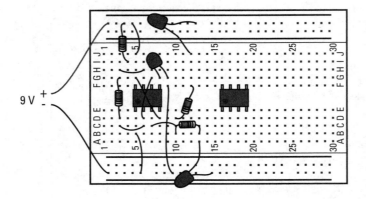

Using a Set/Reset Switch

In this section, you modify the circuit that you built in Project 2-2 so that the circuit is controlled by two pushbuttons that function as a set/reset switch. When you connect the power to this circuit, LED1 turns on and stays on. When you press the set pushbutton, the two LEDs start flashing alternately and continue to flash until you press the reset pushbutton.

The circuit for this project uses two 555 timer chips. The first is configured in bistable mode with two pushbuttons acting as set and reset switches. The second is configured in astable mode, almost identical to the 555 timer that was used in Project 2-2. The difference is that instead of connecting the astable 555 Timer chip's supply voltage pin (pin 8) directly to the battery, it is connected to the output of the first 555. Thus, the first 555 controls the power to the second 555, so the second 555 flashes the LEDs only when the output of the first 555 is high.

This project requires that you use two pushbutton switches that are not designed to be inserted into a solderless breadboard. To make the switches easier to use with the breadboard, solder short lengths of 20-gauge solid wire to the switch terminals.

Project 2-3 shows you how to build this circuit, and the completed project is pictured in Figure 2-17.

Figure 2-17:
The
completed
circuit for
Project 2-3.

Project 2-3: An LED Flasher with a Set/Reset Switch

In this project, you'll expand the circuit you built in Project 2-2 to add a set/reset switch that controls the flashing of the LEDs. You should build Project 2-2 before you attempt this project.

The only tools you'll need to complete this project are wire cutters and wire strippers.

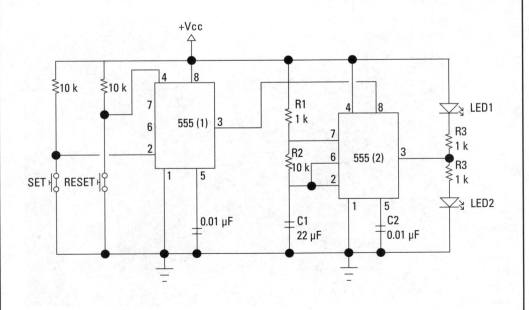

Steps

1. **If you haven't already done so, build the LED flasher circuit described in the preceding section.**

   The 555 timer chip in that project is the one designated as 555 (2) for this project.

2. **Remove the jumper wire you inserted from hole J5 to the positive bus.**

   This step disconnects the Vcc supply (pin 8) of 555 (2). Voltage for 555 (2) is provided by the output of 555 (1).

3. **Insert 555 (1).**

   Insert the chip so that it straddles the gap in the center of the solderless breadboard, with pin 1 in hole E17 and pin 8 in hole F17.

4. **Insert the additional jumper wires.**

   If you're using precut jumper wires, choose an appropriate length for each segment. Otherwise, cut your own jumper wires as needed. You need a total of ten jumper wires, inserted into the solderless breadboard as follows:

From	To
A17	Any hole in the ground bus
I5	I15
E15	F15
D15	D19
C18	C23
D20	D27

F21	Any hole in the ground bus
E23	F23
E27	F27
J17	Any hole in the positive voltage bus

5. **Insert the resistors.**

   Insert the two resistors as follows:

Resistor	From	To
10 kΩ	J23	Any hole on the positive bus
10 kΩ	J27	Any hole on the positive bus

6. **Insert the capacitor.**

   The 0.01 µF capacitor should be inserted in holes G20 and G21.

7. **Insert the pushbuttons.**

   Insert the two pushbutton switches as follows:

Button	From	To
Set	A23	Any hole in the ground bus
Reset	A27	Any hole in the ground bus

8. **Connect the battery.**

   Plug the 9 V battery into the battery snap connector, and then connect the red lead to the positive voltage bus and the black lead to the ground bus. LED1 should immediately light up. (If it doesn't, double-check all your connections and make sure the battery isn't dead.)

9. **You're done!**

Press the set button to start the LEDs flashing. Let them flash for awhile, and then press the reset button to make the flashing stop.

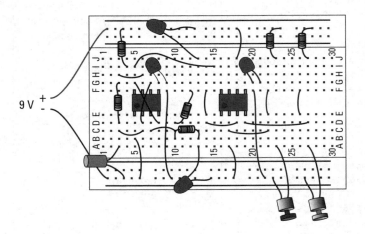

Making a Beeper

In this section, you use two 555 timer chips to build an audible beeper, with both timers configured in astable mode. One timer generates an audible square-wave tone of approximately 500 Hz; the output of this 555 is sent to a speaker so the tone can be heard. The other has a much lower frequency of about 1.5 Hz. You connect its output to the reset of the 500 Hz timer to turn the tone on and off, which creates the beeping effect.

Project 2-4 shows how to build this circuit. Before you start, have a look for a moment at the project's schematic diagram. For the first timer chip — designated *555 (1)* in the schematic — the RC network uses resistors of 1 kΩ and 470 kΩ along with a 1 µF capacitor to produce the 1.5Hz output. The second timer — *555 (2)* — uses two 100 kΩ resistors and a 0.01 µF capacitor to create the 500 Hz output. The output of the first timer is sent to the reset of the second timer, and the output of the second timer is sent through a 22 µF capacitor to an 8 Ω speaker.

Note: You could easily build this circuit using a single 556 dual-timer chip. You'd have to adjust the pin designations on the schematic accordingly.

The speaker used in this circuit can be any 8 Ω speaker. If you have an old unamplified computer speaker, you can use it. I recommend you solder 2–3" lengths of 20-gauge solid wire to the speaker terminals so you can easily connect the speaker to the breadboard.

Figure 2-18 shows the completed project.

Here are some extra-credit assignments for this project, in case you want to experiment with the circuit a bit:

✦ Try building the circuit with a single 556 chip instead of two, 555 chips.

✦ Replace the 100 kΩ R3 with a 1 kΩ resistor, and then add a 1 MΩ potentiometer in series with the resistor. As you turn the potentiometer, the tone changes.

✦ Add a 1 MΩ potentiometer in series with R1. Then, as you turn the potentiometer, the beeping rate changes.

Figure 2-18:
The
completed
beeper
project.

Project 2-4: An Audible Beeper

In this project, you'll use a pair of 555 Timer ICs to build a circuit that produces an audible beep on a small speaker. Both of the 555 Timer ICs are configured in astable mode. The first 555 provides the beeping interval, which is about one beep every 1/10th second. The second 555 generates a 1.5KHz frequency that's fed to the speaker to produce the audible output.

To build this project, you'll need wire cutters and wire strippers in addition to the parts listed below.

Parts List

1	9 V battery
1	9 V battery snap holder (RadioShack 270-325)
1	Small solderless breadboard (RadioShack 276-003)
2	555 timer chip (RadioShack 276-1718)
2	5 mm red LED (RadioShack 276-209)
1	1 k Ω resistor (brown-black-red)
1	470 kΩ resistor (yellow-violet-yellow)
2	100 kΩ resistor (brown-black-yellow)
1	1 μF electrolytic capacitor
1	22 μF electrolytic capacitor
2	0.01 μF ceramic disk capacitor
15	Jumper wires (various lengths)

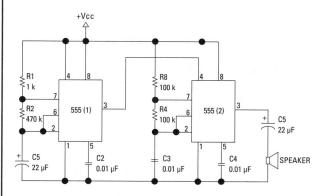

**Book III
Chapter 2**

**The Fabulous
555 Timer Chip**

Steps

1. Insert the 555 timer chips.

Insert the chips so that they straddle the gap in the center of the solderless breadboard. For the first chip, insert pin 1 in hole E5 and pin 8 in hole F5. For the second chip, insert pin 1 in hole E17 and pin 8 in hole F17.

2. Insert the jumper wires.

If you're using precut jumper wires, choose an appropriate length for each segment. Otherwise, cut your own jumper wires as needed. You need a total of 15 jumper wires, inserted into the solderless breadboard as follows:

From	To
A5	Any hole in the ground bus
A17	Any hole in the ground bus
J5	Any hole in the positive bus
J17	Any hole in the positive bus
C3	C6
H3	H6
C15	C18
G15	G18
F9	Any hole in the ground bus
E10	Any hole in the positive bus
D8	D10
D6	G7 (This wire crosses over the top of the first 555 chip.)
D18	G19 (This wire crosses over the top of the second 555 chip.)
F21	Any hole in the ground bus

3. Insert the resistors.

Insert the four resistors as follows:

Designation	Resistor	From	To
R1	470 Ω	J3	Any hole on the positive bus
R2	KΩ	E3	F3
R3	100K Ω	E15	F15
R4	100KΩ	J15	Any hole in the positive bus

4. Insert the capacitors.

Insert the five capacitors as follows:

Designation	Resistor	From	To
C1	1 µF Electrolytic	A3	Any hole in the ground bus
C2	0.01 µF Ceramic Disk	G8	G9
C3	0.01 µF Ceramic Disk	A15	Any hole in the ground bus
C4	0.01 µF Ceramic Disk	G20	G21
C5	22 µF Electrolytic	D19	D25

Note that for C1, the negative lead should be inserted into the ground bus. For C5, the negative lead should be in D25.

5. **Connect the speaker.**
   One lead from the speaker should be inserted in hole C25. The other can be inserted in any hole in the ground bus.
   Note that speakers aren't sensitive to polarity, so it doesn't matter which lead goes in the ground bus and which goes in A25.

6. **Connect the battery.**
   Plug the 9 V battery into the battery snap connector, and then connect the red lead to the positive voltage bus and the black lead to the ground bus. You should immediately hear the speaker beeping.

7. **You're finished!**
   Remove the battery when you can no longer stand the beeping.

**Book III
Chapter 2**

**The Fabulous
555 Timer Chip**

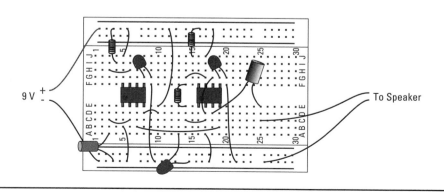

9 V

To Speaker

Chapter 3: Working with Op-Amps

In This Chapter

✔ Getting familiar with op amps

✔ Exploring how feedback circuits work with op amps

✔ Looking at summing amplifiers and comparators

✔ Exploring several popular op-amp packages

Did you ever play *Operation,* the game in which you had to use electrified tweezers to remove plastic body parts from little holes in a body? The edges of the holes were metal conductors, so if you touched the edge of the hole with the tweezers while trying to remove the plastic piece inside, a buzzer would sound, and the patient's nose (which was a red light bulb) would light up.

I was never very good at it, because it required nerves of steel and precise control of the tweezers. The slightest jiggle of the tweezers was amplified into a flashing light, a loud buzzer, and the mocking laughter of my brother, who was a much better *Operation* surgeon than I was.

An operational amplifier (op amp for short) is kind of like the game *Operation.* Well, actually, it isn't really except for the part about the slightest variations in the input (your hand holding the tweezers) being amplified into a huge variation in the output (the flashing red nose, the jarring buzzer, and the ridicule of your brother).

Op amps are among the most common types of integrated circuits — probably second in popularity only to the 555 timer chip, described in the previous chapter. In this chapter, you learn what an op amp is and how to build several useful circuits with it. So put on your scrubs, and let's start the operation!

Looking at Operational Amplifiers

An *op amp* is a super-sensitive amplifier circuit that's designed to amplify the difference of two input voltages. Thus, an op amp has two inputs and one output. The output voltage is often tens or even hundreds of thousands of times greater than the difference in the input voltages. Thus, a very small difference in the two input voltages — perhaps a few hundredths or even a few thousandths of a volt — can result in a large output voltage.

Although an op amp is a type of integrated circuit, op amps were invented long before integrated circuits. Op-amp circuits are a natural for integrated circuits, however, so it wasn't long after the introduction of the first integrated circuits that IC versions of op amps became available. Today, op amps are among the most popular types of integrated circuits.

The name *operational amplifier* may be a bit confusing. Originally, the op-amp circuit was created for use as an amplifier in telephone distribution systems. Later, computer engineers discovered that the circuit could easily be adapted to do mathematical operations such as addition, subtraction, multiplication, and division. It was around that time that the term *operational amplifier* was coined, because the circuits are amplifiers that can perform (mathematical) operations. (For more information, see the nearby sidebar "How the op amp came to be.")

Internally, the simplest op amps consist of several dozen transistors, and more complicated varieties have many more. In this chapter, I completely ignore the internal circuitry of an op amp and treat it as what it is: a handy device you can use without understanding how it works. You can thank the engineers who, many decades ago, worked out all the internal details that make an op amp do its magic.

Many types of op-amp chips are manufactured today, but all have the five connections shown in Figure 3-1. The following paragraphs describe the function of each of these connections:

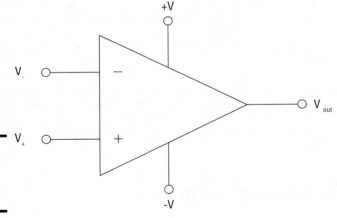

Figure 3-1:
Schematic symbol for an op amp.

✦ **+V and –V:** The power supply for an op amp is provided via two pins usually labeled +V and –V. (These pins may be labeled V_{s+} and V_{s-} instead, but their functions are the same.) Most op amps require both a positive and a negative voltage power supply, with voltages usually ranging from ±6 V to ±18 V. This type of power supply is called a *split supply*. The ± indicates that both positive voltage and negative voltage are required. ±6 V, for example, means that both +6 V and –6 V are required.

You can easily build a split supply by using two batteries connected end to end, as shown in Figure 3-2. Here, two 9 V batteries are connected to create a ±9 V supply. Note that the +9 V and –9 V are measured relative to ground, which is accessed between the two batteries.

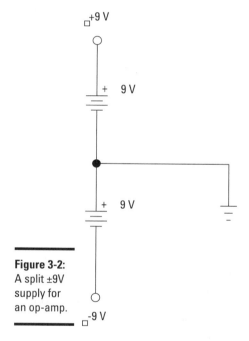

Figure 3-2:
A split ±9V supply for an op-amp.

Note that some op amps don't require split-voltage power supplies. Op amps that use single power supplies have a ground terminal instead of a –V terminal.

✦ **V_{out}:** The output of the op amp is taken from the V_{out} terminal. The voltage at the output terminal can be positive or negative, depending on the voltage difference between the two input terminals. The maximum voltage is usually a few volts less than the supply voltage at the +V and –V terminals. Thus, if the power supply for an op amp is ±9 V, the maximum output will be around ±7 V or 8 V.

Most op amps can handle only a small amount of current through the output terminal — usually, in the neighborhood of 25 mA or less. As Figure 3-3 shows, the output is passed through an external resistance, designated R_L. The other end of this resistance is connected to ground. Thus, the output current that flows through the op amp must eventually end up at ground.

Note that the load resistance isn't necessarily in the form of a simple resistor; it can be any other circuit that provides some load resistance, such as the base-emitter circuit of a transistor or even input of another op amp.

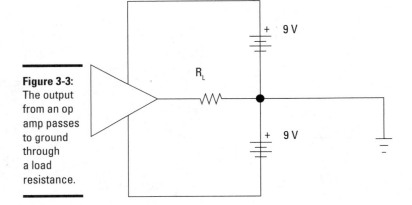

Figure 3-3: The output from an op amp passes to ground through a load resistance.

✦ **V_+ and V_-:** The two inputs of an op amp are the V_+ and V_- terminals. These terminals are sometimes identified by + and – signs inside the triangle. The inputs are called *differential inputs* because the output voltage, which appears on the V_{out} terminal, depends on the difference between the voltage of the + and – terminals.

For most op amps, the maximum allowable input voltage is a bit less than the maximum power supply voltage. ±12 V is a typical limit. Remember, though, that the *difference* between the two input voltages is what an op amp amplifies. In many cases, the two input voltages are very close, so the difference is very small.

I've said it already, but it's worth repeating: The polarity of the op-amp output depends on the polarity of the difference between the V_+ and V_- inputs. Thus, if V_+ is greater than V_-, the output will be a positive voltage, but if V_+ is less than V_-, the output will be a negative voltage.

How the op amp came to be

The modern operational amplifier dates way back to the early 1930s, when Bell Telephone was just starting to run telephone cables throughout the country. In the early days of the telephone, engineers had a difficult problem with phone lines that ran more than a few thousand feet. Long phone lines needed amplifiers to give their signals a boost, but the amplifiers available at the time were very fidgety — too sensitive to weather (temperature and humidity) and unable to work consistently over the range of voltages used in early phone lines.

A Bell engineer named Harry Black was working on the amplifier problem in 1934 when an idea struck him as he was riding the ferry home from work. This idea was a stroke of genius that seems obvious decades later. Instead of trying to design an amplifier that would have the exact amplitude gain needed for the job, Black's idea was to use an amplifier that had far more gain that was needed — thousands of times more gain, in fact — and then feed some of the output back into the input through a resistor. This feedback circuit would reduce the overall gain of the amplifier based on the amount of resistance in the circuit.

The circuit didn't gain the name *operational amplifier* until the computer age began a decade or so later, and computer researchers figured out how to use the amplifier's unique characteristics to perform basic mathematical operations like addition, subtraction, multiplication, and division on the input voltages. Eventually, digital computers replaced the analog computers built from op amps. Op amps still play an important role in computers today, however, primarily to provide an interface with real-world input measuring devices such as voltage sensors and moisture detectors.

The original op-amp circuits were built with vacuum tubes. They were large, required several hundred volts to operate, and generated substantial heat. When transistors replaced vacuum tubes in the 1950s, op amps became smaller, and when integrated circuits were invented in the 1960s, op amps were among the first chips to be designed.

Book III
Chapter 3

Working with Op-Amps

In many op-amp circuits, one input is connected to ground. If the V_+ input is grounded, the output polarity is always the opposite of the polarity of the input voltage on the V_- terminal. In other words, negative voltage on V_- will give positive voltage on V_{out}, and positive voltage on V_- will give negative voltage on V_{out}. For this reason, the V_- input is often called the *inverting input* because its polarity is inverted in the output.

If, on the other hand, the V_- input is connected to ground, the polarity of the output is the same as the polarity of the input voltage applied to V_+. Thus, if V_+ is positive, V_{out} will be positive; if V_+ is negative, V_{out} will be negative. For this reason, the V_+ input is called the *noninverting input* because its polarity is the same in the output — that is, the V_- input voltage is *not* inverted.

Understanding Open Loop Amplifiers

As its name suggests, one of the most basic uses of an op amp is as an amplifier. If you connect an input source to one of the input terminals and ground the other input terminal, an amplified version of the input signal appears on the out terminal.

An important concept in op-amp circuits is *voltage gain,* which simply represents the amount by which the difference between the two input voltages is multiplied to produce the output voltage. If the input voltage difference is 2 V and the output voltage is 12 V, for example, the voltage gain of the amplifier is 6.

If you simply apply an input signal to the V— terminal of an op amp as shown in Figure 3-4, the circuit is called an *open loop amplifier.* The reason it's called this will become more apparent when you get to the next section. For now, just realize that this type of circuit goes by the name *open loop.*

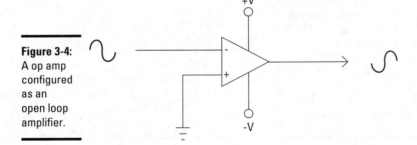

Figure 3-4:
A op amp
configured
as an
open loop
amplifier.

In the open loop op-amp circuit, the V_+ input is connected to ground, and an input signal is placed on the V_- input. In this arrangement, the voltage to be amplified is the same as the voltage of the V_- input. Although Figure 3-4 shows the input as alternating current, the open loop op-amp circuit works for direct current as well.

The voltage gain in an open loop op-amp circuit is extraordinarily high — on the order of tens or even hundreds of thousands. Suppose that you're using an op amp whose open loop voltage gain is 200,000 and that the power supply is ±9 V. In that case, an input voltage of +0.000025 V will result in an output voltage of +5 V. An input voltage of +0.00004 V will give you an output voltage of 8 V.

The ideal op amp

If you read about op amps on the web or in an electronics book, you'll undoubtedly come across the term *ideal op amp*. An *ideal op amp* is a hypothetical op amp with certain characteristics that real op amps strive to achieve. Real op amps come very close to the ideal op amp, but no op amp in existence actually achieves the perfection of an ideal op amp. So you see, in many ways, op amps are almost human.

Depending on which list you read, an ideal op amp has anywhere between two and seven characteristics, the most important of which are

✔ **Infinite open loop gain:** Several times in this chapter, I mention that the open loop gain in an op amp is very large — on the order of tens or even hundreds of thousands. In an ideal op amp, the open loop gain is infinite, which means that *any* voltage differential on the two input terminals will result in an infinite voltage on the output. In real op amps, the output voltage is limited by the power supply voltage. Because the output voltage can't be infinite, the gain can't be infinite either.

✔ **Infinite input impedance:** *Impedance* represents a circuit's opposition to current flow, whether the current is alternating or direct. In an ideal op amp, the impedance of the two input terminals is infinite, which means that no current enters the op amp from the inputs. The inputs are able to see and react to the voltage, but that voltage is unable to push any current into the op amp. What that means in practice is that the op amp has no effect on the input voltage. In an actual op amp, a small amount of current — usually, a few milliamps or less — does leak into the op amp's input circuits.

✔ **Zero output impedance:** In an ideal op amp, the output circuitry has zero internal impedance, which means that the voltage provided from the output is the same regardless of the amount of load placed on it by the circuit to which the output is connected. In reality, most op amps have an output impedance of a few ohms, which means that the actual voltage provided by the output terminal will vary a small amount depending on the load connected to the output.

✔ **Zero offset voltage:** The *offset voltage* is the amount of voltage at the output terminal when the two inputs are exactly the same. If you connect both inputs to ground, for example, there should be exactly 0 V at the output. In reality, real world op amps have a very small voltage on the output even when both inputs are grounded, connected to each other, or not connected to anything at all. For most op amps, this offset voltage is just a few millivolts.

✔ **Infinite bandwidth:** The term *bandwidth* refers to the range of alternating current frequencies within which an op amp can accurately amplify. In an ideal op amp, the frequency of the input signal has no effect on how the op amp behaves. In real-world op amps, the op amp doesn't perform well above a certain frequency — typically, a few megahertz (millions of cycles per second).

The characteristics are often summed up with the following two "golden rules" of op amps:

1. **The output attempts to do whatever is necessary to make the voltage difference between the inputs zero.**

(continued)

(continued)

This rule, which applies only to closed-loop amplifier circuits, means that the feedback sent from the output to the input causes the two input voltages to become the same.

2. The input draws no current.

This rule means that the input terminals look at the voltage placed across them but don't allow any current to flow into the op amp.

Although no actual op amp is able to live up to the standards of the ideal op amp, most come pretty close. Close enough, in fact, that you can safely design an op amp circuit as if the op amps were ideal. In particular, the two golden rules apply: The feedback will equalize the input voltages, and the op amp draws no current from the input.

The output voltage can never exceed the power supply voltage. In fact, the maximum output voltage usually is about 1 V less than the power supply voltage. So if you're using a pair of 9 V batteries to provide a ±9 V power supply, the maximum output voltage is ±8 V. As a result, the most that an open loop op-amp circuit with an open loop gain of 200,000 can reliably amplify is 0.00004 V. If the input voltage difference is any larger than 0.00004 V, the op amp is said to be *saturated,* and the output voltage will go to the maximum.

I guarantee that no matter how much money you invested in a top-quality voltmeter, it isn't sensitive enough to measure voltages that small. Physicists at Cal Tech may be able to measure voltages that size, but for all practical purposes, 0.00004 V is the same as 0 V.

As a result, one of the basic features of an open loop op-amp circuit is that if the input voltage difference is anything other than zero, the op amp will be saturated, and the output voltage will be the same at its maximum. So if the maximum output voltage is ±8 V, the output will be one of only three voltages: +8 V, 0 V, or –8 V.

Open loop op-amp circuits may not sound very useful, but actually, they have many useful applications. You see one example in "Using an Op Amp as a Voltage Comparator," later in this chapter.

Looking at Closed Loop Amplifiers

As I explain in the preceding section, open loop op-amp circuits aren't very useful as amplifiers because they're so easily saturated. To make an op amp useful as an amplifier, you must use it in a *feedback circuit,* which reduces the gain to a more manageable amount so that input voltages that are usable (and even measurable!) can be amplified reliably.

I'm sure that you're already familiar with the concept of feedback. You've probably sat in an auditorium listening to someone talk into a public-address system when suddenly a piercing screech came out of the speakers. That screech was feedback. In the case of the public-address system, the microphone picked up some of the output from the speakers and sent it back through the amplifier again. The result was an annoying high-pitched squeal.

Not all feedback is bad, though. In an op-amp amplifier circuit, feedback is used to reduce the enormous open loop amplification gain to a more manageable gain, such as 10. To do this, the output signal is fed back into the input via the V₊ terminal. A resistor is used to reduce the voltage that is fed back to the input. This type of circuit is called a *closed loop amplifier* because a closed circuit path exists between the output and the input. (Now you understand why an op-amp circuit without the feedback loop is called an *open loop amplifier.*)

The most common op-amp configuration is called an *inverting amplifier* because the voltage of the output is opposite the voltage of the input. Figure 3-5 shows a basic inverting amplifier circuit.

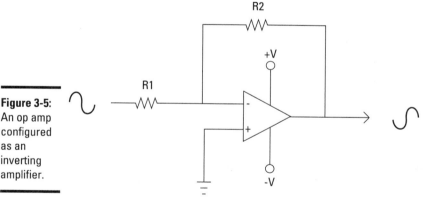

Figure 3-5: An op amp configured as an inverting amplifier.

Book III Chapter 3

Working with Op-Amps

In an inverting amplifier circuit, the input signal works its way through a resistor on its way to the V_ input, and the output is looped back into the V– input through a second resistor. In Figure 3-5, these resistors are designated R1 and R2. You can easily calculate the overall voltage gain of the circuit by using this formula:

$$A_{CL} = -\frac{R2}{R1}$$

Here, the gain is designated A_{CL} (*CL* stands for *closed loop*).

If R1 is 1 kW and R2 is 10 kW, the voltage gain of the circuit will be –10. Then if the input voltage is +0.5 V, the output voltage will be –5 V (0.5 × –10).

Note that the negative sign is required because Figure 3-5 is an inverting amplifier circuit, so positive inputs give negative outputs, and vice versa.

A closed loop amplifier can also be designed as a *noninverting amplifier* in which the output voltage is not reversed. To do that, you simply reverse the inputs, as shown in Figure 3-6. Instead of connecting the input voltage to V_ through a resistor and grounding V_+, you ground V_ through a resistor and connect the input voltage to V_+. The feedback circuit is the same; the output is connected to the V– input through resistor R2.

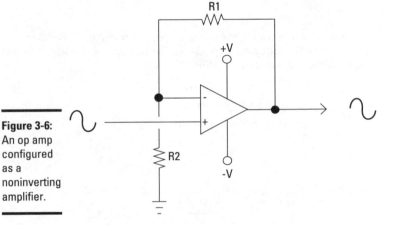

Figure 3-6: An op amp configured as a noninverting amplifier.

The formula for calculating the gain for a noninverting amplifier is a little different from the formula for an inverting amplifier:

$$A_{CL} = 1 + \frac{R2}{R1}$$

If R1 is 1 kW and R2 is 10 kW, the gain is 11. Thus, an input voltage of +0.5 V will result in an output voltage of +5.5 V.

TECHNICAL STUFF

Delving Deeper into Feedback

If you're interested in understanding how the feedback circuit works, remember the rule that I describe in the section "Understanding Open Loop Amplifiers": If the input voltage is anything other than zero, the op amp will be saturated, and the output voltage will be the maximum allowed. The purpose of the feedback circuit is to return some of the output voltage to the inverting input, which results in the input-voltage difference's being driven toward zero. As the voltage approaches zero, the op amp's gain begins to drop to a useful range.

Suppose that the input voltage difference in an inverting op amp circuit is +0.5 V. This difference results in the op amp's becoming saturated, so −8 V appears at the output. A portion of the negative voltage that depends on the voltage divider created by R1 and R2 is returned to the V− input, which has the effect of reducing the input voltage. That results in a smaller voltage difference, but not small enough to prevent the op amp from still being saturated.

Remember, however, that the feedback loop is just that: a loop. As more and more of the saturated output voltage gets fed back through the loop, the input voltage difference gets closer to zero. When it gets oh-so-close to zero, the output voltage drops to a range between zero and the maximum voltage.

The best feature of the closed loop amplifier circuit is that the two resistors, which are outside the op-amp IC, give you precise control of the amount of gain that the circuit will ultimately have. All you have to do to get any gain you want (within the limits of the op amp) is choose the right resistor values.

Using an Op Amp as a Unity Gain Amplifier

A *unity gain amplifier* is an amplifier circuit that doesn't amplify. In other words, it has a gain of 1. The output voltage in a unity gain amplifier is the same as the input voltage.

You may think that such a circuit would be worthless. After all, isn't a simple piece of wire a unity gain circuit? Sure, but a unity gain amplifier provides one important benefit: It doesn't take any current from the input source. (Remember, that's one of the Golden Rules of the ideal op amp.) Therefore, it completely isolates the input side of the circuit from the output side of the circuit. Op amps are often used as unity gain amplifiers to isolate stages of a circuit from one another.

Unity gain amplifiers come in two types: voltage followers and voltage inverters. A *follower* is a circuit in which the output is exactly the same voltage as the input. An *inverter* is a circuit in which the output is the same voltage level as the input but with the opposite polarity.

If you think about it for a moment, you might be able to come up with the circuit for unity gain followers and inverters on your own. The formula for calculating the gain of both an inverting amplifier and a noninverting amplifier requires you to divide R2 by R1, so all you have to do is choose resistor values that will result in a gain of 1.

The following sections explain how to create unity followers and unity inverters.

Configuring a unity follower

A unity gain follower is simply a noninverting amplifier with a gain of 1. Recall that the formula for calculating the value of a noninverting amplifier is this:

$$A_{CL} = 1 + \frac{R2}{R1}$$

To create a unity gain follower, you just omit R2 and connect the output directly to the inverting input, as shown in Figure 3-7. Because R2 is zero, the value of R1 doesn't matter, because zero divided by anything equals zero. So R1 is usually omitted as well, and the V_ input isn't connected to ground.

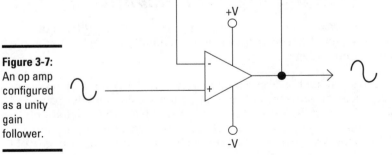

Figure 3-7:
An op amp configured as a unity gain follower.

Configuring a Unity Inverter

The formula for calculating gain for an inverting amplifier is this:

$$A_{CL} = -\frac{R2}{R1}$$

In this case, all you have to do is use identical values for R1 and R2 to make the amplifier gain equal to 1. Figure 3-8 shows a unity gain inverter circuit using 1 kW resistors.

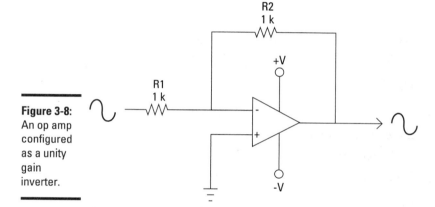

Figure 3-8: An op amp configured as a unity gain inverter.

Using an Op Amp as a Voltage Comparator

A *voltage comparator* is a circuit that compares two input voltages and lets you know which of the two is greater. Suppose that you have a photocell that generates 0.5 V when it's exposed to full sunlight, and you want to use this photocell as a sensor to determine when it's daylight. You can use a voltage comparator to compare the voltage from the photocell with a 0.5 V reference voltage to determine whether or not the sun is shining.

It's easy to create a voltage comparator from an op amp, because the polarity of the op-amp's output circuit depends on the polarity of the difference between the two input voltages. Figure 3-9 shows the basic circuit for an op-amp voltage comparator.

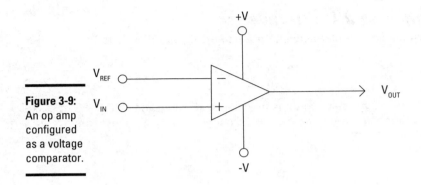

Figure 3-9:
An op amp
configured
as a voltage
comparator.

In the voltage-comparator circuit, first a reference voltage is applied to the inverting input (V_); then the voltage to be compared with the reference voltage is applied to the noninverting input. The output voltage depends on the value of the input voltage relative to the reference voltage, as follows:

Input Voltage	*Output Voltage*
Less than reference voltage	Negative
Equal to reference voltage	Zero
Greater than reference voltage	Positive

Note that the voltage level for both the positive and negative output voltages will be about 1 V less than the power supply. Thus, if the op-amp power supply is ±9 V, the output voltage will be +8 V if the input voltage is greater than the reference voltage, 0 V if the input voltage is equal to the reference voltage, and −8 V if the input voltage is less than the reference voltage.

You can modify the circuit to eliminate the negative voltage if the input is less than the reference by sending the output through a diode, as shown in Figure 3-10. In this circuit, a positive voltage appears at the output if the input voltage is greater than the reference voltage; otherwise, no output voltage exists.

To create a voltage comparator that creates a positive voltage output if the input voltage is *less than* a reference voltage, use the circuit shown in Figure 3-11. Here, the reference voltage is applied to the inverting (V_) input, and the input voltage is applied to the noninverting (V₊) input.

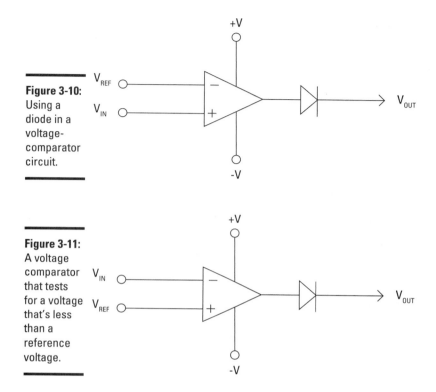

Figure 3-10:
Using a diode in a voltage-comparator circuit.

Figure 3-11:
A voltage comparator that tests for a voltage that's less than a reference voltage.

The final voltage-comparator circuit you should know about is the *window comparator,* which lets you know whether the input voltage falls within a given range. A window comparator requires three inputs: a low reference voltage, a high reference voltage, and an input voltage. The output of the window comparator will be a positive voltage only if the input voltage is greater than the low reference voltage and less than the high reference voltage. If the input voltage is less than the low reference voltage, the output will be zero. Similarly, if the input voltage is greater than the high reference voltage, the output will also be zero.

You need two op amps to create a window comparator, as shown in Figure 3-12. As you can see in the figure, one op amp is configured to produce positive output voltage only if the input is greater than the low reference voltage $(V_{REF(LOW)})$. The other op amp is configured to produce positive output voltage only if the input is less than the high reference voltage $(V_{REF(HIGH)})$.

The input voltage is connected to both op amps; the output voltage is sent through diodes to allow only positive voltage and then combined. The resulting output will have positive voltage only if the input voltage falls between the low and high reference voltages.

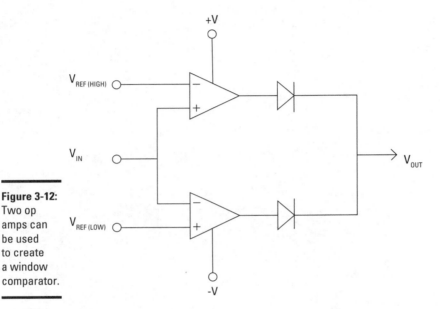

Figure 3-12:
Two op amps can be used to create a window comparator.

 Notice in Figure 3-12 that the power supply connections aren't shown separately for each op amp in the circuit. It's common to omit the power supply connections when multiple op amps are used in a single circuit. If the power supply connections were shown for all of the op amps, the power supply connections would complicate the schematic unnecessarily. I don't know about you, but I don't need any unnecessary complications in my life. I have enough necessary complications as it is.

Adding Voltages

An op amp can be used to add or subtract two or more voltages. A circuit that adds voltages is called a *summing amplifier.* A summing amplifier has two inputs and an output whose voltage is the sum of the two input voltages but with the opposite polarity. If one of the inputs is +1.5 V and the other is +1.0 V, for example, the output voltage will be –2.5 V.

Figure 3-13 shows a basic circuit for a summing amplifier. For the summing amplifier to work, resistors R1, R2, and R3 should all be the same value.

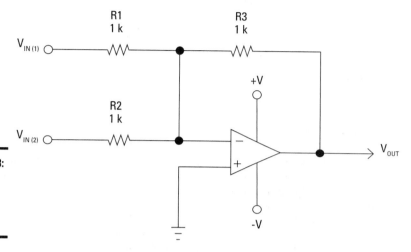

Figure 3-13:
A basic
summing
amplifier
circuit.

If all the resistors in a summing amplifier are the same, the output voltage will be the sum of the input voltages. This is the usual way to configure a summing amplifier, though you can vary the resistor values if you want.

If the resistors have different values, each of the input voltages is weighted according to the value of the resistor on its input circuit, which has the effect of multiplying each input voltage by a certain value before the voltages are summed. The exact value by which each input is multiplied depends on the mix of resistors you use.

If R1 is 1 kW and R2 is 10 kW, for example, the input voltage applied through the 1 kW resistor will be multiplied by 10 before being added to the voltage applied through the 10 kW resistor. Thus, if the input at R1 is +1 V, and the input at R2 is +2 V, the output voltage will be –12 V. (For this formula to work, R3 must also be 10 kW.)

The actual formula for calculating the output voltage based on the input voltages and the resistor values is this:

$$V_{out} = -R3\left(\frac{V1}{R1} + \frac{V2}{R2}\right)$$

I'll leave it to you to work out the math for various combinations of resistor values and input voltages. Here, though, are a few examples that should give you an idea of how the circuit will behave when R1 is 1 kW and both R2 and R3 are 10 kW:

$V_{IN\,(1)}$	$V_{IN\,(2)}$	V_{OUT}
+1 V	+1 V	−11 V
+1 V	+5 V	−15 V
0 V	+5 V	−5 V
+2 V	−5 V	−15 V
−1 V	−5 V	+15 V

One drawback of the summing amplifier is that it inverts the polarity of the input, but you can easily feed the output of a summing amplifier into the input of a unity gain inverter, as shown in Figure 3-14. Here, the second op amp inverts the polarity of the output from the summing amplifier, which has the effect of returning the output voltage polarity to the polarity of the original inputs. (For more information about the voltage-inverter portion of this circuit, refer to "Using an Op Amp as a Unity Gain Amplifier," earlier in this chapter.)

Figure 3-14: A summing amplifier can be combined with a voltage inverter to preserve the input polarity.

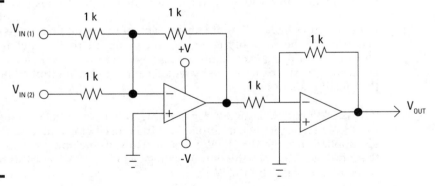

One common use for a summing amplifier circuit is as an audio mixer. When this type of circuit is used as an audio mixer, each input is connected to a microphone. The summing amplifier combines all the microphone inputs by adding the voltages from each microphone, and the resulting output is sent on to another amplifier stage.

The resistors in each input circuit are often potentiometers, which allows you to vary the signal level from each input source. When you increase the resistance on one of the input circuits, less of that input is represented in the output mix — especially useful if one of your singers is a bit off key.

A summing amplifier circuit can be extended with additional inputs. Figure 3-15 shows a circuit with four inputs that uses potentiometers to control the level of each input. You can add as many inputs as you want, but you need to ensure that the total combined voltage from all inputs doesn't exceed the power supply voltage (minus a volt or two).

**Book III
Chapter 3**

**Working with
Op-Amps**

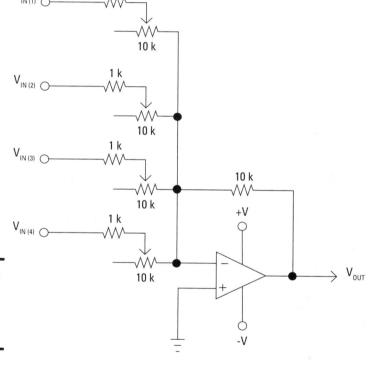

Figure 3-15:
A simple
audio mixer
with four
inputs.

One final variation of the summing amplifier circuit is used in conjunction with a second op amp configured as an inverter. This configuration preserves the polarity of the input voltages.

Working with Op Amp ICs

So far, all the examples in this chapter have assumed that you're using a generic op amp in your circuits. When you get around to building an actual op-amp circuit, of course, you'll need to use a real op amp. Fortunately, op-amp integrated circuits are plentiful, and nearly all stores that sell electronic components sell several types of inexpensive op-amp ICs.

The most popular op-amp IC is the LM741, which comes in a standard eight-pin DIP package. Figure 3-16 shows the pin connections for an LM741 op amp.

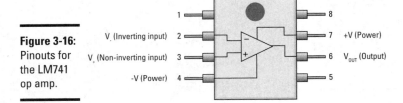

Figure 3-16:
Pinouts for the LM741 op amp.

You can also get integrated circuits that contain two or more op amps in a single package. One of the most common is the LM324 quad op amp, which contains four op amps in a single 14-pin DIP package. Unlike the LM741, the LM324 uses single power supply op amps. Thus, instead of a split + and – voltage supply, you provide just a positive voltage power supply and a ground.

Figure 3-17 shows the pinouts for an LM324. As you can see, the first op amp is accessed via pins 1–3, the second via pins 5–7, the third via pins 8–10, and the fourth via pins 12–14. The positive voltage power supply is connected to pin 4, and pin 11 is connected to ground.

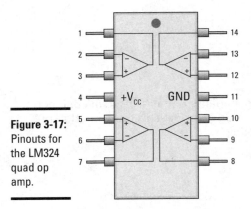

Figure 3-17:
Pinouts for the LM324 quad op amp.

Book IV

Getting into Alternating Current

Good thing I had my laptop with me. Try turning it over now!

Contents at a Glance

Chapter 1: Understanding Alternating Current

In This Chapter

- ✔ Looking at alternating current
- ✔ Glimpsing at the history of how alternate current won out over direct current
- ✔ Examining how an alternator can convert mechanical power to alternating current
- ✔ Discovering how electric motors work
- ✔ Unveiling the magic of transformers

I recently saw the touring production of my favorite musical, *Les Misérables*. It has nothing to do with electricity, of course, but at the end of the first act, as the daring gang of revolutionary college students join the rest of the characters singing "One Day More!," they do a really cool marching step in which they march one pace toward the audience, then one pace back, one pace forward, one pace back, and so on. It creates a convincing effect of a mob on the move, even though the mob isn't going anywhere.

That's how alternating current works. So far in this book, I've focused mostly on working with direct current, in which the electric current flows in one direction and one direction only. If the revolutionaries in *Les Misérables* were demonstrating direct current, they would march straight off the stage, over the orchestra pit, and through the audience, out the back doors of the auditorium.

In *alternating current*, the current flows in both directions — forward and backward — much as the revolutionaries march one step at a time forward and backward.

Alternating current is of vital importance in electronics for one simple reason: The electric current you can access by plugging a circuit into a wall outlet happens to be alternating current. Thus, if you want to free your circuits from the tyranny of batteries, which eventually die, you'll need to learn how to make your circuits work from an alternating current power supply.

In this chapter, take a look at the nature of alternating current and how it can deliver reliable voltage to your home or place of business. You can also look at three fundamental alternating current devices: alternators, which generate alternating current from a source of motion such as a steam turbine or windmill; motors, which turn alternating current into motion; and transformers, which can transfer alternating current from one circuit to another without any physical connection between the circuits.

Incidentally, if you want to see the famous *Les Misérables* march step in action, go to YouTube, search *Les Miserables Tony Awards,* and choose the first hit that comes up. You'll see the marching about three minutes into the video. (The video is also cool because it's introduced by Jerry Orbach, before he landed the part of Lenny Briscoe on *Law and Order.*)

What Is Alternating Current?

As you know, electric current that flows continuously in a single direction is called a *direct current*, or *DC*. In a direct current circuit, current is caused by electrons that all line up and move in one direction. Within a wire carrying direct current, electrons hop from atom to atom while moving in a single direction. Thus, a given electron that starts its trek at one end of the wire will eventually end up at the other end of the wire.

In *alternating current*, the electrons don't move in only one direction. Instead, they hop from atom to atom in one direction for awhile, and then turn around and hop from atom to atom in the opposite direction. Every so often, the electrons change direction. In alternating current, the electrons don't move steadily forward. Instead, they just move back and forth.

When the electrons in alternating current switch direction, the direction of current and the voltage of the circuit reverses itself. In public power distribution systems in the United States, (including household current), the voltage reverses itself 60 times per second. In some countries, the voltage reverses itself 50 times per second.

The rate at which alternating current reverses direction is called its *frequency*, expressed in hertz. Thus, standard household current in the United States is 60 Hz.

In an alternating current circuit, the voltage, and therefore the current, is always changing. However, the voltage doesn't instantly reverse polarity. Instead, the voltage steadily increases from zero until it reaches a maximum voltage, which is called the *peak voltage*. Then, the voltage begins to decrease again back to zero. The voltage then reverses polarity and drops below zero, again heading for the peak voltage but negative polarity. When it reaches the peak negative voltage, it begins climbing back again until it gets to zero. Then the cycle repeats.

The swinging change of voltage is important because of the basic relationship between magnetic fields and electric currents. When a conductor (such as a wire) moves through a magnetic field, the magnetic field induces a current in the wire. But if the conductor is stationary relative to the magnetic field, no current is induced.

Physical movement is not necessary to create this effect. If the conductor stays in a fixed position but then intensity of the magnetic field increases or decreases (that is, if the magnetic field expands or contracts), a current is induced in the conductor the same as if the magnetic field were fixed and the conductor was physically moving across the field.

Because the voltage in an alternating current is always either increasing or decreasing as the polarity swings from positive to negative and back again, the magnetic field that surrounds the current is always either collapsing or expanding. So, if you place a conductor within this expanding and collapsing magnetic field, alternating current will be induced in the conductor.

It seems like magic! With alternating current, it is possible for current in one wire to induce current in an adjacent wire, even though there is no physical contact between the wires.

The bottom line is this: Alternating current can be used to create a changing magnetic field, and changing magnetic fields can be used to create alternating current. This relationship between alternating current and magnetic fields makes three important devices possible:

✦ **Alternator:** A device that generates alternating current from a source of rotating motion, such as a turbine powered by flowing water or steam or a windmill. Alternators work by using the rotating motion to spin a magnet that's placed within a coil of wire. As the magnet rotates, its magnetic field moves, which induces an alternating current in the coiled wire. (Coils of wires are used instead of straight wires simply because coiling up the wire allows a greater length of wire to be exposed to the changing magnetic field.)

✦ **Motor:** The opposite of an alternator. It converts alternating current to rotating motion. In its simplest form, a motor is simply an alternator that's connected backward. A magnet is mounted on a shaft that can rotate; the magnet is placed within the turns of a coil of wire. When alternating current is applied to the coil, the rising and falling magnetic field created by the current causes the magnet to spin, which turns the shaft.

✦ **Transformer:** Consists of two coils of wire placed within close proximity. If an alternating current is placed on one of the coils, the collapsing and expanding magnetic field will induce an alternating current in the other coil.

Measuring Alternating Current

With direct current, it's easy to determine the voltage that's present between two points: You simply measure the voltage with a voltmeter.

With alternating current, however, measuring the voltage isn't so simple. That's because the voltage in an alternating current circuit is constantly changing. So, for example, when we say that the voltage at a wall receptacle is 120 VAC, what does that really mean?

There are actually three ways you can measure voltage in an AC circuit. They are illustrated in Figure 1-1. The three ways are:

✦ **Peak voltage:** A measurement of the largest voltage present between 0 V and the highest point on the AC cycle. It's the maximum voltage that the AC voltage attains.

✦ **Peak-to-peak voltage:** The difference between the highest and lowest peaks of the AC voltage. In most AC voltages, the peak-to-peak voltage is double the peak voltage.

✦ **RMS voltage:** The average voltage of the circuit; also called the *mean voltage. RMS* stands for *root mean square,* but that's important only if you're studying for an exam or something. RMS voltage is far and away the most common way to specify the voltage of an AC circuit. For example, when we say that the voltage at a household electrical outlet is 120 VAC, what we really mean is that the RMS voltage is 120 V.

If the AC voltage follows a true sine wave, the RMS voltage is equal to 0.707 times the peak voltage. Or to turn it around, the peak voltage is equal to about 1.4 times the RMS voltage. Thus, the actual peak voltage at a household electrical outlet is about 168 V.

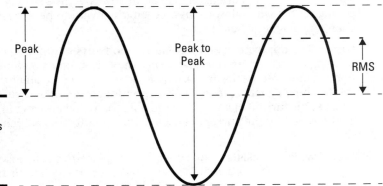

Figure 1-1:
Three ways
to measure
alternating
current.

The current wars

Alternating current is the worldwide standard for power distribution. This wasn't always the case, however. Back in the days when electricity was first being put to practical use, direct current was the normal way to distribute electricity. The biggest champion of direct current was none other than Thomas Edison himself, the Great American Inventor who is credited with inventing just about everything from the light bulb to the phonograph to the motion picture.

In 1880, Edison patented a system of electrical power distribution based on direct current and opened the first public electric utility company in 1882, providing electricity to 59 customers in New York. By 1890, he had more than 100 power plants operating nationwide.

Thomas Edison's biggest rival was a fellow named George Westinghouse, who advocated the use of alternating current for power distribution and promoted a system developed by the brilliant but eccentric inventor Nicola Tesla. Westinghouse promoted the benefits of alternating current over direct current — primarily the benefit that alternating current could transmit power efficiently over much larger distances than direct current. Edison's direct current system required that power plants be located within a few miles of customers. But Tesla's alternating current system could deliver power hundreds of miles away from the power plants.

Edison responded to the negative publicity of direct current the way any true-blooded American marketer would: by launching a smear campaign. In 1887, a man was accidentally killed when he touched bare power lines. Edison had one of his employees develop a method of *intentionally* killing people with electricity. Hence, the electric chair was invented.

Of course, the electric chair used alternating current to electrocute its victims. Edison launched a nationwide publicity campaign to convince the public that alternating current was so dangerous that it was used in prisons to kill condemned murderers. He even went so far as to conduct public executions of stray dogs and, in one case, an elephant. The message was clear: You don't want this dangerous stuff in your house.

Fortunately, the smear campaign didn't work. The benefits of alternating current eventually won out. The turning point came when the alternating current generators at Niagara Falls began operating in 1896, delivering power 20 miles away to Buffalo, NY. By the early part of the twentieth century, nearly all power distribution worldwide was done with alternating current.

Direct current distribution lasted much longer than you might think, however. Con Edison — one of the largest electric companies in the world and the direct descendant of Edison's original electric company — did not convert its last few holdout customers over to alternating current until 2007.

The true RMS voltage is a bit tricky to calculate, since it involves some fairly complicated math. RMS is calculated by sampling the actual voltage in very small time increments. Then, the sample voltages are squared, the squares of the voltages are added up, and the average of all the squared values is calculated. Finally, the square root of the average is calculated. This is the actual RMS value.

For a true sine wave, the preceding calculation turns out to be very close to multiplying the peak voltage by 0.707. For AC voltages that aren't true sine waves, however, the actual RMS value can be different than the "multiply by 0.707" shortcut would indicate.

Nearly all AC voltmeters report the RMS voltage, but only more expensive AC voltmeters calculate the actual RMS by sampling the input voltage and doing the sum-of-the-squares thing. Inexpensive voltmeters simply measure the peak voltage and multiply it by 0.707. Fortunately, this is close enough for most purposes.

Understanding Alternators

One good way to get your mind around how alternating current works is to look at the device that's most often used to generate it: the *alternator*. An alternator is a device that converts rotary motion, usually from a turbine driven by water, steam, or a windmill, into electric current. By its very nature, an alternator creates alternating current.

Figure 1-2 shows a simplified diagram of how an alternator works. Essentially, a large magnet is placed within a set of stationary wire coils. The magnet is mounted on a rotating shaft that's connected to a turbine or windmill. Thus, when water or steam flows through the turbine or when wind turns the windmill, the magnet rotates.

As the magnet rotates, its magnetic field moves across the coils of wire. Because of the phenomenon of electromagnetic induction, the moving magnetic field induces an electric current within the wire coils. The strength and direction of this electric current depends on the position and direction of the rotating magnet.

In Figure 1-2, you can see how the current is induced in the wire at four different positions of the magnet's rotation. In part A, the magnet is at its farthest point away from the coils and oriented in the same direction as the coils. At this moment, the magnetic field doesn't induce any electric current at all. Thus, the light bulb is dark.

But as the magnet begins to rotate clockwise, the magnet comes closer to the coils, thus exposing more of its magnetic field to the coils. The moving magnetic field induces a current that gets stronger as the magnet continues to rotate closer to the coils. This causes the light bulb to glow. Soon, the magnet reaches its closest point to the coils, as shown in part B. At this point, the current and the voltage are at their maximum, and the light bulb glows at its brightest.

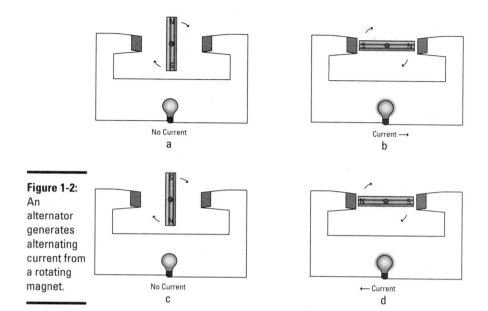

Figure 1-2:
An alternator generates alternating current from a rotating magnet.

As the magnet continues to rotate counterclockwise, it now begins to move away from the coil. The moving electric field continues to induce current in the coil, but the current (and the voltage) decreases as the magnet retreats farther away from the coils. When the magnet reaches its farthest point from the coils, shown in part C, the current stops and the light bulb goes dark.

As the magnet continues to rotate, it now gets closer again to the coils. But this time, the polarity of the magnet is reversed. Thus, the electric current induced in the wire by the moving magnetic field is in the opposite direction, as shown in part D. Once again, the light bulb glows as the current passing through it increases.

And so on. With each revolution of the magnet, voltage starts at zero and rises steadily to its maximum point, then falls until it reaches zero again. Then the process is reversed, with the current flowing in the opposite direction.

Here are a few other interesting tidbits about alternators:

✦ The term *generator* refers to any device that converts mechanical energy into electrical energy. An alternator is a specific type of generator, so it is common — and correct — to refer to an alternator as a generator.

✦ It's possible to generate direct current from rotating magnetic fields. A DC generator is more complex than an alternator, however, and contains additional components that can wear out over time.

✦ The frequency of the alternating current generated by an alternator is dictated by the rate at which the magnet rotates. The faster the magnet rotates, the higher the frequency of the resulting alternating current.

✦ If you place two sets of coils spaced evenly around the magnet, each forming its own complete circuit, alternating current will be induced in each set of coils. However, the polarities of the two voltages will be mirror-images of one another. In other words, when the voltage is positive in one of the circuits, it will be negative in the other. The relationship between the polarity of the circuits is called a *phase*, and a power-generating system with two circuits arranged in this way is called a *two-phase system*. The two circuits are said to be 180° out of phase with one another.

✦ If three sets of coils are used, the system is called a *three-phase system*. In a three-phase system, the three circuits are 120° out of phase. Most public power-generation systems are three-phase systems because that results in the most efficient generation of power from the rotating magnetic fields.

Understanding Motors

An electric *motor* converts electrical energy in the form of electric current to rotating mechanical energy. The simplest type of electric motor is essentially the same thing as an alternator. The difference is that instead of using some other mechanical force such as water or steam to turn the magnet — which in turn induces electric current in the coils — electric current is applied to the coils, which in turn causes the magnet to rotate.

I would show you a diagram depicting the operation of a motor, but it would look pretty much like the alternator diagram shown in Figure 1-2. The only difference would be that the light bulb would be replaced by an AC power source. The same force that causes an electric current to be induced in a coil when the coil passes through a moving magnetic field causes a moving magnetic field to be created when a current is passed through a coil. The moving magnetic field, in turn, causes the magnet to rotate. This rotation is transferred to the shaft to which the magnet is attached.

As with alternators, it's also possible to create motors that work with DC circuits instead of AC. As with alternators, DC motors are more complicated than AC motors. In a DC motor, the polarity of the coils must be reversed every half-revolution of the magnet to keep the magnet moving in complete rotations. Usually, metal brushes are used to do this. In an AC motor, the brushes aren't necessary because the alternating current reverses polarity on its own.

Understanding Transformers

In Book II, Chapter 4, you learn about the basic principles of magnetism and inductance. A transformer is a device that exploits these two principles:

✦ A changing current passing through a wire creates a moving magnetic field around the wire.

✦ A changing current will be induced in a wire that's exposed to a moving magnetic field.

A transformer combines these two principles by placing two coils of wire in close proximity to one another, as shown in Figure 1-3. When a source of AC is connected to one of the coils, that coil creates a magnetic field that expands and collapses in concert with the changing voltage of the AC. In other words, as the voltage increases across the coil, the coil creates an expanding magnetic field. When the voltage reaches its peak and begins to decrease, the magnetic field created around the coil begins to collapse.

Figure 1-3:
A transformer uses magnetic induction to pass current from one circuit to another.

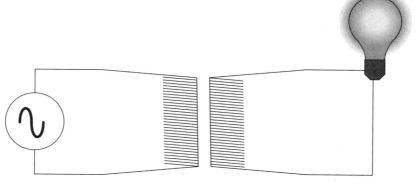

The second coil is located within the magnetic field created by the first coil. As the magnetic field expands, it induces current in the second coil. The voltage across the second coil increases as long as the magnetic field expands. When the magnetic field begins to collapse, the voltage across the second coil begins to decrease.

Thus, the current induced in the second coil mirrors the current that is passed through the first coil. A small amount of energy is lost in the process, but if the transformer is well constructed, the strength of the current induced in the second coil is very close to the strength of the current passed through the first coil.

The first coil in a transformer — the one that's connected to the AC voltage — is called the *primary coil*. The second coil — the one in which an AC voltage is induced — is called the *secondary coil*. All transformers have both a primary and a secondary coil.

One of the most useful characteristics of a transformer is that the voltage induced in the secondary coil is equal to the voltage applied to the primary coil times the ratio of the number of turns in the primary coil and the second coil.

In the simplest case, where both the primary and secondary coils have the same number of turns, the voltage induced in the secondary coil will be the same as the voltage applied to the primary coil. But what if the primary coil has more turns than the secondary coil? In that case, the voltage induced in the secondary coil will be less than the voltage applied to the primary.

How much less depends on the ratio of the turns in the primary and secondary coils. If the secondary coil has half as many turns as the primary coil, the voltage induced in the secondary coil will be half the voltage applied to the primary coil. For example, if you apply 110 VAC to the primary coil, 55 VAC will be induced in the secondary coil.

Likewise, if the secondary coil has more turns than the primary coil, the induced voltage will be more than the voltage applied to the primary coil. For example, suppose the primary coil has 1,000 turns and the secondary coil has 2,000 turns. Then, if you apply 110 V to the primary coil, 220 V will be induced in the secondary coil.

A transformer whose primary coil has more turns than its secondary coil is called a *step-down transformer* because it reduces voltage — that is, the voltage at the secondary coil is less than the voltage at the primary coil. Similarly, a transformer that has more turns in the secondary than in the primary is called a *step-up transformer* because it increases voltage.

Although the voltage increases in a step-up transformer, the current is reduced proportionately. For example, if the primary coil has half as many turns as the secondary coil, the voltage induced in the secondary coil will be twice the voltage that's applied to the primary coil, but the current that flows through the secondary coil will be half the current flowing through the primary coil.

Similarly, when the voltage decreases in a step-down transformer, the current increases proportionately. Thus, if the voltage is cut in half, the current doubles.

This makes perfect sense when you think about it. After all, a transformer can't just conjure up power out of thin air. If it could, we'd have solved the planet's energy problems long ago. But there is no such thing as free energy.

Remember the basic formula for calculating electric power:

$$P = V \times I$$

In other words, power equals voltage times current. A transformer transfers power from the primary coil to the secondary coil. Since the power must stay the same, if the voltage increases, the current must decrease. Likewise, if the voltage decreases, the current must increase.

Transformers are the main reason we use alternating current instead of direct current in large power distribution systems. That's because when you send large amounts of power over a long distance, it's much more efficient to send the power in the form of high voltage and low current. That's why overhead power transmission lines often carry voltages as high as 400,000 VAC. Such high voltages allow the electrical power to be transmitted using much smaller wires than would be required if the same amount of power were transmitted at 120 VAC.

Power distribution systems use large step-up transformers to step voltages generated at power plants up to thousands or hundreds of thousands of volts. Then, as the power gets closer to its final destination (such as your house), a series of step-down transformers drops the voltage down to more manageable levels, until the voltage is dropped to its final level (120 VAC) before it enters your house.

Transformers work only with alternating current. That's because it's the *change* of the magnetic field created by the primary coil that induces voltage in the secondary coil. To create a changing magnetic field, the voltage applied to the primary coil must be constantly changing. Because DC is a steady, fixed voltage, it creates a fixed magnetic field that won't induce voltage in the secondary coil.

Chapter 2: Working with Line Voltage

In This Chapter

✔ Playing it safe with line voltage

✔ Examining how line voltage wiring works

✔ Making safe connections with line-voltage circuits

✔ Using relays to control line-voltage circuits

Line voltage refers to the voltage that's available in standard residential or commercial wall outlets. In the United States, this voltage is almost always in the neighborhood of 120 VAC, though it's commonly referred to as 110 VAC, 115 VAC, or 117 VAC. In other parts of the world, the voltage may be lower or higher.

In Europe, line voltage is often referred to as *mains voltage* or just *mains*.

Unlike the voltage available from household batteries, line voltage is dangerous. In fact, if you're not careful, line voltage can kill you. Thus, you need to take precautions whenever you build a circuit that works with line voltage. In this chapter, you learn how to use line voltage safely so that neither you nor anyone else gets hurt.

Using Line Voltage in Your Projects

So far, none of the projects presented in this book have involved the use of line voltage. But many — if not most — real-world projects do require that you use line voltage.

The most common reason for using line voltage in a project is to eliminate the need for batteries. Batteries are a convenient source of power for your circuits, but they wear out. For many circuits, you want to provide a power source that will last indefinitely. If you use batteries, they'll eventually lose their charge and have to be replaced. If you use line voltage, you can plug the project in and not have to worry about changing batteries.

Of course, most electronic components require direct current rather than alternating current, and at much lower voltage levels than the levels that line voltage supplies. Thus, for your project to use line voltage as its source of power, you need to provide the project with a *power supply* that converts the 120 VAC line voltage to something more useful, such as 5 VDC.

There are at least two ways to accomplish this:

✦ **By using a power adapter:** The easiest way is to use an external *power adapter,* often called a *wall wart* or a *power brick.* Figure 2-1 shows a typical external power adapter. You can purchase power adapters from just about any store that has a consumer electronics department. Just get one that provides the right level of DC voltage and use it instead of batteries.

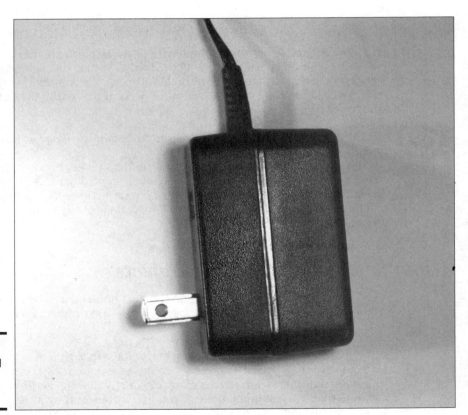

Figure 2-1:
An external
power
adapter.

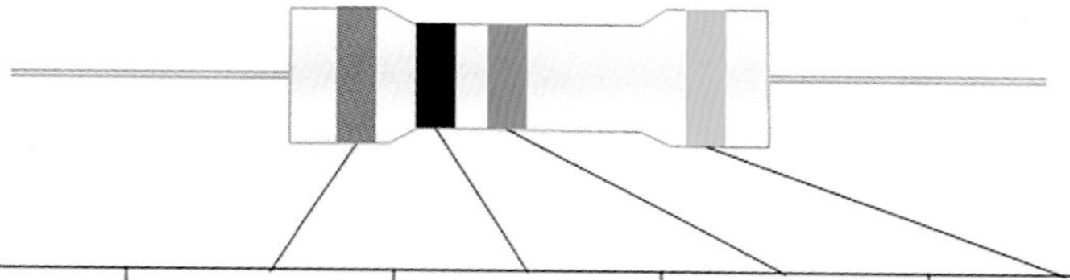

COLOR	1st Digit	2nd Digit	Multiplier	Tolerance
Black	0	0	1	
Brown	1	1	10	± 1%
Red	2	2	100	± 2%
Orange	3	3	1K	
Yellow	4	4	10K	
Green	5	5	100K	± 0.5%
Blue	6	6	1M	± 0.25%
Violet	7	7	10M	± 0.1%
Grey	8	8		± 0.05%
White	9	9		
Gold				± 5%
Silver				± 10%

Color Plate 1: Resistor values are identified by a special color code. You'll find out how to read these codes in Chapter 2 of Book II.

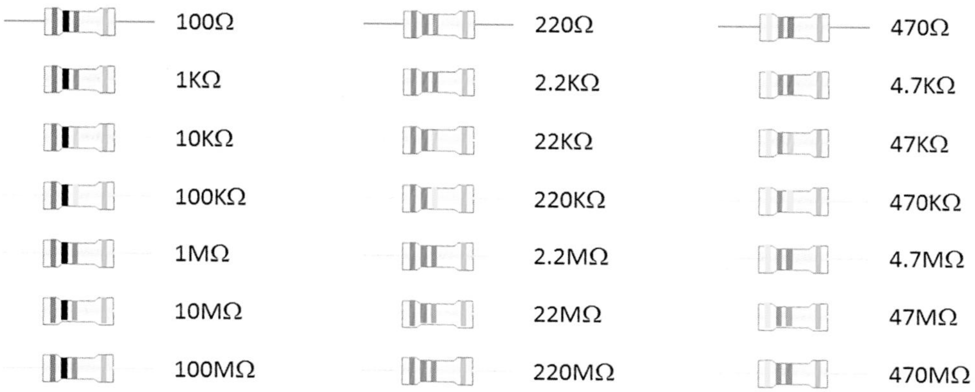

100Ω		220Ω		470Ω
1KΩ		2.2KΩ		4.7KΩ
10KΩ		22KΩ		47KΩ
100KΩ		220KΩ		470KΩ
1MΩ		2.2MΩ		4.7MΩ
10MΩ		22MΩ		47MΩ
100MΩ		220MΩ		470MΩ

Color Plate 2: Colors for common resistor values.

VELLEMAN

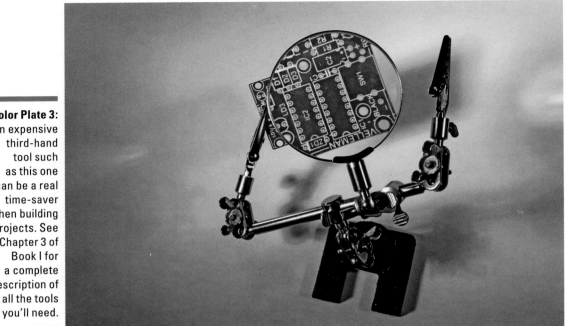

Color Plate 3: An expensive third-hand tool such as this one can be a real time-saver when building projects. See Chapter 3 of Book I for a complete description of all the tools you'll need.

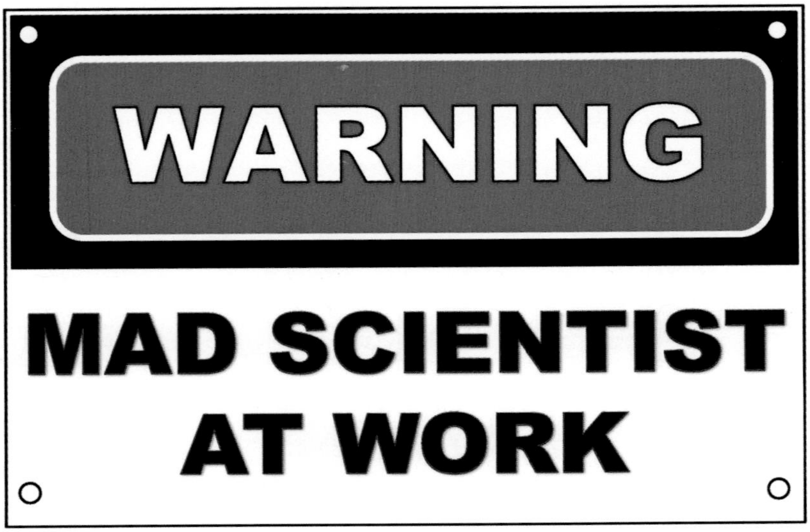

WARNING
MAD SCIENTIST
AT WORK

Color Plate 4: No electronics workbench is complete without a sign such as this.

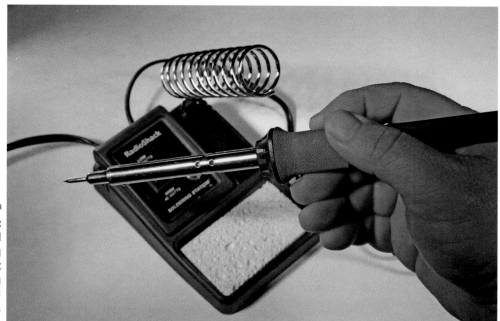

Color Plate 5:
A good soldering iron is a must for building electronic circuits.

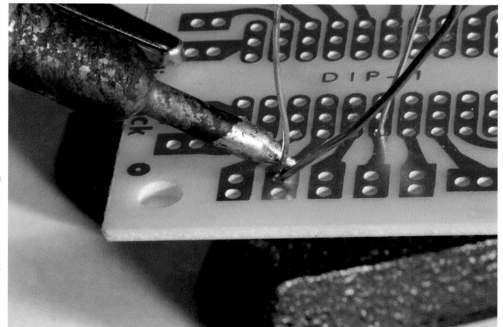

Color Plate 6:
Chapter 7 of Book I explains all about how to apply heat and solder to create a solid solder joint.

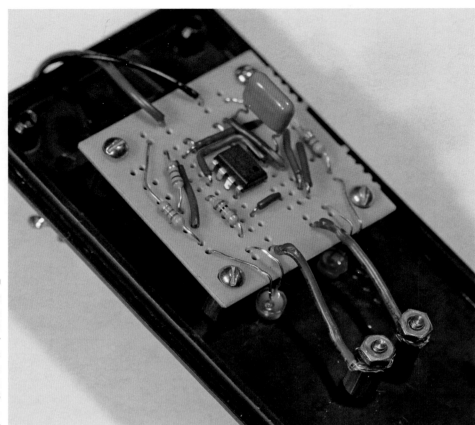

Color Plate 7:
Here is the internal circuitry for the Coin Toss Game which is described in Chapter 6 of Book I.

Color Plate 8:
The completed Coin Toss Game. Your friends will be amazed that you built it yourself.

Color Plate 9:
This nifty circuit, described in Chapter 6 of Book II, uses a pair of transistors to alternately flash a pair of LEDs.

Color Plate 10:
Here's a circuit that uses a 555 Timer IC to flash a pair of LEDs. You learn how to build this circuit in Chapter 2 of Book III.

Color Plate 11:
You learn how to program a Parallax Basic STAMP HomeWork board in Book VII.

Color Plate 12:
A simple LED flasher circuit assembled on a Parallax Basic STAMP HomeWork board.

Color Plate 13: The circuitry for the Color Organ project, which can create great lighting effects such as a thunderstorm or a beating heart. Chapter 2 of Book VIII explains how to build it.

Color Plate 14: The completed Color Organ project.

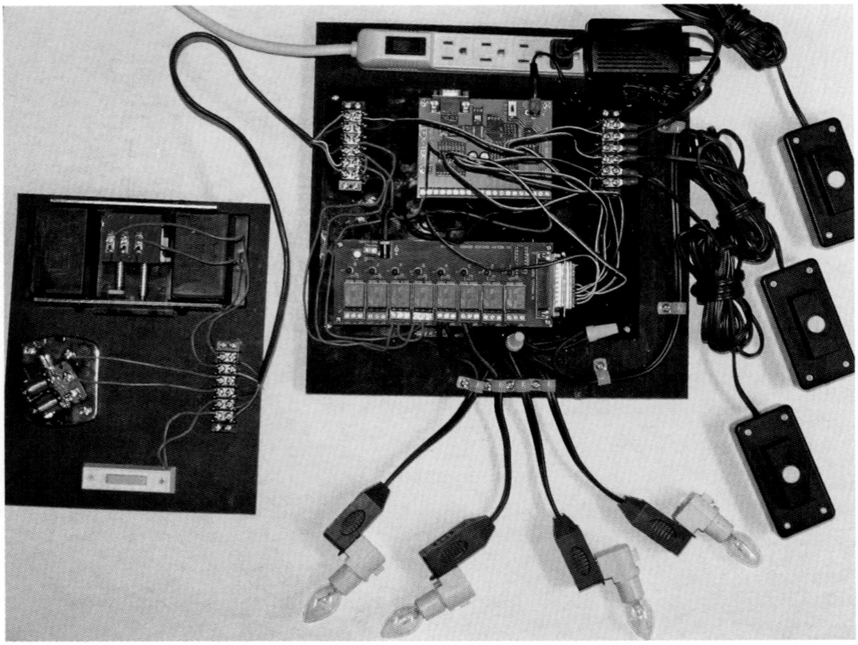

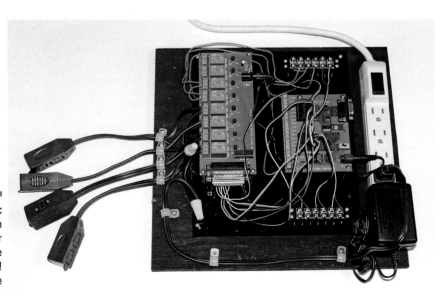

Color Plate 15: The Quiz-O-Matic! game-show controller lets players buzz in to answer questions in the classic quiz-show game format. For complete instructions to build this system, see Chapter 1 in Book VIII.

Color Plate 16: The main controller board for the Quiz-O-Matic! quiz-game controller.

✦ **By building your own power supply:** The alternative to purchasing a power adapter is to build your own power supply circuit. This circuit must accomplish two things. First, it must step the voltage down from 120 VAC to whatever voltage your circuit requires, and second, it must convert the AC voltage to DC voltage. You learn how to design and build a power supply circuit in the next chapter.

The second common reason for using line voltage in a project is if the project needs to control some external device that runs on line voltage, such as a flood lamp or a pump. In that case, your project needs to be able to turn the line voltage on and off.

The most common way to turn a line voltage device on and off from an electronic circuit is to use a device called a *relay*, which is basically an electronic switch that uses a low-current input to control a high-current output. For example, a relay can let you use a 12 VDC circuit to control a separate line voltage circuit. You learn how to use a relay for this purpose later in this chapter, in the section "Using Relays to Control Line Voltage Circuits."

Being safe with line voltage

Whenever you build an electronics project that uses line voltage, you must take extra precautions to ensure your safety and the safety of anyone who may come in contact with your project. Line voltage is potentially deadly, so these precautions are absolutely mandatory.

Many people are under the mistaken belief that line voltage is not sufficient to cause serious injury or death. That simply isn't true; 120 VAC is more than enough voltage to kill, given the right conditions. In fact, you should treat any voltage above 50 V as potentially lethal.

When you work with line voltage, be sure to take the following precautions:

✦ *Never* work on the circuit when the power plug is plugged in.

✦ *Never* leave exposed line-voltage connections anywhere that you or anyone who comes into contact with your project might accidentally touch. *All* line-voltage connections must be completely insulated or contained within an insulated project box or a grounded metal project box.

✦ *Always* enclose projects that use line voltage in a sealed project box so that stray hands can't accidentally come in contact with bare wires or other components.

✦ *Always* use grounded power cords if your project is contained in a metal box, and *always* connect the metal box itself to the power cord's ground lead.

**Book IV
Chapter 2**

**Working with
Line Voltage**

✦ *Always* use the correct gauge of wire for the amount of current your circuit will be carrying. For more information, see the section, "Wires and Connectors for Working with Line Voltage."

✦ *Always* ensure that all line-voltage connections are tight and secure. If you're using stranded wire instead of solid wire, always check for stray strands at your connections.

✦ *Always* provide some form of strain-relief for wires that carry line voltage. The most common way to do this is to pass the wire through a grommet-protected hole in the project box and tie the wire into a knot inside the box. The knot will prevent the cord from sliding through the hole in the case and possibly pulling loose.

✦ *Always* incorporate a fuse in the primary side of your line-voltage circuit. The fuse will automatically detect when too much current is flowing and immediately break the circuit.

✦ *Never* use a fuse that's rated for more than the maximum current your circuit is designed to bear. For example, if you're using a relay that can switch 5 A of current, use a fuse rated for 5 A or less. (For more on fuses, see the section, "Using Fuses to Protect Line-Voltage Circuits," later in this chapter.)

✦ *Never* arrange the wiring for your project in a way that causes wires to move or rub against one another. The rubbing will eventually wear off the insulation and create a shock hazard.

✦ *Always* be aware of heat sinks that may be hot.

✦ *Never* use a three-prong to two-prong adapter to plug a three-prong power connector into a two-prong extension cord. This disables the safety provided by proper grounding and can result in a potentially fatal electric shock if a short circuit occurs.

Understanding hot, neutral, and ground

Before you start working with line voltage in your circuits, you need to understand a few details about how most residential and commercial buildings are wired. The following description applies only to the United States; if you're in a different country, you'll need to determine the standards for your country's wiring.

Standard line voltage wiring in the United States is done with plastic-sheathed cables, which usually have three conductors, as shown in Figure 2-2. This type of cable is technically called *NMB cable,* but most electricians refer to it using its most popular brand name, Romex.

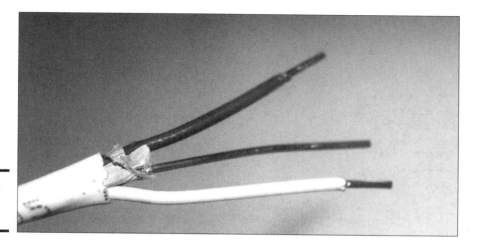

Figure 2-2:
NMB
cabling.

Two of the conductors in NMB cable are covered with plastic insulation (one white, the other black). The third conductor is bare copper. These conductors are designated as follows:

✦ **Hot:** The black wire is the *hot wire*, which provides a 120 VAC current source.

✦ **Neutral:** The white wire is called the *neutral wire*. It provides the return path for the current provided by the hot wire. The neutral wire is connected to an earth ground.

✦ **Ground:** The bare wire is called the *ground wire*. Like the neutral wire, the ground wire is also connected to an earth ground. However, the neutral and ground wires serve two distinct purposes. The neutral wire forms a part of the live circuit along with the hot wire. In contrast, the ground wire is connected to any metal parts in an appliance such as a microwave oven or coffee pot. This is a safety feature, in case the hot or neutral wires somehow come in contact with metal parts. Connecting the metal parts to earth ground eliminates the shock hazard in the event of a short circuit.

Note that some circuits require a fourth conductor. When a fourth conductor is used, it's covered with red insulation and is also a hot wire.

The three wires in a standard NMB cable are connected to the three prongs of a standard electrical outlet (properly called a *receptacle*) as shown in Figure 2-3. As you can see, the neutral and hot wires are connected to the two vertical prongs at the top of the receptacle (neutral on the left, hot on the right) and the ground wire is connected to the round prong at the bottom of the receptacle.

**Book IV
Chapter 2**

**Working with
Line Voltage**

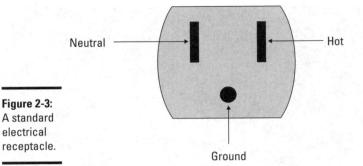

Neutral — Hot

Figure 2-3:
A standard
electrical
receptacle.

Ground

You can plug a two-prong or three-prong plug into a standard three-prong receptacle. Two-prong plugs are designed for appliances that don't require grounding. Most nongrounded appliances are *double-insulated*, which means that there are two layers of insulation between any live wires and any metal parts within the appliance. The first layer is the insulation on the wire itself; the second is usually in the form of a plastic case that isolates the live wiring from other metal parts.

Three-prong plugs are for appliances that require the ground connection for safety. Most appliances that use a metal chassis require a separate ground connection.

There is only one way to insert a three-prong plug into a three-prong receptacle. But regular two-prong plugs, which lack the ground prong, can be connected with either prong on the hot side. To prevent that from happening, the receptacles are *polarized*, which means that the neutral prong is wider than the hot prong. Thus, there's only one way to plug a polarized plug into a polarized receptacle. That way, you can always keep track of which wire is hot and which is neutral.

You should always place switches or fuses on the hot wire rather than on the neutral wire. That way, if the switch is open or the fuse blows, the current in the hot wire will be prevented from proceeding beyond the switch or fuse into your circuit. This minimizes any risk of shock that might occur if a wire comes loose within your project.

Wires and Connectors for Working with Line Voltage

When working with line voltage, you must always use wire that's designed specifically to handle line-voltage currents. Depending on your needs, you may choose to use solid or stranded wire. Stranded wire is usually easier to work with because it's more flexible.

When choosing wire, make sure you get the right gauge for the current your circuit will be carrying. For circuits that are designed to carry 15 A or less (which is the maximum current limit for devices plugged into most household electrical outlets), you can use 14-gauge wire. If the circuit will carry no more than 13 A, 16-gauge wire is sufficient. For less than 10 A, 18-gauge wire is sufficient.

Lamp wire, also known as *zip cord,* is the wire that lamp cords and indoor extension cords are made of. It's usually two-conductor, nongrounded 16- or 18-guage stranded wire in which the two conductors are joined in a way that lets you easily peel them apart. Lamp wire is the easiest wire to use for short connections within your project. You can buy lamp wire in bulk at most hardware stores, or you can buy an inexpensive indoor extension cord at a discount dollar store and cut the ends off.

Make sure all connections you make with wires carrying line voltage are secure. The easiest way to connect the wires is to use *wire nuts,* illustrated in Figure 2-4. To use wire nuts, strip off 3/8" or so of insulation from the two wires to be connected and loosely twist the two ends together. Then, slip the wire nut over the twisted end and tighten the nut onto the connection by pushing down and twisting. When the wire nut is as tight as you can get it, check to make sure that none of the stripped wire extends below the base of the wire nut. For good measure, you can wrap a short strip of black electrical tape around the connection.

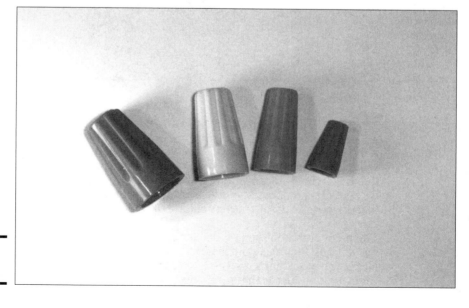

Figure 2-4:
Wire nuts.

Another way to make connections is to use a *barrier strip*, as shown in Figure 2-5. Barrier strips come in various sizes and shapes. The small one pictured in the figure can make four connections; the larger one can make eight. To use a barrier strip, simply strip away a short length of insulation from the end of the wires you want to connect and secure them under the terminal screws. If you're using stranded wire, make sure that all of the strands are held by the screw. Loose strands can cause short circuits.

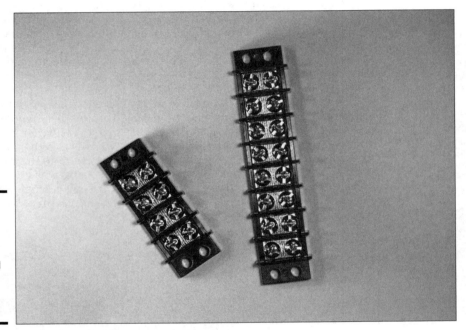

Figure 2-5:
Barrier strips are useful for connecting wires that carry AC power.

Using Fuses to Protect Line-Voltage Circuits

A *fuse* is an inexpensive device that can carry only a certain amount of current. If the current exceeds the rated level, the fuse melts (blows), thus breaking the circuit and preventing the excessive current from flowing. Fuses are an essential component of any electrical system that uses line voltage and has the possibility of short-circuiting or overheating and causing a fire.

The most common type of fuse is the *cartridge fuse,* which consists of a cylindrical body that's usually made of glass, plastic, or ceramic, with two metal ends. The metal ends are the two terminals of the fuse. Inside the body is a thin wire conductor that's designed to melt away if the current exceeds the rated threshold. As long as the current stays below the maximum level, the conductor passes the current from one metal end to the other. But when the current exceeds the rated maximum, the conductor melts, and the circuit is broken.

Figure 2-6 shows an *AGC fuse,* which is a small fuse made of glass, 1-1/4" in length and ¼" in diameter. This particular fuse is rated at 2 A, but you can get AGC fuses in larger ratings, up to 15 A. (*AGC* stands for *Automotive Glass Cartridge,* but that won't be on the test.)

Figure 2-6:
A 2 A,
1-1/4" x ¼"
AGC fuse.

Fuses should always be connected to the hot wire and should be placed before any other component in the circuit. In most projects, the fuse should be the first thing the hot wire connects to after it enters your project enclosure. Figure 2-7 shows how a fuse is depicted in a schematic diagram. Here, the fuse is placed on the hot wire before the lamp.

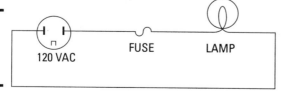

Figure 2-7:
A fuse in a
schematic
diagram.

120 VAC   FUSE   LAMP

If you plan on using a fuse in your circuit, you'll need to purchase a fuse holder to hold the fuse. For AGC cylindrical fuses, there are two distinct types of fuse holders to choose from. If the fuse will be mounted inside your project's enclosure, you can use a chassis-type fuse holder. If you want the fuse to be accessible from outside the project's enclosure, you should choose a panel-mount holder instead. Figure 2-8 shows both types of holders.

Figure 2-8:
Fuse
holders.

Using Relays to Control Line-Voltage Circuits

In many projects, you need to turn line-voltage circuits on and off using circuits that use low-voltage DC power supplies. For example, suppose you want to flash a 120 VAC flood lamp on and off at regular intervals. You could build a circuit to provide the necessary timing using a 555 timer IC as described in Book III, Chapter 2, but the 555 timer IC requires just a small DC power supply, in the range of 5 to 15 V. And the output current can't exceed 200 mA, not nearly enough to light a flood lamp.

Relays to the rescue!

A *relay* is an electromechanical device that uses an electromagnet to open or close a switch. The circuit that powers the electromagnet's coil is completely separate from the circuit that is switched on or off by the relay's switch, so it's possible to use a relay whose coil requires just a few volts to turn a line voltage circuit on or off.

Figure 2-9 shows a typical relay. For this relay, the coil requires just 12 VDC to operate and pulls just 75 mA, well under the current limit that can be sourced by the 555 timer IC's output pin. But the switch part of this relay can handle up to 10 A of current at 120 VAC, more than enough to illuminate a flood lamp.

Figure 2-9:
A relay is a switch that is controlled by an electro-magnet.

The switch part of a relay is available in different configurations just like manual switches. The most common switch configuration is double pole, double throw (DPDT), which means that the relay actually controls two separate switches that operate together, and that each switch has both normally open and normally closed contacts.

Figure 2-10 shows a schematic diagram for a simple circuit that uses a 9 VDC circuit with a handheld pushbutton to turn a 120 VAC lamp on and off. The relay in the circuit has a coil rated for 9 VDC and a switch rating of 10 A at 117 VAC. Thus, only 9 VDC passes through the handheld pushbutton. If the person holding the switch decides to take it apart, he or she won't be exposed to dangerous voltage.

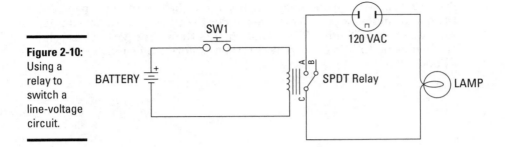

Figure 2-10:
Using a
relay to
switch a
line-voltage
circuit.

Figure 2-11 shows a more complicated circuit, in which a 555 timer IC controls a flood lamp via a relay. Here, one end of the relay coil is connected to the 555 timer IC's output pin (pin 3), and the other end is connected to ground. When the 555's output switches on, the relay closes, and the flood lamp circuit is completed.

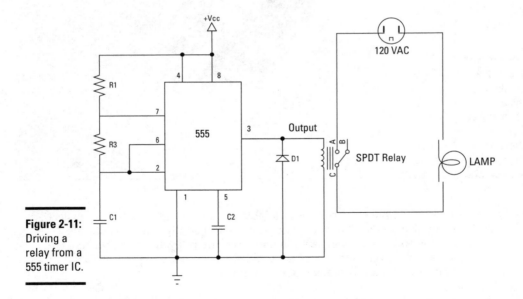

Figure 2-11:
Driving a
relay from a
555 timer IC.

Note the diode that's placed across the relay coil in this circuit. This diode is required to protect the 555 timer IC from any back-current that might be created within the relay's coil when the coil is energized. Because of electromagnetic induction, relay coils are prone to this problem.

When the coil is energized, it creates a magnetic field that causes the relay's switch contacts to move. However, this magnetic field has a subtle side effect. In the instant that the voltage on the coil goes from zero to the Vss supply voltage, the magnetic field surrounding the coil expands from nothing to its maximum strength. During this expansion, the magnetic field is moving relative to the coil itself. Because of the principal of induction, this moving magnetic field induces a current in the coil itself, in the opposite direction as the current that is energizing the coil.

Depending on the circumstances, this back current can be powerful — powerful enough to overwhelm the output current coming from the 555 timer IC and possibly powerful enough to send current into the output pin, which can damage or destroy the 555 chip. D1 prevents this from happening by providing the equivalent of a short circuit across the coil for current flowing back toward the output pin.

Whenever you drive a relay from a circuit that has delicate components such as integrated circuits or transistors, you should always include a diode across the relay coil to prevent the relay from damaging your circuits.

Chapter 3: Building Power Supplies

In This Chapter

✔ Looking at how power supplies work

✔ Stepping down the voltage

✔ Converting AC to DC

✔ Filtering wavy DC

✔ Making the voltage level more reliable through regulation

*W*ith very few exceptions, every electronic circuit requires a power supply of some sort. Although some projects run off of solar power or more exotic power sources such as wind turbines, fuel cells, or nuclear reactors, most of the projects you build will get their power from one of two sources: batteries or an electrical outlet.

So far in this book, I assume that all circuits get their power from batteries. In this chapter, you look at how you can get power from an electrical outlet instead. Electrical outlets have a compelling advantage over batteries: Unless there's a power outage, electrical outlets don't go dead like batteries do. However, electrical outlets have an equally compelling disadvantage over batteries: Unless you use really long extension cords, you can't take a project powered from an electrical outlet very far from the outlet.

Most electronic circuits require a relatively low DC voltage, typically in the range of 3 to 12 V. It's easy to get that range of voltage out of batteries. Since each battery provides about 1.5 V, you just team up two or more batteries to get the right voltage. For example, if your circuit needs 6 V, you can use four batteries connected in series.

Powering a project from an electrical outlet is a little more challenging. First, the 120 V provided at the electrical outlet is much more than most circuits require, so you have to step the voltage down to a more appropriate level. Second, electronic circuits usually require direct current — and the wall outlet provides alternating current — so you have to convert the AC to DC. And third, circuits that run directly on 120 VAC are inherently more dangerous than circuits that run on lower voltages because of the shock danger that accompany higher voltages.

The circuit that converts 120 VAC to direct current at a lower voltage is called a *power supply*. In this chapter, you learn the basics of creating your own power supplies so you can power your projects from a wall outlet instead of from batteries.

Using a Power Adapter

Before I show you how to build your own power supply circuit, I want to let you know that you can probably purchase a preassembled *power adapter* that will provide the voltage you need for just a few dollars more than you could build the circuit yourself. A power adapter, also called a *wall wart*, is a self-contained power supply circuit that plugs into a wall outlet and provides a specified level of AC or DC voltage as its output. As long as the power adapter supplies the correct voltage, you can use the power adapter instead of batteries in just about any circuit.

When you purchase a power adapter, you need to check the specifications to make sure you're purchasing the correct adapter. The specifications are usually printed on the adapter itself. Look for the following important specifications:

+ **AC or DC:** Not all power adapters supply direct current; some are made to power low-voltage AC devices. So make sure that you get an adapter that provides direct-current output.

+ **Voltage level:** Next, check the output voltage. Some power adapters have a switch that lets you choose from among several output voltages. If you use such an adapter, make sure you set the switch to the correct output level for your circuit.

+ **Current capacity:** Most power adapters will have a maximum current rating expressed in milliamps. Smaller adapters can handle a few hundred milliamps, whereas larger adapters may be able to handle an ampere or more. Make sure that the adapter you use can handle the current requirements of your project. (Although some power adapters can handle several, few can handle more than that.)

+ **Polarity:** Most power adapters use a *barrel connector* to plug the power adapter into the circuit. In nearly all modern power adapters, the center connection of the barrel connector is positive, and the outer connection is negative. However, some power adapters are wired just the opposite, with negative in the center and positive on the outside. The polarity of the connector should be printed on the adapter along with the voltage and current specifications.

✦ **Connector size:** Unfortunately, there are far too many different sizes and styles of connectors used for power adapters. Once you've purchased a power adapter, you can go to a local electronics store such as RadioShack and purchase a jack that is compatible with the connector on the power adapter. Then, you can use the jack to connect the power adapter to your circuit. (Note that some power adapters have different interchangeable plugs in different sizes.)

Using a power adapter instead of building your own power supply can make your project safer to build and use. That's because the part of your project that is potentially dangerous — the part that works directly with 120 VAC line voltage — is fully contained inside the preassembled power adapter.

However, as you'll soon discover, you get what you pay for when it comes to power supplies. Inexpensive wall warts convert AC to DC and step down the voltage, but most do not provide power that is very clean (that is, a pure level of DC) or stable (that is, with a predictable voltage). Thus, even if you use a wall wart to power your project, you may still need to add circuitry that will improve the quality of the DC supplied by the wall wart.

Understanding What a Power Supply Does

If you want to add your own power supply circuit to a project to convert 120 VAC line voltage to a DC voltage that's suitable for your circuit, you'll have to design a power supply circuit that provides at least three distinct functions:

✦ **Voltage transformation:** Reduces the 120 VAC line voltage to the voltage your circuit needs.

✦ **Rectification:** Converts the reduced AC voltage to DC voltage. Note that the DC voltage produced by a rectifier circuit is technically direct current, but it isn't steady direct current. Instead, a rectifier produces pulsating direct current in which the voltage fluctuates in sync with the 60 Hz alternating current that's fed into it from the transformation stage.

✦ **Filtering:** Smoothes out the ripples in the DC voltage produced by the rectification stage.

Transforming Voltage

You already know that a *transformer* is a device that uses the principal of electromagnetic induction to transfer voltage and current from one circuit to another. The transformer uses a *primary coil* that's connected to line voltage and a *secondary coil* that provides the output voltage.

In most power supplies, the transformer reduces the voltage. The amount of the voltage reduction depends on the ratio of the number of turns in the primary coil versus the number of turns in the secondary coil. For example, if the secondary coil has half as many turns as the primary coil, the primary coil voltage will be cut in half at the secondary coil. In other words, if 120 VAC is applied to the primary coil, 60 VAC will be available at the secondary coil.

Common secondary voltages for transformers used in low-voltage power supplies range from 6 to 24 VAC. Note that because some voltage will be lost in the rectifier and filtering stages, you'll want to choose a secondary coil voltage that's a few volts higher than the final DC voltage your circuit actually needs. (Note, however, that the actual DC voltage level used for most circuits isn't all that critical. So if you're designing a power supply for a circuit that calls for 6 VDC and you use a transformer that provides 6 VAC in its secondary coil, the output from the power supply after it's rectified to DC voltage will be closer to 5 VDC. Most likely, 5 VDC will be close enough, and the circuit will work just fine.

Note that many transformers have more than one *tap* in the secondary coil. A tap is simply a wire connected somewhere in the middle of a coil, effectively dividing a single coil into two smaller coils. Multiple taps let you access several different voltages in the secondary coil. The most common arrangement is a *center-tapped* transformer, which provides two voltages as shown in Figure 3-1.

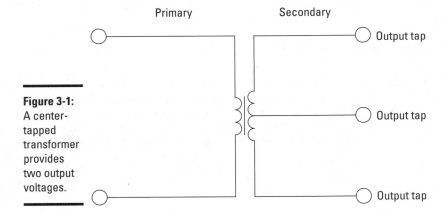

Primary   Secondary

Output tap

Figure 3-1:
A center-tapped transformer provides two output voltages.

Output tap

Output tap

In a center-tapped transformer, the voltage measured across the two outer taps is double the voltage measured from the center tap to either one of the two outer taps. Thus, if the voltage across the two outer taps is 24 VAC, the voltage across the center tap and either of the outer taps is 12 VAC.

It's important to note that when a transformer reduces voltage, it increases current. Thus, if a transformer cuts the voltage in half, the current will double. As a result, the overall power in the system (defined as the voltage multiplied by the current) remains the same.

If the current didn't increase as the voltage decreased, the transformer would violate a basic law of physics — the one about conservation of energy, which says that energy can't just disappear. That's a good thing. You don't want to be violating the laws of physics unless you know what you're doing or you're in a science fiction movie, in which case you can violate the laws of physics at will.

A transformer is strictly an alternating current device. That means:

✦ Transformers work only when alternating current is applied to the primary coil. If you apply direct current to the primary coil, no voltage will appear across the secondary coil. (Actually, there will be a brief spike of voltage across the secondary coil the moment voltage is applied to the primary coil, but in most circuits this fleeting voltage is insignificant.)

✦ A step-down transformer reduces the voltage from the primary to the secondary coils but *doesn't* convert alternating current to direct current. The voltage at the secondary coil is always AC.

✦ A transformer isolates the circuit attached from the secondary coil from the circuit connected to the primary coil. Thus, you can use a transformer to isolate your project from line voltage.

Turning AC into DC

The task of turning alternating current into direct current is called *rectification,* and the circuit that does the job is called a *rectifier.* The most common way to convert alternating current into direct current is to use one or more *diodes,* those handy electronic components that allow current to pass in one direction but not the other. Diodes are covered in detail in Book II, Chapter 5. You may want to briefly review that chapter before reading further if the concept doesn't sound familiar.

Although a rectifier converts alternating current to direct current, the resulting direct current isn't a steady voltage. It would be more accurate to refer to it as "pulsating DC." Although the pulsating DC current always moves in the same direction, the voltage level has a distinct ripple to it, rising and falling a bit in sync with the waveform of the AC voltage that's fed into the rectifier. For many DC circuits, a significant amount of ripple in the power supply can cause the circuit to malfunction. Therefore, additional filtering is required to "flatten" the pulsating DC that comes from a rectifier to eliminate the ripple. (For more on filtering, see the section, "Filtering Rectified Current," later in this chapter.)

There are three distinct types of rectifier circuits you can build: half-wave, full-wave, and bridge. The following sections describe each of these three rectifier types.

Half-wave rectifier

The simplest type of rectifier is made from a single diode, as shown in Figure 3-2. This type of rectifier is called a *half-wave rectifier* because it passes just half of the AC input voltage to the output. When the AC voltage is positive on the cathode side of the diode, the diode allows the current to pass through to the output. But when the AC current reverses direction and becomes negative on the cathode side of the diode, the diode blocks the current so that no voltage appears at the output.

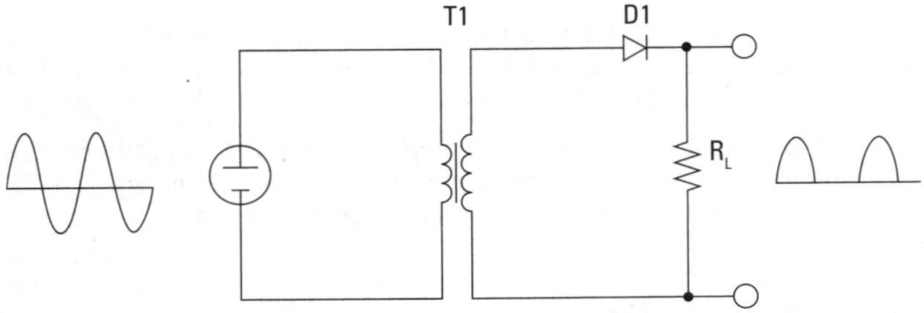

Figure 3-2:
A half-wave rectifier uses just one diode.

Half-wave rectifiers are simple enough to build but aren't very efficient. That's because the entire negative cycle of the AC input is blocked by a half-wave rectifier. As a result, output voltage is zero half of the time. This causes the average voltage at the output to be half of the input voltage.

Note the resistor marked R_L in Figure 3-2. This resistor isn't actually a part of the rectifier circuit. Instead, it represents the resistance imposed by the load that will ultimately be placed on the circuit when the power supply is put to use.

Full-wave rectifier

A *full-wave rectifier* uses two diodes, which enables it to pass both the positive and the negative side of the alternating current input. The diodes are connected to the transformer as shown in Figure 3-3.

Figure 3-3:
A full-wave
rectifier
uses two
diodes.

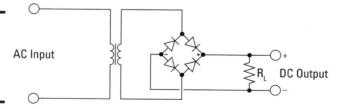

Notice that the full-wave rectifier requires that you use a center-tapped transformer. The diodes are connected to the two outer taps, and the center tap is used as a common ground for the rectified DC voltage. The full-wave rectifier converts both halves of the AC sine wave to positive-voltage direct current. The result is DC voltage that pulses at twice the frequency of the input AC voltage. In other words, assuming the input is 60 Hz household current, the output will be DC pulsing at 120 Hz.

Bridge rectifier

The problem with a full-wave rectifier is that it requires a center-tapped transformer, so it produces DC that's just half of the total output voltage of the transformer. A *bridge rectifier*, shown in Figure 3-4, overcomes this limitation by using four diodes instead of two. The diodes are arranged in a diamond pattern so that, on each half phase of the AC sine wave, two of the diodes pass the current to the positive and negative sides of the output, and the other two diodes block current. A bridge rectifier doesn't require a center-tapped transformer.

Figure 3-4:
A bridge
rectifier
uses four
diodes.

The output from a bridge rectifier is pulsed DC, just like the output from a full-wave rectifier. However, the full voltage of the transformer's secondary coil is used.

You can construct a bridge rectifier using four diodes, or you can use a bridge rectifier IC that contains the four diodes in the correct arrangement. A bridge rectifier IC has four pins: two for the AC input and two for the DC output.

**Book IV
Chapter 3**

**Building Power
Supplies**

Filtering Rectified Current

Although the output from a rectifier circuit is technically direct current because all of the current flows in the same direction, it isn't stable enough for most purposes. Even full-wave and bridge rectifiers produce direct current that pulses in rhythm with the 60 Hz AC sine wave that originates with the 120 VAC current that's applied to the transformer. And that pulsing current isn't suitable for most electronic circuits.

That's where filtering comes in. The filtering stage of a power supply circuit smoothes out the ripples in the rectified DC to produce a smooth direct current that's suitable for even the most sensitive of circuits.

Filtering is usually accomplished by introducing a capacitor into the power supply circuit, as shown in Figure 3-5. Here, the capacitor is simply placed across the DC output.

Figure 3-5:
A capacitor can be used to filter the output from the rectifier.

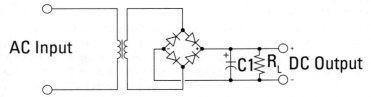

As you learn in Book II, Chapter 3, a capacitor has the useful characteristic of resisting changes in voltage. It accomplishes this magic feat by building up a charge across its plates when the input voltage is increasing. When the input voltage decreases, the voltage across the capacitor's plates decreases as well, but more slowly than the input voltage decreases. This has the effect of leveling out the voltage ripple, as shown in Figure 3-6.

Figure 3-6:
A filter circuit smooths the output voltage.

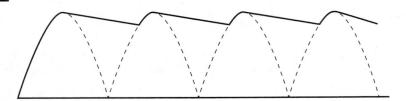

The difference between the minimum DC voltage and the maximum DC voltage in the filtering stage is called the *voltage ripple*, or just *ripple,* which is usually measured as a percentage of the average voltage. For example, a 10% ripple in a 5 V power supply means that the actual output voltage varies by 0.5 V.

The filter capacitor must usually be large to provide an acceptable level of filtering. For a typical 5 V power supply, a 2,200 µF electrolytic capacitor will do the job. The bigger the capacitor, the lower the resulting ripple voltage.

Don't forget to watch the polarity on electrolytic capacitors. The positive side of the capacitor must be connected to the positive voltage output from the rectifier, and the negative side must be connected to ground.

One way to improve the filter circuit is to use two capacitors in combination with a resistor, as shown in Figure 3-7. In this circuit, the first capacitor acts like the capacitor in Figure 3-6, eliminating a large portion of the ripple voltage. The resistor and second capacitor work as an RC network that eliminates the ripple voltage even further.

The advantages of this circuit are that the resulting DC has a smaller ripple voltage and the capacitors can be smaller. The disadvantage is that the resistor drops the DC output voltage. How much depends on the amount of current drawn by the load. For example, if you use a 100 Ω resistor and the load draws 100 mA, the resistor will drop 10 V (100×0.1). Thus, to provide a final output of 5 V, the rectifier circuit must supply 15 V because of the 10 V drop introduced by the resistor.

Figure 3-7:
Two capacitors and a resistor cut ripple voltage but also reduce the DC output voltage.

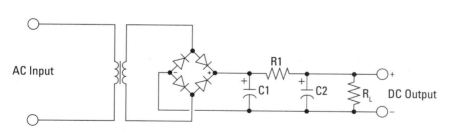

You can also use an inductor in a filter circuit, as shown in Figure 3-8. Unlike a resistor-capacitor filter, an inductor-capacitor filter doesn't significantly reduce the DC output voltage. Although inductor-capacitor filter circuits create the smallest ripple voltage, inductors in the range needed (typically 10 henrys) are large and relatively expensive. Thus, most filter circuits use a single capacitor or a pair of capacitors coupled with a resistor.

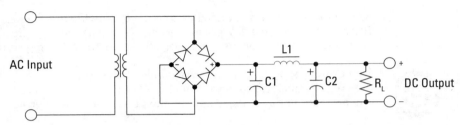

Regulating Voltage

The purpose of a power supply is to provide power for an electronic circuit. There's a basic formula for calculating the amount of power circuit uses:

$$P = I \times V$$

Power, measured in watts, is equal to current measured in amperes times voltage measured in volts.

If you know any two of these three elements for a circuit, you can easily calculate the third. For example, if you know that the current is 0.5 A and the voltage is 10 V, you can calculate that the circuit consumes 5 W of power by multiplying 0.5 by 10.

For a given amount of power, there's an inverse relationship between voltage and current. Whenever current increases, voltage must decrease, and whenever current decreases, voltage must increase. This simple fact, unfortunately, has an adverse effect on power supply circuits. When you connect a voltmeter to the output terminals of a power supply, the meter itself draws an almost insignificant amount of current, so the meter reads very close to the voltage you expect to obtain from the power supply.

However, if you connect a circuit that draws significant current from the power supply, the voltage from the power supply will drop in proportion to the current. Depending on the nature of the circuit you're connecting to the power supply, this voltage drop may or may not be a bad thing. Some circuits designed for 12 VDC will work fine if only given 9 VDC. But other circuits are sensitive to the input voltage, so the power supply needs to work harder to make sure it delivers the desired voltage.

To maintain a steady voltage level regardless of the amount of current drawn from a power supply, the power supply can incorporate a *voltage regulator* circuit. The voltage regulator monitors the current drawn by the load and increases or decreases the voltage accordingly to keep the voltage level constant.

A power supply that incorporates a voltage regulator is called a *regulated power supply*.

You can, if you want, design your own voltage regulator circuit using a couple of transistors, some resistors, and a Zener diode. However, it's far too easy to buy one of the many available integrated circuit voltage regulators. Voltage regulator ICs are inexpensive (under two dollars) and, with just three pins to connect, easy to incorporate into your circuits.

The most popular type of voltage regulator IC is the *78XX* series, sometimes called the *LM78XX* series. These voltage regulators combine 17 transistors, three Zener diodes, and a handful of resistors into one handy package with three pins and a heat sink that helps dissipate the excess power consumed by the regulator as it compensates for increases or decreases in current draw to keep the voltage at a constant level.

The last two digits of the 78XX ID number indicate the output voltage regulated by the IC. The most popular models are:

Model	*Voltage*
7805	5
7806	6
7809	9
7810	10
7812	12
7815	15
7818	18
7824	24

Of these, the most common are the 7805 (5 V) and 7812 (12), which are available at most RadioShack stores.

To use a 78XX voltage regulator, you just insert it in series on the positive side of the power supply circuit and connect the ground lead to the negative side as shown in Figure 3-9. As this figure shows, it's also a good idea to place a small capacitor (typically 1µF) after the regulator.

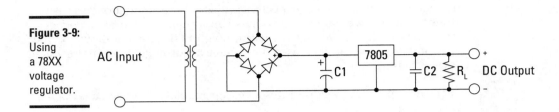

Figure 3-9:
Using
a 78XX
voltage
regulator.

You must supply a voltage regulator with about 3 V more than the regulated output voltage. Thus, for a 7805 regulator, you should give it at least 8 V. The maximum input voltage for a 7805 is 30 V. Remember that the diodes in a bridge rectifier will drop about 3 V from the transformer output, so you'll need a transformer whose secondary delivers at least 11 V to produce 5 V of regulated output.

Eleven-volt transformers are rare, but 12 V transformers are readily available. Thus, a 5 V regulated power supply starts with a 12 VAC transformer that delivers 12 V to the bridge rectifier, which converts the AC to DC and drops the voltage down to about 9 V and then delivers the voltage to the filter circuit, which smoothes out the ripples and passes the voltage on to the 7805 voltage regulator, which holds the output voltage at 5 V.

Another popular voltage regulator IC is the LM317, which is an adjustable voltage regulator. An LM317 regulator works much like a 78XX regulator, except that instead of connecting the middle lead directly to ground, you connect it to a voltage divider built from a pair of resistors, as shown in Figure 3-10. The value of the resistors determines the regulated voltage. In Figure 3-10, I used a potentiometer so that the user can vary the output voltage by adjusting the potentiometer.

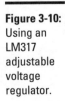
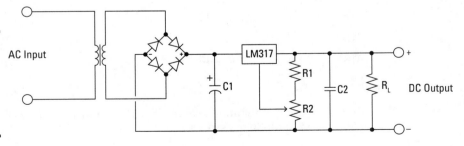

Figure 3-10:
Using an
LM317
adjustable
voltage
regulator.

Book V

Working with Radio and Infrared

"The sensor in my running shoe is transmitting information and encouragements to my iPod. Right now, Lance Armstrong is encouraging me to stop running like a girl."

Contents at a Glance

Chapter 1: Understanding Radio

In This Chapter

✔ How radio waves work

✔ How transmitters and receivers work

✔ The difference between AM and FM radio

✔ Some interesting radio history

*1*t's pet peeve time! One of my pet peeves is the phrase, "The Golden Age of Radio," which implies that the height of radio's popularity was in the 1930s and '40s and that radio has declined in popularity since then.

In reality, radio — the technology if not the audio programming — has never been more popular than it is right now. In the 1930s and '40s, there was only one use for radio: broadcast audio signals. Today, audio broadcast over radio is as commonplace as ever, but the list of other types of information being broadcast by radio technology has skyrocketed.

First, there was broadcast television, which is nothing more than the combination of audio and video broadcast over radio. Then, there were cellphones, which use radio to extend the world's telephone networks to places that phone cables can't reach. Then there was wireless networking, which replaced bulky computer network cables with data transmitted over radio. And now there are cellular data plans, which transmit Internet data over radio. And there are many other popular uses for radio technology, including radar, GPS navigation systems, and wireless Bluetooth devices.

In this chapter, you learn some of the basic concepts of radio, including what it is, how it works, and how it was discovered. Along the way, you learn a few interesting — and in some cases sad — stories about the early pioneers of radio.

This chapter lays the important foundation for the next two chapters, in which you learn how to build circuits that receive and play radio signals broadcast in the AM radio band.

Throughout this chapter, I occasionally violate my own pet peeve and refer to radio as if it's only for the broadcast of sound. Whenever I do, just keep in mind that radio is also used for broadcasting video and digital data as well as other types of information.

Pet peeves are interesting things, aren't they? I think peeves do make interesting pets. They're not as cute as puppies or kittens, but if you get a pet peeve when it's young, it can bring you a lifetime of pleasure. Be careful, however. Like dogs that bark late at night or cats that roam in other people's yards, a pet peeve may annoy your friends and neighbors.

I have always wondered why cities and counties require that you register pet dogs and cats but don't require that you register pet peeves. Just think of the revenue local governments are missing out on! Even a fee as low as $10 per peeve could raise enough money to fill in the potholes on all those roads paved with good intentions, and a portion of the money could go to care for homeless peeves.

Understanding Radio Waves

Most people think of radio as wireless broadcast of sound, most often music and speech. But the term *radio* is actually much broader than that; the broadcast of sound is actually just one application of the extremely useful electrical phenomenon that is called radio.

Radio takes advantage of one of the most interesting of all electrical phenomena: *electromagnetic radiation* (often abbreviated *EMR*), which is a type of energy that travels in waves at the speed of light. EMR travels freely through the air and even in the vacuum of space.

EMR waves can oscillate at any imaginable frequency. The rate of the oscillation is measured in cycles per second, also known as *hertz* (abbreviated Hz). The term *hertz* here does *not* refer to the car rental company. Instead, it honors the great German physicist Heinrich Hertz, who was the first person to build a device that could create and detect radio waves.

Radio is simply a specific range of frequencies of EMR waves. The low end of this range is just a few cycles per second, and the upper end is about 300 billion cycles per second (also known as *gigahertz*, abbreviated *GHz*.) That's a pretty big range, but EMR waves with much higher frequencies exist as well, and are in fact commonplace. EMR waves with frequencies higher than radio waves go by various names, including infrared, ultraviolet, X-rays, gamma rays, and — most importantly — visible light.

That's right; what we call light is exactly the same thing as what we call radio, but at higher frequencies. The frequency of visible light is measured in billions of hertz, also called *terahertz* and abbreviated *THz*. The low end of visible light (red) is around 405THz and the upper end (violet) is around 790 THz.

Who really invented radio?

The history of radio technology is plagued by controversy over the question of who actually invented the thing. The answer most often given is Italian inventor Guglielmo Marconi, but many others made important discoveries that give them good claim to the title.

Here's a rundown on the contenders for the title of The Father of Radio:

- **Marconi:** He was the first person to demonstrate radio successfully and exploit it commercially. In 1901, Marconi sent a message via radio across the Atlantic from England to Canada, though the message was faint, consisted of nothing but the letter *S*, and reception of the message wasn't independently confirmed. Nevertheless, Marconi's accomplishment was astonishing, and he made many important contributions to the technology and business of radio.

- **Tesla:** In 1943, the U.S. Supreme Court ruled that many of Marconi's important radio patents were invalid because Nikola Tesla had already described the devices covered by Marconi's patents. Tesla was a brilliant engineer who is best known for being the champion of alternating current over direct current for power distribution. He publicly demonstrated wireless communication devices as early as 1893. Tesla believed that wireless technology would be used not only for communication, but for power distribution as well.

- **Lodge:** In England, Sir Oliver Lodge was building wireless telegraph systems in the mid-1890s.

- **Popov:** In Russia, Alexander Stepanovich Popov was demonstrating wireless telegraph transmissions around the same time as Lodge.

- **Bose:** In India, Sir Jagadish Chandra Bose was also demonstrating wireless telegraph transmissions in the early 1890s. Whether these demonstrations occurred before, after, or at the same time as other demonstrations by Lodge, Popov, and others is under dispute.

- **Many others:** The list of names of others who did important research and made important discoveries in the last decades of the nineteenth century is long: Heinrich Hertz, Edouard Branly, Roberto de Moura, Ernest Rutherford, Ferdinand Braun, Julio Baviera, and Reginald Fessenden are just a few of the many individuals who made important contributions.

So it seems that no one person has a clear-cut claim to being the first to invent radio. Work was being done all around the world and discoveries were being made it seems every day.

It may be that the best answer is that no one "invented" radio. Radio is a natural phenomenon. It was *discovered*, not *invented*.

What was invented were ways to exploit the phenomenon of radio by building devices that could generate radio waves and modulate them to add information, as well as devices that could receive radio waves and extract the information that was added.

So here's an interesting thought to ponder: Radio stations broadcast on a specific frequency. For example in San Francisco, there's a popular radio station called KNBR, which has been broadcasting on the frequency 680 kHz since 1922. There are plenty of other radio stations in the area, but only KNBR broadcasts at 680 kHz.

The term *channel* is often used to refer to a radio station broadcasting at a particular frequency. For example, if someone asks me what radio channel I like to listen to when I'm in San Francisco, I would tell them KNBR.

I might listen to KNBR on my portable radio, which happens to be made of purple plastic. *Purple* is the color we perceive when we see light whose frequency is right around 680 THz. There are many other colors, but only the color purple is at 680 THz. So in a way, color is the same thing as channel. If EMR waves are vibrating at 680 kHz, they are KNBR radio. If those same EMR waves are vibrating a million times faster, at 680 THz, they are the color purple.

An important concept that's related to frequency is the idea of wavelength. The term *wavelength* refers to the distance between the crests of each cycle of EMR at a particular frequency. Because EMR waves travel at the speed of light, you can calculate the wavelength of a given frequency by dividing the distance that light travels in a single second by the number of cycles per second.

Light is pretty fast: It scoots along at 186,282 miles per second. Thus, the wavelength of an EMR wave oscillating at 100 kHz is about 1.86 miles: 186,282 divided by 100,000.

The higher the frequency, the shorter the wavelength. The wavelength of most AM broadcast radio stations is a few hundred feet. The wavelength of visible light is a very small fraction of an inch.

Sound waves are not a type of EMR. Sound waves are created when particles of matter bump against each other. Thus, you must have matter — such as air or water — to transmit sound waves. Radio waves don't require particles of matter to travel. In fact, radio waves travel best in the vacuum of space, where there's no matter to get in the way.

Transmitting and Receiving Radio

There are many natural sources of radio waves. But in the later part of the 19th century, scientists figured out how to generate radio waves using electric currents. In a nutshell, if you pass an alternating current into a length of wire, radio waves at the same frequency as the alternating current are generated.

Two components are required for radio communication: a *transmitter* and a *receiver*. The transmitter generates radio waves, and the receiver detects them. The following sections describe the basic operation of radio transmitters and receivers.

Understanding radio transmitters

A radio transmitter consists of several elements that work together to generate radio waves that contain useful information such as audio, video, or digital data. These components are shown in Figure 1-1 and described here:

✦ **Power supply:** Provides the necessary electrical power to operate the transmitter.

✦ **Oscillator:** Creates alternating current at the frequency on which the transmitter will transmit. The oscillator usually generates a sine wave, which is referred to as a *carrier wave.*

✦ **Modulator:** Adds useful information to the carrier wave. There are two main ways to add this information. The first, called amplitude modulation or AM, makes slight increases or decreases to the intensity of the carrier wave. The second, called frequency modulation or FM, makes slight increases or decreases the frequency of the carrier wave. For more information about AM and FM, see the sections "Understanding AM Radio" and "Understanding FM Radio" later in this chapter.

(Actually, there is a third method of adding information to a radio signal: by simply turning the signal on and off in a pattern that represents the information. For example, radio signals can send Morse code in this way.)

✦ **Amplifier:** Amplifies the modulated carrier wave to increase its power. The more powerful the amplifier, the more powerful the broadcast.

✦ **Antenna:** Converts the amplified signal to radio waves.

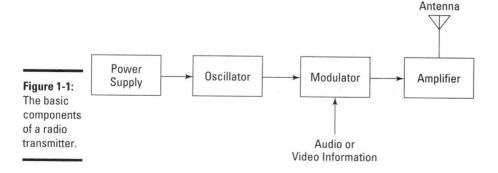

Figure 1-1:
The basic components of a radio transmitter.

Understanding radio receivers

A radio receiver is the opposite of a radio transmitter. It uses an antenna to capture radio waves, processes those waves to extract only those waves that are vibrating at the desired frequency, extracts the audio signals that were added to those waves, amplifies the audio signals, and finally plays them on a speaker. Figure 1-2 shows these components, and the following paragraphs explain how each works:

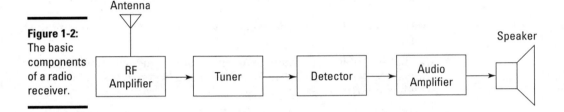

Figure 1-2: The basic components of a radio receiver.

+ **Antenna:** Captures the radio waves. Typically, the antenna is simply a length of wire. When this wire is exposed to radio waves, the waves induce a very small alternating current in the antenna.

+ **RF amplifier:** A sensitive amplifier that amplifies the very weak radio frequency (RF) signal from the antenna so that the signal can be processed by the tuner.

+ **Tuner:** A circuit that can extract signals of a particular frequency from a mix of signals of different frequencies. On its own, the antenna captures radio waves of all frequencies and sends them to the RF amplifier, which dutifully amplifies them all. Unless you want to listen to every radio channel at the same time, you need a circuit that can pick out just the signals for the channel you want to hear. That's the role of the tuner.

The tuner usually employs the combination of an inductor (for example, a coil) and a capacitor to form a circuit that resonates at a particular frequency. This frequency, called the *resonant frequency,* is determined by the values chosen for the coil and the capacitor. This type of circuit tends to block any AC signals at a frequency above or below the resonant frequency.

You can adjust the resonant frequency by varying the amount of inductance in the coil or the capacitance of the capacitor. In simple radio receiver circuits such as the one you learn about in the next chapter, the tuning is adjusted by varying the number of turns of wire in the coil. More sophisticated tuners use a variable capacitor (also called a *tuning capacitor*) to vary the frequency.

Looking at the radio spectrum

The term *spectrum* simply means a range of frequencies. Radio is generally considered to be frequencies between 3 Hz to 300 GHz. That broad range of frequencies is carved up into smaller pieces that are used for specific types of radio, as described in the following table:

Frequency	Abbreviation	Description
3–30 Hz	ELF	Extremely low frequency, used for communications with submarines.
30–300 Hz	SLF	Super low frequency, also used for communications with submarines.
300 Hz–3 kHz	ULF	Ultra low frequency, used for underground communications within mines.
3–30 kHz	VLF	Very low frequency, also for submarine communications and a few other unusual applications.
30–300 kHz	LF	Low frequency, used for navigation, RFID, and a few other applications.
300–3,000 kHz	MF	Medium frequency, used for AM radio.
3–30 MHz	HF	High frequency, used for shortwave and CB radio.
30–300 MHz	VHF	Very high frequency, used for FM radio and television.
300–3,000 MHz	UHF	Ultra high frequency, used for television, mobile phones, wireless networking, Bluetooth, and so on.
3–30 GHz	SHF	Super high frequency, used for high-speed wireless networking, radar, and communication satellites.
30–300 GHz	EHF	Extreme high frequency, used for microwave communications.
300–3,000 THz	THF	Tremendously high frequency (no I didn't make that up), used for exotic applications that border on science fiction.
Above 3,000 GHz		You're off the edge of the spectrum map, mate. Here there be dragons.

✦ **Detector:** Responsible for separating the audio information from the carrier wave. For AM signals, this can be done with a diode that just rectifies the alternating current signal. What's left after the diode has its way with the alternating current signal is a direct current signal that can be fed to an audio amplifier circuit. For FM signals, the detector circuit is a little more complicated.

✦ **Audio amplifier:** This component's job is to amplify the weak signal that comes from the detector so that it can be heard. This can be done using a simple transistor amplifier circuit as described in Book II, Chapter 6. You can also use an op-amp IC as described in Book 3, Chapter 3.

Of course, there are many variations on this basic radio receiver design. Many receivers include additional filtering and tuning circuits to better lock on to the intended frequency — or to produce better-quality audio output — and exclude other signals. Still, these basic elements are found in most receiver circuits.

Understanding AM Radio

The original method of encoding sound information on radio waves is called *amplitude modulation*, or *AM*. It was developed in the first few decades of the twentieth century. AM is a relatively simple way to add audio information to a carrier wave so that sounds can be transmitted.

One of the simplest forms of AM modulators simply runs the power supply for an oscillator circuit through an audio transformer that is coupled to a microphone or other sound source. Figure 1-3 shows this arrangement.

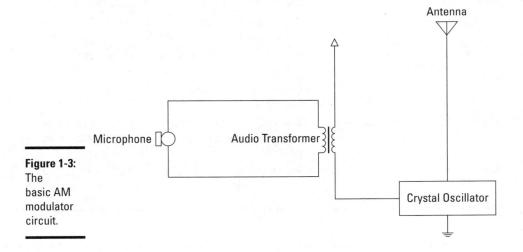

Figure 1-3:
The basic AM modulator circuit.

The circuit in Figure 1-3 uses a 1 MHz *crystal oscillator*, which is often used to generate the clock frequencies for microprocessor circuits. 1 MHz is perfect for a simple AM transmitter circuit because 1 MHz falls right in the middle of the band that's used for AM radio transmissions.

Although you can't buy a crystal oscillator at your local RadioShack, you can get it on the Internet. Use Google to search for *1 MHz crystal oscillator,* and you should find several online sources that will sell you one for under $2.

The crystal oscillator is contained in a metal can that has three pins. One pin is for ground, the second pin is the supply voltage (typically 9 VDC), and the third is the oscillator output.

By running the Vss supply through the secondary coil of a transformer whose primary coil is connected to an audio input source such as a microphone, the actual voltage supplied to the oscillator will fluctuate based on the variations in the input signal. Because crystal oscillators are very stable, these voltage variations won't affect the frequency generated by the oscillator, but they will affect the voltage of the oscillator output. Thus, the audio input signal will be reflected as voltage changes in the oscillator's output signal.

A better AM modulation circuit uses a transistor as shown in Figure 1-4. In this circuit, the carrier-wave generated by an oscillator that isn't shown in the circuit is applied to the base of a transistor. Then, the audio input is applied to the transistor's emitter through a transformer. The AM signal is taken from the transistor's collector.

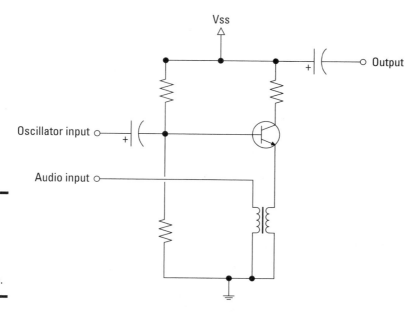

Figure 1-4:
Using a transistor for amplitude modulation.

So how does this circuit work? The transistor amplifies the input from the oscillator through the emitter-collector circuit. However, as the audio input varies, it induces a small current in the secondary coil of the transformer. This, in turn, affects the amount of current that flows through the collector-emitter circuit. In this way, the intensity of the output varies with the audio input.

Figure 1-5 shows how a carrier wave is combined with an audio signal to produce an AM radio waveform. As you can see, the carrier wave is a constant frequency and amplitude. In other words, each cycle of the sine wave is of the same intensity. The current of the audio wave varies, however. When the two are combined by the modulator circuit, the result is a signal with a steady frequency, but the intensity of each cycle of the sine wave varies depending on the intensity of the audio signal.

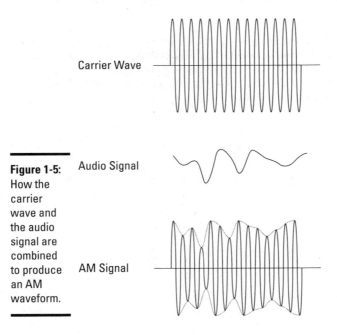

Carrier Wave

Audio Signal

AM Signal

Figure 1-5: How the carrier wave and the audio signal are combined to produce an AM waveform.

Understanding FM Radio

AM radio is relatively simple. However, it has several weaknesses. The main drawback of AM radio is that it's difficult, if not impossible, for an AM radio receiver to distinguish between a signal broadcast by a radio transmitter and spurious signals at the same frequency generated by other sources. The most obvious example of this is lightning. When lightning strikes, it generates a brief but powerful burst of electromagnetic radiation with a very large spectrum of frequencies. The noise generated by a lightning strike includes just about the entire range of frequencies used by AM radio. If you're listening to an AM radio station when the lightning strikes, the sudden burst of radio energy on the frequency you're listening to will be interpreted as sound. Thus, when lightning strikes, you can hear it on the radio.

Signals that interfere with an intentional broadcast are called *static*, and static is the main drawback of AM radio. To counteract static, a better method of superimposing information on a radio wave, called *frequency modulation* or *FM*, was developed in 1933. (See the sidebar titled "The tragic genius behind FM radio" for the fascinating and sad story about the inventor of FM radio.)

In frequency modulation, the intensity of the carrier wave isn't varied. Instead, the exact frequency of the carrier wave is varied in sync with the audio signal. When the audio signal is higher, the frequency of the broadcast signal goes up a little. When the audio signal is lower, the frequency slows down a bit.

Figure 1-6 shows how this appears in a graph. At the top of the figure, you can see the carrier wave that clocks the specific frequency of the broadcast station. In the middle, you can see the audio signal that is to be superimposed on the carrier wave. And at the bottom, you can see the resulting modulated signal. As you can see, the frequency decreases when the input signal gets lower and increases when the input signal is higher.

Carrier Wave

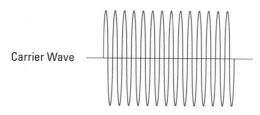

Figure 1-6: How the carrier wave and the audio signal are combined to produce an FM waveform.

Audio Signal

FM Signal

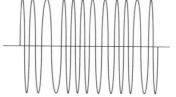

The tragic genius behind FM radio

One of the great inventors in the history of radio was a brilliant engineer named Edwin H. Armstrong. Born in 1890, he was fascinated with electrical technology from a very young age. At the age of 14, he started experimenting with wireless radio circuits, building an antenna in his family's backyard that was more than 100' tall.

He made his first major contribution to radio technology while he was a junior at Columbia University in 1912. His invention was a circuit that amplified incoming radio signals by feeding them back though the amplifier tube in what came to be called a *regenerative circuit*. It was an important early breakthrough in radio technology that for the first time allowed radio to be heard through a speaker rather than with headphones.

During World War I, Armstrong invented another type of radio receiver, which he called the *superheterodyne circuit*. The basic principal of the superheterodyne circuit is that a radio signal broadcasting at a high frequency — say 1,500 kHz, can be combined with a nearby frequency from an oscillator — say, 1,560 kHz, in such a way that the original signal could also be detected at 60 kHz — the difference between the original signal's frequency (1,560 kHz) and the oscillator's frequency (1,500 kHz). The superheterodyne circuit may be one of the most important electronic circuits ever invented. It's still used in nearly all radio receivers to this day.

His third great invention came in 1933, when he created a method for transmitting radio signals that wasn't subject to interference from atmospheric disturbances like lightning. His new system was called *frequency modulation*. We know it as FM radio.

Armstrong patented his inventions, but his patents were challenged or ignored by the titans of radio. He lost his lawsuit to protect his patent for his regenerative circuit in 1934 because the justices of the Supreme Court didn't understand how the circuit worked, and the industry challenged his FM radio patents and used his technology freely throughout the 1940s and 1950s.

Finally, in 1954, ill and broke from his legal battles, Armstrong committed suicide by jumping from his high-rise apartment window.

Eventually his widow, Marion, won a series of patent lawsuits and was awarded damages of $10 million.

Note that the frequency variations in a frequency-modulated signal are all within a small proportion of the carrier-wave frequency. Typically, the frequency stays within 100 kHz of the base frequency.

FM radio stations broadcast at frequencies in the range of 88 to 108 MHz, but the base frequency for each station always ends in 0.1, 0.3, 0.5, 0.7, or 0.9. That's why FM radio stations have frequencies such as 89.3 or 107.5, but never 92.0 or 98.6.

Assigning base frequencies in increments of 0.2 MHz gives each station 100 kHz of room on either side of the center frequency for its frequency modulation. Thus, a station broadcasting at 103.1 actually sends signals whose frequencies range from 103.0 to 103.2. Most stations limit the variation from the base frequency to ±75 kHz to leave some margin for error. This helps prevent adjacent stations from interfering with one another.

FM modulators usually use a type of electronic component called a *varactor*, which is a type of diode that has an unusual characteristic: It has capacitance like a capacitor, and its capacitance increases when voltage is applied across the diode. In essence, a varactor is a voltage-controlled variable capacitor. The schematic symbol for a varactor, shown in the margin, looks like a cross between a diode and a capacitor.

Varactors can be used in oscillator circuits to create an oscillator that vibrates faster when voltage increases. This ability makes it ideal for an FM radio modulator. As the voltage of the audio input increases, the capacitance of the varactor increases and thus the frequency of the oscillator increases. When the voltage decreases, the capacitance of the varactor decreases and so does the oscillator's frequency. Figure 1-7 shows a sample of an FM modulator circuit that uses a varactor.

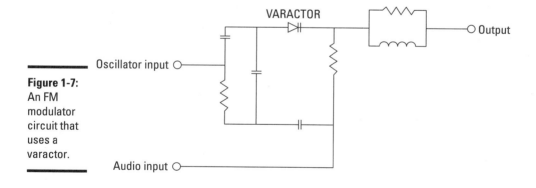

Figure 1-7:
An FM modulator circuit that uses a varactor.

Chapter 2: Building a Crystal Radio

In This Chapter

✔ **Understanding how crystal radios work**

✔ **Gathering materials to make a crystal radio**

✔ **Winding up your coil**

✔ **Building the circuit**

✔ **Creating an antenna and a ground**

✔ **Listening to your radio**

*I*n this chapter, you learn how to build one of the simplest of all useful electronic circuits: a *crystal radio*, which is a radio receiver that can receive AM radio broadcasts. It's not a particularly good radio receiver. Only one person at a time can listen to it because it uses a headphone instead of a speaker. And it's not very sensitive; you'll be lucky if you can receive two or three different stations even if dozens of AM radio stations broadcast in your area.

But what makes the crystal radio unique is that, unlike every other electronic circuit described in this book, a crystal radio has no obvious source of power, no batteries or other power supply. The only source of power used by a crystal radio is the power present in the radio waves themselves.

Crystal radios have been around since the very beginning of radio broadcasting. In the 1920s, it was common for people to build their own crystal radio receivers. Newspapers and magazines published articles telling readers how to construct crystal radios using mostly household items such as scraps of wood and metal and empty oatmeal boxes, plus a few specialty items including a crystal and a telephone headset.

The total cost for a 1920s crystal radio was around $10. That sounds cheap, but adjusted for inflation, that's more like $125 today. Fortunately, the total cost for a crystal radio today is still around $10.

If you prefer, you can buy a kit to build your own crystal radio. Amazon sells a nice kit for about $25, and you can find crystal radio kits at many local hobby or school-supply stores. Although you can probably round up the parts separately for less than the cost of a kit, a few of the parts are relatively hard to find outside of a kit. (One option you may want to consider is to buy a kit to get these few hard-to-get parts, but build the radio using the instructions found in this chapter.)

Looking at a Simple Crystal Radio Circuit

Figure 2-1 shows a basic crystal radio receiver circuit. As you can see, this circuit consists of just a few basic components: an antenna and a ground connection, a coil, a variable capacitor, a diode, and an earphone.

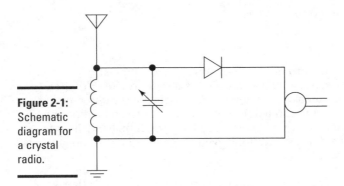

Figure 2-1:
Schematic
diagram for
a crystal
radio.

The antenna, of course, captures the radio waves travelling through the air and converts them into alternating current. In order for current to flow, a complete circuit is required. The ground connection is what completes the circuit, allowing current to flow.

The combination of the coil and the capacitor form the tuning circuit. The inductance of the coil combines with the capacitance of the variable capacitor to create a circuit that resonates at a particular frequency, allowing that frequency to pass but blocking other frequencies. In a basic crystal radio such as the one shown in Figure 2-1, the tuning circuit isn't very precise. As a result, you'll probably hear several stations at once. However, it is possible to build more sensitive tuning circuits that can hone in on individual stations.

The diode forms the detector part of the circuit. It simply converts the alternating current signal that comes from the antenna and tuning circuit to direct current. This direct current is extremely small, but it is enough to drive a sensitive piezoelectric earphone, which converts the current to sound.

So that, in a nutshell, is how a crystal radio works. The rest of this chapter shows you how to build a crystal radio of your own.

Figure 2-2 shows the finished crystal radio that you can build in this project. Note that there are many ways to build a crystal radio, so the instructions that follow in this chapter are by no means definitive. Use your imagination when looking for materials to build your radio.

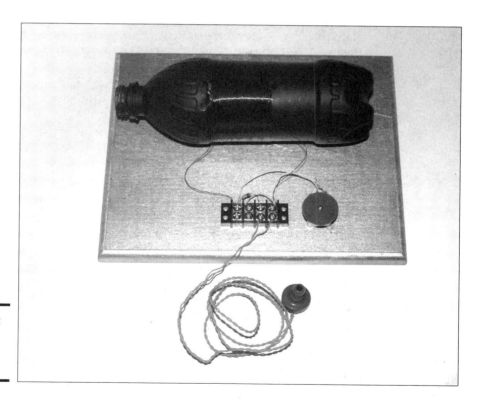

Figure 2-2:
A finished crystal radio.

Gathering Your Parts

You'll need a handful of parts to build your crystal radio. The following is a recommended list:

✦ At least 50 feet of **antenna wire.** You can use almost any wire for the antenna. I prefer 18-gauge solid hook-up wire, which you can buy at RadioShack.

✦ A few feet of **hook-up wire** to connect the radio to a ground connection.

✦ At least 50 feet of 30-gauge, enamel-coated **magnet wire.** You can buy this at RadioShack.

✦ Something to wrap the coil on. I used an empty soda bottle.

✦ A **variable capacitor,** also called a *tuning capacitor.* These are getting hard to get, as RadioShack no longer carries them. However, you can easily harvest one out of an old radio that doesn't work, or you can purchase them online.

Note that the variable capacitor is an optional component. If you can't find one, you can still build your crystal set without one; you just won't be able to tune out competing stations.

✦ A **germanium diode.** You can't buy these at RadioShack, but you can order them over the Internet. Just use your favorite search engine to search for *1N34A*, and you'll find several suppliers.

✦ A **piezoelectric earphone.** Regular earphones like the kind used with an iPod or cell phone won't work. Search for *piezoelectric earphone* and you'll find several suppliers that sell them for about $3.

✦ A **board** to mount the radio on. About 6 by 9" should be sufficient.

✦ Something to make your electrical connections. I like to use a four-pole barrier strip from RadioShack (part number 274-658).

If you have an old radio that doesn't work lying around, feel free to open it up and harvest its parts. In particular, look for the tuning capacitor. You'll be able to spot it easily because it will be connected to the radio's tuning knob.

The germanium diode, tuning capacitor, and piezoelectric earphone are the three parts that are a bit difficult to find. You may want to purchase a crystal radio kit from a hobby or school-supply store and harvest those three parts from the kit.

Building the Coil

When you look at a crystal radio, the first thing you're likely to notice is the large coil. The coil usually consists of 100 turns or more of small-gauge magnet wire wrapped around a non-conductive tube anywhere from one to five inches in diameter. The coil is an essential part of the radio's tuning circuit.

Many different types of materials can be used to wrap the coil around. Here are a few ideas:

✦ An empty 16 oz soda bottle. That's what I used for the radio built in this chapter. To make the coil look better, you may first want to spray-paint the bottle with black paint. See Figure 2-3. (Don't use metallic paint!)

✦ An empty toilet-paper roll.

✦ An empty oatmeal container.

✦ An empty bottle of contact-lens fluid or another similarly sized plastic bottle.

✦ A 6" length of 2 or 3" diameter PVC sprinkler pipe.

✦ A 6" length of a wooden closet rod.

✦ A 6" length of a cardboard mailing tube.

In short, any sturdy cylindrical object that is made of an insulating material can be used as the core of your coil. As long as it's cylindrical and not made of metal, you can use it.

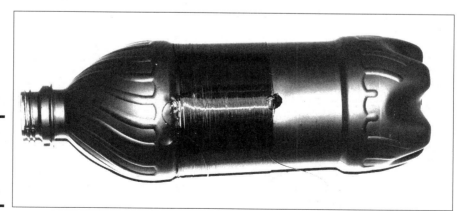

Figure 2-3:
A coil
wound on
an empty
soda bottle.

As for the choice of wire to use for the coil, the most common is magnet wire, which is coated with thin enamel insulation rather than encased in plastic insulation. Wire wrapped with plastic insulation will work, but enamel insulation is thinner and thus allows the turns to be spaced closer together.

The number of turns in the coil and the diameter of the cylinder you wrap the coil around will determine how much wire you need. You'll want to wrap at least 100 turns. To determine how much wire you'll need for each turn, multiply the diameter of the cylinder by 3.14. Then, multiply the result by the number of turns, and divide by 12 to determine how many feet of wire you'll need.

For example, suppose you're wrapping the coil around a 2" cylinder and you want to wrap 100 turns. In this case, each turn will require 6.28" of wire (2" × 3.14). So you'll need just over 52' of wire (6.28" × 100 ÷ 12"). Allowing a foot or so of extra wire at each end of the coil to connect the coil to the radio circuit, you'll need about 54' of wire for the coil.

For the simplest type of crystal radio, the exact number of turns doesn't affect the operation of the radio significantly. So in the example here, if you have a 50' spool of wire, you can just wind the coil a few turns short of 100, and the radio will work just as well.

The easiest way to wind a coil is to place the tube you're winding the coil around on a screwdriver blade or other long narrow object so that the tube will spin freely. That way, you can turn the tube and slowly feed wire from its spool onto the tube. This keeps the wire from becoming twisted as you wind the coil. If you have a vice on your workbench, you can clamp the screwdriver horizontally in the vice, and then slide the tube onto the screwdriver so that the tube will spin freely.

To start the coil, you must first attach one end of the magnet wire to one end of the tube. The easiest way to do that is with a dab of hot glue. If you prefer, you can punch a hole through the tube and feed the wire through. Either way, be sure to leave six inches or more of wire free. This will give you plenty of wire to connect the coil to the circuit when you finish winding the coil.

Once you've attached one end of the coil, turn the tube slowly while feeding wire from the spool onto the tube. Each half turn or so, use your fingers to carefully scoot the wire you just fed up against the turns you've already wound. The goal is for each turn of wire to be adjacent to the previous turn, with no gaps between the turns.

You'll find that you need to keep a bit of tension on the wire as you feed it onto the tube in order to keep the windings nice and tight. If you slip and

TIP

let go of the tension, several turns may unravel, and you'll have to untangle them to restore the coil's tightness.

It helps if you wrap the coil in sections of about ten turns each. When you finish each section, dab a little hot glue on it to hold it in place.

When you reach the end of the tube (or run out of wire), use a little hot glue to secure the last turn of the coil, or cut a slit in the tube and slide the wire through it. Be sure to leave about six inches of free wire after the last turn.

When you're done, the coil should have a nice, tight appearance with no major gaps between the turns, and you should have about six inches of free wire on each end of the coil.

Assembling the Circuit

Once you have your coil, the next step is to assemble the various parts of the radio on a base. I recommend you use a piece of wood about 6 by 9". To make your radio look good, consider painting or staining the wood before you assemble the circuit.

Here's a list of the parts you'll need to assemble the circuit:

+ The **coil** you assembled according to the instructions in the previous section of this chapter
+ A four-position **barrier strip**
+ A **germanium diode** (1N34A or similar)
+ A **tuning capacitor** (optional)
+ One length of **hook-up wire,** approximately 1.5" long
+ Two lengths of **hook-up wire,** approximately 3" long

You'll need the following tools to build the crystal radio circuit:

+ **A hot glue gun and some glue sticks**
+ **A Phillips-head screwdriver**
+ **Wire cutters**
+ **Wire strippers**
+ **Soldering iron and some solder**

Figure 2-4 shows the layout for the assembled crystal radio circuit.

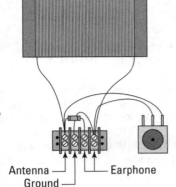

Figure 2-4:
Layout for
the crystal
radio circuit.

Antenna ——
Ground ——

—— Earphone

In the instructions here, I refer to the individual terminals of the barrier strip by numbering them from left to right, 1 through 4. Terminal connectors in the top row are given the letter A, and those in the bottom row are given the letter B. Thus, the terminal at the top left of the barrier strip is terminal 1A, and the terminal at the bottom right is 4B.

To assemble the crystal radio circuit, follow these steps:

1. **Glue the barrier strip, tuning capacitor, and coil to the board.**

   Use Figure 2-4 to judge the placement of each of these parts. Be sure to give the glue enough time to cool and harden before you continue.

2. **Connect the diode between terminals 1A and 3A.**

   Interestingly enough, the direction in which you connect the diode doesn't matter in a crystal radio circuit.

3. **Strip about ⅜" of insulation from both ends of all three lengths of hook-up wire.**

4. **Connect one end of the 1.5" length of hook-up wire to terminal 2A on the barrier strip, and then connect the other end to terminal 4A.**

5. **Use some sandpaper to gently scrape the enamel insulation off the ends of the wire.**

6. **Connect the two wires from the coil to terminals 1A and 4A.**

7. **Solder one end of one of the 3" hook-up wires to the center lead of the capacitor and one end of the other 3" wire to either one of the other leads.**

   It doesn't matter which of the two outside leads you use.

Foxhole radios

In World War II, GIs would often build their own crystal radios from whatever materials they could scrounge up. These were called *foxhole radios*, although I'm not certain how many of them were actually built in foxholes. I think if I were in a foxhole and the enemy was launching mortar shells at me, I wouldn't be too interested in listening to my crystal radio. Nevertheless, that's what they're called.

Wire for making antennas and coils wasn't too hard to come by, and super-sensitive headphones weren't hard to scrounge. But crystals for making the detector part of the circuit were another story. So the GIs came up with a cleverly improvised solution: They used razor blades, pencil lead, and safety pins.

To build the detector for a foxhole radio, the razor blade must be made of blue steel. Most modern blades aren't, but you can fix that by placing the blade in a metal vice and blasting it with a propane torch until it glows red hot. Let it cool before you handle it!

Note that it also doesn't hurt if the razor blade is a bit rusty; the oxide in the rust actually helps.

To build the detector, first glue the razor blade to a piece of wood. Sharpen the pencil, then cut it short (½" is long enough). Bend open the safety pin to about 90° and jam the pointed end of the pin into the lead at the end of the pencil that you cut off. Then nail or screw the flat end of the safety pin to the board, positioned so that the tip of the pencil sits on the razor blade.

Connect one wire to the razor blade and the other to the safety pin and wire it into your circuit right where the germanium diode would go. Then, hook your radio up to the antenna and ground, put the earphone in your ear, and drag the pencil tip around to different parts of the razor blade until you hear a signal.

This type of detector is very finicky, so you might have to try different angles and positions, and you might have to try different razor blades or pencils. But once you get it to work, you'll be delighted that you were able to make a radio out of an old razor blade, a safety pin, and a pencil.

8. **Connect the free ends of the wires you soldered in Step 7 to terminals 1A and 1D of the barrier strip.**

You're done!

When the radio circuit is assembled, look it over to make sure all the pieces are connected as shown in Figure 2-4.

Stringing Up an Antenna

A good, long antenna is vital to the successful operation of a crystal radio. In general, the longer the antenna, the better. If possible, try make your antenna at least 50'. Longer is better.

You can make your antenna from just about any type of wire, insulated or not. RadioShack sells a 60' roll of 18-gauge, solid hook-up wire that's perfect.

The best configuration for a crystal radio antenna is to run the wire horizontally between two supports as high off the ground as you can get them, as shown in Figure 2-5. For your antenna, you probably won't find actual poles as shown in the figure. However if you look around, you should be able to find two suitable points to which you can connect the ends of your antenna. Fence posts, trees, basketball hoops, flag poles, swing sets, or almost any other tall structure will do the trick.

Figure 2-5: Stringing your antenna.

to radio

Notice that one end of the antenna wire must run to the ground to a convenient place where you can connect it to your crystal radio. You'll need to run this wire to the location at which you intend to operate your radio.

Wood isn't a great insulator, and most metals, of course, are excellent conductors. It's vital that your antenna is well insulated from the ground. Thus, you must be careful about how you support the ends of the antenna wire to make sure you don't inadvertently ground the antenna.

If you use insulated wire for the antenna, you can secure the ends to wood by using ½" eye screws available from any hardware store. Screw the eye screw into the wood, and then simply tie the end of the antenna wire to it.

If the wire is uninsulated, you'll need to support it with something that doesn't conduct electricity. I suggest browsing the sprinkler parts department of your local hardware store to find a PVC pipe fitting. You can screw this fitting into wood or use duct tape or zip ties to secure it to metal, then loop your antenna wire through the fitting and tie it off.

A crystal radio is a relatively safe electronics project, but there are a few dangers associated with the antenna. Here are some things to be careful of:

✦ Don't string up your antenna in a thunderstorm! Lightning loves wires, and you don't want to tempt Mother Nature by providing her with a convenient path to discharge her fury through.

✦ Likewise, don't operate your crystal radio in a thunderstorm!

✦ Do not under any circumstances run your antenna wire anywhere near a power cable or other utility line. That's a sure way to become a statistic.

✦ Be very careful if you must climb a ladder to string your antenna. I don't have the statistics to back it up, but I bet that the most common way to seriously injure yourself building a crystal radio is to fall off a 10" extension ladder while putting up your antenna.

Connecting to Ground

A good ground connection is every bit as important as a good antenna. The best way to create a good ground connection is to use a metal cold water pipe. Assuming you placed your antenna outdoors, you may be lucky enough to find an outdoor water faucet near the end of the antenna. Then, you can connect one end of a length of hookup wire to the water pipe and the other end to your crystal radio. (Note that this won't work if the home uses plastic pipe; it will only work with metal pipe.)

If you can't find a water pipe, get a length of metal rebar and pound it into the ground. The deeper you go, the better the ground connection.

The easiest way to connect a wire to a water pipe (or a piece of rebar) is to use a pipe clamp, which you can find in the plumbing section of any hardware store. Use some coarse sandpaper to sand the pipe where you attach the clamp to improve the electrical connection, especially if the pipe has been painted or varnished. Strip an inch or two of insulation from the end of your ground wire and wrap it around the clamp, then slip the clamp around the water pipe and tighten it down as shown in Figure 2-6.

Figure 2-6:
A good ground connection.

Using the Crystal Radio

Once your crystal radio circuit is built, your antenna is up, and your ground wire is connected, it's time to put your crystal radio to the test. Follow these steps:

1. **Connect the two leads of the piezoelectric earphone to terminals 3B and 4B of the barrier strip.**

2. **Connect the antenna lead to terminal 1B of the barrier strip.**

3. **Connect the ground lead to terminal 2B of the barrier strip.**

4. **Put the earphone in your ear.**

   You'll probably immediately hear a radio station.

5. **Turn the knob on the tuning capacitor to hear other stations.**

   The tuner on this crystal radio circuit isn't very sensitive, so you'll probably be able to distinguish just two or three different stations.

   Note that if you didn't add a tuning capacitor, you won't be able to tune to a specific station at all. Instead, you'll probably hear several stations at once. (Even with a tuning capacitor, you may still hear several stations at once. As I said, the tuning circuit for a simple crystal radio like this isn't very discriminating.)

Chapter 3: Working with Infrared

*I*n this chapter, you learn how to work with circuits that detect the invisible light that's commonly called *infrared*. Infrared light has all sorts of applications for wireless communication, the most common of which is the remote control for your television. Other uses for infrared include night-vision goggles and cameras, and temperature detection.

Have fun!

Introducing Infrared Light

Infrared light is light whose frequency is just below the range of visible red light. Specifically, infrared is light whose frequency falls between 1 THz to 400 THz (one THz is one trillion cycles per second). The infrared spectrum falls right between microwaves and visible light.

Remember from Chapter 1 of this minibook that there's an inverse relationship between *frequency* and *wavelength*. In other words, the lower the frequency, the longer the wavelength. If you describe infrared in terms of its wavelength rather than its frequency, infrared waves are longer than the waves of visible light, but shorter than microwaves. The wavelength of infrared light is between 0.75 to 300 micrometers, which is a millionth of a meter. Thus, at the very bottom edge of the infrared spectrum, the infrared waves are about one third of a millimeter long. At the upper end, the waves are about one thousandth of a millimeter long. If the waves get any shorter than that, they become visible light.

Figure 3-1 shows the entire spectrum of electromagnetic radiation, so you can see where infrared falls within the grand scheme of things radiation-wise.

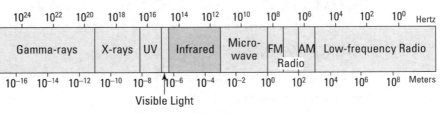

Figure 3-1:
Infrared
light falls
between
visible
light and
microwaves.

Infrared light isn't visible to human eyes. I guess Mother Nature decided that we didn't need to be looking at things in infrared. It's too bad because more than half of all the light energy emitted from our sun is in the form of infrared light. If our eyes could see infrared light as well as the visible light we can see, a sunny day would seem twice as bright.

Sometimes infrared light is confused with heat. That's because we can't see infrared light waves, but we can feel them in the form of heat. In other words, infrared light waves heat surfaces that absorb them. Visible light does this too. That's why it's cooler in the shade than it is in the sun. Because heat is an effect of infrared light, infrared light can be used as a heat source. But infrared light and heat aren't the same thing.

Infrared light is often used to detect objects that we can't see in visible light. One common application of this is night vision. According to a principal of physics called *Planck's law*, all matter emits electromagnetic radiation if its temperature is above absolute zero. Some of that radiation is in the infrared spectrum, so devices that can detect infrared light can literally see in the dark.

To enhance the effect, some night-vision devices actually illuminate an area with infrared light. Because the human eye can't see the infrared light, the area illuminated still appears dark to us, but to a detector sensitive to infrared light, the area is lit up and fully visible.

Another common application of infrared light is for wireless communications across short distances. The best known infrared devices are television remote controls. The remote control unit contains a bright infrared light source, and the television itself includes an infrared detector. When you point the remote control at the television and push a button, the remote control turns on the infrared light source and encodes a message on it. The receiver picks up this signal, decodes the message, and does whatever the message directs it to do — turns up the volume, changes the channel, and so on.

Like visible light, infrared light can be blocked by solid objects and it can bounce off of reflective objects. That's why the remote won't work if your spouse is standing between you and the television. But it's also why you can get around your spouse by pointing the remote at a window. The infrared waves bounce off the glass and, if the angle is right, arrive at the television.

The first wireless remote control was developed by Zenith in 1955. It used ordinary visible light, could turn the TV on or off, and could change channels. It had one nasty defect: You had to position your television in the room so that light from an outside source (such as the setting sun shining through a window) didn't hit the light sensor. Otherwise, the TV might shut itself off right in the middle of the evening news when the sun reached just the right angle and hit the sensor.

Remotes today use complicated encoding schemes to avoid such random misfirings. You're probably familiar with the procedure you must go through when programming a remote control to work with a particular television. This programming is necessary because there's no widely accepted standard for how the codes on a remote control should work, so each manufacturer uses its own encoding scheme.

Detecting Infrared Light

There are several ways to detect infrared light in an electronic circuit, but the most common is with a device called a *phototransistor*. You can buy a phototransistor for less than a dollar at RadioShack or any other store that stocks electronic components.

To understand how a phototransistor works, first review how a transistor works. A transistor has three terminals, known as the *base, collector,* and *emitter.* Within the transistor, there's a path between the collector and emitter. How well this path conducts depends on whether voltage is applied across the base and the emitter. If voltage is applied, the collector-emitter path conducts well. If there's no voltage on the base, the collector-emitter path doesn't conduct.

In a phototransistor, the base isn't a separate terminal that's connected to a voltage source in your circuit. Instead, the base is exposed to light. When infrared light hits the base, the energy in the light is converted to voltage, and the emitter-collector path conducts.

Thus, infrared light hitting the base has the same effect as voltage on the base of a traditional transistor: The infrared light turns the transistor on. The brighter the infrared light, the better the emitter-collector path conducts.

Figure 3-2 shows a simple circuit that uses an IR phototransistor to detect infrared light. When infrared light is present, the collector-emitter circuit conducts, and the LED lights up. Thus, the LED lights when the phototransistor is exposed to infrared light.

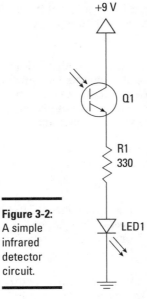

+9 V

Q1

R1
330

Figure 3-2:
A simple
infrared
detector
circuit.

LED1

Project 3-1 shows you how to build this circuit on a solderless breadboard, and Figure 3-3 shows the assembled circuit.

Once you have assembled this circuit, try exposing the phototransistor to different light sources to see whether they emit infrared light. One sure source of infrared is a TV remote control. Point the remote at the phototransistor and press any button on the remote. You should see the LED flash on and off quickly as it responds to the infrared signals being sent by the remote.

Another interesting source of infrared is an open flame. Be very careful, of course; I don't want you burning down your house just to see if the flames produce infrared light. If you have a small gas lighter, light it up and hold it near the phototransistor.

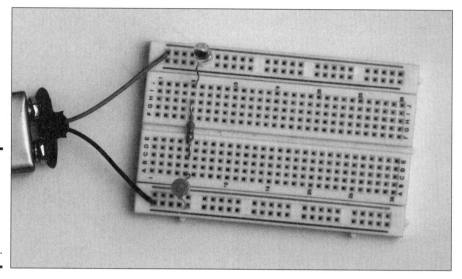

Figure 3-3:
The assembled infrared detector circuit (Project 3-1).

Creating Infrared Light

The easiest way to create infrared light is by using a special light-emitting diode (LED) that operates in the infrared spectrum. Infrared LEDs (often called *IR LEDs*) are readily available at RadioShack or any other store that sells electronic parts.

IR LEDs are similar to regular LEDs, except that you can't see the light they emit. The LED itself is usually a dark purple or blue color. Like other LEDs, the cathode lead is shorter than the anode lead.

Project 3-1: A Simple IR Detector

In this project, you can build a simple infrared light detector using a phototransistor. When the phototransistor is exposed to infrared light, the LED lights up.

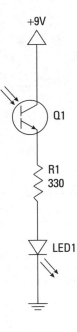

Parts List

1   9 V battery connector
1   9 V battery
1   IR phototransistor
1   Red LED
1   330 Ω ¼ W resistor
(orange-orange-brown)

Steps

1. **Insert the photodiode.**

   Collector (short lead): Positive bus
   Emitter (long lead): J5

2. **Insert the resistor.**

   C5 to H5

3. **Insert the LED.**

   Cathode (short lead): Ground bus
   Anode (long lead): A5

4. **Connect the battery.**

   Red lead: Positive bus
   Black lead: Negative bus

5. **Expose the phototransistor to an infrared light source.**

   Try a variety of sources, including a TV remote control, a flame (be careful!), and sunlight. Try other sources that don't emit infrared, such as an LED flashlight.

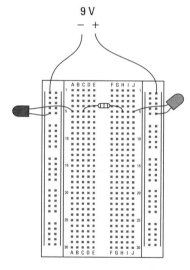

As with any LED, you must use a resistor in series with an IR LED to prevent excess current from burning out the LED. To calculate the size of the resistor, you need to know three things:

✦ **The supply voltage:** For example, 9 V.

✦ **The LED forward-voltage drop:** For most infrared LEDs, the forward-voltage drop is 1.3 V.

✦ **The desired current through the LED:** Usually, the current flowing through the IR LED should be kept under 50 mA. However, IR LEDs are typically rated for more current than regular LEDs. The ones I buy from RadioShack can handle up to 150 mA.

With these three facts in hand, you can calculate the correct resistor size by using Ohm's law:

1. **Calculate the resistor voltage drop.**

   To do that, subtract the voltage drop of the IR LED (typically 1.3 V) from the total supply voltage. For example, if the total supply voltage is 9 V and the LED drops 1.3 V, the voltage drop for the resistor is 7.7 V.

2. **Convert the desired current to amperes.**

   In Ohm's law, the current must be expressed in amperes. You can convert milliamperes to amperes by dividing the milliamperes by 1,000. Thus, if your desired current through the IR LED is 50 mA, you must use 0.05 A in your Ohm's law calculation.

3. **Divide the resistor voltage drop by the current in amperes.**

   This gives you the desired resistance in ohms. For example, if the resistor voltage drop is 7.6 V and the desired current is 50 mA, you need a 152 Ω resistor.

4. **Choose a standard resistor size that's close to the calculated resistance.**

   A 150 Ω resistor is close enough for a 9 V circuit. If you don't have a 150 Ω resistor, a 220 Ω will do the job.

Once you've chosen the correct resistor size, just wire it in series with the IR LED, as shown in the schematic in Figure 3-4.

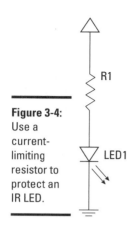

Figure 3-4:
Use a
current-
limiting
resistor to
protect an
IR LED.

Building a Proximity Detector

The combination of an IR LED and a photodiode is often used as a *proximity detector,* a gadget that detects when an object is nearby. There are two ways to build a proximity detector. One is to mount the IR LED and the phototransistor so that they face each other. Then, the infrared light from the IR LED is detected by the phototransistor. If an object comes between the IR LED and the phototransistor, the light is blocked, and the phototransistor turns off.

The other way to build a proximity detector is to mount the IR LED and the IR photodiode next to each other facing the same direction. When an object comes near the IR LED, some infrared light will bounce off the object and be detected by the phototransistor.

You can learn how to build two types of proximity detector circuits in the next two sections of this chapter.

Building a Common-Emitter Proximity Detector

A schematic for a simple proximity circuit is shown in Figure 3-5. Although it isn't shown in the schematic, this circuit assumes that IR LED and Q1 are oriented so that Q1 can detect the infrared light emitted by IR LED, either indirectly (for a proximity detector) or directly (for an interrupter).

This circuit is called a *common-emitter* circuit because the phototransistor's emitter is common between the phototransistor side of the circuit and the output side of the circuit that's connected to the IR LED. In a common-emitter circuit, the output voltage is on when infrared light is detected by the photo-transistor. Thus, the red LED lights up when the path between the IR LED and phototransistor isn't obstructed. If you block the path between the IR LED and phototransistor, the red LED goes dark.

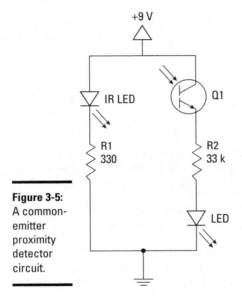

Figure 3-5:
A common-emitter proximity detector circuit.

Project 3-2 shows how to build this circuit configured as an interrupter, and Figure 3-6 shows the finished project. When you connect this circuit to power, the red LED will come on. If you pass an object such as a piece of paper between the IR LED and the phototransistor, the red LED will go off.

Figure 3-6:
Using an
IR LED and
a photo-
transistor as
a proximity
detector
(Project 3-2).

The output in this circuit is simply a red LED. However, you could just as easily connect the output to other circuit components. For example, the output could drive a mechanical relay if you want to use the proximity detector to turn on a floodlight or other 120 VAC device, or you could connect the output to a digital logic circuit as described in Book VI, Chapter 4.

Project 3-2: A Common-Emitter Proximity Detector

In this project, you can build a common-emitter proximity detector that lights a red LED whenever the path between an IR LED and a phototransistor is clear. If anything blocks the path, the red LED goes dark.

Parts List

1   9 V battery connector
1   9 V battery
1   IR phototransistor
1   IR LED
1   Red LED
2   330 Ω ¼ W resistor
(orange-orange-brown)

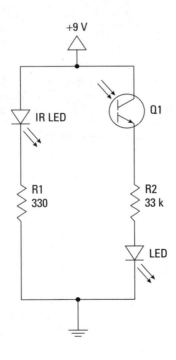

Steps

1. **Insert the photodiode.**
   Collector (short lead): Positive bus
   Emitter (long lead): A3

2. **Insert the red LED.**
   Cathode (short lead): Ground bus
   Anode (long lead): J3

3. **Insert the IR LED.**
   Cathode (short lead): A15
   Anode (long lead): Positive bus

4. **Insert the resistors.**
   R1: C3 to H3
   R2: E15 to ground bus

5. **Connect the battery.**
   Red lead: Positive bus
   Black lead: Negative bus
   The red LED will illuminate when
   the power is connected.

6. **Use a piece of paper or other flat
   object to block the light path
   between the IR LED and the
   phototransistor.**
   The red LED will go dark when
   you interrupt the light path
   between the IR LED and the
   phototransistor.

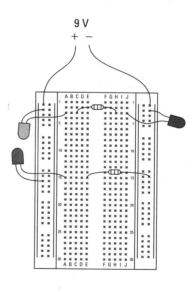

Building a Common-Collector Proximity Detector

Figure 3-7 shows a circuit that uses a *common-collector* circuit, in which the collector is the common point between the phototransistor circuit and the LED output circuit. When wired in this way, the LED is dark whenever the path between the IR LED and the phototransistor is clear. When something blocks the path and the phototransistor stops detecting IR light, the red LED comes on.

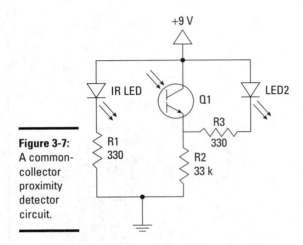

Figure 3-7:
A common-collector proximity detector circuit.

Project 3-3 shows how to build this circuit configured as an interrupter, and Figure 3-8 shows the finished project. When you connect this circuit to power, the red LED stays dark. But if you pass an object between the IR LED and the phototransistor, the red LED turns on.

Figure 3-8:
This circuit lights a red LED when the path between an IR LED and a photo-transistor is blocked (Project 3-3).

Project 3-3: A Common-Collector Proximity Detector

In this project, you can build a common-collector proximity detector that lights a red LED whenever something comes between an infrared LED and a phototransistor.

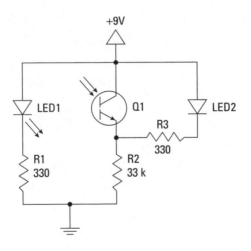

Parts List

1 9 V battery connector
1 9 V battery
1 IR phototransistor
1 IR LED
1 Red LED
2 330 Ω ¼ W resistor
(orange-orange-brown)
1 33 k Ω ¼ W resistor
(orange-orange-orange)
2 Short lengths of jumper wire

Steps

1. **Insert the red LED.**
   Anode (long lead): D5
   Cathode (short lead): F5

2. **Insert the photodiode.**
   Collector (short lead): E10
   Emitter (long lead): F10

3. **Insert the IR LED.**
   Anode (long lead): E22
   Cathode (short lead): F22

4. **Insert the resistors.**
   R1 (330 Ω):   J22 to ground bus
   R2 (33 k Ω):   J10 to ground bus
   R3 (330 Ω):   H10 to H5

5. **Insert the jumper wires.**
   A5 to positive bus
   A10 to positive bus
   A22 to positive bus

6. **Connect the battery.**
   Red lead: Positive bus
   Black lead: Negative bus

7. **Use a piece of paper or other flat object to block the light path between the IR LED and the phototransistor.**
   The red LED will light up when you interrupt the light path between the IR LED and the phototransistor. When you remove the obstacle, the red LED will go dark.

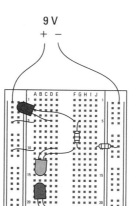

Book VI

Doing Digital Electronics

A Kit 74 relay

Contents at a Glance

Chapter 1: Understanding Digital Electronics

In This Chapter

✔ **Looking at the difference between analog and digital circuits**

✔ **Exploring binary codes**

✔ **Learning about logical operators**

✔ **Building some very simple logic circuits with switches and lamps**

*W*elcome to the world of digital electronics. Beginning with this chapter, you start to learn about the fundamental building-block circuits that you'll ultimately use to create computers and other advanced electronic devices.

Digital electronics is a complex, almost-overwhelming topic, but the building blocks are actually quite simple. In this chapter, you learn some basic principles, such as how to tell the difference between analog and digital circuits, how the binary system works, and how basic logic operations work. In the remaining chapters in this part, you build on that knowledge as you explore the amazing world of digital electronics.

Hang on!

Distinguishing Analog and Digital Electronics

All of electronics can be divided into two broad categories: analog and digital.

Analog refers to circuits in which quantities such as voltage or current vary at a continuous rate. When you turn the dial of a potentiometer, for example, you change the resistance by a continuously varying rate. The resistance of the potentiometer can be any value between the minimum and maximum allowed by the pot.

If you create a voltage divider by placing a fixed resistor in series with a potentiometer, the voltage at the point between the fixed resistor and the potentiometer increases or decreases smoothly as you turn the knob on the potentiometer.

In *digital* electronics, quantities are counted rather than measured. There's an important distinction between counting and measuring. When you *count* something, you get an exact result. When you *measure* something, you get an approximate result.

Consider a cake recipe that calls for 2 cups of flour, 1 cup of milk, and 2 eggs. To get 2 cups of flour, you scoop some flour into a 1-cup measuring cup, pour the flour into the bowl, and then do it again. To get a cup of milk, you pour milk into a liquid measuring cup until the top of the milk lines up with the 1-cup line printed on the measuring cup and then pour the milk into the mixing bowl. To get 2 eggs, you count out 2 eggs, crack them open, and add them to the mixing bowl.

The measurements for flour and milk in this recipe are approximate. A teaspoon too much or too little won't affect the outcome. But the eggs are precisely counted: exactly 2. Not 3, not 1, not 1½, but 2. You can't have a teaspoon too many or too few eggs. There will be exactly 2 eggs, because you count them.

One of the most common examples of the difference between analog and digital devices is a clock. Figure 1-1 shows two clocks: one analog and the other digital. On the analog clock, the time is represented by hands that spin around a dial and point to a location on the dial that represents the approximate time. On a digital clock, a numeric display indicates the exact time.

Figure 1-1:
Analog
and digital
clocks.

Another example is a thermometer. A traditional glass–mercury thermometer has a small amount of liquid mercury inside a glass column. Mercury expands when it gets warm, so the warmer the mercury is, the higher it climbs in the glass column. Little tick marks are printed on the column to help you read the temperature. On a digital thermometer, the exact temperature is indicated by a numeric display.

So which is more accurate — analog or digital? In one sense, digital circuits are more accurate because they count with complete precision. You can precisely count the number of jelly beans in a jar, for example. But if you weigh the jar by putting it on an analog scale, your reading may be a bit imprecise because you can't always judge the exact position of the needle. Say that the needle on the scale is about halfway between 4 pounds and 5 pounds. Does the jar weigh 4.5 pounds or 4.6 pounds? You can't tell for sure, so you settle for approximately 4.5 pounds.

On the other hand, digital circuits are inherently limited in their precision because they must count in fixed units. Most digital thermometers, for example, have only one digit to the right of the decimal point. Thus, they can indicate a temperature of 98.6 or 98.7 but can't indicate 98.65.

Here are a few other thoughts to ponder concerning the differences between digital and analog systems:

+ The term *digital* is actually a reference to your digits — that is, your fingers. The most natural way for us humans to count is with our fingers.

+ The term *analog* is a variation on the word *angler* — a reference to the ancient practice of estimating the size of a fish by spreading out your arms while saying, "He was at least this big!" That's the original analog measurement.

+ Saying that a system is digital isn't the same as saying that it's binary. *Binary* is a particular type of digital system in which the counting is all done with the binary number system. Nearly all digital systems are also binary systems, but the two words aren't interchangeable. (For more information about the binary number system, see the next section, "Understanding Binary.")

+ Many systems are a combination of binary and analog systems. In a system that combines binary and analog values, special circuitry is required to convert from analog to digital, or vice versa. An input voltage (analog) might be converted to a sequence of pulses, one for each volt; then the pulses can be counted to determine the voltage.

+ I made up that item about the angler, of course. Made you think about it for a second, though, didn't I?

Understanding Binary

Most digital electronic circuits work with the binary number system. Thus, before you can understand the details of how digital circuits work, you need to understand how the binary numbering system works.

Knowing your number systems

A *number system* is simply a way of representing numeric values. Number systems use symbols called *numerals* to represent numeric quantities. The numerals 1, 2, and 3 represent the numeric quantities commonly known as one, two, and three.

In most number systems, numerals can be strung together to create larger numeric values, and the position of each numeral in the string determines its relative value. In the number 12, the numeral 1 represents the quantity ten, and the numeral 2 represents the quantity two. In the number 238, the numeral 2 represents the quantity two hundred, the numeral 3 represents the quantity thirty, and the numeral 8 represents the quantity eight.

You learned all this stuff in grade school, so it's completely intuitive. Unfortunately, it's so intuitive that it's easy to miss its brilliance — and to overlook the fact that it's completely arbitrary.

The fact that we have ten numerals in our everyday counting system — which is called the *decimal system,* or *base 10* — is a simple result of the fact that humans have ten fingers. If humans had evolved with 12 fingers, we would've learned how to count in base 12, and we would've invented two more numerals.

Different number bases may seem strange, but we actually encounter them every day without thinking about it. One common example is the system we use for keeping time. Our system for measuring time actually works in several number bases. An hour contains 60 minutes, for example. We recognize that 1:30 is halfway between 1:00 and 2:00 without even thinking about it. You may not realize it, but you're thinking in base 60 when you tell time.

Counting by ones

Binary is one of the simplest of all number systems because it has only two numerals: 0 and 1. In the decimal system (with which most people are accustomed), you use 10 numerals: 0 through 9. In an ordinary decimal number, such as 3,482, the rightmost digit represents ones; the next digit to the left, tens; the next, hundreds; the next, thousands; and so on. These digits represent powers of ten: first 10^0 (which is 1); next, 10^1 (10); then 10^2 (100); then 10^3 (1,000); and so on.

In binary, you have only two numerals rather than ten, which is why binary numbers look somewhat monotonous, as in 110011, 101111, and 100001.

The positions in a binary number (called *bits* rather than *digits)* represent powers of two rather than powers of ten: 1, 2, 4, 8, 16, 32, and so on. To figure the decimal value of a binary number, you multiply each bit by its corresponding power of two and then add the results. The decimal value of binary 10111, for example, is calculated as follows:

$$1 \times 2^0 = 1 \times 1 = 1$$
$$+1 \times 2^1 = 1 \times 2 = 2$$
$$+1 \times 2^2 = 1 \times 4 = 4$$
$$+0 \times 2^3 = 0 \times 8 = 0$$
$$+1 \times 2^4 = 1 \times 16 = \underline{16}$$
$$23$$

Fortunately, converting a number between binary and decimal is something that a computer is good at — so good, in fact, that you're unlikely ever to need to do any conversions yourself. The point of learning binary is not to be able to look at a number such as 1110110110110 and say instantly, "Ah! Decimal 7,606!" (If you could do that, Barbara Walters would probably interview you, and Hollywood would even make a movie about you. You'd probably be played by Dustin Hoffman.)

Instead, the point is to have a basic understanding of how computers store information and — most important — of how the binary counting system works, which I describe in the following section.

If you do find that you need to convert binary numbers to decimal, or vice versa, and you have access to a computer, you can use the Calculator program that comes free with Windows to do the conversion for you. For more information, see the nearby sidebar "Using Windows Calculator for binary conversions."

Here are some of the most interesting characteristics of binary, which explain how the system is similar to and different from the decimal system:

✦ **In decimal, the number of decimal places allotted for a number determines how large the number can be.** If you allot six digits, for example, the largest number possible is 999,999. Because 0 is itself a number, however, a 6-digit number can have any of 1 million different values.

Similarly, the number of bits allotted for a binary number determines how large that number can be. If you allot 8 bits, the largest value that number can store is 11111111, which happens to be 255 in decimal. Thus, a binary number that is 8 bits long can have any of 256 different values (including 0).

✦ **To quickly figure how many different values you can store in a binary number of a given length, use the number of bits as an exponent of two.** An 8-bit binary number, for example, can hold 2^8 values. Because 2^8 is 256, an 8-bit number can have any of 256 different values. This is why a *byte* — 8 bits — can have 256 different values.

♦ **This "powers of two" thing is why digital systems don't use nice, even round numbers for measuring such values as memory capacity.** A value of 1k, for example, isn't an even 1,000 bytes: It's actually 1,024 bytes, because 1,024 is 2^{10}. Similarly, 1MB isn't an even 1,000,000 bytes, but 1,048,576 bytes, which happens to be 2^{20}.

One basic test of digital geekdom is knowing your powers of two because they play such an important role in binary numbers. Just for the fun of it, but not because you really need to know, Table 1-1 lists the powers of two up to 32.

Table 1-1 also shows the common shorthand notation for various powers of two. The abbreviation *k* represents 2^{10} (1,024). The *M* in *MB* stands for 2^{20}, or 1,024k, and the *G* in *GB* represents 2^{30}, which is 1,024MB.

Table 1-1			Powers of Two		
Power	*Bytes*	*Kilobytes*	*Power*	*Bytes*	*k, MB, or GB*
2^1	2	2^{17}	131,072		128k
2^2	4	2^{18}	262,144		256k
2^3	8	2^{19}	524,288		512k
2^4	16	2^{20}	1,048,576	1MB	
2^5	32	2^{21}	2,097,152	2MB	
2^6	64	2^{22}	4,194,304	4MB	
2^7	128	2^{23}	8,388,608	8MB	
2^8	256	2^{24}	16,777,216	16MB	
2^9	512	2^{25}	33,554,432	32MB	
2^{10}	1,024	1k	2^{26}	67,108,864	64MB
2^{11}	2,048	2k	2^{27}	134,217,728	128MB
2^{12}	4,096	4k	2^{28}	268,435,456	256MB
2^{13}	8,192	8k	2^{29}	536,870,912	512MB
2^{14}	16,384	16k	2^{30}	1,073,741,824	1GB
2^{15}	32,768	32k	2^{31}	2,147,483,648	2GB
2^{16}	65,536	64k	2^{32}	4,294,967,296	4GB

Doing the logic thing

One of the great things about binary is that it's very efficient at handling special operations called logical operations. *Logical operations* compare two binary bits and render a third binary bit as a result. There are 16 possible logical operations, and you learn about all 16 of them in Book VI, Chapter 2. For now, I want to introduce you to three of them: AND, OR, and XOR.

The following list summarizes these three basic logical operations:

✦ **AND:** An AND operation compares two binary values. If both values are 1, the result of the AND operation is 1. If one value is 0 or both of the values are 0, the result is 0.

✦ **OR:** An OR operation compares two binary values. If at least one of the values is 1, the result of the OR operation is 1. If both values are 0, the result is 0.

✦ **XOR:** An XOR operation compares two binary values. If exactly one of them is 1, the result is 1. If both values are 0 or if both values are 1, the result is 0.

Table 1-2 summarizes how AND, OR, and XOR work.

Table 1-2	Logical Operations for Binary Values			
First Value	*Second Value*	*AND*	*OR*	*XOR*
0	0	0	0	0
0	1	0	1	1
1	0	0	1	1
1	1	1	1	0

You can apply logical operations to binary numbers that have more than one binary digit by applying the operation one bit at a time. The easiest way to do this manually is to line the two binary numbers on top of one another and then write the result of the operation below each binary digit. The following example shows how you'd calculate 10010100 AND 11011101:

```
        10010100
AND  11011101
        10010100
```

As you can see, the result is 10010100.

Using Windows Calculator for binary conversions

If you have a computer, you can use the free Calculator program that comes with all versions of Windows to work with binary numbers. The Calculator program has a special Programmer mode that many users don't know about. When you flip the Calculator into this mode, you can do instant binary and decimal conversions, which occasionally come in handy when you're working with IP addresses.

To use the Windows Calculator in Programmer mode, launch the Calculator by choosing Start⇨ All Programs⇨Accessories⇨Calculator. Then choose View⇨Programmer from the Calculator menu. The Calculator changes to a fancy programmer model — the kind that I paid $150 for when I was a rookie computer programmer 30 years ago. All kinds of buttons appear:

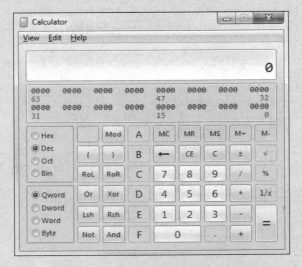

You can select the Hex, Dec, Oct, and Bin radio buttons to switch from decimal to the different numbering systems commonly used in digital electronics: hexadecimal, octal, and binary. For example, to find the binary equivalent of decimal 155, enter **155** and then select the Bin radio button. The value in the display changes to 10011011.

Here are a few other things to note about the Programmer mode of the Calculator:

✔ The Programmer Calculator has several features that are designed specifically for binary calculations, such as AND, XOR, NOT, and NOR.

✔ The Programmer Calculator can also handle hexadecimal conversions. Hexadecimal doesn't come into play when you're dealing with IP addresses, but it's used for other types of binary numbers, so this feature sometimes proves to be useful.

✔ The calculator in earlier versions of Windows (prior to Windows 7) doesn't have Programmer mode. However, the older versions do have a Scientific mode which included features for working with binary numbers.

Using Switches to Build Gates

To give you an idea of how basic gates work, Projects 1-1, 1-2, and 1-3 show you how to assemble an AND gate, an OR gate, and an XOR gate using simple DPDT knife switches. In actual practice, gate circuits are built with transistors or with integrated circuits. But these three projects will give you a good idea of how gates work.

Figure 1-2 shows these three projects fully assembled. The AND gate (Project 1-1) is shown at the top of the photo. As you can see, the AND gate is implemented by connecting two switches together in series. Both switches must be closed for the lamp to light.

The circuit for the OR gate (Project 1-2) is shown in the middle of the figure. To implement an OR gate with switches, you simply wire the switches in parallel. Then, the lamp lights if either of the switches is closed.

The XOR gate circuit (Project 1-3) is shown at the bottom of the figure. This wiring is a bit trickier. The DPDT switches are cross-connected in such a way that the circuit to the lamp is completed if one of the switches is in the A position and the other is in the B position. If both switches are in the same position, the circuit is not completed so the lamp does not light.

**Book VI
Chapter 1**

**Understanding
Digital Electronics**

Figure 1-2:
Imple-
menting
AND, OR,
and XOR
gates
with knife
switches.

 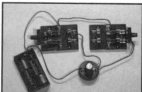 

AND gate   OR gate   XOR gate

Project 1-1: Building a Simple AND Circuit

In this project you use two switches and a lamp to create a simple circuit that performs a logical AND operation.

The two switches represent the two binary values that are input to the AND operation. A closed switch represents the binary value 1, and an open switch represents the binary value 0.

The lamp represents the binary output of the AND operation. When lit, the lamp represents the binary value 1. When it's not lit, the lamp represents the binary value 0.

Parts List

2 AA batteries
1 Battery holder
1 Lamp holder
1 3 V flashlight lamp
2 DPDT knife switches
2 5" lengths of 22-gauge stranded
   wire, stripped ½" on each end
1 Small Phillips-head screwdriver
1 Wire cutter
1 Wire stripper

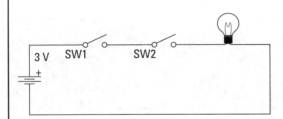

Steps

1. **Open both switches.**

   Move the handles to the upright position so that none of the contacts is connected.

2. **Attach the red lead from the battery holder to terminal 1X of one of the switches.**

3. **Attach the black lead to one of the terminals on the lamp holder.**

4. **Connect the two 5" wires as follows:**

Connect From	Connect To
Terminal 1A of the first switch	Terminal 1X of the second switch
Terminal 1A of the second switch	The unused terminal of the lamp holder

5. **Insert the batteries into the holder.**

6. **Manipulate the switches to verify the correct operation of the AND circuit.**

   The switches should operate as follows:

Switch 1	Switch 2	Lamp
Open	Open	Off
Closed	Closed	Off
Closed	Open	Off
Closed	Closed	On

   Notice that the lamp comes on only when both switches are closed, which is exactly how an AND circuit should work.

7. **You're done!**

   Give yourself a pat on the back, because you've just built your first logic circuit.

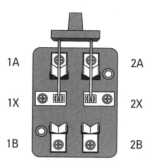

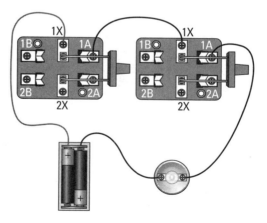

Project 1-2: Building a Simple OR Circuit

In this project, you build a simple OR circuit by wiring two switches in parallel to control a lamp. In this circuit, a lamp will be lit with either of two switches in closed. The switches represent the two binary inputs to the OR operation, and the lamp represents the binary output.

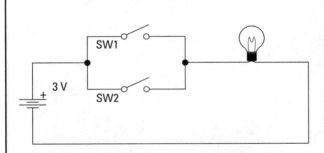

Parts List
2 AA batteries
1 Battery holder
1 Lamp holder
1 3 V flashlight lamp
2 DPDT knife switches
3 5" lengths of 22-gauge stranded wire
1 Small Phillips-head screwdriver
1 Wire cutter
1 Wire stripper

Steps

1. **Open both switches.**

   Move the handles to the upright position so that none of the contacts is connected.

2. **Attach the red lead from the battery holder to terminal 1X of the first switch.**

3. **Attach the black lead to the first terminal on the lamp holder.**

4. **Connect the first 5" wire from terminal 1X of the first switch to terminal 1X of the second switch.**

5. **Connect the second 5" wire from terminal 1A of the first switch to the second terminal on the lamp holder.**

6. **Connect the third 5" wire from terminal 1A of the second switch to the second terminal on the lamp holder.**

7. **Insert the batteries into the holder.**

8. **Manipulate the switches to verify the correct operation of the AND circuit.**

   The switches should operate as follows:

Switch 1	*Switch 2*	*Lamp*
Open	Open	Off
Closed	Closed	On
Closed	Open	On
Closed	Closed	On

   Notice that the lamp is on whenever at least one of the two switches is closed. The lamp is off only when both switches are open. That's the way an OR circuit should operate.

9. **You're done!**

   Congratulate yourself on a job well done. You've completed your second logic circuit.

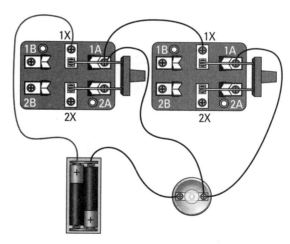

**Book VI
Chapter 1**

**Understanding
Digital Electronics**

Project 1-3: Building a Simple XOR Circuit

In this project, you build a simple XOR circuit by wiring two switches to control a lamp. In this circuit, a lamp will light when one or the other switch is closed. If both switches are open or both are closed, the lamp won't light. The switches represent the two binary inputs to the XOR operation, and the lamp represents the binary output.

Unlike the other two projects in this chapter, the switches in this project use both the A and B positions. Thus, a switch represents a binary 1 when it's in the A position. In the B position, the switch represents a binary 0.

Notice how in the schematic diagram the A and B terminals are cross-connected between the two switches. That way, the circuit to the lamp will be closed only if the switches are in opposite positions. In other words, the lamp will light if SW1 is in the A position and SW2 is in the B position, or if SW1 is in the B position and SW2 is in the A position. If both switches are in the A position or if both switches are in the B position, the lamp won't light.

Parts List
2 AA batteries
1 Battery holder
1 Lamp holder
1 3 V flashlight lamp
2 DPDT knife switches
3 5" lengths of 22-gauge stranded wire
1 Small Phillips-head screwdriver
1 Wire cutter
1 Wire stripper

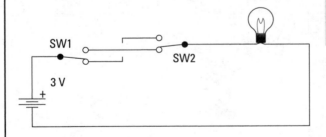

Steps

1. **Open both switches.**

   Move the handles to the upright position so that none of the contacts is connected.

2. **Attach the black lead from the battery holder to terminal 2X of the first switch.**

3. **Attach the red lead to the first terminal on the lamp holder.**

4. **Connect the first 5" wire from the second terminal on the lamp holder to terminal 1X on the second switch.**

5. **Connect the second 5" wire from terminal 2A on the first switch to terminal 1B on the second switch.**

6. **Connect the third 5" wire from terminal 2B on the first switch to terminal 1A on the second switch.**

7. **Insert the batteries into the holder.**

8. **Manipulate the switches to verify the correct operation of the XOR circuit.**

   The switches should operate as follows:

Switch 1	*Switch 2*	*Lamp*
Open	Open	Off
Open	Closed	On
Closed	Open	On
Closed	Closed	Off

9. **You're done!**

   Congratulate yourself on a job well done. You've completed your third logic circuit!

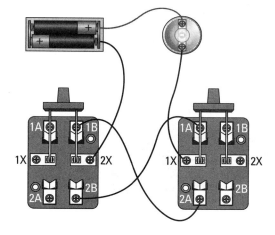

Chapter 2: Getting Logical

In This Chapter
✓ Learning about Boolean logic

✓ Examining the different types of logic gates

✓ Seeing how two or more gates can be used in combination

✓ Building logic circuits with only NAND or NOR gates

✓ Using software to simulate logic gate circuits

1 grew up on *Star Trek.* Captain Kirk and his valiant crew sparked my imagination as they journeyed through the galaxy on their amazing adventures. So it's only logical that I'm inclined to begin a chapter on logic with a reference to Mr. Spock. In fact, it would be illogical to begin this chapter in any other way.

In this chapter, I look at the basic principles of logic, which are the underpinnings of digital electronics. In particular, I explore *logic gates,* which are handy devices that perform a logical operation on two binary inputs and produce a single binary output result.

It's incredible to realize that the overwhelming complexity of modern computer systems is built on the simple concept of logic gates. A modern computer processor consists of millions of individual logic gates connected in a way that enables the processor to perform complicated operations at amazing speed.

So back to Mr. Spock: My favorite Spock quote concerning logic is in an episode entitled *The Changeling.* At the very end of the episode, Spock congratulates Captain Kirk for getting the *Enterprise* crew out of a real bind by using sophisticated logic on a destruction-bent robot:

Spock: My congratulations, Captain. A dazzling bit of logic.

Kirk: You didn't think I had it in me, did you, Spock?

Spock: No.

Well, Kirk did have it in him, and I think you have it in you, too. So let's get started!

Introducing Boolean Logic and Logic Gates

In digital electronics, *Boolean logic* refers to the manipulation of binary values in which a 1 represents the concept of *true* and a 0 represents the concept of *false*. In electronic circuits that implement logic, binary values are represented by voltage levels. In the most common convention, a binary value of one is represented by +5 V (also called *HIGH*), and a binary zero is represented by 0 V (also called *LOW*).

This type of logic is called *Boolean* because it was invented in the 19th century by George Boole, an English mathematician and philosopher. In 1854, he published a book titled *An Investigation of the Laws of Thought,* which laid out the initial concepts that eventually came to be known as Boolean algebra, also called Boolean logic. Boolean logic is among the most important principles of modern computers. Thus, most people consider Boole to be the father of computer science.

As I mention at the start of this section, in Boolean logic, *true* is represented by the binary digit 1 and *false* by the binary digit 0. *Logical operations* (also called *logical functions)* are functions that can be applied to one or more logic inputs and produce a single logic output. One of the most common types of logic operations is NOT, which simply inverts the state of its input. In other words, with the NOT operation, if the input is true, the output is false; if the input is false, the output is true.

A *gate* is a circuit or device that implements a logical function. Thus, a NOT gate is a circuit or device that implements the logical NOT operation. NOT gates are very common in digital circuits.

You can create gate switches in a variety of ways. The most common method uses transistors as switches, arranged in such a way that the correct output is generated based on the logical inputs and the type of gate being implemented. In Chapter 3, you see how gates are implemented. For the rest of this chapter, I discuss what the various types of gates do and how they can be combined in real-world circuits.

Regardless of the method used to create gate circuits, all logic circuits depend on different voltage ranges to represent 1 and 0. As I've already mentioned, the most common voltage convention is to represent 1 by approximately +5 V and 0 by approximately 0 V. The +5 V signal is usually referred to as *HIGH,* and the 0 V signal is usually called simply *LOW.*

In the rest of this chapter, I look at seven of the most common types of logic gates: NOT, AND, OR, NAND, NOR, XOR, and NXOR. All these gates except NOT use at least two inputs; the NOT gate has just one input. To help you get your bearings, Table 2-1 provides a brief overview of the distinctions among the gate types:

Table 2-1	The Most Common Types of Logic Gates
Gate	*Description*
NOT	Inverts the input (HIGH becomes LOW, LOW becomes HIGH)
AND	Outputs HIGH if all the inputs are HIGH; otherwise, outputs LOW
OR	Outputs HIGH if at least one of the inputs is HIGH; otherwise, outputs LOW
NAND	Outputs HIGH if all the inputs are LOW; otherwise, outputs LOW
NOR	Outputs HIGH if at least one of the inputs is LOW; otherwise, outputs LOW
XOR	Outputs HIGH if one, and only one, of the inputs is HIGH; otherwise, outputs LOW
NXOR	Outputs HIGH if one, and only one, of the inputs is LOW; otherwise, outputs LOW

Looking at NOT Gates

The simplest of all gates is the *NOT gate,* which is also called an *inverter.* A NOT gate has just one input, and its output is the opposite of the input. If the input is LOW, the output is HIGH. If the input is HIGH, the output is LOW.

Table 2-2 shows the truth table of an inverter. A *truth table* is simply a table that lists every possible combination of input values and shows the resulting output for each combination. For an inverter, the truth table is simple. Because only one input exists, only two possibilities exist: The input is LOW, or it is HIGH. As you can see in the table, the output is simply the opposite of the input.

Note that in truth tables, it's common to use 0s and 1s to represent logic values rather than the terms *HIGH* and *LOW.*

Table 2-2	The Truth Table for an Inverter
Input	*Output*
0	1
1	0

Figure 2-1 shows the standard logic symbol for a NOT gate. Symbols such as this are often used in schematic diagrams for circuits that use gates. The NOT symbol is simply a triangle with the input at one end and the output at the other. The small circle on the output is called a *negation bubble,* which indicates that the output is inverted.

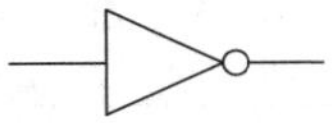

Figure 2-1:
The symbol
for a NOT
gate.

Looking at AND Gates

A *two-input AND gate* is a gate that has two inputs and one output. The output is HIGH only if both of the inputs are HIGH. Any other combination of inputs results in the output's being LOW. Table 2-3 shows the truth table for an AND gate with two inputs.

Table 2-3	The Truth Table for a Two-Input AND Gate	
Input A	*Input B*	*Output*
0	0	0
0	1	0
1	0	0
1	1	1

The standard symbol for an AND gate is shown in Figure 2-2. The inputs are on the left, and the output is on the right.

Figure 2-2:
The symbol
for a two-
input AND
gate.

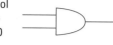

Notice that you can create AND gates with more than two inputs. For each additional input that you add to a gate, the number of possible input combinations doubles. A two-input gate has four possible input combinations; for a three-input gate, there are eight possible combinations; a four-input gate has 16 possible input combinations, and so on.

Table 2-4 shows the truth table for a three-input AND gate. As you can see, the output is HIGH only if all the inputs are HIGH. Any other combination of inputs produces LOW output.

Table 2-4	The Truth Table for a Three-Input AND Gate		
Input A	*Input B*	*Input C*	*Output*
0	0	0	0
0	0	1	0
0	1	0	0
0	1	1	0
1	0	0	0
1	0	1	0
1	1	0	0
1	1	1	1

You can combine gates in a circuit to create logic networks that are more complicated than a single gate could produce. You can create a three-input AND gate by using two two-input AND gates as shown in Figure 2-3, for example. In that figure, the first AND gate produces a HIGH output only if the inputs A and B are true. Then the output of the first AND gate is used as one of the inputs to the second AND gate; the other input is the input C.

Figure 2-3:
A pair of
two-input
AND gates
can be used
to create
a logic
network that
operates
like a three-
input AND
gate.

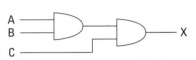

Because the output of the second gate will be HIGH only if both of its inputs are HIGH, and because the first input to the second gate is the output from the first gate, which is HIGH only if both of its inputs are HIGH, the output of the entire circuit (designated as X) is HIGH only if all three inputs (A, B, and C) are HIGH.

If you want to, you can think of an AND gate as being a form of multiplication. To understand why, first consider that when you multiply any quantity of single-bit binary numbers, only two possible results exist: zero or one. Look back at the truth tables in Table 2-3 and Table 2-4, and try multiplying the binary inputs in each row. In each case, the answer will be zero for any combination that contains a zero in any of the inputs. The answer will be one only if all the inputs are one.

The following paragraphs describe just two of the many situations in which you might use an AND gate in a real-life electronic circuit:

✦ Figure 2-4 shows how AND gates might be used in a home alarm system. Here, the inputs for the various sensors placed on the home's doors and windows are processed by the sensor circuit, which sends a HIGH signal to one of the inputs of the AND gate if any of the sensors indicates an intrusion anywhere in the house. Then the arming circuit sends a 1 to the other input of the AND gate if the system is armed. Finally, the alarm circuit sounds an audible alarm if the AND gate's output is 1. As a result, the alarm will sound if an intrusion is detected and the system is armed.

When used in this way, an AND gate is often called an *enable input,* as one of the inputs to the AND gate enables the other input to be processed. When the enable input is HIGH, the controlled input is allowed to pass through the AND gate. When the enable is LOW, the controlled input is inhibited.

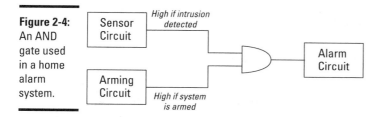

Figure 2-4:
An AND gate used in a home alarm system.

Figure 2-5 shows a more developed version of the home alarm system that uses an enable input. In this version, the alarm doesn't sound immediately when an intrusion is detected. Instead, the alarm sounds 30 seconds after it is triggered. This gives you time to disable the alarm before waking up the neighbors. When the sensor circuit detects an intrusion, it sends a trigger pulse to a 555 timer circuit, which then generates a 30-second HIGH pulse. The HIGH signal from the 555 timer output is routed through a NOT gate, which inverts the signal to LOW. The LOW signal is sent to one of the inputs to the second AND gate. The other input to the second AND gate is the output of the first AND gate, which indicates that an intrusion has been detected, and the system is armed.

For the first 30 seconds after the intrusion is detected, the first input to the second AND gate is LOW, and the second input is HIGH, so the output from the second AND gate is LOW. Thus, the alarm doesn't sound. In effect, the timer output inhibits the alarm from sounding. But when the output from the 555 timer circuit goes LOW after the 30-second pulse ends, the NOT gate inverts the signal, sending a HIGH output to the second AND gate. This situation causes the second AND gate's output to go HIGH, so the alarm sounds. Thus, the inverted pulse from the timer circuit enables the alarm circuit.

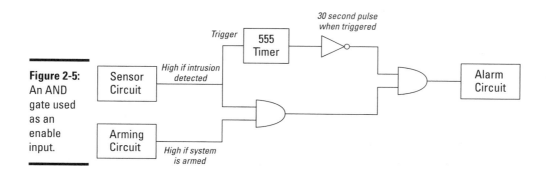

Figure 2-5:
An AND gate used as an enable input.

Looking at OR Gates

An *OR gate* produces a HIGH output if any of the inputs is HIGH. The output from an OR gate is LOW only if all the inputs are LOW. Table 2-5 shows the truth table for a two-input OR gate.

It's important to note that in an OR gate, it doesn't matter how many of the inputs are HIGH. If at least one input is HIGH, the output will be HIGH. Thus, in a two-input OR gate, the output will be HIGH if either input is HIGH or both inputs are HIGH. In a three-input OR gate, the output will be HIGH if any one, any two, or all three of the inputs are HIGH.

Table 2-5	The Truth Table for a Two-Input OR Gate	
Input A	*Input B*	*Output*
0	0	0
0	1	1
1	0	1
1	1	1

The standard symbol for an OR gate is shown in Figure 2-6. The inputs are on the left, and the output is on the right.

Figure 2-6:
The symbol for a two-input OR gate.

Like AND gates, OR gates with more than two inputs are easy to create. No matter how many inputs the OR gate has, the output is HIGH if any of the inputs is HIGH.

Multiple-input OR gates are easy to create by combining two input OR gates. Figure 2-7 shows a logic network with three OR gates, effectively acting like a four-input OR gate: If any of the four inputs is HIGH, the output will be HIGH.

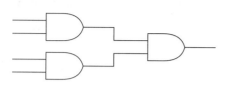

Figure 2-7:
Three OR
gates used
to create a
four-input
OR gate.

Book VI
Chapter 2

Getting Logical

You can combine 2-Input OR gates in a circuit to create OR networks that have more than two inputs. For example, Figure 2-8 shows how OR gates might be used in the sensor circuit of a home alarm system that has more than two inputs. In this circuit, eight separate alarm sensors are fed into a network of OR gates. If any one of the inputs is HIGH, the output from the sensor circuit is HIGH.

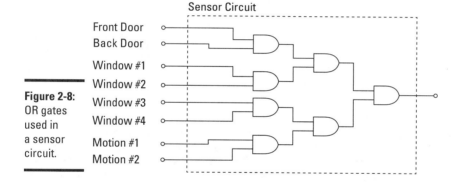

Figure 2-8:
OR gates
used in
a sensor
circuit.

The sensor circuit uses seven OR gates to create an eight-input OR gate. Study the way that the OR gates are arranged in this network for a few moments to make sure you understand how it works. Each of the eight inputs is routed to one of the four OR gates in the first tier of gates. These four OR gates reduce the eight inputs to four outputs, which are then sent to the two OR gates in the second tier. These two OR gates reduce their four inputs to two outputs, which are sent to the final OR gate. Then the output of the last OR gate becomes the output of the entire sensor circuit.

Looking at NAND Gates

A *NAND* gate is a combination of an AND gate and a NOT gate. In fact, the name *NAND* is a contraction of *NOT* and *AND*. As you can see in Table 2-6, the output of a NAND gate is LOW when both of the inputs are HIGH. Otherwise, the output of the NAND gate is HIGH.

Table 2-6	The Truth Table for a Two-Input NAND Gate	
Input A	*Input B*	*Output*
0	0	1
0	1	1
1	0	1
1	1	0

The standard symbol for a NAND gate is shown in Figure 2-9. This symbol is the same as the symbol for an AND gate, with the addition of a circle at the output. As in the symbol for a NOT gate, the circle indicates that the output is inverted. In other words, a NAND gate is an AND gate whose output is inverted.

Figure 2-9:
The symbol for a two-input NAND gate.

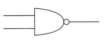

The NAND gate is special because you can use various combinations of NAND gates to create AND, OR, or NOT gates. Thus, a logic network that consists of a combination of NOT, AND, and OR gates can be created with an equivalent combination of just NAND gates. For this reason, the NAND gate is called a *universal gate*. You learn more about this characteristic of NAND gates in "Exploring Universal Gates," later in this chapter.

Another interesting thing about the NAND gate is that it can be used as a kind of OR gate that tests for LOW inputs instead of HIGH inputs. In other words, the output of a NAND gate is HIGH whenever either of the inputs are LOW.

This characteristic of NAND gates is useful in many situations. For example, consider the alarm sensor circuit that was presented in the previous section and in Figure 2-8. Suppose all of the alarm sensors produce a HIGH signal when there is no intrusion, then go LOW when there is an intrusion. In the real world, many alarm sensors work exactly in this way. For example, a sensor that detects whether a door is open is essentially a simple switch that is closed when the door is closed and open when the door is open. When the door is closed, current flows through the switch so the signal from the sensor is HIGH. When the door is opened, current stops flowing and the signal from the sensor goes LOW.

Figure 2-10 shows how NAND gates can be used to test for a LOW input on any of eight sensors. This circuit is identical to the circuit that was shown in Figure 2-8, except that all of the OR gates have been replaced by NAND gates.

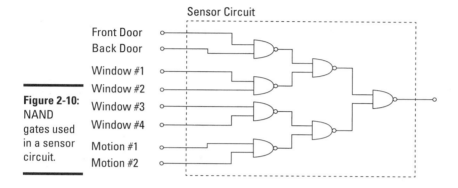

Figure 2-10: NAND gates used in a sensor circuit.

Looking at NOR Gates

A *NOR* gate is a combination of an OR gate and a NOT gate. As with NAND, the name NOR is a contraction of NOT and OR. The truth table for a NOR gate is shown in Table 2-7. As you can see, the output of a NOR gate is LOW when any of its input are HIGH. Otherwise, the output of the NAND gate is HIGH.

Table 2-7	The Truth Table for a Two-Input NOR Gate	
Input A	*Input B*	*Output*
0	0	1
0	1	0
1	0	0
1	1	0

As you can see in Figure 2-11, the standard symbol for a NOR gate is the same as the symbol for an OR gate, with a negation circle added to the output. The circle simply indicates that the output is inverted. A NOR gate is essentially a combination of an OR gate and a NOT gate

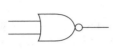

Like the NAND gate, the NOR gate is a universal gate. Any logic network that consists of NOT, AND, and OR gates can be recreated with just NOR gates. For more information, see the section "All You Need is NAND (or NOR)".

Just as a NAND gate is like an OR gate for LOW inputs, a NOR gate is like an AND gate for LOW inputs. In other words, the output of a NAND gate is HIGH when both of the inputs are LOW.

Figure 2-5, earlier in this chapter, shows an alarm circuit that sounds an alarm 30 seconds after an input sensor detects an intrusion. In that circuit, the output pulse from a 555 timer circuit is inverted by a NOT gate and then sent to an AND gate, which sends a HIGH output to an audible alarm circuit when the output from a sensor circuit is HIGH and when the 30-second timer pulse ends.

Figure 2-12 shows a version of the same circuit that uses a NOR gate instead of an AND gate to feed the audible alarm circuit. In this circuit, the alarm sounds when the output from the sensor circuit is LOW and the output from the timer circuit is also LOW. A NOT gate isn't needed for the 555 timer output, but now a NOT gate is required on the output from the AND gate to invert its signal, so that it emits LOW when the system is armed and the alarm is tripped.

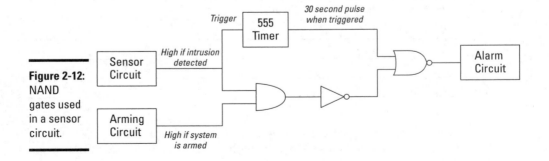

Figure 2-12:
NAND
gates used
in a sensor
circuit.

Looking at XOR and XNOR Gates

I have but two gate types remaining for your approval in this chapter: *XOR,* which stands for *Exclusive OR,* and *XNOR,* which stands for *Exclusive NOR.*

In an XOR gate, the output is HIGH if one, and only one, of the inputs is HIGH. If both inputs are LOW or both are LOW, the output is LOW. The truth table for an XOR gate is shown in Table 2-8.

Table 2-8	The Truth Table for a Two-Input XOR Gate	
Input A	*Input B*	*Output*
0	0	0
0	1	1
1	0	1
1	1	0

Another way to explain an XOR gate is as follows: The output is HIGH if the inputs are different; if the inputs are the same, the output is LOW.

The XOR gate has a lesser-known cousin called the XNOR gate. An *XNOR gate* is an XOR gate whose output is inverted. Table 2-9 lists the truth table for an XNOR gate.

Table 2-9	The Truth Table for a Two-Input XNOR Gate	
Input A	*Input B*	*Output*
0	0	1
0	1	0
1	0	0
1	1	1

Figure 2-13 shows the symbols used for both XOR and XNOR gates. As you can see, the only difference between these two symbols is that the XNOR has a circle on its output to indicate that the output is inverted.

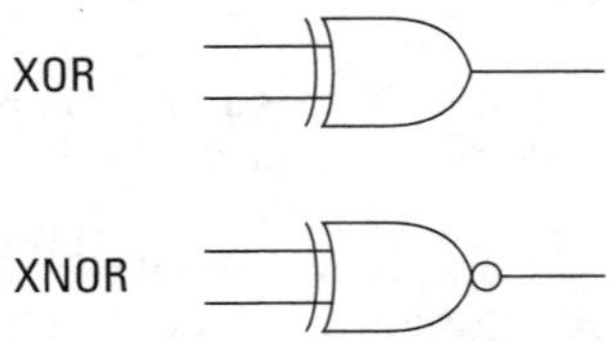

Figure 2-13:
The symbols for two-input XOR and XNOR gates.

XOR

XNOR

One of the most common uses for XOR gates is to add two binary numbers. For this operation to work, the XOR gate must be used in combination with an AND gate, as shown in Figure 2-14.

Figure 2-14:
An XOR gate and an AND gate can be used to add two binary numbers.

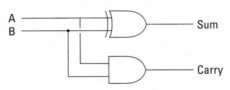

To understand how the circuit shown in Figure 2-14 works, review how binary addition works:

$$0+0=0$$
$$0+1=1$$
$$1+0=1$$
$$1+1=10$$

If you wanted, you could write the results of each of the preceding addition statements by using two binary digits, like this:

$$0+0=00$$
$$0+1=01$$
$$1+0=01$$
$$1+1=10$$

When results are written with two binary digits, as in this example, you can easily see how to use an XOR and an AND circuit in combination to perform binary addition. If you consider just the first binary digit of each result, you'll notice that it looks just like the truth table for an AND circuit and that the second digit of each result looks just like the truth table for an XOR gate.

The adder circuit shown in Figure 2-14 has two outputs. The first is called the Sum, and the second is called the Carry. The Carry output is important when several adders are used together to add binary numbers that are longer than 1 bit.

De Marvelous De Morgan's Theorem

It's time for a theorem! Although I've presented some pretty sophisticated logic concepts in this chapter, I've avoided mentioning any that go by the name *theorem* — until now, anyway.

De Morgan's Theorem was created by Augustus De Morgan, a 19th-century British mathematician who developed many of the concepts that make Boolean logic work. Among De Morgan's most important work are two related theorems that have to do with how NOT gates are used in conjunction with AND and OR gates:

+ An AND gate with inverted output behaves the same as an OR gate with inverted inputs.

+ An OR gate with inverted output behaves the same as an AND gate with inverted inputs.

An AND gate with inverted output is also called a NAND gate, of course, and an OR gate with inverted output is also called a NOR gate. Thus, De Morgan's laws can also be stated like this:

+ A NAND gate behaves the same as an OR gate with inverted inputs.

+ A NOR gate behaves the same as an AND gate with inverted inputs.

An OR gate with inverted inputs is called a *negative OR gate,* and an AND gate with inverted inputs is called a *negative AND gate.*

In case you're not persuaded, review for a moment the truth table for a NAND gate:

A	B	X
0	0	1
0	1	1
1	0	1
1	1	0

Now look at the truth table for an OR gate, with an extra set of columns added to show the inverted inputs:

A	B	NOT A	NOT B	X
0	0	1	1	1
0	1	1	0	1
1	0	0	1	1
1	1	0	0	0

Here, the A and B columns represent the inputs. The NOT A and NOT B columns are the inputs after they've been inverted. Finally, the X column represents an OR operation applied to the NOT A and NOT B values.

As you can see, the final output column of these truth tables is the same. Thus, a NAND gate is equivalent to a negative OR gate. Any time you see a NAND gate in a circuit diagram, you can substitute a negative OR gate.

Now take a look at the other side of De Morgan's Theorem. Here's a truth table for a NOR gate:

A	B	X
0	0	1
1	0	0
0	1	0
1	1	0

And here's the output of a negative AND gate:

A	B	NOT A	NOT B	X
0	0	1	1	1
0	1	1	0	0
1	0	0	1	0
1	1	0	0	0

Again, you can see that these two truth tables give the same output.

Just as a circle is used on the output of a NAND or NOR gate to indicate that the output is inverted, you can use a circle on the inputs to an OR or AND gate to indicate that the inputs are inverted. Figure 2-15 shows these symbols. The figure also shows that the negative OR and AND gates are interchangeable with the NAND and NOR gates.

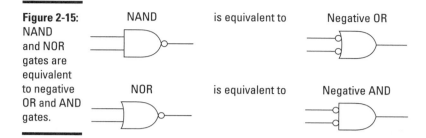

Figure 2-15: NAND and NOR gates are equivalent to negative OR and AND gates.

All You Need Is NAND (Or NOR)

In "Looking at NAND Gates" and "Looking at NOR Gates," earlier in this chapter, I mention that the NAND gate is a universal gate because any other type of gate (such as AND, OR, XOR, and NXOR) can be constructed solely from combinations of NAND gates.

This fact is incredibly useful because it enables you to build any logic circuit, simple or complex, by using just NAND gates. When you begin to build your own digital circuits, you can stock up on integrated circuits that contain just NAND gates and be assured that you can build even the most complex circuits with your stock of NAND gates.

In "Looking at NOR Gates," I also mention that the NOR gate is a universal gate. Thus, you can also build any logic circuit by using nothing but NOR gates.

In the following sections, I first explain how you can use NAND gates to build other types of gates; then I show you how to do the same thing with NOR gates.

Universal NAND Gates

Figure 2-16 shows how you can use NAND gates in various combinations to create NOT, AND, OR, and NOR gates. The following paragraphs describe how the circuits work:

✦ **NOT:** You can create a NOT gate from a NAND gate simply by tying the two inputs of the NAND gate together. Because the two inputs of the NAND gate are tied together, only two input combinations are possible: both HIGH or both LOW. If both inputs are HIGH, the NAND gate will output a LOW. If both inputs are LOW, the NAND gate will output HIGH. Thus, the circuit behaves exactly as a NOT gate would.

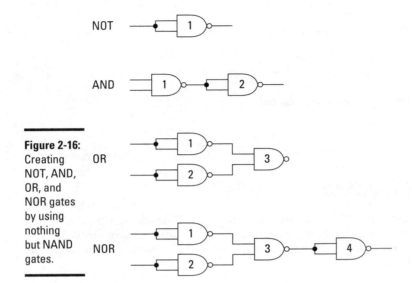

Figure 2-16:
Creating NOT, AND, OR, and NOR gates by using nothing but NAND gates.

✦ **AND:** You can create an AND gate by using two NAND gates. The first NAND gate does what NAND gates do: returns LOW if both inputs are HIGH and returns HIGH if both inputs are anything else. Then the second NAND gate is configured as a NOT gate to invert the output from the first NAND gate.

One of the basic rules of NOT gates is that if you invert a signal twice, you end up with the same signal. If the original input is HIGH, and you invert it, the signal becomes LOW. Invert the input again, and it returns to HIGH. With this rule in mind, you should be able to see how the two NAND gates work together to create an AND gate.

✦ **OR:** You need three NAND gates to create an OR gate. First, you use a pair of NAND gates configured as NOT gates to invert the two inputs. Then the third NAND gate produces a LOW output if both of the original inputs are LOW. If one of the original inputs is HIGH, or if both of the original inputs are HIGH, the output of the third gate is HIGH.

✦ **NOR:** It takes four NAND gates to create a NOR gate. As you can see in Figure 2-16, this circuit is the same as the OR gate circuit, with the addition of another NOT gate to invert the output from the third NAND gate. Inverting the output changes the overall function of the circuit from OR to NOR.

Universal NOR Gates

Like NAND, NOR is a universal gate. Figure 2-17 shows how NOR gates can be combined in various and sundry ways to create NOT, AND, OR, and NAND gates.

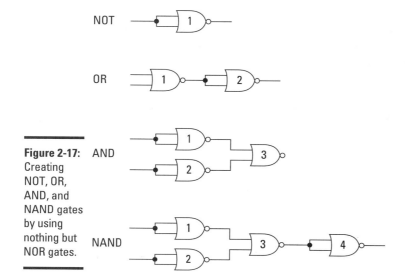

Figure 2-17:
Creating
NOT, OR,
AND, and
NAND gates
by using
nothing but
NOR gates.

The following paragraphs describe how the circuits work:

✦ **NOT:** Creating a NOT gate from a NOR gate is the same as creating a NOT gate from a NAND gate: You simply tie the two inputs of the NOR gate together. If both inputs are LOW, the NOR gate will output a HIGH. Otherwise, the output is LOW.

✦ **OR:** You need two NOR gates to create an OR gate. The first NOR gate returns LOW if either input is HIGH or both inputs are HIGH. Then the second NOR gate is configured as a NOT gate to invert the output of the first NOR gate.

✦ **AND:** You need three NOR gates to create an AND gate. The first two are configured as NOT gates, so they invert the inputs. Then the third NOR gate produces a HIGH output if both of the original inputs are HIGH.

✦ **NAND:** It takes four NOR gates to create a NAND gate. The first three NOR gates are configured just as they are for an AND gate. Then a fourth NOR gate configured as a NOT gate inverts output from the third NOR gate.

Using Software to Practice with Gates

If you really want to learn how logic gates work, one of the best ways is to download one of the many software logic gate simulators that are available free on the Internet. You can find these programs by firing up Google and searching for keywords such as *"logic gate simulator."*

One of my favorite programs is Logic Circuit Designer, written by Ivan Andrei. You can download it from `http://download.cnet.com/Logic-Circuit-Designer/3000-2054_4-10840569.html`. Figure 2-18 shows this useful program in action.

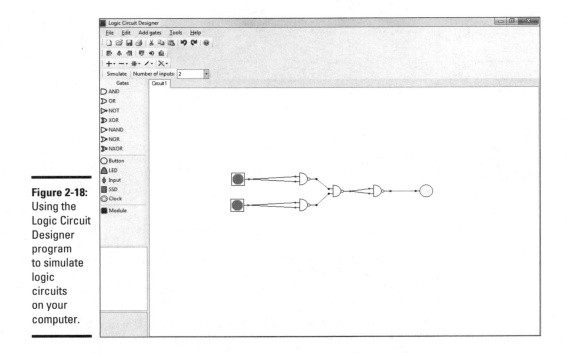

Figure 2-18: Using the Logic Circuit Designer program to simulate logic circuits on your computer.

Here are just a few of the features of the Logic Circuit Designer program:

✦ You can add AND, OR, NOT, XOR, NAND, NOR, and NXOR gates to your circuit simply by dragging them from a toolbar.

✦ You can connect the gate inputs and outputs by clicking an input or output and then dragging it to another gate input or output.

✦ You can add simple switches to the circuit for external inputs.

✦ You can add LEDs at any point in the circuit to indicate the status of an output or an input.

Chapter 3: Working with Logic Circuits

In This Chapter

✔ Implementing logic gates with transistors

✔ Looking at TTL and CMOS integrated circuits

✔ Building some simple logic circuits

*I*n Chapter 2, you learn all about logic gates, including the seven most popular kinds of gates: NOT, AND, OR, NAND, NOR, XOR, and XNOR.

In this chapter, you learn how to create actual circuits that use logic gates. I start you off by showing you the basics of how logic gates can be created from simple transistor circuits. Then I look at two popular integrated circuit families that provide prebuilt logic gates.

If you haven't already read Chapter 2, I suggest that you go back and do so now before you read this chapter. Little of this chapter will make sense if you aren't familiar with the different types of gates described in that chapter.

Creating Logic Gates with Transistors

Way back in Book II, Chapter 6, you saw how transistors can be used as switches. In a nutshell, a voltage applied to the base of a transistor allows current to flow from the collector to the emitter. Thus, by applying an input signal to a transistor's base, you can control an output signal taken from the collector-emitter path.

You can build any logic gate you want by cobbling together a few transistors and resistors in just the right way. In this section, I look at simple transistor circuits for five gate types: NOT, AND, OR, NAND, and NOR.

If you need a refresher on how transistors work, refer to Book II, Chapter 6.

All the circuits in this chapter assume that a HIGH signal (logical one) is represented by a DC voltage of +5 V or more. LOW (logical zero) is represented by near-zero voltage.

Note that you won't often actually build your own logic gates by using transistors and resistors. Instead, you'll use integrated circuits that contain prefabricated logic gates. Before you use logic ICs in your circuits, however, you should have a basic understanding of how the gates that they contain work. After you take a look at simple transistor circuits for basic logic gates, you'll take a look at the logic ICs.

A transistor NOT gate circuit

A NOT gate simply inverts its input. If the input is HIGH, the output is LOW, and if the input is LOW, the output is HIGH. Such a circuit is easy to build, using a single transistor and a pair of resistors. Figure 3-1 shows the schematic.

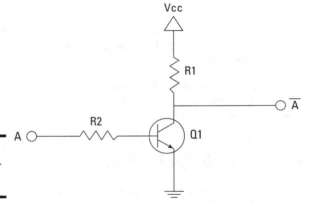

Figure 3-1:
A transistor
NOT gate.

The operation of this circuit is simple. The input is connected through resistor R2 to the transistor's base. When no voltage is present on the input, the transistor turns off. When the transistor is off, no current flows through the collector-emitter path. Thus, current from the supply voltage (V_{cc} in the schematic, typically between +5 V and +9 V) flows through resistor R1 to the output. In this way, the circuit's output is HIGH when its input is LOW.

When voltage is present at the input, the transistor turns on, allowing current to flow through the collector-emitter circuit directly to ground. This ground path creates a shortcut that bypasses the output, which causes the output to go LOW.

In this way, the output is HIGH when the input is LOW and LOW when the input is HIGH.

Project 3-1 shows how to assemble a simple transistor NOT gate on a solderless breadboard. For this project, a normally open pushbutton is used as the input. When the button isn't pressed, the input is LOW and the output is HIGH, which causes the LED to light. When you press the button, the input goes HIGH, the output goes LOW, and the LED goes out. The assembled project is shown in Figure 3-2.

Figure 3-2: A transistor NOT gate assembled on a breadboard (Project 3-1).

Project 3-1: A Transistor NOT Gate

In this project, you build a simple NOT gate by using a bipolar transistor. A NOT, also known as an inverter, reverses the logic level of its input. Thus, if the input is HIGH, the output of the NOT gate is LOW. If the input is LOW, the output is HIGH.

The input to this gate is controlled by the pushbutton switch SW1. When the switch is open, the input is LOW. When the switch is closed, the input is HIGH.

The output from this gate is sent through an LED, so the LED is on when the output is HIGH and off when the output is LOW.

Parts List

1	Four-cell AAA battery holder
4	AAA batteries
1	NPN transistor (2N2222 or equivalent)
1	Red LED
2	1 kΩ ¼ W resistors (Brown-Black-Red)
1	Normally open pushbutton Miscellaneous jumper wires

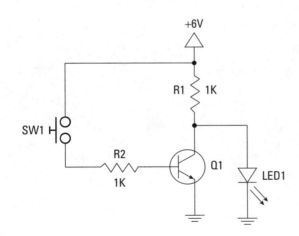

NOT Gate Truth Table

Input	Output
0	1
1	0

Steps

1. **Insert transistor Q1.**
   Collector: G9
   Base: G10
   Emitter: G11

2. **Insert resistors R1 and R2.**
   R1 - 1 kΩ: F9 to positive bus
   R2 - 1 kΩ: F10 to D12

3. **Insert LED1.**
   Cathode (short lead): Ground bus
   Anode (long lead): J9

4. **Insert the jumper wire.**
   From: J11
   To: Ground bus

5. **Insert pushbutton SW1.**
   From: A12
   To: Positive bus

6. **Connect the batteries.**
   Red lead: Positive bus
   Black lead: Negative bus

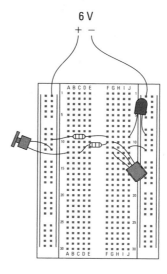

6 V
+ −

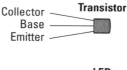

Collector
Base   **Transistor**
Emitter

Cathode   **LED**
Anode

A transistor AND gate circuit

A two-input AND gate produces a HIGH output if both of its inputs are HIGH. You can create a two-input AND gate by using two transistors and three resistors, as shown in Figure 3-3. In this circuit, the output current must flow from the Vcc supply voltage through the collector-emitter circuits of both transistors to reach the output. Thus, current will flow to the output only if both of the transistors are on.

The bases of both transistors are fed through R2 and R3 from the two inputs. Thus, if both inputs are HIGH, current will flow through the base-emitter path of both transistors, turning both transistors on and allowing current to flow through to the output. If either input is LOW, the corresponding transistor turns off, and the output goes LOW.

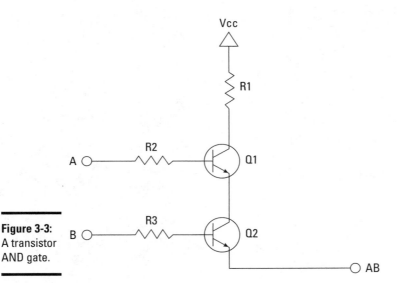

Figure 3-3:
A transistor
AND gate.

A transistor NAND gate circuit

A two-input NAND gate produces a LOW output if both of its inputs are HIGH. You could create a NAND gate by combining the circuits shown in Figure 3-1 and Figure 3-3 so that the output from the AND gate is used as input to the NOT gate, but that combination would require three transistors. It's easy enough to create a NAND gate by using just two transistors, as shown in Figure 3-4.

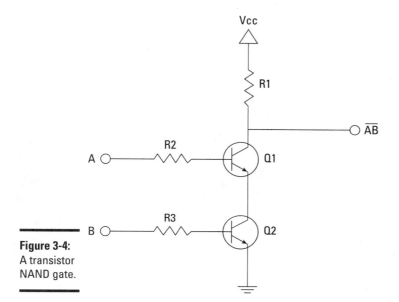

Book VI
Chapter 3

Working with
Logic Circuits

Figure 3-4:
A transistor
NAND gate.

This NAND gate circuit is almost identical to the AND gate circuit shown in Figure 3-3, earlier in this chapter. The only difference is that instead of connecting the output to the emitter of the second transistor, the output is obtained before the collector of the first transistor. If both of the inputs are HIGH, both of the transistors will conduct through their collector-emitter paths, which creates a short circuit to ground. This causes the current to bypass the output altogether, which in turn causes the output to go LOW.

If either transistor turns off, however, the supply current can't flow through the transistors to ground, so it flows through the output circuit instead. Thus, the output is HIGH if either one of the inputs is LOW. If both inputs are HIGH, the output is LOW.

Project 3-2 shows how to assemble a simple transistor NAND gate on a solderless breadboard. Normally open pushbuttons are used for the two inputs. The LED will be on until you press both of the pushbuttons. This action causes both inputs to go HIGH, which causes the output to go LOW and the LED to go dark. The completed project is shown in Figure 3-5.

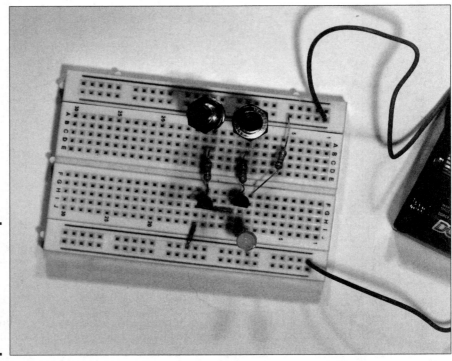

Figure 3-5:
A two-transistor NAND gate on a breadboard (Project 3-2).

A transistor OR gate circuit

A two-input OR gate produces a HIGH output if either of its inputs is HIGH or both of its inputs are HIGH. Figure 3-6 shows a schematic for an OR gate created with two transistors and three resistors.

In the OR gate circuit, the supply voltage is connected separately to the collector of each transistor. Then the emitters of both transistors are connected to the output. That way, if voltage is applied to the base of either one of the transistors, that transistor will turn on and pass current through to the output.

Thus, the output is HIGH if one input is HIGH or both inputs are HIGH. The output is LOW only if both inputs are LOW.

A transistor NOR gate circuit

A NOR gate is an inverted OR gate. If at least one of the inputs is HIGH, the output is LOW. If both inputs are LOW, the output is HIGH.

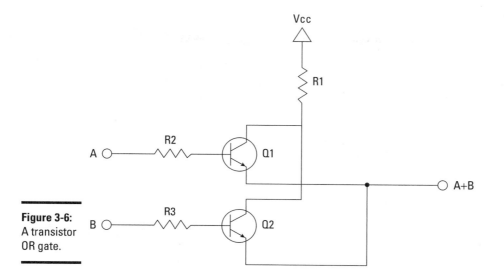

Figure 3-6:
A transistor
OR gate.

Figure 3-7 shows a schematic for a NOR gate. This circuit is similar to the circuit shown in Figure 3-6, earlier in this chapter, except that the output is connected to the collector of both transistors and the emitter of each transistor is connected to ground. If either one of the transistors is on, current from Vcc will be short-circuited to ground, bypassing the output. As a result, the output will be HIGH only when both inputs are LOW. If either input is HIGH or both inputs are HIGH, the output will be LOW.

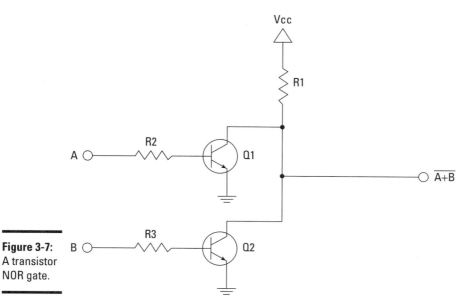

Figure 3-7:
A transistor
NOR gate.

Project 3-2: A Transistor NAND Gate

In this project, you build a simple NAND gate by using two bipolar transistors. The output of a NAND gate is LOW if both inputs are HIGH; otherwise, the output is HIGH.

The output from this gate is sent through an LED, so the LED is on when the output is HIGH and off when the output is LOW.

The inputs to this gate are controlled by the pushbutton switches SW1 and SW2. When a switch is open, the corresponding input is LOW. When the switch is depressed, the corresponding input is HIGH.

Parts List

1 Four-cell AAA battery holder
4 AAA batteries
2 NPN transistors (2N2222 or equivalent)
1 Red LED
3 1 kΩ 1/4 W resistors (Brown-Black-Red)
2 Normally open pushbuttons
Miscellaneous jumper wires

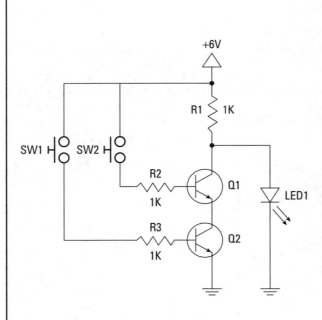

NAND Gate Truth Table

Input A	Input B	Output
0	0	1
0	1	1
1	0	1
1	1	0

Steps

1. **Insert transistors Q1 and Q2.**

Lead	Q1	Q2
Collector:	G9	G13
Base:	G10	G14
Emitter:	G11	G15

2. **Insert resistors R1, R2, and R3.**

   R1 - 1 kΩ: F9 to positive bus
   R2 - 1 kΩ: F10 to C10
   R3 - 1 kΩ: F14 to C14

3. **Insert LED1.**

   Cathode (short lead): Ground bus
   Anode (long lead): J9

4. **Insert the jumper wires.**

   Jumper 1: From H11 to H13
   Jumper 2: From J15 to ground bus

5. **Insert switches SW1 and SW2.**

   SW1: From A10 to positive bus
   SW2: From A14 to positive bus

6. **Connect the batteries.**

   Red lead: Positive bus
   Black lead: Negative bus

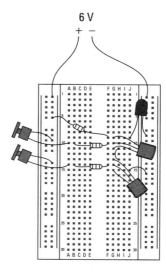

6 V
+ −

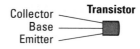

Transistor
Collector
Base
Emitter

LED
Cathode
Anode

Book VI
Chapter 3

Working with
Logic Circuits

You can build a two-transistor NOR gate by following the steps outlined in Project 3-3. As with the other projects in this chapter, Project 3-3 uses normally open pushbuttons to control the input circuits. When power is applied to this circuit, both inputs initially will be LOW, and the output will be HIGH. Pressing either one of the switches causes that switch's input to go HIGH, which in turn causes the output to go LOW. The assembled project is shown in Figure 3-8.

Figure 3-8: A two-transistor NOR gate on a breadboard (Project 3-3).

Introducing Integrated Circuit Logic Gates

Although you can build your own logic gates by using transistors and resistors as described so far in this chapter, it's far easier to buy prepackaged integrated circuits that implement logic gates. The main advantage of using integrated circuit logic gates is that you don't have to design the individual gates yourself or waste time assembling them.

The logic gate circuits you've learned about so far in this book are among the simplest circuits for creating logic gates, but they're by no means the only — and not necessarily the best — ways to create logic gates. Over the 50 years or so that circuit designers have been working on semiconductor-based logic circuits, many designs have been developed for creating logic gates.

Because each approach to designing logic circuits results in an entire family of logic circuits for the various types of gates (NOT, AND, OR, NAND, NOR, XOR, and XNOR), the different designs are often referred to as *design families.* Also, because these design families were developed by engineers far nerdier than you and I, each family has a nice three- or four-letter acronym. Here are the most popular:

✦ **RTL:** *Resistor-Transistor Logic,* which uses resistors and bipolar transistors. The circuits presented so far in this chapter are examples of RTL circuits.

✦ **DTL:** *Diode-Transistor Logic,* which is similar to RTL but adds a diode to each input circuit.

✦ **TTL:** Not Tinkertoy Logic, but *Transistor-Transistor Logic.* It uses two transistors, one configured to work as a switch and the other configured to work as an amplifier. The switching transistor is used in the input circuits, and the amplifier transistor is used in the output circuits. The amplifier allows the gate's output to be connected to a larger number of inputs than RTL or DTL circuits.

In a TTL circuit, the switching transistors are actually special transistors that have two or more emitters. Each input is connected to one of these emitters so that the separate inputs all control the same collector-emitter circuit. The switching transistor's base is connected to the Vcc supply voltage, and the collector is connected to the base of the amplifying transistor. Figure 3-9 shows a typical TTL circuit.

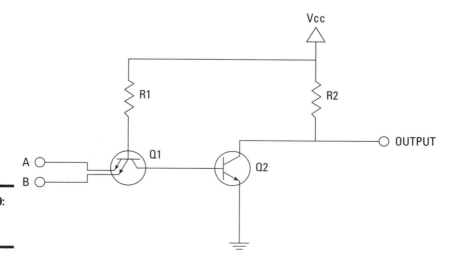

Figure 3-9:
A typical
TTL gate.

Project 3-3: A Transistor NOR Gate

In this project, you build a simple NOR gate by using two bipolar transistors. The output of a NOR gate is HIGH if both inputs are LOW. Otherwise the output is LOW.

The output from this gate is sent through an LED, so the LED is on when the output is HIGH and off when the output is LOW.

The inputs to this gate are controlled by normally open pushbutton switches. When a switch is open, the corresponding input is LOW. When the switch is depressed, the corresponding input is HIGH.

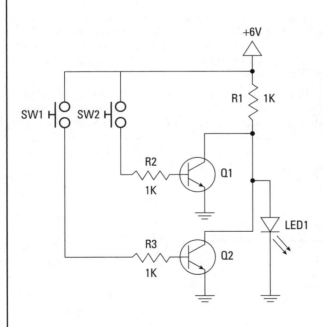

Parts List

1 Four-cell AAA battery holder
4 AAA batteries
2 NPN transistors (2N2222 or equivalent)
1 Red LED
3 1 kΩ ¼ W resistors (Brown-Black-Red)
2 Normally open pushbuttons Miscellaneous jumper wires

NOR Gate Truth Table

Input A	Input B	Output
0	0	1
0	1	0
1	0	0
1	1	0

Steps

1. **Insert transistors Q1 and Q2.**

Lead	Q1	Q2
Collector:	G9	G13
Base:	G10	G14
Emitter:	G11	G15

2. **Insert resistors R1, R2, and R3.**

   R1 - 1 kΩ: F9 to positive bus
   R2 - 1 kΩ: F10 to C10
   R3 - 1 kΩ: F14 to C14

3. **Insert LED1.**

   Cathode (short lead): Ground bus
   Anode (long lead): J9

4. **Insert the jumper wires.**

   Jumper 1: From J10 to ground bus
   Jumper 2: From J15 to ground bus
   Jumper 3: From I9 to I13

5. **Insert switches SW1 and SW2.**

   SW1: From A10 to positive bus
   SW2: From A14 to positive bus

6. **Connect the batteries.**

   Red lead: Positive bus
   Black lead: Negative bus

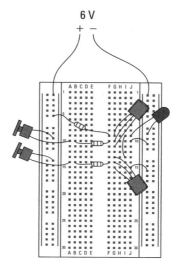

6 V

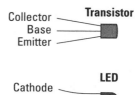

Collector — **Transistor**
Base
Emitter

LED
Cathode
Anode

Although you can build TTL circuits by using individual transistors, ICs with TTL circuits are readily available. The most popular types of TTL ICs are designated by numbers in the form 74*nn*. In all, a few hundred types of 7400-series integrated circuits are available. Many of them provide advanced logic circuits that you aren't likely to use for home electronics projects. The ICs listed in Table 3-1 provide several basic logic gates in a single package.

Table 3-1	7400-Series TTL Logic Gates
Number	*Description*
7400	Quad two-input NAND gate (four NAND gates)
7402	Quad two-input NOR gate (four NOR gates)
7404	Hex inverter (six NOT gates)
7408	Quad two-input AND gate (four AND gates)
7432	Quad two-input OR gate (four OR gates)
7486	Quad two-input XOR gate (four XOR gates)

There actually is such a thing as Tinkertoy Logic. In fact, a team of MIT students actually built a complete computer that plays tic-tac-toe by using Tinkertoys. To learn more, search the Internet for *"Tinkertoy Computer."*

✦ **CMOS:** *Complementary Metal-Oxide Semiconductor Logic,* which refers to logic circuits built with a special type of transistor called a MOSFET. *MOSFET* stands for *Metal Oxide Semiconductor Field Effect Transistor,* but that won't be on the test. The physics of how a MOSFET differs from a standard bipolar transistor aren't all that important unless you want to become an IC designer. What is important is that MOSFETs use much less power, can switch states much faster, and are significantly smaller than bipolar transistors. These differences make MOSFETs ideal for modern integrated circuits, which often require millions of transistors on a single chip.

Apart from drawing less power and operating faster than TTL circuits, CMOS circuits work much like TTL circuits. In fact, CMOS chips are designed to be interchangeable with comparable TTL chips.

CMOS logic chips have a four-digit part number that begins with the number 4 and are often called 4000-series chips. As with the 7400 series of TTL logic chips, several hundred types of 4000-series chips are available. Table 3-2 lists the 4000-series chips that provide basic logic gates.

Table 3-2	4000-Series CMOS Logic Gates
Number	*Description*
4001	Quad two-input NOR gate (four NOR gates)
4009	Hex inverter (six NOT gates)
4011	Quad two-input NAND gate (four NAND gates)
4030	Quad two-input XOR gate (four XOR gates)
4071	Quad two-input OR gate (four OR gates)
4077	Quad two-input XNOR gate (four XNOR gates)
4081	Quad two-input AND gate (four AND gates)

CMOS logic circuits are very sensitive to static electricity. As a result, you should take special precautions when you handle them. Make sure that you discharge yourself properly by touching a grounded metal surface before you touch a CMOS chip. For maximum protection, wear an antistatic wrist band. For more information about static precautions, refer to Book I, Chapter 4.

The remaining sections of this chapter describe several popular 4000-series ICs and present a few projects that use them.

Introducing the Versatile 4000-Series Logic Gates

The 4000-series CMOS logic circuits include several ICs that provide several logic gates in a single package. Figure 3-10 shows the pinout connections for six popular 4000-series chips. Each of these 6 chips contains 4 2-input logic gates in a 14-pin DIP package. Power, which can range from +3 V to +15 V, is connected to pin 14, and ground is connected to pin 7.

Note that the 4001 and 4011 chips contain four NOR and NAND gates. Because NOR and NAND are universal gates, you can use them in combination to create other types of gates. As a result, if you stock up on 4001 or 4011 chips, you'll be able to create any type of logic circuit you may need.

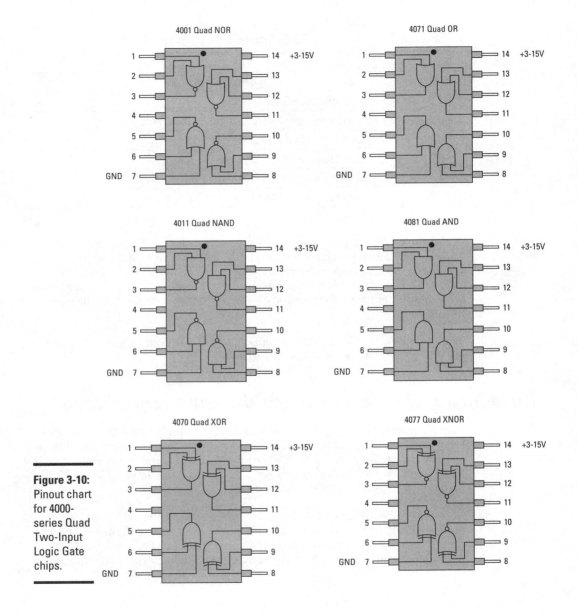

Figure 3-10:
Pinout chart for 4000-series Quad Two-Input Logic Gate chips.

Before you start building circuits with CMOS logic chips, here are a few tips for working with 4000-series chips:

✦ The outputs can source as much as 10 mA with a 9 V power supply. At 6 V, the maximum is about 5 mA, which is just enough to light up an LED. If your output circuit requires more current, you can always use a transistor. Just connect the transistor's collector to the positive voltage source and the base to the logic gate output. Then connect your output circuit to the transistor's emitter. The output circuit should eventually lead to ground, of course, to complete the circuit.

✦ The input pins of CMOS logic chips are notorious for picking up stray signals in the form of electrical noise. Although it isn't necessary to do so in experimental breadboard circuits, in an actual circuit you should connect all unused input pins to the positive supply voltage or to ground. (You don't have to connect the unused outputs to anything — just the inputs.)

✦ In addition to connecting unused inputs to positive voltage or ground, it's a good idea to place a small capacitor (47 mF is typical) across the power leads (pins 7 and 14). The capacitor helps ensure that the input voltage stays constant.

✦ Don't forget that CMOS chips are very susceptible to damage from small amounts of static electricity. Be sure to ground yourself by touching a metal object before you touch CMOS chips.

**Book VI
Chapter 3**

**Working with
Logic Circuits**

Building Projects with the 4011 Quad Two-Input NAND Gate

The 4011 Quad Two-Input NAND Gate is a popular CMOS logic gate IC. As its name suggests, this IC contains four two-input NAND gates. The pinouts for this IC are shown in Figure 3-10, earlier in this chapter, along with the pinouts for several other quad two-input gate chips.

Unlike most of the components used in this book, a 4011 IC isn't available at your local RadioShack. Thus, you'll have to find a local electronics store that carries parts such as CMOS logic chips, or you'll have to order them from an online source. I list several reputable sources in Book I, Chapter 3.

In Book VI, Chapter 2, you learned that NAND gates (along with NOR gates) are *universal gates,* which means that you can construct any other type of gate by using nothing but NAND gates combined in various ways. Projects 3-4 through 3-7 walk you step by step through the process of building various types of gate circuits by using only NAND gates.

Project 3-4 uses just one of the NAND gates in a 4011. The two inputs of the NAND gate are connected to pushbuttons, and the output is connected to an LED. When you build this project, you'll be able to visualize the operation of a NAND gate: The LED will be on unless you press both buttons.

Figure 3-11 shows Project 3-4 assembled on a solderless breadboard. This figure should give you a good idea of how to connect the components to the breadboard. The project description itself provides detailed instructions. The circuits for projects 3-5, 3-6, and 3-7 are similar enough in appearance that you can use Figure 3-11 as a guide for the overall appearance of your finished projects.

Figure 3-11:
A NAND
gate circuit
that uses
a CMOS
logic chip.
(Project 3-4)

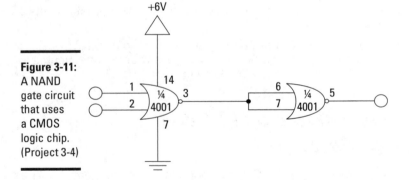

Project 3-5 uses two of the NAND gates on the 4011 to create an AND gate. Because a NAND gate is nothing more than an AND gate whose output is inverted, you can create an AND gate from a NAND gate by inverting the NAND gate's output. This inversion works because of one of the fundamental rules of logic: If you invert a value twice, you get the original value. Thus, if you invert an AND gate once, you get a NAND gate; if you invert it again, you're back to an AND gate.

As luck would have it, you can easily turn a NAND gate into a single-input inverter (that is, a NOT gate) by connecting the single input to both inputs of the NAND gate. This connection causes the two inputs to always be the same: Either both are HIGH, or both are LOW. In a NAND gate, if both inputs are HIGH, the output is LOW, and if both inputs are LOW, the output is HIGH. Thus, wiring the inputs of a NAND gate together has the effect of inverting the input.

In Project 3-6, you see how to create an OR gate by using three NAND gates. Book VI, Chapter 2 shows that a NAND gate is the same as an OR gate whose inputs have been inverted. Thus, to create an OR gate by using NAND gates, you invert the two inputs with NAND gates configured as inverters (that is, with their inputs wired together). The output from these inverters is sent to the inputs of the third NAND gate.

The final project in this chapter, Project 3-7, uses all four of the NAND gates on the 4011 chip to create a NOR gate. A NOR gate is nothing more than an OR gate whose output has been inverted. Thus, you first create an OR gate by using the technique in Project 3-6; then you configure the fourth NAND gate on the 4011 chip as an inverter to invert the output of the OR gate.

Because these projects build on one another, I recommend that you don't tear down your breadboard after completing each project. Instead, you can use each assembled project as a starting point for the next one. If you choose to build the projects in this way, you can just scan the steps to see what resistors and jumper wires need to be moved for each project. (The 4011 IC itself, the LED, and the two switches are in the same locations for all four projects.)

**Book VI
Chapter 3**

**Working with
Logic Circuits**

Project 3-4: A CMOS NAND Gate

In this project, you use a 4011 CMOS chip to build a NAND gate. The output of a NAND gate is HIGH unless both inputs are HIGH. When both inputs are HIGH, the output goes LOW.

The 4011 IC has four independent NAND gate circuits. In this project, you use just one of the four.

The inputs to this gate are fed through two normally open pushbutton switches. When a switch isn't pressed, the corresponding input is LOW. When the switch is pressed, the corresponding input goes HIGH.

The output from this gate is sent through an LED, so the LED is on when the output is HIGH and off when the output is LOW.

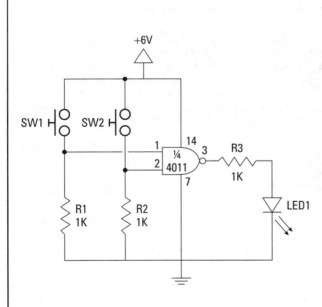

Parts List

1	Four-cell AAA battery holder
4	AAA batteries
1	4011 CMOS Quad Two-Input NAND Gate
1	RED LED
3	1 kΩ ¼ W resistors (Brown-Black-Red)
2	Normally open pushbuttons
	Miscellaneous jumper wires

NAND Gate Truth Table

Input A	Input B	Output
0	0	1
0	1	1
1	0	1
1	1	0

Steps

1. **Insert the 4011 IC.**
   Pin 1 should be in hole E10.

2. **Insert resistors R1, R2, and R3.**
   R1 - 1 kΩ: B4 to ground bus
   R2 - 1 kΩ: B8 to ground bus
   R3 - 1 kΩ: D12 to D20

3. **Insert LED1.**
   Cathode (short lead): Ground bus
   Anode (long lead): A20

4. **Insert the jumper wires.**
   Jumper 1: E4 to F4
   Jumper 2: E8 to F8
   Jumper 3: D4 to D10
   Jumper 4: C8 to C11
   Jumper 5: A16 to ground bus
   Jumper 6: J10 to positive bus

5. **Insert switches SW1 and SW2.**
   SW1: J4 to positive bus
   SW2: J8 to positive bus

6. **Connect the batteries.**
   Red lead: Positive bus
   Black lead: Negative bus

6 V

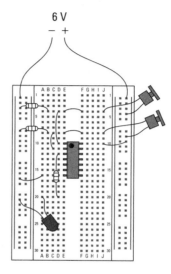

LED
Cathode
Anode

Book VI
Chapter 3

**Working with
Logic Circuits**

Project 3-5: A CMOS AND Gate

In this project, you use two of the NAND gates in a 4011 Quad NAND Gate IC to build an AND gate. The output of an AND gate is HIGH if both inputs are HIGH. If either input is LOW or both inputs are LOW, the output is LOW.

To create the AND gate, you send the output from the first NAND gate to the second NAND gate, which you configure as an inverter by tying its inputs together. Inverting the output of a NAND gate creates an AND gate.

The inputs to this gate are fed through two normally open pushbutton switches. When a switch isn't pressed, the corresponding input is LOW. When the switch is pressed, the corresponding input goes HIGH.

The output from this gate is sent through an LED, so the LED is on when the output is HIGH and off when the output is LOW.

Parts List

1 Four-cell AAA battery holder
4 AAA batteries
1 4011 CMOS Quad Two-Input NAND Gate
1 RED LED
3 1 kΩ ¼ W resistors (Brown-Black-Red)
2 Normally open pushbuttons
  Miscellaneous jumper wires

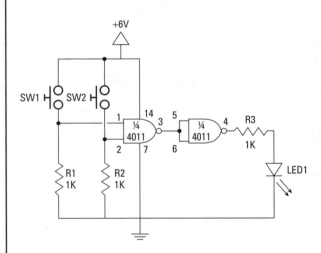

AND Gate Truth Table

Input A	Input B	Output
0	0	0
0	1	0
1	0	0
1	1	1

Steps

1. **Insert the 4011 IC.**
   Pin 1 should be in hole E10.

2. **Insert resistors R1, R2, and R3.**
   R1 - 1 kΩ: B4 to ground bus
   R2 - 1 kΩ: B8 to ground bus
   R3 - 1 kΩ: B13 to B20

3. **Insert LED1.**
   Cathode (short lead): Ground bus
   Anode (long lead): A20

4. **Insert the jumper wires.**
   Jumper 1: E4 to F4
   Jumper 2: E8 to F8
   Jumper 3: D4 to D10
   Jumper 4: C8 to C11
   Jumper 5: D12 to D14
   Jumper 6: C14 to C15
   Jumper 7: A16 to ground bus
   Jumper 8: J10 to positive bus

5. **Insert switches SW1 and SW2.**
   SW1: J4 to positive bus
   SW2: J8 to positive bus

6. **Connect the batteries.**
   Red lead: Positive bus
   Black lead: Negative bus

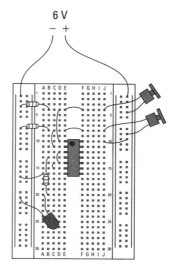

6 V
− +

LED
Cathode
Anode

Project 3-6: A CMOS OR Gate

In this project, you use three of the NAND gates in a 4011 Quad NAND Gate IC to build an OR gate. The output of an OR gate is HIGH if either input is HIGH or both of the inputs are HIGH. If both inputs are LOW, the output is LOW.

An OR gate can be built from NAND gates because an OR gate is the same thing as a NAND gate whose inputs are inverted. Thus, each of the inputs for this circuit is connected to a NAND gate configured as an inverter. Then the outputs from

these inverters are sent to a third NAND gate.

The inputs to this gate are fed through two normally open pushbutton switches. When a switch isn't pressed, the corresponding input is LOW. When the switch is pressed, the corresponding input goes HIGH.

The output from this gate is sent through an LED, so the LED is on when the output is HIGH and off when the output is LOW.

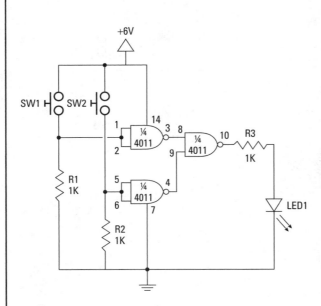

Parts List

1 Four-cell AAA battery holder
4 AAA batteries
1 4011 CMOS Quad Two-Input
   NAND Gate
1 RED LED
3 1 kΩ ¼ W resistors (Brown-
   Black-Red)
2 Normally open pushbuttons
   Miscellaneous jumper wires

OR Gate Truth Table

Input A	Input B	Output
0	0	0
0	1	1
1	0	1
1	1	1

Steps

1. **Insert the 4011 IC.**
   Pin 1 should be in hole E10.

2. **Insert resistors R1, R2, and R3.**
   R1 - 1 kΩ: B4 to ground bus
   R2 - 1 kΩ: B8 to ground bus
   R3 - 1 kΩ: B20 to G20

3. **Insert LED1.**
   Cathode (short lead): Ground bus
   Anode (long lead): A20

4. **Insert the jumper wires.**
   Jumper 1: E4 to F4
   Jumper 2: E8 to F8
   Jumper 3: D4 to D10
   Jumper 4: C8 to C14
   Jumper 5: B10 to B11
   Jumper 6: B14 to B15
   Jumper 7: D12 to G16
   Jumper 8: D13 to G15
   Jumper 9: H14 to H20
   Jumper 10: A16 to ground bus
   Jumper 11: J10 to positive bus

5. **Insert switches SW1 and SW2.**
   SW1: J4 to positive bus
   SW2: J8 to positive bus

6. **Connect the batteries.**
   Red lead: Positive bus
   Black lead: Negative bus

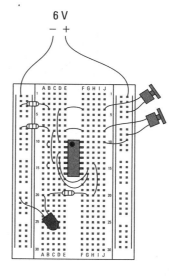

6 V
− +

LED
Cathode
Anode

**Book VI
Chapter 3**

**Working with
Logic Circuits**

Project 3-7: A CMOS NOR Gate

In this project, you use all four of the NAND gates in a 4011 Quad NAND Gate IC to build a NOR gate. The output of a NOR gate is LOW if either input is HIGH or both of the inputs are HIGH. If both inputs are LOW, the output is HIGH.

A NOR gate is simply an OR gate whose output is inverted by a NOT gate. Thus, you use the first three NAND gates of the 4011 to create an OR gate, following the steps in Project 3-6. Then you use the fourth NAND gate to invert the OR gate's output.

The inputs to this gate are fed through two normally open pushbutton switches. When a switch isn't pressed, the corresponding input is LOW. When the switch is pressed, the corresponding input goes HIGH.

The output from this gate is sent through an LED, so the LED is on when the output is HIGH and off when the output is LOW.

Parts List

1 Four-cell AAA battery holder
4 AAA batteries
1 4011 CMOS Quad Two-Input NAND Gate
1 RED LED
3 1 kΩ 1/4 W resistors (Brown-Black-Red)
2 Normally open pushbuttons
   Miscellaneous jumper wires

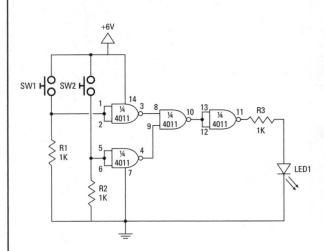

NOR Gate Truth Table

Input A	Input B	Output
0	0	1
0	1	0
1	0	0
1	1	0

Steps

1. **Insert the 4011 IC.**
   Pin 1 should be in hole E10.

2. **Insert resistors R1, R2, and R3.**
   R1 - 1 kΩ: B4 to ground bus
   R2 - 1 kΩ: B8 to ground bus
   R3 - 1 kΩ: B20 to G20

3. **Insert LED1.**
   Cathode (short lead): Ground bus
   Anode (long lead): A20

4. **Insert the jumper wires.**
   Jumper 1: E4 to F4
   Jumper 2: E8 to F8
   Jumper 3: D4 to D10
   Jumper 4: C8 to C14
   Jumper 5: B10 to B11
   Jumper 6: B14 to B15
   Jumper 7: D12 to G16
   Jumper 8: D13 to G15
   Jumper 9: G11 to G12
   Jumper 10: H12 to H14
   Jumper 11: I13 to I20
   Jumper 12: A16 to ground bus
   Jumper 13: J10 to positive bus

5. **Insert switches SW1 and SW2.**
   SW1: J4 to positive bus
   SW2: J8 to positive bus

6. **Connect the batteries.**
   Red lead: Positive bus
   Black lead: Negative bus

6 V
− +

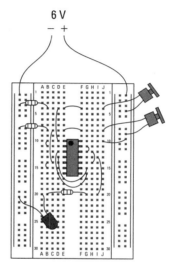

Cathode
Anode
LED

Chapter 4: Interfacing to Your Computer's Parallel Port

In This Chapter

✔ Getting to know the parallel port

✔ Building circuits you can connect to your computer's parallel port

✔ Programming the parallel port

✔ Using a kit to connect to the parallel port

Every year at Halloween, I build an elaborate animatronic band that features Frankenstein's Creature on the drums, Dracula on guitar, a severed hand that plays a mean keyboard, and a skeleton front man that sings and dances. I call them Grateful Undead. They're programmed to sing songs such as "Monster Mash," "Dead Man's Party," and "Walk Like an Egyptian."

My party guests are amazed by the sight of these creatures moving in sync with the music. They think I'm a technological genius, but actually, the system that controls the band's movement is pretty simple. It's all driven by a 15-year-old laptop computer and a single circuit board that activates mechanical relays in response to output from the computer. A simple software program runs on the computer to send signals to the relays via the computer's parallel printer port, causing the band's characters to sing and dance in sync with the music.

In this chapter, you learn how to create your own circuits that can be controlled via a parallel printer port so that you can build your own contraptions to amaze your friends. They'll think you're a genius too (which, of course, you are).

Understanding the Parallel Port

Until a few years ago, all computers came equipped with a *parallel port*, which was used mostly to connect to a printer. Today, most printers connect to computers via USB (Universal Serial Bus) ports. USB has many advantages over parallel port, the most significant being faster data transfer and smaller cables.

The lowly parallel port has one advantage over USB ports, however: It makes it easy to create your own circuits that interface directly with the port. These circuits can control low-current devices such as LEDs, or they can activate transistors or even mechanical relays that in turn can activate high-current devices such as motors, incandescent lamps, or sound systems.

After you've created the circuits to connect to a parallel port, it's a simple matter to create a software program on the computer that sends data to the parallel port. When you run this program, your circuit can detect the data sent to the parallel port to control such things as LEDs or other low-power circuits.

The makeup of a parallel port

A standard parallel port has eight data pins, which are essentially TTL logic outputs, with +5 V HIGH representing 1 and 0 V LOW representing 0. In fact, the first parallel printer ports designed for the IBM PC back in 1980 actually used 7400-series logic chips.

Because of the TTL logic levels used by the parallel port, it's a simple matter to create logic circuits that interface with the output from a parallel port. And after you build a circuit that connects to a parallel port, it's a relatively simple matter to use software on the computer to send data to the parallel port. Then your circuit can respond to the data you send.

Unfortunately, few computers today come with a built-in parallel port. Before you waste your time building circuits to interface with a parallel port, you need to find a computer that has one. Your best bet is to scavenge for an old laptop computer. Almost any laptop computer 10 years old or so will do the job, because most computers that age have a parallel port.

If you can't find a computer with a parallel port, you can buy an inexpensive parallel-port card that adds a parallel port to any computer. Online retailers such as Newegg (www.newegg.com) sell them for less than $15 apiece. In fact, it's a good idea to use an add-on parallel port card even if your computer has a built-in parallel port. That's because if you incorrectly wire the circuit that interfaces with the parallel port, you risk damaging the computer's internal circuitry. Better to fry a $15 add-on card that your computer's motherboard.

The DB25 connector and its pins

A parallel printer port uses a standard type of connector called a *DB25 connector,* which has 25 pins, each of which serves a different purpose in the port's communication with a printer.

Like most data connectors, DB25 connectors come in both male and female variants. The male DB25 connectors consist of 25 pins, and the female connectors have 25 holes. Figure 4-1 shows how the pins of a DB25 female and male connector are numbered. As you can see, the connector consists of two rows of pins. The top row has 13 pins, and the bottom row has 12. The pin in the top-right corner of the female connector is designated as pin 1; the pin at top left is pin 13; pin 14 is at bottom left; and pin 25 is at bottom right.

Note that the pin numbers for a male connector are the mirror images of the numbers of the pins in the female connector. This arrangement is necessary so that when the male connector is plugged into the female connector the pins connect properly (e.g., pin 1 connects to pin 1, pin 2 connects to pin 2, and so on.)

**Book VI
Chapter 4**

Interfacing to Your
Computer's Parallel
Port

DB25 Parallel Port Pins

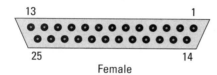

Female

Figure 4-1:
Pinouts
for a DB25
connector.

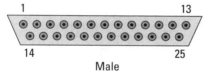

Male

Pinout assignments

Table 4-1 lists the pinout assignments for a standard parallel port.

Table 4-1		Pinouts for a Standard Parallel Printer Port	
Pin	*Name*	*Input or Output*	*Description*
1	STROBE	Output or Input	LOW when data is present on the data pins
2	D0	Output	Data bit 0
3	D1	Output	Data bit 1
4	D2	Output	Data bit 2

(continued)

Table 4-1 *(continued)*

Pin	Name	Input or Output	Description
5	D3	Output	Data bit 3
6	D4	Output	Data bit 4
7	D5	Output	Data bit 5
8	D6	Output	Data bit 6
9	D7	Output	Data bit 7
10	ACK	Input	LOW when data has been read
11	BUSY	Input	HIGH when the printer is busy
12	PE	Input	HIGH when the printer is out of paper
13	SEL	Input	HIGH when the printer is ready
14	LINEFEED	Output or Input	Advances the printer
15	ERROR	Input	HIGH when an error condition exists
16	RESET	Output or Input	HIGH when the printer is reset
17	SELECT	Output or Input	HIGH when the printer is offline
18	GND0	Neither	Ground connection
19	GND1	Neither	Ground connection
20	GND2	Neither	Ground connection
21	GND3	Neither	Ground connection
22	GND4	Neither	Ground connection
23	GND5	Neither	Ground connection
24	GND6	Neither	Ground connection
25	GND7	Neither	Ground connection

For purposes of this chapter, the pins you're most interested in are pins 2 through 9 — the eight data pins that are collectively called the *data port*. When the data port is connected to a printer, its eight pins are capable of sending 1 byte of data at a time to the printer. When the data port is connected to a circuit of your own design, its pins operate as eight separate logic outputs, which you can use as inputs to your own logic circuits.

A parallel port also features four additional output pins called the *control port*, which you can also use for output. When the control port is connected to a printer, these pins are used to control the operation of the printer. One of them, called the *strobe*, indicates that a new byte of data is available on the data pins; when the strobe pin goes HIGH, the printer reads a byte of data from the data pins. Another control-port pin resets the printer.

Finally, the five pins that make up the *status port* allow the printer to send information back to the computer. One of the status-port pins lets the printer tell the computer that it's ready to receive data via the data port. Another pin lets the printer know that it has finished reading data from the data port. A third pin informs the computer that the printer is out of paper. The other status pins have similar functions.

When the status port is connected to a circuit of your own design, its pins can be used to send information to the computer. You could use the status pins to tell the computer that a switch has been closed, that a light sensor detects light, or that a water-level sensor has reached a certain level.

Book VI
Chapter 4

Interfacing to Your
Computer's Parallel
Port

Unfortunately, the programming required to detect the presence of input on one of the status-port pins is a bit too complex for this chapter. If you're an experienced C or C++ programmer, you'll have no trouble writing programs to read the status-port pins. C and C++ programming are beyond the scope of this book, however, so I'll focus for the rest of this chapter on sending output to the parallel-port data pins.

As I've already mentioned, the output pins of a parallel port use a +5 V HIGH signal to represent 1 and 0 V to represent 0. The amount of current that each pin can source is relatively small — typically, around 10–12 mA. That current is enough to drive an LED, but for anything more demanding, you need a way to isolate the output load from the parallel port itself. To do that, you can use individual transistors or an IC designed specifically for this purpose. For details, see "Using Darlington Arrays to Drive High-Current Outputs," later in this chapter.

Designing a Parallel-Port Circuit

Each of the eight data output pins provides 5 V DC, which can source about 10 mA or 12 mA — enough to drive an LED directly. Alternatively, you can connect the data output to the base of a switching transistor, which allows you to control circuits that require more current.

Figure 4-2 shows these two methods for connecting output devices to a parallel-port data pin. The first circuit in this figure drives an LED directly. A current-limiting resistor is required to prevent the LED from pulling too much current and damaging the LED, the parallel port itself, or both.

The second circuit in Figure 4-2 shows how to use a transistor to switch a circuit by using the output from a parallel-port data pin. As you can see, the data pin is connected through a resistor to the transistor's base. When the data pin goes high, the transistor is turned on, allowing current to flow through the collector-emitter circuit.

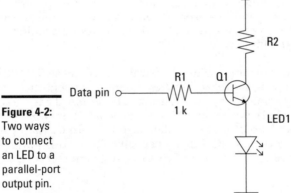

Figure 4-2:
Two ways
to connect
an LED to a
parallel-port
output pin.

Using a transistor driver this way provides two benefits:

✦ The transistor can switch more than the 10–12 mA that the parallel port's data pin can source directly.

✦ The collector-emitter circuit can operate at a different voltage from the one that the data pin's +5 V HIGH signal provides.

When you build Project 4-1 later in this chapter, you'll see an example of a circuit that uses the first method to connect to parallel-port output. This project connects eight LEDs to the eight data pins. With the cable to connect this circuit to a computer's parallel port and the right software in place on the computer, you can write simple scripts that will flash the LEDs on and off in just about any sequence you desire.

Working with DB25 Connectors

To build your own circuits that interface with a parallel port, you need to use a DB25 connector that attaches to your circuit. The standard parallel port on a computer uses a DB25 female connector, which means that you need to provide a DB25 male connector for your circuits.

You have several ways to fashion a DB25 connector for your circuits. The simplest way is to find an old parallel printer cable, cut it a few inches past the male end, strip back the insulation, and separate the 25 wires that run inside the cable. Strip the ends of each of these wires and then use your ohmmeter to match up wire with the individual pins on the connector. Keep good notes so that you'll know which wires to use for your circuits.

Alternatively, you can purchase a male DB25 connector and solder wires to each of the pins. Figure 4-3 shows a connector that I made to use along with Project 4-1 and Project 4-2, which appear later in this chapter. I soldered eight red wires to pins 2 through 9 (the data pins) and one black wire to pin 25 (one of the ground pins).

If you plan to complete Project 4-1 or Project 4-2, you need to assemble a similar cable yourself. All you need are a male DB25 connector, some 18-gauge solid wire, some solder, a soldering iron, and a vise or third-hand tool to hold the connector while you solder. You may also want to get some magnifying goggles. Although they'll make you look like Christopher Lloyd in *Back to the Future,* they make the job much easier by helping you see the project more closely.

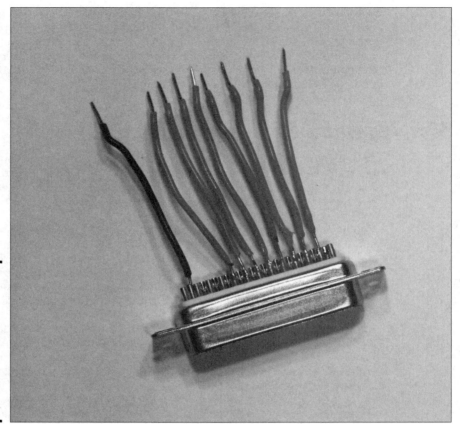

Figure 4-3:
A male DB25 connector with leads soldered to the data pins and one of the ground pins.

Controlling Parallel-Port Output from an MS-DOS Prompt

Project 4-1 (see "Building a Parallel-Port LED Flasher," later in this chapter) covers the details of building a simple breadboard project that lets you control eight LEDs from a computer's parallel port.

You can plug the project directly into the parallel port on the back of your computer, or you can use a short male-to-female DB25 printer cable to connect the project to the computer. Either way, you need some special software on the computer to send data to the parallel port to turn on the LEDs. Fortunately, you don't have to be a programmer to create this software, because software that does this job is readily available on the Internet.

The simplest software available is designed to work with a popular kit that lets you connect relays to the parallel port. This kit, called *Kit 74,* is described in "Using a Kit 74 Relay Controller," later in this chapter. To find the software, just search for *"Kit 74 software."* The software is contained in an archive file named k74_dos.zip. Download it to your computer and extract its contents to a folder on your hard drive named C:\k74_dos.

To use the commands, click the Windows Start button, type cmd, and press the Enter key. (In versions prior to Windows 7, you must click Run after you click the start button; then type cmd.) An MS-DOS command window opens (see Figure 4-4).

Figure 4-4:
You can run an MS-DOS command to send data to the parallel port.

Type the following command, and press Enter:

```
cd \k74_dos
```

This command transfers you to the folder in which you saved the Kit 74 commands. Then you can type and run the commands to test your parallel-port circuits.

The Kit 74 DOS software consists of three commands — RELAY, DELAY, and WAITFOR — that you can run from a command prompt. I explain these three commands in the following sections.

Using the RELAY command

The RELAY command sends a single byte of data to the parallel port. Each of the eight output pins is set HIGH or LOW, depending on the byte you send. This command sets all eight pins to HIGH:

```
RELAY FF
```

And the following command sets all eight outputs to LOW:

```
RELAY 00
```

Unfortunately, most versions of the RELAY command available on the Internet have a bug that requires you to issue the command twice to get it to work. Thus, you must actually enter the command RELAY FF twice in sequence to turn on all the output pins.

You must specify the output data as a single hexadecimal number. For your reference, Table 4-2 shows the hexadecimal numbers you should use for each of the eight output pins.

Table 4-2	Hex values for data output pins
Data Pin	*Hex Value*
1	01
2	02
3	04
4	08
5	10
6	20
7	40
8	80

To turn all the pins on, use the value FF. To turn them all off, use the value 00.

To turn more than one pin on or off, you must first calculate the eight-bit binary number equivalent of the pins you want to set. To turn on pins 1, 2, 3, and 8, for example, you'd use the binary value 100000111. (Notice that pin 1 is represented by the rightmost bit of the binary number and that pin 8 is the leftmost bit.)

After you've concocted the binary number for the pins you want to set, split the binary number in half so that you have two four-bit numbers. In the example that sets pins 1, 2, 3, and 8, the first binary number is 1000, and the second is 0111.

Finally, look up each four-bit number in Table 4-3 to determine the single hexadecimal digit to use. For this example, the first four-bit number converts to 8, and the second four-bit number converts to 7. Combining these two numbers gives you the hexadecimal number 87. Thus, the command to turn on pins 1, 2, 3, and 8 is

RELAY 87

You must enter this command twice to get it to work.

Table 4-3	Converting Binary Values to Hexadecimal Digits		
Binary Value	*Hexadecimal Digit*	*Binary Value*	*Hexadecimal Digit*
0000	0	1000	8
0001	1	1001	9
0010	2	1010	A
0011	3	1011	B
0100	4	1100	C
0101	5	1101	D
0110	6	1110	E
0111	7	1111	F

Creating a command script

You can easily create command scripts that can execute a series of RELAY commands in sequence. Just use Notepad (the simple text editor that comes free with all versions of Windows) to enter a series of RELAY commands, one on each line. Save the file to the C:\k74_dos folder, using the file extension .bat. Then you can run your script by typing the name of your script file (without the .bat extension) in the MS-DOS command window.

Figure 4-5 shows a simple script created with Notepad. Notice that this script contains two identical RELAY commands because of the bug in the RELAY command that requires you to run the command twice to get it to work.

Figure 4-5:
A simple script in a Notepad window.

On an older Windows XP computer (the kind that's likely to have a parallel port to play with), you'll find Notepad in the Start menu under Accessories.

Besides the RELAY command (and the DELAY and WAITFOR commands, which I explain in just a moment), you often use two special MS-DOS commands in command scripts. The first command is called a *label;* it lets you give a name to a line in your script. Labels are indicated by a colon followed by a short word. :LOOP is a typical label.

The second command, called GOTO, creates a program loop by telling the script to jump to a label. Labels and GOTO commands are always used together, like this:

```
:LOOP
RELAY FF
RELAY FF
GOTO LOOP
```

This sequence of commands causes the two RELAY commands to be executed. Then the GOTO command sends the script back to the :LOOP label command, which executes the RELAY commands again. The commands between the GOTO command and the label will be executed again and again until you stop the script by pressing Ctrl+C or closing the command window.

Listing 4-1 shows a simple script that quickly flashes the LEDs in sequence, beginning with LED1. When the script gets to LED8, it reverses the direction and then flashes the LEDs back to LED1. Then a GOTO command sends the script back to the :LOOP label to repeat the flashing. The resulting effect is that the LEDs sweep back and forth indefinitely.

This script is named CYLON.BAT because it resembles the flashing eyes of the evil Cylons from the classic science-fiction TV series *Battlestar Galactica*.

Listing 4-1:   The CYLON.BAT Script

```
:LOOP
RELAY 01
RELAY 01
RELAY 02
RELAY 02
RELAY 04
RELAY 04
RELAY 08
RELAY 08
RELAY 10
RELAY 10
RELAY 20
RELAY 20
RELAY 40
RELAY 40
RELAY 80
RELAY 80
RELAY 40
RELAY 40
RELAY 20
RELAY 20
RELAY 10
RELAY 10
RELAY 08
RELAY 08
RELAY 04
RELAY 04
RELAY 02
RELAY 02
RELAY 01
RELAY 01
GOTO LOOP
```

Seeing why timing is everything

Besides the RELAY command, the Kit 74 software includes two commands that let you add delays to your scripts. By incorporating delays, you can control the timing of the devices controlled by your parallel-port circuit. You could turn pin 1 on, wait 5 minutes, and then turn it off again, for example.

The most useful of the timing commands is DELAY, which simply causes your script to pause for a certain number of seconds. To delay your script for 10 seconds, for example, use this command:

```
DELAY 10
```

The following sequence shows how to turn all outputs on and off at 1-second intervals:

```
:LOOP
RELAY FF
RELAY FF
DELAY 1
RELAY 00
RELAY 00
DELAY 1
GOTO LOOP
```

This sequence starts by turning on all the output pins. Then it waits 1 second, turns all the outputs off, waits another second, and jumps to the LOOP label to start the sequence all over again.

You must always specify the delay period in seconds. To wait 1 minute, use this command:

```
DELAY 60
```

An hour contains 3,600 seconds, so the following command delays the script for 1 hour:

```
DELAY 3600
```

The second timing command is WAITFOR, which waits to execute until a certain time of day arrives. To stop your script until 10:30 AM, for example, use this command:

```
WAITFOR 10:30
```

Here's a sequence that turns all outputs on at 10:30 AM every day, leaves them on for an hour, and turns them off:

```
:LOOP
WAITFOR 10:30
RELAY FF
RELAY FF
DELAY 3600
```

```
RELAY 00
RELAY 00
GOTO LOOP
```

Building a Parallel-Port LED Flasher

Project 4-1 presents a simple breadboard circuit that connects eight LEDs to the eight output pins of a parallel port. To complete this project, you need to have a computer with a parallel port, and you need to install the Kit 74 parallel-port software (refer to "Controlling Parallel-Port Output from an MS-DOS Prompt," earlier in this chapter).

You also need to build a parallel-port connector that you can attach to your circuit. Refer to Figure 4-3 in "Working with DB-25 Connectors," earlier in this chapter, for information on building the required connector.

After you assemble the circuit, you can test it by connecting it to your computer's parallel port, opening a command prompt, and using the RELAY command to send data to the port. To turn on all eight LEDs, for example, run this command:

```
RELAY FF
```

Book VI
Chapter 4

Interfacing to Your
Computer's Parallel
Port

You may need to run the command twice to make the LEDs turn out. This requirement is due to a flaw in the RELAY command itself, not to a problem with your circuit or your computer's parallel port.

When the circuit works properly when you enter the RELAY command from the command prompt, try running the CYLON.BAT script, presented in Listing 4-1 earlier in this chapter. This script flashes the LEDs sequentially from left to right and back again. The LEDs run repeatedly until you terminate the batch file by pressing Ctrl+C or closing the command window.

Figure 4-6 shows the finished parallel-port LED flasher circuit. You can see in the photo that I've connected the male DB-25 connector to one end of a parallel printer cable. The other end of this cable is connected to the computer.

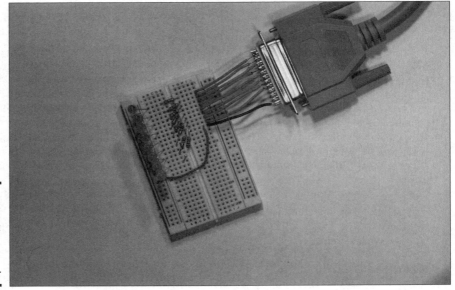

Figure 4-6:
The finished
parallel-port
LED flasher
circuit
(Project 4-1).

Introducing Seven-Segment Displays

A *seven-segment display* is an array of seven LEDs arranged in a way that can display numerals as well as some alphabetic characters. You see this type of display most often on calculators, which have several seven-segment displays arranged side by side to display numbers. You can purchase an inexpensive (less than $2) seven-segment display at your local RadioShack or other electronics-parts stores.

Before I go any further, I want to acknowledge that you may be wondering why I'm talking about seven-segment displays in a chapter on parallel-port interfacing. It's true that these two topics don't have a lot in common, except that to use a seven-segment display for any practical purpose, you have to connect the display to a digital circuit that's capable of controlling the individual segments to display meaningful information such as numerals or alphabetic letters. You *can* do that with logic gates (as described in Book VI, Chapter 3), but the resulting circuits are very complicated. It's much easier to use the power of a computer to control the individual segments via a parallel-port connection.

Figure 4-7 shows how a single-digit seven-segment display module is usually wired up. As you can see, the segments themselves are referred to by the letters *a* through *g*. This particular display module is contained in a 14-pin DIP package, but only 8 of the pins are actually used. The anode of each LED segment is connected to one of the pins. The cathodes for all the segments are connected at pin 4. (This is called *common-cathode* wiring. You can also get 7-segment displays in which the anodes are connected to a common pin; this arrangement is called *common-anode* wiring.)

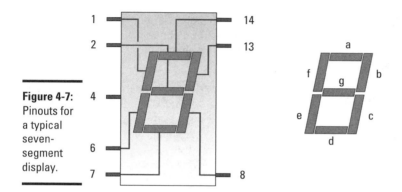

Figure 4-7: Pinouts for a typical seven-segment display.

**Book VI
Chapter 4**

**Interfacing to Your
Computer's Parallel
Port**

To control a seven-segment display, you must first connect a positive voltage source to the anode of each of the seven segments. The cathode should be connected to ground. Be sure to use a current-limiting resistor in series with each anode to limit the current that flows through the LEDs.

To drive a seven-segment display from your computer's parallel port, just connect the anode of each segment through a current-limiting resistor (1 kΩ is typical) to one of the data output pins. The most straightforward way to do that is to connect DATA1 (pin 2) to the *a* segment, DATA1 (pin 3) to the *b* segment, and so on until DATA7 is connected to the *g* segment.

After you've connected a seven-segment display to the parallel port, you can form numerals or some alphabetic characters by sending the right data to the parallel port. Table 4-4 shows the data byte you should send to display the numerals 0 through 9.

Project 4-1: A Parallel-Port LED Circuit

In this project, you build a circuit that connects an LED to each of the eight data pins on a parallel port. Then you write a program on the computer that flashes the LEDs to make sure that the circuit is working properly.

To complete this project, you need a male DB25 connector with solid 18-gauge wire soldered to pins 2 through 9 and pin 25. You can build a connector like this yourself (refer to Figure 4-3, earlier in this chapter), or you can cut an old parallel printer cable apart, separate the wires, and use an ohmmeter to single out the wires connected to required pins. For more information, refer to "Working with DB-25 Connectors," earlier in this chapter.

You also need to locate and download the Kit74 software. You can get it on several websites; just search for "Kit74 software" to find it. For more information, refer to "Controlling Parallel-Port Output from an MS-DOS Prompt," earlier in this chapter

Parts List
1 Male DB25 connector with 18-gauge wire on pins 2 through 9 and 25
8 1 kΩ resistors (brown-black-red)
8 Red LEDs
1 2" jumper wire
1 Computer with a parallel printer port

Steps

1. **Insert resistors R1 through R8**

R1 - 1 kΩ:	D3 to F3
R2 - 1 kΩ:	D5 to F4
R3 - 1 kΩ:	D7 to F5
R4 - 1 kΩ:	D9 to F6
R5 - 1 kΩ:	D11 to F7
R6 - 1 kΩ:	D13 to F8
R7 - 1 kΩ:	D15 to F9
R8 - 1 kΩ:	D17 to F10

2. **Insert LEDs LED1 through LED8.**
   Cathode (short lead): Ground bus
   Anode (long lead): A3, A5, A7, A9, A11, A13, A15, A17

3. **Insert the DB25 wires.**

Pin 2	J3
Pin 3	J4
Pin 4	J5
Pin 5	J6
Pin 6	J7
Pin 7	J8
Pin 8	J9
Pin 9	J10
Pin 25	J11

4. **Insert the jumper wire.**
   I11 to ground bus

5. **Plug the male DB25 connector into the parallel port on the computer.**

6. **Turn the computer on.**

7. **Click the Windows Start button, choose Run, type CMD, and press the Enter key.**
   An MS-DOS command window opens

8. **Type the command CD \K74_DOS.**

This command switches you to the folder that contains the Kit 74 software.

9. **Enter the command RELAY FF and press the Enter key; then do it again.**
   All eight LEDs should come on, verifying that your circuit works.

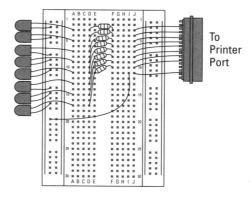

To
Printer
Port

LED
Cathode
Anode

Table 4-4	Data Values to Display Numerals on a Seven-Segment Display	
Numeral	Data Value	
0	3F	
1	06	
2	5B	
3	4F	
4	66	
5	6D	
6	7D	
7	07	
8	7F	
9	6F	

Thus, to display the numeral 5, use this RELAY command:

```
RELAY 6D
```

To understand why the data values in Table 4-4 are required to display numerals on a seven-segment display, remember that each of the segments in the display is connected to one of the data output pins of the parallel port. Thus, to light a particular combination of segments to form a numeral, you must set the parallel port's output so that the data pins corresponding to the segments you want lit are HIGH and the remaining pins are LOW.

To form the numeral 3, for example, segments *a, b, c, d,* and *g* should be turned on. Those segments are connected to data output pins 1, 2, 3, 4, and 7. Thus, you must send a byte of data to the parallel port with the bit positions corresponding to pins 1, 2, 3, 4, and 7 set to the binary value 1 and the other bit positions set to binary 0.

In a binary number, the bit positions are numbered right to left, so the binary pattern you need to send to the parallel port to form the numeral 3 is

```
01001111
```

**Book VI
Chapter 4**

Interfacing to Your
Computer's Parallel
Port

The hexadecimal equivalent for this binary number is 4F. Thus, the following command displays the numeral 3:

```
RELAY 4F
```

Listing 4-2 shows a script called COUNTDOWN.BAT that displays a NASA-style countdown from 9 to 0 at 1-second intervals. When the script reaches 0, that numeral flashes repeatedly until you cancel the batch file by pressing Ctrl+C or closing the command window.

Listing 4-2:   The COUNTDOWN.BAT Script

```
RELAY 6F
RELAY 6F
DELAY 1
RELAY 7F
RELAY 7F
DELAY 1
RELAY 07
RELAY 07
DELAY 1
RELAY 7D
RELAY 7D
DELAY 1
RELAY 6D
RELAY 6D
DELAY 1
RELAY 66
RELAY 66
DELAY 1
```

(continued)

Listing 4-2 *(continued)*

```
RELAY 4F
RELAY 4F
DELAY 1
RELAY 5B
RELAY 5B
DELAY 1
RELAY 06
RELAY 06
DELAY 1
:LOOP
RELAY 3F
RELAY 3F
RELAY 00
RELAY 00
GOTO LOOP
```

Notice in this script that the `:LOOP` label appears near the end of the listing, not at the beginning. You can place labels anywhere you want in a script. The `GOTO LOOP` command at the end of the script causes the script to repeat the last four commands over and over until you interrupt the script by pressing Ctrl+C or closing the command window.

Do *not* use the script shown in Listing 4-2 to launch any actual rockets or missiles. This script has not been certified for actual space flight by NASA or any other space agency.

Building a Seven-Segment Display Countdown Timer

Project 4-2 presents a breadboard circuit that connects a seven-segment display to seven of the eight output pins of a parallel port. As with Project 4-1, you need a computer with a parallel port and the Kit 74 software installed. You also need a parallel-port connector with wires soldered to the data pins and one of the common data ground pins. You can find information about the required software and the required cable earlier in this chapter, in the section "Controlling Parallel-Port Output from an MS-DOS Prompt."

To test your assembled circuit, connect it to your computer's parallel port, open a command prompt, and use the following `RELAY` command:

```
RELAY FF
```

All seven of the display's segments should light up. If one doesn't light, carefully double-check your wiring. Also verify that your seven-segment display uses the same pinouts as the one listed in the project; some display modules have different packaging with different pin connections. Finally, remember that you may have to run the RELAY command twice to get it to work.

When the circuit checks out, run the COUNTDOWN.BAT script (refer to Listing 4-2, earlier in this chapter). The seven-segment display should count down the digits 9 through 0 at 1-second intervals. When it gets to zero, the display should flash until you terminate the batch file by pressing Ctrl+C or closing the command window.

Figure 4-8 shows the finished seven-segment-display circuit.

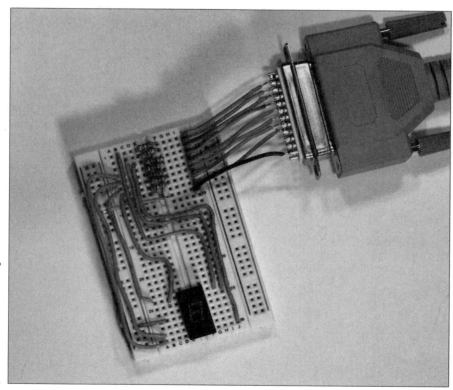

Figure 4-8:
The finished
parallel-port
seven-
segment
display
(Project 4-2).

Project 4-2: A Seven-Segment Display Circuit

In this project, you build a circuit that connects a seven-segment LED display to seven of the eight data pins on a parallel port. Then you write a program on the computer that flashes the segments in the correct sequence to count down the digits 9 through 0.

To complete this project, you need a male DB25 connector with solid 18-gauge wire soldered to pins 2 through 9 and pin 25. You can build a connector like this yourself (refer to Figure 4-3, earlier in this chapter), or you can cut an old parallel printer cable apart, separate the wires, and use an ohmmeter to single out the wires connected to required pins. For more information, refer to "Working with DB-25 Connectors," earlier in this chapter.

You also need to locate and download the Kit74 software. You can get it on several websites; just search for "Kit74 software" to find it. For more information, refer to "Controlling Parallel-Port Output from an MS-DOS Prompt," earlier in this chapter.

Parts List
1 Male DB25 connector with 18-gauge wire on pins 2–9 and 25
7 1 kΩ resistors (brown-black-red)
1 Seven-segment display module
9 Jumper wires (various lengths)
1 Computer with a parallel printer port

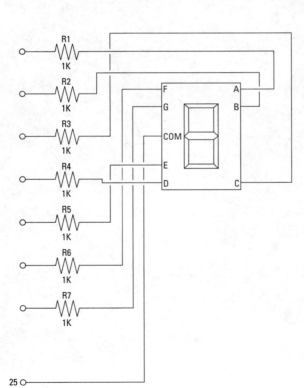

Steps

1. **Insert resistors R1 through R8.**
   R1 - 1 kΩ: E1 to F1
   R2 - 1 kΩ: DE2 to F2
   R3 - 1 kΩ: E3 to F3
   R4 - 1 kΩ: E4 to F4
   R5 - 1 kΩ: E5 to F5
   R6 - 1 kΩ: E6 to F6
   R7 - 1 kΩ: E7 to F7

2. **Insert the seven-segment display.**
   Pin 1 should go in E24, and pin 14 should go in F24.

3. **Insert the wires from the DB25 connector.**

Pin 2	J1
Pin 3	J2
Pin 4	J3
Pin 5	J4
Pin 6	J5
Pin 7	J6
Pin 8	J7
Pin 9	J8
Pin 25	J9

4. **Insert the jumper wires.**
   F9 to the ground bus
   B27 to the ground bus
   B1 to H24
   C2 to I25
   D3 to J30
   A4 to A30
   A5 to A29
   A6 to A24
   A7 to C25

5. **Plug the male DB25 connector into the parallel port on the computer.**

6. **Turn the computer on.**

7. **Click the Windows Start button, choose Run, type CMD, and press the Enter key.**
   An MS-DOS command window opens.

8. **Type the command CD \K74_DOS.**
   This command switches you to the folder that contains the Kit 74 software.

9. **Enter the command RELAY FF and press the Enter key; then do it again.**
   All eight LEDs should come on, verifying that your circuit works.

10. **Use Notepad or any other text editor to create a bath file with the program shown in Listing 4-2. Then, run it to verify that the program and your circuit work properly.**

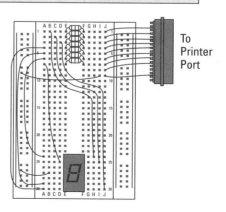

To
Printer
Port

Using Darlington Arrays to Drive High-Current Outputs

In Book II, Chapter 6, you see that two transistors can be connected to form a *Darlington transistor* (sometimes called a *Darlington pair*), which can switch much more current than the collector-emitter circuit of a standard transistor can. You can use darlington transistors to switch up to 500 mA from the output of a parallel-port data pin, which is enough current to drive a mechanical relay or a small electric motor.

Rather than use individual darlington transistors, you can use an integrated circuit specially designed for driving high-current loads from TTL-level inputs. The most common ICs of this type are the ULN2003, which has 7 darlington drivers in a 16-pin DIP package, and the ULN2803, which has 8 drivers in an 18-pin DIP package.

You won't find these ICs at your local RadioShack store, but if you have access to an electronics-parts store, it probably will have this useful chip in stock. If not, you can easily find it on the Internet by searching for *"ULN2003"* or *"ULN2803."*

Table 4-5 lists the pinouts for the ULN2003. As you can see, pins 1 through 7 are the input pins, which you can connect directly to the output pins from the parallel port. Pins 10 through 16 are the output pins, which you can connect to the circuit you want to control. Pin 8 connects to ground, and pin 9 connects to a voltage source.

Table 4-5 also shows the pinouts for the ULN2803, which are similar to the ULN2003 pinouts.

Table 4-5		**Pinouts for a ULN2003 and ULN2803 Darlington Array IC**	
ULN2003		*ULN2803*	
Pin	*Description*	*Pin*	*Description*
1	Input 1	1	Input 1
2	Input 2	2	Input 2
3	Input 3	3	Input 3
4	Input 4	4	Input 4
5	Input 5	5	Input 5

ULN2003		ULN2803	
Pin	*Description*	*Pin*	*Description*
6	Input 6	6	Input 6
7	Input 7	7	Input 7
8	Common ground	8	Input 8
9	Vss	9	Common ground
10	Output 1	10	Vss
11	Output 2	11	Output 1
12	Output 3	12	Output 2
13	Output 4	13	Output 3
14	Output 5	14	Output 4
15	Output 6	15	Output 5
16	Output 7	16	Output 6
		17	Output 7
		18	Output 8

The output circuit for a ULN2003/2803 is a little different from what you may expect. Rather than sourcing current for the load, the darlington array sinks the current. Thus, the output pin is on the ground side of the load circuit, as shown in Figure 4-9. As you can see, the voltage source (Vss) feeds both the load circuit (in this case, a small motor) and the ULN2003.

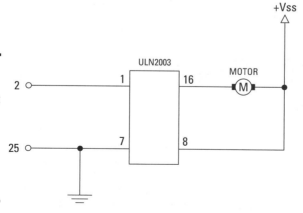

Figure 4-9:
Driving a
high-current
load from
the parallel
port by
using a
darlington
array.

If you'll be using a ULN2003 or ULN2803 to drive an inductive load such as a relay or motor, you should use a Zener diode on pin 10. This diode prevents current from flowing in the wrong direction into the ULN2803 in case the relay or motor coil creates a large backward voltage spike, as coils are apt to do.

Building a Motor Driver

Project 4-3 presents a breadboard circuit that drives a small 3 V DC motor from a parallel port. Because this motor uses much more current than a parallel port can handle, a ULN2003 darlington array IC is used to drive the motor.

As with the preceding projects in this chapter, you need a computer with a parallel port and the Kit 74 software installed, plus a parallel-port connector with wires soldered to the data pins and one of the common data ground pins. Please refer to those earlier projects for more information.

When the circuit is assembled and connected to your computer, you can test it by running the following command twice in a command window:

```
RELAY 01
```

The motor should start running. To stop the motor, use this command twice:

```
RELAY 00
```

You can easily write a script to run the motor for a certain interval. Listing 4-3 shows a simple example. Here, the MOTOR.BAT file simply runs the motor for 30 seconds, turns it off for 30 seconds, and then jumps to the :LOOP label to repeat the cycle.

Listing 4-3:   The MOTOR.BAT Script

```
:LOOP
RELAY 01
RELAY 01
DELAY 30
RELAY 00
RELAY 00
DELAY 30
GOTO LOOP
```

Figure 4-10 shows the finished seven-segment-display circuit.

Figure 4-10:
The finished
parallel-port
motor driver
(Project 4-3).

Using a Kit 74 Relay Controller

One of my favorite ways to interface to a computer via the parallel port is with an inexpensive build-it-yourself kit called the Kit 74. You can purchase this kit from various sources on the Internet; just search for *"Kit 74"* to find it. The cost of the unassembled kit is $35 or $40.

You can also find fully assembled versions of the kit for about $45 or $50, and for about $20 more, you can find complete versions that are contained in a plastic case.

Project 4-3: A Parallel-Port Motor Driver

In this project, you build a circuit that connects a small 3 V motor to one of the eight data pins on a parallel port. Then you write a program on the computer that turns the motor on and off at 30-second intervals.

To complete this project, you need a male DB25 connector with solid 18-gauge wire soldered to pins 2 through 9 and pin 25. You can build a connector like this yourself (refer to Figure 4-3, earlier in this chapter), or you can cut an old parallel printer cable

apart, separate the wires, and use an ohmmeter to single out the wires connected to required pins. For more information, refer to "Working with DB-25 Connectors," earlier in this chapter.

You also need to locate and download the Kit74 software. You can get it on several websites; just search for "Kit74 software" to find it. For more information, refer to "Controlling Parallel-Port Output from an MS-DOS Prompt," earlier in this chapter.

Parts List
1 Male DB25 connector with 18-gauge wire on pins 2 and 9 and 25
1 ULN2003 darlington array
1 1.5 and 3 V DC motor
4 Jumper wires (various lengths)
1 Two-AA-battery holder
2 AA batteries
1 Computer with a parallel printer port

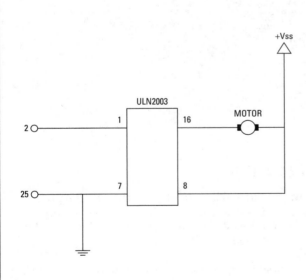

Steps

1. **Insert the ULN2003.**
   Pin 1 should go in hole E15, and pin 16 should go in hole F15.

2. **Insert the wires from the DB25 connector.**

Pin 2	J1	Pin 7	J6
Pin 3	J2	Pin 8	J7
Pin 4	J3	Pin 9	J8
Pin 5	J4	Pin 25	J9
Pin 6	J5		

3. **Insert the jumper wires.**
   D1 to D15
   E1 to F1
   I9 to the ground bus
   B27 to the ground bus

4. **Insert the Zener diode.**
   J22 to the positive bus.
   Note that the banded end goes in the positive bus.

5. **Connect the motor.**
   One lead goes in the positive bus; the other lead goes in J15.

6. **Connect the batteries.**
   Positive goes anywhere in the positive bus; negative goes anywhere in the ground bus.

7. **Plug the male DB25 connector into the parallel port on the computer.**

8. **Turn the computer on.**

9. **Click the Windows Start button, choose Run, type CMD, and press the Enter key.**
   An MS-DOS command window opens.

10. **Type the command CD \K74_DOS.**
    This command switches you to the folder that contains the Kit 74 software.

11. **Enter the command RELAY 01 and press the Enter key; then do it again.**
    The motor should start running.

12. **Enter the command RELAY 00 to stop the motor.**

13. **Use Notepad or any other text editor to create a bath file with the program shown in Listing 4-3. Then, run it to verify that the program and your circuit work properly.**

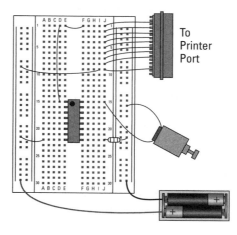

To Printer Port

Figure 4-11 shows an assembled Kit 74 board. As you can see, one end of the board features a DB25 connector. To connect the Kit 74 board to a computer, you need a printer cable with a DB25 male connector on one end and a DB25 female connector on the other end.

Figure 4-11:
An assembled Kit 74 relay controller.

Along the long edge of the board are eight relays that can switch high-current loads, including 120 VAC circuits for lighting or small motors. The maximum current rating for these relays is 5 A at 120 VAC. Each of the connectors for the relays has three screw terminals. The terminal in the middle is for a common wire. One of the two remaining terminals is normally open; the other is normally closed. When the data pin connected to the relay goes HIGH, these two terminals reverse themselves. Because of this arrangement, you can use the relay to turn a circuit on, to turn it off, or to switch between two circuits.

The Kit 74 doesn't provide voltage for the devices that you connect to the relays, so you must do that work yourself. If you're controlling lights or other devices that require 120 VAC, the easiest method is to cut up an inexpensive household extension cord. Using a utility knife, carefully cut through the insulation between the two wires at a convenient point along the cord. Separate the wires a bit so that you have some slack to work with; then use your wire cutters to cut through one of the wires. Strip off about 3/8" of insulation from the two ends, and insert one end into the center terminal of one of the Kit 74 relay connectors. Finally, insert the other end into either the normally open or the normally closed terminal, depending on whether you want the relay to turn the circuit on (normally open) or to turn it off (normally closed).

When your connections are secure, and you've checked for loose or bare wires, you can plug the male end of the extension cord into a wall outlet and then plug the lamp or other device you're controlling into the other end of the cord. Figure 4-12 shows an extension cord connected to a Kit 74 relay connector in this manner.

Figure 4-12:
Using an extension cord to connect to a Kit 74 relay.

Use extreme caution when using a Kit 74 to control 120 VAC circuits! Household current is strong enough to kill you, so make sure that you don't leave any dangerous exposed wires. Also, *never* work on the circuit while power is applied.

One particular risk of cutting up an extension cord to connect to a Kit 74 is that the screw terminals on the relay connectors aren't all that secure. For safety, you should secure the Kit 74 to a board, and use wire holders to secure both ends of the extension cord to the board adjacent to the relay connector. This method prevents the wire from pulling out of the connector, which could create a hazardous situation.

Using Vixen for parallel-port control

If you want to create elaborate scripts for controlling devices connected via a computer's parallel port, you should look into Vixen, a free software program that makes it easy to sequence your parallel port with music. You can download Vixen from `http://vixenlights.com`. I use this software to control the animatronic rock band that I mention at the start of this chapter.

Vixen is a very powerful program that's designed to work with many types of output devices — not just parallel ports. Its main use is for creating elaborate Christmas lighting displays. When used with a parallel port, Vixen can control up to eight output channels — one for each data pin on the parallel port. It can handle as many as 3 parallel ports on a single computer, so in theory, you can control up to 24 separate circuits by using parallel ports. Vixen also interfaces with other output devices, some of which are capable of supporting hundreds of ports.

The user interface is essentially a big grid, with each row in the grid representing an output port and each column representing a time interval. Depending on how much precision you want in your sequence, the time intervals can be as small as a fraction of a second or as large as a minute. After the grid is set up, you program the ports simply by clicking the grid to indicate when each port should be on or off.

Chapter 5: Working with Flip-Flops

In This Chapter

▶ **Understanding latches and gated latches**

▶ **Looking at flip-flops**

▶ **Building some basic latch and flip-flop circuits**

*T*his chapter is about flip-flops, but not the kind you wear on your feet. I wish it were about the kind you wear on your feet, especially if I happened to be wearing a pair while I wrote this because that might mean I'd be writing this at the beach, parked in a beach chair watching children play with the waves.

Don't you love watching children play with waves? They laugh uncontrollably as they chase after a retreating wave. But when the next wave comes in, the children turn and run, screaming with delight for fear that the wave might catch them, and they might — of all things — get wet.

Alas, this chapter isn't about the kind of flip-flops you wear to the beach. Instead, it's about the electronic kind of flip-flop. A *flip-flop* is a circuit that stores data. As such, flip-flops are the basis of modern computers.

Don't you think it's odd that one of the fundamental building blocks of modern thinking machines has a name that suggests it can't make up its mind.

In this chapter, you learn how to work with simple flip-flop circuits. Before I tell you about flip-flops, however, I first show you a simpler type of circuit called a *latch*.

Looking at Latches

A *latch* is a logic circuit that has two inputs and one output. One of the inputs is called the *SET input;* the other is called the *RESET input.*

Latch circuits can be either active-high or active-low. The difference is determined by whether the operation of the latch circuit is triggered by HIGH or LOW signals on the inputs. (Refer to Chapter 2 of this minibook for an explanation of concepts such as HIGH and LOW.)

- ✦ **Active-high circuit:** Both inputs are normally tied to ground (LOW), and the latch is triggered by a momentary HIGH signal on either of the inputs.

- ✦ **Active-low circuit:** Both inputs are normally HIGH, and the latch is triggered by a momentary LOW signal on either input.

In an active-high latch, both the SET and RESET inputs are connected to ground. When the SET input goes HIGH, the output also goes HIGH. When the SET input returns to LOW, however, the output remains HIGH. The output of the active-high latch stays HIGH until the RESET input goes HIGH. Then, the output returns to LOW and will go HIGH again only when the SET input is triggered once more.

In other words, the latch remembers that the SET input has been activated. If the SET input goes HIGH for even a moment, the output goes HIGH and stays HIGH, even after the SET input returns to LOW. The output returns to LOW only when the RESET input goes HIGH.

On the other hand, in an active-low latch the inputs are normally held at HIGH. When the SET input momentarily goes LOW, the output goes HIGH. The output then stays HIGH until the RESET input momentarily goes LOW.

Note that most latch circuits actually have a second output that is simply the first output inverted. In other words, whenever the first output is HIGH, the second output is LOW, and vice versa. These outputs are usually referred to as Q and $\overline{Q}$.

The notation $\overline{Q}$ is usually pronounced either "bar Q" or "Q bar," though some people pronounce it "not Q." The horizontal bar symbol over a label is a common logical shorthand for inversion. That is, $\overline{Q}$ is the inverse of Q. If Q is HIGH, $\overline{Q}$ is LOW, and if Q is LOW, $\overline{Q}$ is HIGH.

You can easily create an active-high latch from a pair of NOR gates, as shown in Figure 5-1. (Recall from Chapter 2 in this minibook that the output of a NOR gate is HIGH if both inputs are LOW; otherwise, the output is LOW.) In this circuit, the SET input is connected to one of the inputs of the first NOR gate, and the RESET input is connected to one of the inputs of the second NOR gate. The trick of the latch circuit is that the output of the NOR gates are cross-connected to the remaining NOR gate inputs. In other words, the output from the first NOR gate is connected to one of the inputs of the second NOR gate, and the output from the second NOR gate is connected to one of the inputs of the first NOR gate.

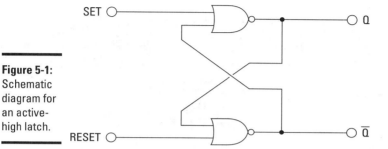

Figure 5-1:
Schematic diagram for an active-high latch.

The schematic for an active-low latch is shown in Figure 5-2. As you can see, the only difference between this schematic and the one shown in Figure 5-1 is that the active-low latch uses NAND gates instead of NOR gates. Notice also in this diagram that the inputs are referred to as $\overline{\text{SET}}$ and $\overline{\text{RESET}}$ rather than SET and RESET, which indicates that the inputs are active-low.

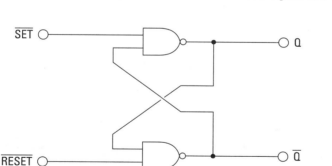

Figure 5-2:
Schematic diagram for an active-low latch.

Projects 5-1 and 5-2 show you how to build simple active-high and active-low latch circuits using a 4001 quad 2-input NOR gate IC and a 4011 quad 2-input NAND gate IC. Both the Q and $\overline{\text{Q}}$ outputs are used to drive LEDs so you can see the state of the latch, and both inputs are controlled by normally-open pushbuttons so that you can trigger the latch by momentarily pressing the buttons. Figure 5-3 shows the assembled active-high latch, and Figure 5-4 shows the assembled active-low latch.

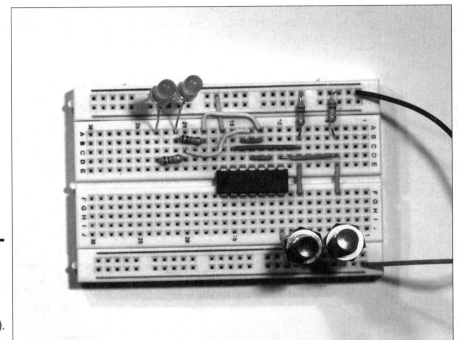

Figure 5-3:
The assembled active-high latch (Project 5-1).

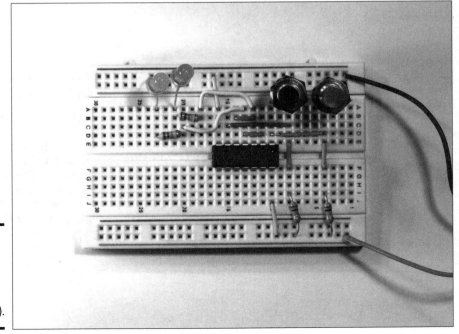

Figure 5-4:
The assembled active-low latch (Project 5-2).

If you compare the schematic diagrams between these two projects, you'll see that there are only these two differences between them:

✦ **Gates:** The active-high circuit uses a 4001 IC, which contains NOR gates, while the active-low uses a 4011 IC, which contains NAND gates.

✦ **Resistor and switch positions:** The positions of R1 and R2 and SW1 and SW2 are reversed. In the active-high circuit, the resistors connect the two gate inputs to ground and the switches short the gate inputs to +6 V. In the active-low circuit, the resistors connect the gate inputs to +6 V and the switches short the gate inputs to ground.

Both of these circuits use simple pushbutton switches to provide the trigger inputs. However, you can easily imagine other sources for the trigger pulse. For example in a home alarm system, the $\overline{SET}$ input in an active-low latch might come from a window switch that breaks contact when the window is open, and the $\overline{RESET}$ input may come from a key lock on the alarm system's control panel.

Here are a few other things you should know about latches before we move on:

✦ A latch with a SET and RESET input is often called an *SR latch*. The term *RS latch* is also used.

✦ In some cases, you may need a latch in which one of the inputs is active-high and the other is active-low. For example in the alarm system described in the previous paragraph, the key lock may send a HIGH signal when the alarm should be reset. Thus, the $\overline{SET}$ input for the alarm latch is active-low, but the RESET input is active high.

You can easily accomplish that by adding an inverter to one of the inputs, as shown in Figure 5-5. Here, I've used NAND gates to create an active-low latch, but I've added a NOT gate to invert the RESET input. Thus, the $\overline{SET}$ input of this inverter is active-low, and the RESET input is active-high.

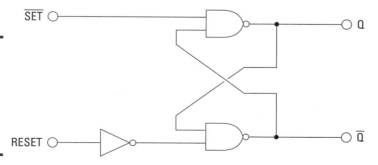

Figure 5-5:
A latch in
which $\overline{SET}$
is active-
low and
RESET is
active-high.

Project 5-1: An Active-High Latch

In this project, you build a simple active-high latch circuit using a pair of NOR gates from a 4001 Quad 2-Input NOR Gate IC. You connect switches to the SET and RESET inputs and LEDs to the Q and Q̄ outputs so you can see the operation of the circuit.

Parts List

1 4001 Quad 2-Input NOR Gate
2 1 kΩ resistors (brown-black-red)
2 10 kΩ resistors (brown-black-orange)
2 Red LEDs
2 Normally open pushbuttons
4 AAA batteries
1 Four AAA-battery holder
10 Jumper wires (various lengths)

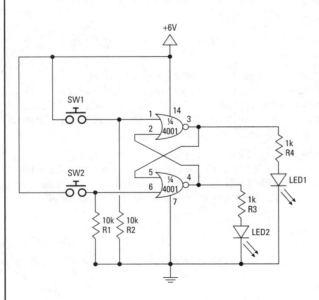

Steps

1. **Insert the 4001 IC.**
   Pin 1 goes in E10; pin 14 in F10

2. **Insert R1 through R4.**
   R1 (10 kΩ): B4 to ground bus
   R2 (10 kΩ): B8 to ground bus
   R3 (1 kΩ): B18 to B21
   R4 (1 kΩ): D20 to D23

3. **Insert the LEDs.**
   LED1: A21 to ground bus
   LED2: A23 to ground bus
   Connect the cathode (short lead)
   to the ground bus.

4. **Insert the jumper wires.**
   D4 to D10
   E4 to F4
   C8 to C15
   E8 to F8
   J10 to positive bus
   D11 to D13
   B12 to B14
   A12 to A18
   A13 to C20
   A16 to ground bus

5. **Insert the switches.**
   SW1: J4 to positive bus
   SW2: J8 to positive bus

6. **Connect the battery holder.**
   Red lead to positive bus
   Black lead to negative bus

7. **Insert the batteries.**

8. **Press the pushbuttons and
   observe the operation of the LEDs.**

6 V
− +

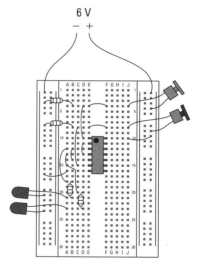

LED

Cathode

Anode

Project 5-2: An Active-Low Latch

In this project, you build a simple active-low latch circuit using a pair of NAND gates from a 4011 Quad 2-Input NAND Gate IC. Connect switches to the $\overline{\text{SET}}$ and $\overline{\text{RESET}}$ inputs and LEDs to the Q and $\overline{\text{Q}}$ outputs so you can see the operation of the circuit.

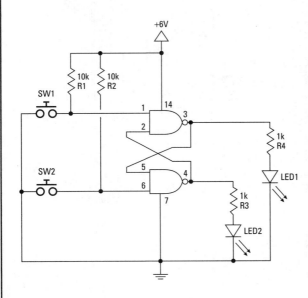

Parts List

1	4011 Quad 2-Input NAND Gate
2	1 kΩ resistors (brown-black-red)
2	10 kΩ resistors (brown-black-orange)
2	Red LEDs
2	Normally open pushbuttons
4	AAA batteries
1	Four AAA-battery holder
10	Jumper wires (various lengths)

Steps

1. **Insert the 4011 IC.**
   Pin 1 goes in E10; pin 14 in F10

2. **Insert R1 through R4.**
   R1 (10 kΩ): I4 to positive bus
   R2 (10 kΩ): I8 to positive bus
   R3 (1 kΩ): B18 to B21
   R4 (1 kΩ): D20 to D23

3. **Insert the LEDs.**
   LED1: A21 to ground bus
   LED2: A23 to ground bus
   Connect the cathode (short lead)
   to the ground bus.

4. **Insert the jumper wires.**
   D4 to D10
   E4 to F4
   C8 to C15
   E8 to F8
   J10 to positive bus
   D11 to D13
   B12 to B14
   A12 to A18
   A13 to C20
   A16 to ground bus

5. **Insert the switches.**
   SW1: A4 to ground bus
   SW2: A8 to ground bus

6. **Connect the battery holder.**
   Red lead to positive bus
   Black lead to negative bus

7. **Insert the batteries.**

8. **Press the pushbuttons and
   observe the operation of the LEDs.**

6 V
− +

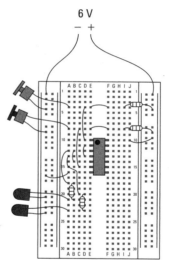

Cathode
Anode

LED

Looking at Gated Latches

A *gated latch* is a latch that has a third input that must be active in order for the SET and RESET inputs to take effect. This third input is sometimes called ENABLE because it enables the operation of the SET and RESET inputs.

The ENABLE input can be connected to a simple switch. Then, when the switch is closed, the SET and RESET inputs are enabled; when the switch is open, any changes in the SET and RESET inputs are ignored.

Alternatively, the ENABLE input can be connected to a clock pulse. For example, you could connect the output of a 555 timer circuit to the ENABLE input. Then, the latch inputs will be operational only when the 555 timer's output is HIGH. Note that the ENABLE input is often called the *CLOCK input*. (For more information about 555 timer circuits, refer to Book III, Chapter 2.)

You can easily add an ENABLE input to a latch by adding a pair of NAND gates as shown in Figure 5-6. Here, the SET and RESET inputs (SR latch) are connected to one input of each of the two NAND gates. The ENABLE input is connected to the other input of each NAND gate. Then, the output from these gates are used as the inputs to the basic latch circuit.

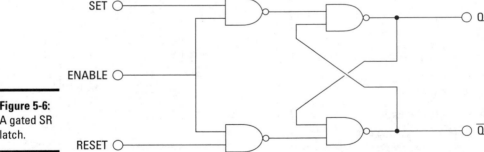

Figure 5-6:
A gated SR
latch.

Another common type of gated latch is called a *gated D latch*, which has just two inputs: DATA and ENABLE. When a HIGH is received at the ENABLE input, the DATA input is copied to the output. Even if the ENABLE input then goes low, the output remains unchanged. The output cannot be changed until the ENABLE input goes high.

To create a gated D latch from a gated SR latch, you simply connect the SET and RESET inputs together through an inverter, as shown in Figure 5-7. Thus, the SET and RESET inputs will always be opposite of one another. When the DATA input is HIGH, the SET input is HIGH and the RESET input is LOW. When the DATA input is LOW, the SET input is LOW and the RESET input is HIGH.

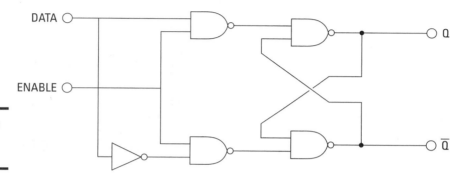

Figure 5-7:
A gated D
latch.

Project 5-3 shows how to build a gated D latch using two 4011 Quad 2-Input NAND gates. Two 4011 chips are required because the NAND gate requires a total of five gates (four NAND gates and one NOT gate), and each 4011 provides just four gates. In Chapter 2 of this minibook, you learn that you can create a NOT gate from a NAND gate simply by tying the two inputs of the NAND gate together. In this project, you use that technique to create the NOT gate.

Figure 5-8 shows the assembled gated D latch. To operate it, use the first button (the one in row 4) as the DATA input and the second button (the one in row 8) as the ENABLE input, as follows:

✦ **Set the Q output to HIGH.** First, press and hold the DATA input button, and then press and release the ENABLE input button to activate the latch. The first LED lights to indicate that the output is HIGH.

✦ **Set the Q output to LOW (which sets the $\overline{Q}$ output to HIGH).** Just press and release the ENABLE button without pressing the DATA button. The first LED goes out to indicate that the Q output is LOW, and the second LED lights to indicate that the $\overline{Q}$ output is HIGH.

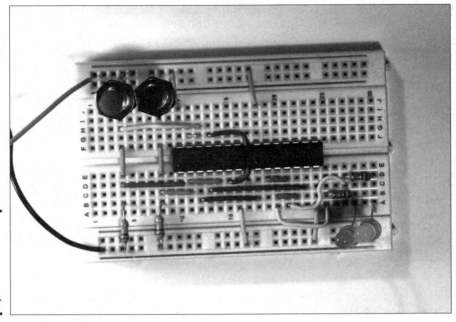

Figure 5-8:
The
assembled
gated
D latch
(Project 5-3).

Introducing Flip-Flops

A *flip-flop* is a special type of gated latch. The difference between a flip-flop and a gated latch is that in a flip-flop, the inputs aren't enabled merely by the presence of a HIGH signal on the CLOCK input. Instead, the inputs are enabled by the *transition* of the CLOCK input. Thus, at the moment that the clock input transitions from low to high, the inputs are briefly enabled. Once the clock stabilizes at the HIGH setting, the output state of the flip-flop is latched until the next clock pulse.

Flip-flops are often said to be *edge-triggered* because it's the edge of the clock signal that triggers the flip-flop. When used in clock-driven computer circuits, edge-triggering is an important characteristic because it helps circuit designers maintain better control over the timing in circuits that contain hundreds or perhaps thousands of flip-flops.

The circuitry that enables a flip-flop to respond to just the leading edge can be pretty complicated. One of the simplest methods is to feed the clock input into a NAND gate, passing one of the legs through an inverter as shown in Figure 5-9. This works because in all logic gates, there is a very small delay between the time a signal arrives at the input and the correct signal arrives at the output.

Figure 5-9:
The clock transitions from LOW to HIGH.

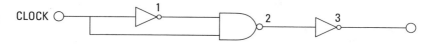

Here I step you through what happens when the clock transitions from LOW to HIGH in Figure 5-9:

1. Initially, the clock input is LOW. The inverter causes the first input to the NAND gate (marked *1* in the figure) to be HIGH, while the second input is LOW. Because the inputs aren't both HIGH, the output from the NAND gate at point 2 in the figure is HIGH. The second inverter inverts the NAND gate output so the final output from the circuit at point 3 is LOW, just like the clock input.

2. When the clock input goes high, the second input to the NAND gate goes high immediately. However, it takes a few milliseconds for the inverter to respond, so for those few milliseconds, the output from the inverter is still HIGH. Thus, both inputs to the NAND gate are HIGH for a few milliseconds, which causes the output from the NAND gate at point 2 in the figure to go LOW. Then, the second NOT gate inverts the NAND gate output, causing the output at point 3 in the signal to go HIGH for a brief moment.

3. Once the first NOT gate catches up and its output goes LOW (at point 1 in the figure), the NAND gate responds to the LOW and HIGH input by setting its output to HIGH at point 2 in the figure. The second NOT gate then inverts that output at point 3 in the figure.

The net result of the circuit in Figure 5-9 is that long clock pulses are turned into short clock pulses. The duration between the pulses remains the same, but the HIGH part of the pulse becomes much shorter.

Flip-flops are designed for use in circuits that use steady clock pulses. An easy way to provide clock pulses for a flip-flop circuit is to use a 555 timer IC, as described in Book IV, Chapter 2. However, the input source for the CLOCK input of a flip-flop doesn't have to be an actual clock; it can also be a one-shot input triggered by a pushbutton.

Project 5-3: A Gated D Latch

In this project, you build a simple active-low latch circuit using a pair of NAND gates from a 4011 Quad 2-Input NAND Gate IC. Connect switches to the $\overline{SET}$ and $\overline{RESET}$ inputs and LEDs to the Q and $\overline{Q}$ outputs so you can see the operation of the circuit.

Parts List

2 4011 Quad 2-Input NAND Gate
2 1 kΩ resistors (brown-black-red)
2 10 kΩ resistors (brown-black-orange)
2 Red LEDs
2 Normally open pushbuttons
4 AAA batteries
1 Four AAA-battery holder
18 Jumper wires (various lengths)

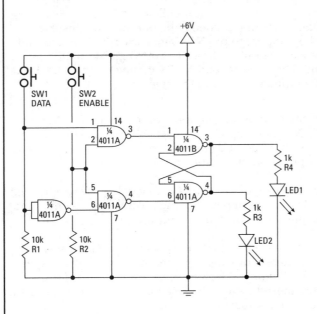

Steps

1. **Insert the two 4011 ICs.**

   4011A: Pin 1 in E10; pin 14 in F10
   4011B: Pin 1 in E18; pin 14 in F18

2. **Insert R1 through R4.**

   R1 (10 kΩ): I4 to positive bus
   R2 (10 kΩ): I8 to positive bus
   R3 (1 kΩ): B25 to B28
   R4 (1 kΩ): D27 to D30

3. **Insert the LEDs.**

   LED1: A28 to ground bus
   LED2: A30 to ground bus
   Connect the cathode (short lead)
   to the ground bus.

4. **Insert the jumper wires.**

   D4 to D10
   E4 to F4
   H4 to H12
   C8 to C11
   E8 to F8
   J10 to positive bus
   D11 to D14
   G11 to G12
   B12 to B18
   C13 to C23
   G13 to D15
   A16 to ground bus
   J18 to positive bus
   D19 to D21
   A20 to A25
   B20 to B22
   A21 to C27
   A24 to ground bus

5. **Insert the switches.**

   SW1: J4 to positive bus
   SW2: J8 to positive bus

6. **Connect the battery holder.**

   Red lead: positive bus
   Black lead: negative bus

7. **Insert the batteries.**

8. **Press the pushbuttons and
   observe the operation of the LEDs.**

6 V

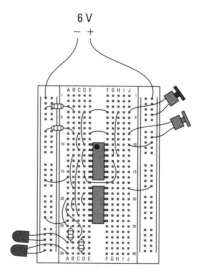

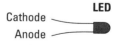

LED
Cathode
Anode

As with latches, there are several different types of flip-flops. The most common are:

✦ **SR flip-flop:** Is similar to an SR latch. Besides the CLOCK input, an SR flip-flop has two inputs, labeled SET and RESET. If the SET input is HIGH when the clock is triggered, the Q output goes HIGH. If the RESET input is HIGH when the clock is triggered, the Q output goes LOW.

Note that in an SR flip-flop, the SET and RESET inputs shouldn't both be HIGH when the clock is triggered. This is considered an invalid input condition, and the resulting output isn't predictable if this condition occurs.

✦ **D flip-flop:** Has just one input in addition to the CLOCK input. This input is called the DATA input. When the clock is triggered, the Q output is matched to the DATA input. Thus, if the DATA input is HIGH, the Q output goes HIGH, and if the DATA input is LOW, the Q output goes LOW.

Most D-type flip-flops also include S and R inputs that let you set or reset the flip-flop. Note that the S and R inputs in a D flip-flop ignore the CLOCK input. Thus, if you apply a HIGH to either S or R, the flip-flop will be set or reset immediately, without waiting for a clock pulse.

✦ **JK flip-flop:** A common variation of the SR flip-flop. A JK flip-flop has two inputs, labeled *J* and *K*. The J input corresponds to the SET input in an SR flip-flop, and the K input corresponds to the RESET input.

The difference between a JK flip-flop and an SR flip-flop is that in a JK flip-flop, both inputs can be HIGH. When both the J and K inputs are HIGH, the Q output is *toggled*, which means that the output alternates between HIGH and LOW. For example, if the Q output is HIGH when the clock is triggered and J and K are both HIGH, the Q output is set to LOW. If the clock is triggered again while J and K both remain HIGH, the Q output is set to HIGH again, and so forth, with the Q output alternating from HIGH to LOW at every clock tick.

✦ **T flip-flop:** This is simply a JK flip-flop whose output alternates between HIGH and LOW with each clock pulse. Toggles are widely used in logic circuits because they can be combined to form counting circuits that count the number of clock pulses received.

You can create a T flip-flop from a D flip-flop by connecting the $\overline{Q}$ output directly to the D input. Thus, whenever a clock pulse is received, the current state of the Q output is inverted (that's what the $\overline{Q}$ output is) and fed back into the D input. This causes the output to alternate between HIGH and LOW.

You can also create a T flip-flop from a JK flip-flop simply by hard-wiring both the J and K inputs to HIGH. When both J and K are HIGH, the JK flip-flop acts as a toggle.

Although you can construct your own flip-flop circuits using NAND gates, it's much easier to use ICs that contain flip-flops. One common example is the 4013 Dual D Flip-Flop. This chip contains two D-type flip-flops in a 14-pin DIP package. The pinouts are listed in Table 5-1.

Table 5-1		Pinouts for the 4013 Dual D Flip-Flop IC			
Pin	*Name*	*Explanation*	*Pin*	*Name*	*Explanation*
1	Q1	Flip-flop 1 Q output	8	SET2	Flip-flop 2 SET input
2	$\overline{Q1}$	Flip-flop 1 $\overline{Q}$ output	9	DATA2	Flip-flop 2 DATA input
3	CLOCK1	Flip-flop 1 CLOCK input	10	RESET2	Flip-flop 2 RESET input
4	RESET1	Flip-flop 1 RESET input	11	CLOCK2	Flip-flop 2 CLOCK input
5	DATA1	Flip-flop 1 DATA input	12	$\overline{Q2}$	Flip-flop 2 $\overline{Q}$ output
6	SET	Flip-flop 1 SET input	13	Q2	Flip-flop 2 Q output
7	GND	Ground	14	V_{DD}	+3 to 15 V

When you use one of the flip-flops in a 4013 IC, be sure to connect any unused inputs to ground. All unused inputs in CMOS logic chips should be connected to ground, but for simple breadboard circuits, the ground connections aren't usually required. However, the DATA and CLOCK inputs of a 4013 flip-flop won't work properly if you don't ground the SET and RESET inputs.

Project 5-4 shows how to use a 4013 IC to create a basic D flip-flop. This circuit works much the same as the D-type latch you create in Project 5-3. However, it requires only one IC rather than two, and the wiring is much simpler. That's because the engineers who designed the 4013 IC crammed all the wiring between the individual NAND gates in the IC, so you don't have to wire the gates together on the breadboard. Instead, all you have to do is hook up the inputs and the outputs and watch the circuit work. Figure 5-10 shows the assembled circuit.

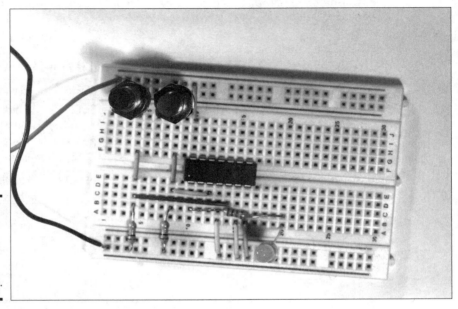

Figure 5-10:
The
assembled
D flip-flop
circuit
(Project 5-4).

Project 5-5 shows how to build a T flip-flop in which each press of a button causes an output LED to alternate between on and off. For this project, the $\overline{Q}$ output from the flip-flop is connected to the DATA input. Then, each time the Clock input goes HIGH, the inverted output from the $\overline{Q}$ output is fed into the DATA input. This causes the Q output to invert. Figure 5-11 shows the assembled circuit.

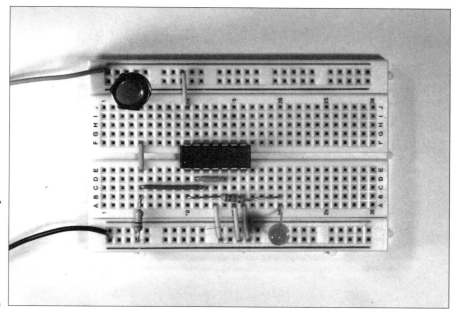

Figure 5-11: The assembled T flip-flop circuit (Project 5-5).

Debouncing a Clock Input

When you use a mechanical switch to trigger the clock input of a flip-flop, the switch will very likely have some mechanical bounce. This bounce happens when the switch contacts don't close completely cleanly; instead, the contacts bounce a little bit when they first touch each other. Even though these bounces are usually just a few milliseconds apart, they can end up confusing the flip-flop, as it thinks that each bounce of the switch contacts is actually a separate press of the button. So instead of just turning the LED attached to the Q output from off to on, a single press of the button might turn it from off to on, and then back off, then on, then off again, and so on until the switch settles down into its fully-closed position.

There are several different ways you can *debounce* a mechanical switch — that is, eliminate the bounce effect. The easiest is to connect the mechanical switch to a one-shot timer circuit that uses an RC network to create a very short time interval such as 10 or 20 ms. Though short, this interval is enough to eliminate the negative bouncing effect.

For more information about how to build a one-shot circuit using a 555 timer IC, refer to Chapter 2 of Book III.

Project 5-4: A D Flip-Flop

In this project, you build a circuit that demonstrates the operation of a type-D flip-flop. This circuit uses one of the two flip-flops on a 4013 Dual D Flip-Flop IC. The Data and Clock inputs are connected to pushbuttons, and the $\overline{Q}$ output is connected to an LED. The Q output isn't connected.

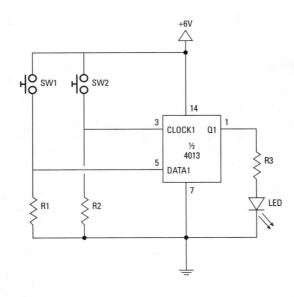

Parts List

- 1   4013 Dual D Flip-Flop
- 1   1 kΩ resistor (brown-black-red)
- 2   10 kΩ resistors (brown-black-orange)
- 1   Red LED
- 2   Normally open pushbuttons
- 4   AAA batteries
- 1   Four AAA-battery holder
- 7   Jumper wires (various lengths)

Steps

1. **Insert the 4013.**

   44013: Pin 1 in E10; pin 14 in F10

2. **Insert R1 through R3.**

   R1 (10KΩ): B4 to negative bus
   R2 (10KΩ): B8 to negative bus
   R3 (1KΩ): B10 to B20

3. **Insert the LED.**

   A20 to ground bus
   Connect the cathode (short lead)
   to the ground bus.

4. **Insert the jumper wires.**

   C4 to C14
   E4 to F4
   D8 to D12
   E8 to F8
   A13 to ground bus
   A15 to ground bus
   A16 to ground bus
   J10 to positive bus

5. **Insert the switches.**

   SW1: J4 to positive bus
   SW2: J8 to positive bus

6. **Connect the battery holder.**

   Red lead: positive bus
   Black lead: negative bus

7. **Insert the batteries.**

8. **Press the pushbuttons and
   observe the operation of the LEDs.**

   Press and hold the DATA button,
   then momentarily press the CLOCK
   button. The LED should come on.
   Release the DATA button, and
   then momentarily press the Clock
   button. The LED should go out.

6 V
− +

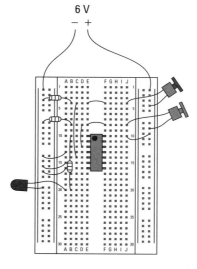

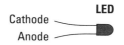

LED

Cathode
Anode

Project 5-5: A Toggle Flip-Flop

In this project, you build a toggle flip-flop using a 4013 Dual D Flip-Flop IC. The Clock input is connected to a pushbutton, and the $\overline{Q}$ output is connected to an LED. The Q output is connected to the DATA input. As a result, each time the button is pressed, the $\overline{Q}$ output is read into the DATA input, which causes the Q output to invert.

Parts List	
1	4013 Dual D Flip-Flop
1	1 kΩ resistor (brown-black-red)
1	10 kΩ resistors (brown-black-orange)
1	Red LED
1	Normally open pushbuttons
4	AAA batteries
1	Four AAA-battery holder
7	Jumper wires (various lengths)

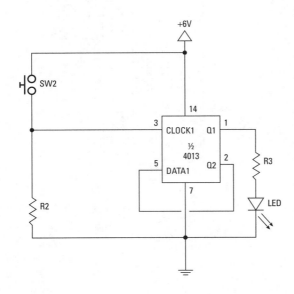

Steps

1. **Insert the 4013.**
   4013: Pin 1 in E10; pin 14 in F10

2. **Insert the resistors.**
   R1 (10 kΩ): B5 to ground bus
   R2 (1 kΩ): B10 to B20

3. **Insert the LED.**
   A20 to ground bus
   Connect the cathode (short lead) to the ground bus.

4. **Insert the jumper wires.**
   C5 to C12
   E5 to F5
   D11 to D14
   J10 to positive bus
   A13 to ground bus
   A15 to ground bus
   A16 to ground bus

5. **Insert the switch.**
   J5 to positive bus

6. **Connect the battery holder**
   Red lead: positive bus
   Black lead: negative bus

7. **Insert the batteries.**

8. **Press the pushbutton and observe the operation of the LED.**
   Each time you press the pushbutton, the LED switches from on to off or from off to on. Note, however, that it's sometimes difficult to press the switch in a way that makes a single, clean contact. Often the switch bounces, sending several very short pulses to the clock input. When that happens, the flip-flop may change state two or more times with a single push of the button. This is to be expected when a mechanical switch is used as the clock source.

6 V
− +

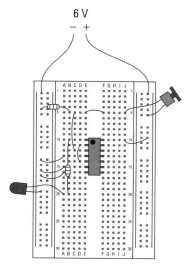

LED
Cathode
Anode

Book VII

Working with BASIC Stamp Processors

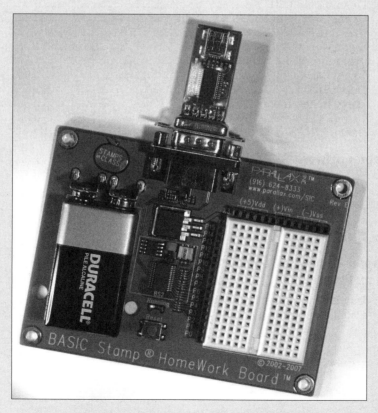

A BASIC Stamp HomeWork board with a USB to serial
adapter connected

Contents at a Glance

Chapter 1: Introducing Microcontrollers and the BASIC Stamp

In This Chapter

✔ Looking at microcontrollers and how they work

✔ Examining the BASIC Stamp microcontroller

✔ Learning how to download a program to a BASIC Stamp microcontroller

✔ Building some simple BASIC Stamp projects

I first learned about microcontrollers years ago, when I wanted to create an automated prop for a Halloween display. The prop consisted of a Frankenstein creature that was attached to a pneumatic (air-powered) pop-up mechanism. Unsuspecting trick-or-treaters would walk into Frankenstein's lab, where the creature was sitting on a big work table. After a few moments, a voice would cry out, "It's alive! It's *alive*! It's *alive*!" and lights would flash on and off. Then the creature would start to twitch, and then pop up screaming "puttin' on the Ritz!" As the creature popped up, a bright floodlight would illuminate him. After a few seconds, the light would go out and the creature would lie back down.

I pondered how I could create the electronic circuit to control my prop. It could be done with a small computer and a parallel-port relay interface. You learn how to create such circuits in Book VI, Chapter 4. But I then realized it would be more fun and cost about the same to do it with a microcontroller instead of an old laptop computer and a parallel-port relay board. So I decided to build the prop using a BASIC Stamp microcontroller to handle all of the sequencing of lights and motion for the prop, and it could even play the sounds when the creature begins to awake and then finally pop up.

In a nutshell, a *microcontroller* is a small computer on a single chip, which you can purchase for $50 or less. This chapter introduces you to the BASIC Stamp microcontroller, one of the popular microcontrollers available today. In subsequent chapters in this minibook, you learn the details of using the BASIC Stamp microcontroller, including how to write the programs that control the BASIC Stamp's operation.

Note that there are many other different kinds of microcontrollers available on the market today. I chose the BASIC Stamp for this minibook primarily because its programming language is easier to learn and use than the programming languages that other types of microcontrollers use. And as an added plus, you can purchase a BASIC Stamp starter kit from most RadioShack stores for under $100. Thus, the BASIC Stamp is an easy way to start learning about microcontrollers.

Introducing Microcontrollers

A *microcontroller* is a complete computer on a single chip. Like all computer systems, microcomputers consist of several basic subsystems:

- ✦ **Central processing unit (CPU):** The brains of the microprocessor. A CPU carries out the instructions provided to it by a program. The CPU can do basic arithmetic as well as other operations necessary to the proper functioning of the computer, such as moving data from one location of memory to another or receiving data as input from the outside world.

  The CPU of a microcontroller is usually much simpler than the CPU found in a desktop computer. However, it is conceptually very similar. In fact, the CPUs found in many modern microcontrollers are as advanced as CPUs used in desktop computers just a few years ago.

- ✦ **Clock:** The CPU and other components of the microcontroller are driven by a clock that provides timing pulses that control the pacing of program instructions as they are executed one at a time by the CPU. For most microcontrollers, the clock ticks along at a pace of a few million ticks per second. In contrast, the clock that drives a typical desktop computer ticks along at a few *billion* ticks per second.

- ✦ **Random access memory (RAM):** Provides a scratchpad area where the computer can store the data it's working on. For example, if you want the computer to determine the result of a calculation (such as two plus two), you need to provide a location in RAM where the computer can store the result.

  In a desktop computer, the amount of available RAM is measured in billions of bytes (GB for gigabytes). In a microcontroller, the RAM is often measured just bytes. That's right: not billions (GB), millions (MB; megabytes), or even thousands (KB; kilobytes) of bytes, but plain old bytes. For example, the popular BASIC Stamp 2, which will be used in this chapter and throughout the remaining chapters in this book has a total of 32 bytes of RAM.

- ✦ **EEPROM:** A special type of memory that holds the program that runs on a microcontroller. *EEPROM* stands for *Electrically Erasable Programmable Read-Only Memory*, but that won't be on the test.

EEPROM is *read-only,* which means that once data has been stored in EEPROM, the data can't be changed by a program running on the microcontroller's CPU. However, it's possible to write data to EEPROM memory by connecting the EEPROM to a computer via a USB port. Then, the computer can send data to the EEPROM.

This is how microcontrollers are programmed. You use special software on a PC to create the program that you want to run on the microcontroller. Then, you connect the microcontroller to the PC and transfer the program from the PC to the microcontroller. The microcontroller then executes the instructions set forth in the program.

Most microcontrollers have a few thousand bytes of EEPROM memory, which is enough to store relatively complicated programs downloaded from a PC.

One of the most important features of EEPROM memory is that it doesn't lose its data when you turn off the power. Thus, once you transfer a program from a PC to a microcontroller's EEPROM, the program remains in the microcontroller until you replace it with some other program. You can turn the microcontroller off and put it on a closet shelf for years, and when you turn the microcontroller back on, the program that was recorded years ago will run again.

✦ **I/O pins:** One of the most important features of a microcontroller is its *I/O pins,* which enable the microcontroller to communicate with the outside world. Although some microcontrollers have separate input pins and output pins, most have shared I/O pins that can be used for both input and output.

I/O pins usually use the basic TTL logic interface that I describe throughout Book VI: HIGH (logic 1) is represented by +5 V, and LOW (logic 0) is represented by 0 V.

Most microcontrollers can handle only a small amount of current directly through the I/O pins. 20–25 mA is typical. That's enough to light up an LED, but circuits that require more current should isolate the higher current load from the microcontroller I/O pins. This is usually done by using a transistor driver.

Introducing the BASIC Stamp

The BASIC Stamp is a microcontroller made by Parallax, a company started by two kids fresh out of high school in 1986. A *BASIC Stamp* is an essentially complete, self-contained computer system. Depending on the model, the Stamp (as it's usually called) is either a single chip or a small circuit board.

The key feature that sets BASIC Stamp microcontrollers apart from other microcontrollers is that BASIC Stamps include a built-in programming language called PBASIC (Parallax's modified version of BASIC). This programming language makes it easy to create programs that run on the BASIC Stamp. With most other microcontrollers, programming is much more difficult because you must write the programs in a more complicated programming language.

Figure 1-1 shows one of the more popular BASIC Stamps, called the Basic Stamp 2 Module. This is a complete computer system on a 24-pin DIP package that you can solder directly to a circuit board or (more likely) insert into a 24-pin DIP socket.

Figure 1-1:
A BASIC
Stamp 2
Module.

Here are the basic specifications for the BASIC Stamp 2 microcontroller:

✦ **Clock speed:** 20 MHz

✦ **RAM:** 32 bytes

✦ **EEPROM:** 2,048 bytes

✦ **Number of I/O pins:** 16

✦ **Power supply:** 5.5 to 15 V

✦ **Current draw:** 3 mA

✦ **Maximum I/O pin current:** 20 mA. (However, there is a limit of to the total amount of current the I/O pins combined can handle: 40mA for the first group of 8 I/O pins and another 40mA for the second group.

I show you how to work with BASIC Stamp 2 microcontrollers throughout this chapter and the remaining chapters in this minibook.

Buying a BASIC Stamp

Although you can purchase a BASIC Stamp microcontroller by itself, the easiest way to get into BASIC Stamp programming is to purchase a starter kit that includes a BASIC Stamp along with the software that runs on your PC for programming the BASIC Stamp and the USB cable that connects your PC to the Stamp. In addition, most starter kits come with a prototype board that makes it easy to design and test simple circuits that interface with the Stamp.

One such starter kit is the BASIC Stamp Activity Kit, which you can purchase at most RadioShack stores for about $100. This kit comes with the following components, which are pictured in Figure 1-2:

✦ **BASIC Stamp HomeWork board:** This is a prototyping board that includes a BASIC Stamp 2 microcontroller, a small solderless breadboard, a clip for a 9 V battery, and a connector for the USB programming cable.

✦ **USB cable:** Connect the HomeWork board to a computer so that you can program the Stamp.

✦ **A handful of useful electronic components:** Use these components for building circuits that connect to the Stamp, such as resistors, LEDs, capacitors, a seven-segment display, some pushbuttons and a potentiometer, and plenty of jumper wires.

**Book VII
Chapter 1**

**Introducing
Microcontrollers and
the BASIC Stamp**

+ **Servo:** This is a fancy motor that the Stamp can control.

+ ***What's a Microcontroller?*:** This book gives a nice detailed overview of programming the BASIC Stamp II.

Figure 1-2:
The BASIC
Stamp
Activity Kit.

Working with the BASIC Stamp HomeWork Board

Figure 1-3 shows the BASIC Stamp HomeWork board. The chips near the center of this board constitute the BASIC Stamp 2 module. On the left is a battery clip to which you can connect a 9 V battery for power. On the right is a small solderless breadboard, consisting of 17 rows of solderless connectors.

Immediately to the left of the breadboard is a row of 16 connectors that provide access to the BASIC Stamp 2's 16 I/O pins, and immediately above the breadboard is a row of connectors that provide access to +5 V (identified as *Vdd* and *Vin* on the board) and ground (identified as *Vss*). This breadboard is designed to allow you to construct simple circuits that connect to the 16 I/O pins. This allows you to connect pushbuttons and other devices that can send input to the program running on the Stamp or LEDs or other output devices that the Stamp program can control.

Figure 1-3:
The BASIC
Stamp
HomeWork
board.

You can construct all of the projects described in this minibook on the
BASIC Stamp HomeWork board. However, if you want to create more elabo-
rate projects, you may want to consider using an alternative called the
Board of Education, as shown in Figure 1-4. It's about the same price as the
HomeWork board, and it provides a few handy features. In particular, the
Board of Education includes the following:

✦ A prototype board similar to the one on the HomeWork board.

✦ A special 20-pin connector that allows you to permanently connect
   external circuits to the BASIC Stamp's 16 I/O pins. This is useful if you
   want to build circuits that are more permanent than the solderless
   breadboard allows.

✦ Special connectors for I/O pins 12, 13, 14, and 15 that are designed for
   connecting the BASIC Stamp to servos.

✦ A 9 V battery clip and an external power connector that allow you to
   power the BASIC Stamp with either a 9 V battery or an external power
   supply.

✦ A removable 24-pin BASIC Stamp 2 Module. This allows you to use the Board of Education to program the Stamp. Then when you're certain that the program is working properly, you can remove the BASIC Stamp 2 Module from the Board of Education and plug it into your own circuit.

Figure 1-4:
The
Board of
Education.

Connecting to BASIC Stamp I/O Pins

Before you start writing programs to run on a BASIC Stamp, you must first build the circuit(s) that will connect to the I/O pins on the Stamp. Once the circuit is constructed, you can then write a program that controls the circuits that you've connected to each pin.

The BASIC Stamp 2 has a total of 16 separate I/O pins, which means you can connect as many as 16 separate circuits. These 16 I/O circuits are more than enough for most BASIC Stamp projects.

For now, start with a basic circuit that simply connects an LED to one of the Stamp's output pins. The program that you write and load onto a Stamp can turn each of the 16 I/O pins on (HIGH) or off (LOW) by using simple programming commands. When an I/O pin is HIGH, +5 V is present at the pin. When it's LOW, no voltage is present.

Each I/O pin can handle as much as 20 mA of current, which is more than enough to light an LED. As with any LED circuit, you need to provide a current-limiting resistor in the circuit. If you forget to include this resistor, you'll destroy the LED, and you might fry the Stamp too — so always be sure to include the current-limiting resistor. A 470Ω resistor is usually appropriate for LED circuits connected to Stamp I/O pins.

Figure 1-5 shows a simple schematic diagram for a circuit that drives an LED from pin 15 of a BASIC Stamp. Notice in this schematic that the pin output is represented by a simple five-sided shape. It's common in schematics to *not* draw the Stamp as a single rectangle as you would other integrated circuits. Instead, each I/O connection in the circuit is shown using a connector shape.

Figure 1-5:
Schematic for an LED circuit connected to a BASIC Stamp I/O pin.

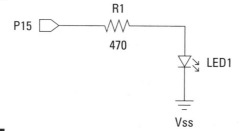

Figure 1-6 shows how you might build this circuit on a BASIC Stamp HomeWork board. Notice that the resistor is inserted into the P15 connection and the sixth hole in the second row. The LED's anode is inserted in the seventh hole in row 2, and the cathode is in one of the Vss connectors. Project 1-1, which appears later in this chapter, walks you through the creation of this circuit.

Figure 1-6:
Assembling an LED circuit on a BASIC Stamp HomeWork board.

Installing the BASIC Stamp Windows Editor

The BASIC Stamp Windows Editor is the software that you use on your computer to create programs that can be downloaded to a BASIC Stamp microcontroller. This software is available free from the Parallax website.

The easiest way to find the right page from which you can download the software is to use Google to search for *BASIC Stamp Windows Editor.*

Or you can do it this way: To download the software, open a web browser such as Internet Explorer, browse to `www.parallax.com`, and follow these instructions:

1. **Click the Resources tab.**
2. **Click Downloads in the menu on the left side of the page.**
3. **Click BASIC Stamp Software.**
4. **Select the version of the software for your computer and download and install the software.**

Once you get the software installed, run it by choosing BASIC Stamp Editor from the Windows Start menu. Figure 1-7 shows how the editor appears when you first run it.

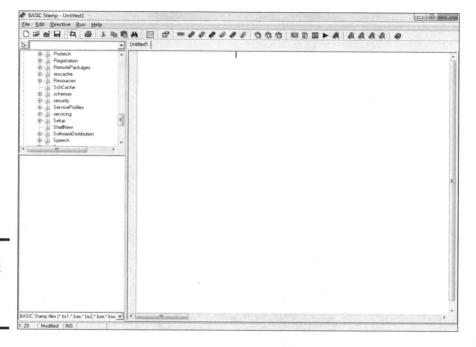

Figure 1-7:
The BASIC
Stamp
Windows
Editor.

Connecting to a BASIC Stamp

Before you can use the Stamp Editor to program a BASIC Stamp, you must first connect the Stamp to your computer, and then configure the Stamp Editor so that it can communicate with the Stamp. The following steps describe the procedure for doing that with a BASIC Stamp HomeWork board. The procedure for connecting a Board of Education is very similar, so you shouldn't have any trouble following along in case you have a Board of Education instead of a HomeWork board.

1. **Insert a 9 V battery in the HomeWork board.**

You'll see a green LED light up on the board when the battery is inserted. If this light doesn't come on, check the battery — it may be dead.

2. **Plug the USB to serial adapter into the DB9 connector on the HomeWork board.**

This step is necessary because the HomeWork board uses an older-style serial connector to connect to your computer, but most computers don't have serial ports. The adapter converts the serial port connection on the HomeWork board to a USB port.

Figure 1-8 shows how the HomeWork board appears with the USB to serial adapter plugged in.

Figure 1-8:
The
HomeWork
board with
the USB
to serial
adapter
connected.

3. **Plug the smaller end of the USB cable into the USB to serial adapter and the larger end into an available USB port on your computer.**

   After a moment, your computer will chirp happily to acknowledge the presence of the BASIC Stamp. A pop-up bubble may appear informing you that Windows is installing the driver needed to access the device. If so, just wait for the bubble to disappear.

   In addition to the activity on your computer, you'll also notice a green LED on the USB to serial adapter. This indicates that the adapter is powered up and ready to adapt.

4. **In the BASIC Stamp Editor, choose Run⇨Identify.**

   This brings forth the Identification window, as shown in Figure 1-9.

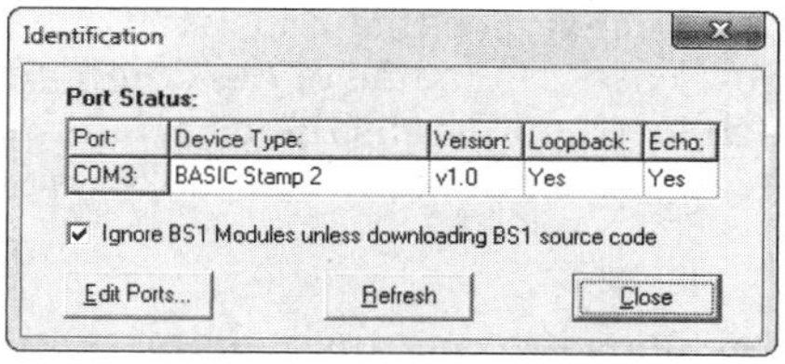

Figure 1-9:
Identifying
your BASIC
Stamp.

This window should indicate that a BASIC Stamp 2 is connected. If it doesn't, consult the "Connection Troubleshooting" section of the Stamp Editor's Help command (Help⇨BASIC Stamp Help).

5. **Click Close to dismiss the Identification window.**

6. **You're done!**

You're ready to write your first BASIC Stamp program.

Here are a few other tidbits you should know about connecting a BASIC Stamp to the Stamp Editor:

✦ If you're having trouble connecting, the most likely cause (other than a loose connection or forgetting to insert the battery) is an incorrect driver for the USB to serial adapter. To install the right driver software, browse to the web page at `www.parallax.com/usbdrivers`. Follow the instructions that appear on that page to install the correct driver.

✦ Parallax makes a version of the Board of Education that has a USB port directly on the board. If you're using that board instead of the HomeWork board, you don't need a USB to serial adapter. Instead, you can plug the USB cable directly into the Board of Education.

Writing Your First PBASIC Program

If you've never done any form of computer programming before, you're in for a fun and fascinating adventure, during which you learn how computers really work. In a nutshell, a computer program is a set of written instructions that a computer knows how to read, interpret, and carry out. The instructions are written in a language that both humans and computers can read. The instructions aren't quite English, but they resemble English enough that we English-speaking people can understand what they mean. (Of course, there are non-English programming languages as well, but PBASIC happens to be an English-based programming language.)

Computer programs are stored in text files that consist of one or more lines of written instructions. In most cases, each line of the computer program contains one instruction. Each instruction tells the computer to do something specific, such as add two numbers together or make one of the output pins go HIGH.

The trick of computer programming is to put the right instructions together in the right sequence to get the program to do exactly what you want it to do. Of course, to do that, you need to have a solid understanding of what you want the program to do, and you need to have a solid knowledge of the variety of instructions that are available to you. The PBASIC programming language consists of about 70 different types of instructions. But don't be discouraged; you can write useful programs using only a handful of these commands.

In just about every book on programming languages, the first program presented is called Hello World. This simple program displays the string "Hello, World!" as a way of demonstrating what the simplest possible program looks like.

In PBASIC (the official name of the BASIC language that's used on BASIC Stamps), the Hello World program consists of three lines:

```
' {$STAMP BS2}
' {$PBASIC 2.5}
DEBUG "Hello, World!"
```

The first two lines are called *directives*. They don't tell the BASIC Stamp to actually do anything; instead, they provide information that the Stamp Editor needs to know to prepare your program so that it can be downloaded to the Stamp. The first line indicates that the microcontroller you'll run the program on is a BASIC Stamp 2 (BS2). The second line indicates that this program uses version 2.5 of PBASIC for this program. (That's the current version.)

Every program you write must include these two lines. Fortunately, you don't have to type them yourself. Instead, you can use menu commands or toolbar buttons to insert the directives automatically:

✦ **Directive➪Stamp➪BS2:** Inserts the $STAMP BS2 directive to indicate that you're using BASIC Stamp 2.

✦ **Directive➪PBASIC➪Version 2.5:** Inserts the $PBASIC 2.5 directive to indicate that you're using version 2.5 of PBASIC.

The third line of the Hello World program is the only line that actually tells the BASIC Stamp to do something. This command, called DEBUG, tells the BASIC Stamp to send a bit of text to the computer connected via the USB port (The DEBUG command always consists of two parts: The word *DEBUG* followed by some text that must be enclosed in quotation marks. For example:

```
DEBUG "Hello, World!"
```

This line sends the message "Hello, World!" to the computer. The message is displayed in a window called the Debug Terminal window within the Stamp Editor.

Project 1-1 shows you how to run the Hello World program on a HomeWork board. Figure 1-10 shows the resulting output as displayed in the Debug Terminal window.

Figure 1-10:
The Debug Terminal window displays the output from the Hello World program.

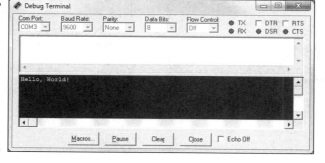

Book VII Chapter 1

Introducing Microcontrollers and the BASIC Stamp

Flashing an LED with a BASIC Stamp

A BASIC Stamp is serious overkill for a circuit that simply flashes an LED on and off: You can do that for a few bucks with a 555 timer IC, a capacitor, and a couple of resistors.

But learning how to flash an LED on and off with a BASIC Stamp is an important step toward completing more complex projects. To flash an LED on and off, you first have to connect an LED to an output pin in such a way that the Stamp can turn the LED on by taking the output pin HIGH. You learn how to do that earlier in this chapter, in the section "Connecting to BASIC Stamp I/O Pins." So all that remains is learning how to write a PBASIC program that will flash the LED.

Project 1-1: Hello, World!

In this project, you run a simple Hello World program on a BASIC Stamp. The output is displayed in the Debug Terminal window with the Stamp Editor program.

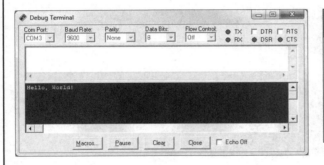

Parts List

1 Computer with BASIC Stamp Editor software installed.
1 BASIC Stamp HomeWork board
1 9 V Battery
1 USB cable
1 USB to serial adapter

Steps

1. **Open the BASIC Stamp Editor.**
2. **Connect your BASIC Stamp to the computer and identify it in the Stamp Editor.**

   For more information about how to do this, refer to the section "Connecting to a BASIC Stamp."
3. **Type the Hello World program into the Editor program window.**

   The Hello World program is shown in the accompanying listing.
4. **Choose File⇨Save.**

   This brings up a Save As dialog box.
5. **Navigate to the folder where you want to save the program, type a filename, and click Save.**

   You can use any filename you wish, but BASIC Stamp 2 program files must have the extension .bs2.
6. **Choose Run⇨Run.**

   If you prefer, you can click the Run button on the toolbar or press F9. When you choose the Run command, the program is downloaded to the BASIC Stamp. Once the program is downloaded, it automatically starts to run on the Stamp.
7. **View the Hello, World! message displayed in the Debug Terminal window.**

Hello World Program

```
'{$STAMP BS2}
'{$PBASIC 2.5}
DEBUG "Hello, World!"
```

Book VII
Chapter 1

Introducing
Microcontrollers and
the BASIC Stamp

To write such a program, you need to know the following five PBASIC instructions:

✦ **HIGH** — Sets one of the Stamp's I/O pins to HIGH. You use this instruction to turn the LED on.

✦ **LOW** — Sets one of the Stamp's I/O pins to LOW. You use this instruction to turn the LED off.

✦ **PAUSE** — Causes the Stamp to sit idle for a specified period of time. You use this instruction to delay the program a bit between HIGH and LOW commands so that the LED stays on for awhile before you turn it off, and then stays off for awhile before you turn it back on.

✦ **GOTO** — Causes the program to loop back to a previously designated location. You use this to cause the program to repeatedly flash the LED on and off instead of flashing the LED on and off only once.

✦ **Label** — Marks the location that you want the GOTO statement to loop to.

Here's the complete program that flashes the LED:

```
' {$STAMP BS2}
' {$PBASIC 2.5}
Main:
 HIGH 15
 PAUSE 1000
 LOW 15
 PAUSE 1000
 GOTO Main
```

Take a look at how this program works, one line at a time:

Program Line	What It Does
' {$STAMP BS2}	Indicates that the program will run on a BASIC Stamp 2.
' {$PBASIC 2.5}	Indicates that the program uses version 2.5 of PBASIC.
Main:	Creates a label named Main that marks the location that the GOTO command will loop back to.
HIGH 15	Makes I/O pin 15 HIGH, which turns the LED on.

Program Line	*What It Does*
PAUSE 1000	Pauses the program for 1,000 ms, which is the same as one second. This allows the LED to stay on for one full second.
LOW 15	Makes I/O pin 15 LOW, which turns the LED off.
PAUSE 1000	Pauses the program for 1,000 ms. This allows the LED to stay off for one full second.
GOTO Main	Causes the program to skip back to the Main label, which causes the program to loop through the HIGH, PAUSE, LOW, and PAUSE instructions over and over again.

The net effect of this program is that the LED on pin 15 flashes on and off at one-second intervals.

Project 1-2 shows how to build a simple circuit that connects an LED to pin 15 and then download and run the LED Flasher program so that the LED flashes on and off. The completed circuit for this project was shown back in Figure 1-6.

Project 1-2: An LED Flasher

In this project, you build a circuit that connects an LED to pin 15 of the BASIC Stamp processor. Then, you download and run a program that flashes the LED on and off at one second intervals.

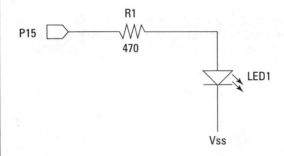

Parts List

1 Computer with BASIC Stamp Editor software installed.
1 BASIC Stamp HomeWork board
1 9 V Battery
1 USB cable
1 USB to serial adapter
1 Red LED
1 460 Ω omega resistor (yellow-violet-brown)

LED Flasher

```
'{STAMP BS2}
'{$PBASIC 2.5}
Main:
    HIGH 15
    PAUSE 1000
    LOW 15
    PAUSE 1000
    GOTO Main
```

Steps

1. **Insert the resistor and LED in the HomeWork board's solderless breadboard as shown in the accompanying diagram.**

   The resistor should connect the P15 I/O pin to the anode (long lead) of the LED, and the cathode (short lead) of the LED should connect to one of the Vss connections.

2. **Open the BASIC Stamp Editor.**

3. **Connect your BASIC Stamp to the computer and identify it in the Stamp Editor.**

   For more information about how to do this, refer to the section "Connecting to a BASIC Stamp."

4. **Type the LED Flasher program into the Editor program window.**

   The LED Flasher program is shown in the accompanying listing.

5. **Choose File⇨Save.**

   This brings up a Save As dialog box.

6. **Navigate to the folder where you want to save the program, type a filename, and click Save.**

   You can use any filename you wish, but BASIC Stamp 2 program files must have the extension .bs2.

7. **Choose Run⇨Run.**

   If you prefer, you can click the Run button on the toolbar or press F9. When you choose the Run command, the program is downloaded to the BASIC Stamp. Once the program is downloaded, it automatically starts to run on the Stamp.

8. **Observe the LED flashing on and off.**

   If the LED doesn't flash on and off, recheck your wiring. The most likely error is that you've inserted the LED backward or you've connected the resistor to the wrong I/O pin.

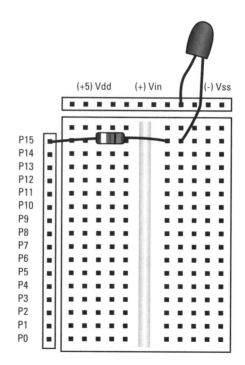

Chapter 2: Programming in PBASIC

*T*his chapter is about the exciting but somewhat intimidating topic of computer programming. Specifically, programming the BASIC Stamp microprocessor using its built-in programming language, which is called PBASIC.

If you've never done any kind of computer programming, you're in for an interesting and fun journey of discovery as you start to learn what makes computers work. This journey may seem rocky at times, as there are impor-tant concepts to be grasped and those concepts can be difficult to get your mind around at first. But trust me on this: Programming a BASIC Stamp microprocessor isn't rocket science. You can and will get your mind around the difficult concepts, and when you do, your imagination is the only limit to what you can get a BASIC Stamp microprocessor to do.

So grab your thinking cap and get started.

Note that there's a lot more to PBASIC programming than this short chapter can cover. For complete information about the PBASIC programming lan-guage, you can download the 500+ page *BASIC Stamp Syntax and Reference Manual* from www.parallax.com.

Introducing PBASIC

As you learn in Chapter 1 of this minibook, you use the BASIC Stamp Windows Editor (hereafter referred to simply as the Stamp Editor) to create your programs and download them to the BASIC Stamp.

If you're unclear on how to do that, please reread the preceding chapter. You can't progress very far in PBASIC programming without writing some actual programs, downloading them to a BASIC Stamp, and observing how they work.

PBASIC is a special version of a very old programming language called *BASIC,* which was developed in the early 1960s by two professors at Dartmouth College (John Kemeny and Thomas Kurtz) as a way to teach computer programming to students who weren't interested in majoring in computer science. In other words, BASIC was originally designed to be a programming language for nonprogrammers.

Back in the 1960s, the names of programming languages were usually acronyms that indicated the intended purpose of the language. For example, FORTRAN, which was designed for the purpose of solving math formulas, stands for *formula translator.* COBOL, which was designed for business problems, stands for *common business-oriented language.* BASIC was no exception: The name *BASIC* stands for *beginners all-purpose symbolic instruction code,* and the *P* in PBASIC stands for *Parallax,* the company that invented the BASIC Stamp.

In this chapter, I focus on the PBASIC statements that you use to control BASIC Stamp output pins as well as the statements that you use to control the execution of your program. With just the statements used in this chapter, you can set up complicated programs that can control output devices connected to the BASIC Stamp.

For information about how to use BASIC Stamp I/O ports with input devices, please see the next chapter.

This chapter focuses on the PBASIC programming techniques you use for Basic Stamp 2 microcontrollers. If you're using a BASIC Stamp 1 microprocessor, please refer to Chapter 5 of this minibook for information about programming this older and simpler microcontroller.

Building a Test Circuit

All of the programs in this chapter assume that an LED is connected to output pins 0, 2, 4, 6, 8, and 10. The programs work with just the even numbered I/O pins simply because the holes on the breadboard in the BASIC Stamp HomeWork board are too close together to easily connect LEDs to adjacent pins.

Project 2-1 shows how you can build a test circuit that has six LEDs connected to pins 0, 2, 4, 6, 8, and 10 using components that come with the

BASIC Stamp Activity Kit, which is available from most RadioShack stores. *Note:* You can also easily assemble this circuit with a Board of Education and your own LEDs and resistors.

Figure 2-1 shows the finished circuit, ready to be used for testing the programs you'll write as you work your way through this chapter.

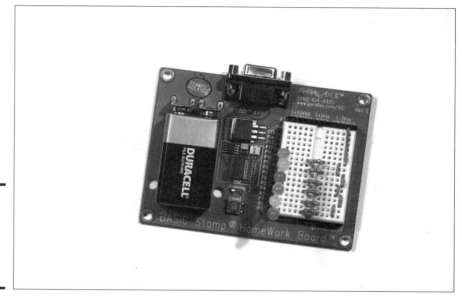

Figure 2-1:
A BASIC
Stamp
HomeWork
Board with
six LEDs.

Flashing the LEDs

In the preceding chapter, you see a program that flashes a single LED on and off. In this chapter, I show you several variants of that program, which flash the six LEDs in the test project in various sequences. Along the way, you can add more PBASIC statements to your repertoire to provide more and more complex ways to control the flashing.

Keep in mind throughout this chapter that if you can turn an LED on or off with a PBASIC program, you can control *anything* that can be connected to a BASIC Stamp I/O port. The BASIC Stamp itself doesn't know or care what kind of circuit you connect to an I/O pin. All it knows is that when you tell it to, the BASIC Stamp makes the I/O port HIGH or LOW. It's the external circuitry connected to the pin that determines what happens when the pin goes HIGH.

Project 2-1: An LED Test Board

In this project, you connect six LEDs to a BASIC Stamp Home Work or Board of Education Board to test the programs presented in this chapter.

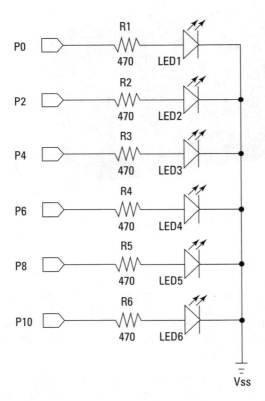

Parts List

1 Computer with BASIC Stamp Editor software installed.
1 BASIC Stamp HomeWork board or Board of Education
1 9 V Battery
1 USB cable
1 USB to serial adapter
6 LEDs
6 470 Ω resistors (yellow-violet-brown)
6 Jumper wires

Test Flash Program

```
'{$PBASIC 2.5}
'{$STAMP BS2}

Main:
    HIGH 0
    HIGH 2
    HIGH 4
    HIGH 6
    HIGH 8
    HIGH 10
    PAUSE 500
    LOW 0
    LOW 2
    LOW 4
    LOW 6
    LOW 8
    LOW 10
    PAUSE 500
    GOTO Main
```

Steps

1. **Connect the six LEDs to the output pins as follows:**

Pin (Cathode)	Breadboard Hole
P0	A17
P2	A15
P4	A13
P6	A11
P8	A9
P10	A7

2. **Connect the resistors across the gap in the center of the brea board as follows:**

Resistor	From	To
R1	E17	F17
R2	E15	F15
R3	E13	F13
R4	E11	F11
R5	E9	F9
R6	E7	F7

3. **Use the jumper cables to connect the resistors to ground.**

   I17 to I15
   J15 to J13
   I13 to I11
   J11 to J9
   I9 to I7
   J7 to Vss strip

4. **Open the BASIC Stamp Editor.**

5. **Connect your BASIC Stamp to the computer and identify it in the Stamp Editor.**

   For more information about how to do this, refer to the preceding chapter.

6. **Type the Test Flash program into** the Editor program window, and then save the file.

   The Test LED program is shown in the accompanying listing.

7. **Choose File⇨Save.**

   This brings up a Save As dialog box.

8. **Navigate to the folder where you want to save the program, type a filename, and click Save.**

   You can use any filename you wish, but BASIC Stamp 2 program files must have the extension .bs2.

9. **Choose Run⇨Run.**

   If you prefer, you can click the Run button in the toolbar or press F9.

   When you choose the Run statement, the program is downloaded to the BASIC Stamp. Once the program is downloaded, it automatically starts to run on the Stamp.

10. **Watch the LEDs flash!**

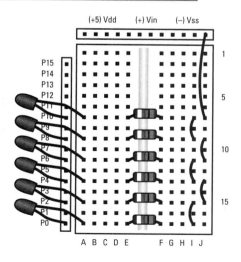

The only limitation is that the BASIC Stamp itself can swing only about 20 mA through its I/O pins. If the circuit connected to the pin requires more current than 20 mA, you must isolate the higher-current portion of the circuit from the BASIC Stamp. The easiest way to do that is to use a transistor driver or a relay.

Listing 2-1 shows a simple program that flashes all six of the LEDs on and off at half-second intervals. This program uses nothing more than the HIGH, LOW, PAUSE, and GOTO statements that are presented in the preceding chapter. The program turns all six LEDs on, pauses 500 ms (half a second), turns the LEDs off, waits another half second, and then jumps back to the Main label to start the whole process over.

Listing 2-1: Flashing LEDs

```
' {$PBASIC 2.5}                                          → 1
' {$STAMP BS2}                                           → 2

Main:                                                    → 4
  HIGH 0                                                 → 5
  HIGH 2
  HIGH 4
  HIGH 6
  HIGH 8
  HIGH 10
  PAUSE 500                                              → 11
  LOW 0                                                  → 12
  LOW 2
  LOW 4
  LOW 6
  LOW 8
  LOW 10
  PAUSE 500                                              → 18
  GOTO Main                                              → 19
```

The following paragraphs summarize the operation of this program:

→ **1** This line indicates that the program is written in version 2.5 of PBASIC. Every program you write for the BASIC Stamp 2 should include this line. You can insert it automatically into a program by choosing Directive⇨PBASIC⇨Version 2.5 or by clicking the PBASIC Version: 2.5 button in the toolbar.

→ **2** This line indicates that the program will run on a BASIC Stamp 2, and it's required for every program you run on a BASIC Stamp 2 microcontroller. You can insert it automatically by choosing Directive⇨Stamp⇨BS2.

→ **4**   The label `Main:` identifies the location that the `GOTO` statement in line 19 jumps to. `Main` is known as a *label*, which is simply a named location in your program. To create a label, you just type a name followed by a colon. For more information about creating value names in PBASIC, see the section "Creating Names" later in this chapter.

→ **5**   This line sets the output of pin 0 to `HIGH`, which in turn lights up the LED. The following lines (6 through 10) similarly turn on pins 2, 4, 6, 8, and 10.

→ **11**   This line pauses the program for 500 ms (one-half of a second).

→ **12**   This line and the five that follow set the outputs of pins 0, 2, 4, 6, 8, and 10 to `LOW`, which in turn extinguishes the LEDs.

→ **18**   This line pauses the program for an additional half second.

→ **19**   This line transfers control of the program back to the `Main` label in line 4 so that the program will repeat.

Using Comments

A *comment* is a bit of text that provides an explanation of your code. PBASIC completely ignores comments, so you can put any text you wish in a comment. Using plenty of comments in your programs to explain what your program does and how it works is a good idea.

A comment begins with an apostrophe. When PBASIC sees an apostrophe on a line, it ignores the rest of the line. Thus, if you place an apostrophe at the beginning of a line, the entire line is considered to be a comment. If you place a comment in the middle of a line (for example, after a statement), everything after the apostrophe is ignored.

It's a common programming practice to begin a program with a group of comments that indicates what the program does, who wrote it, and when. This block of comments should also indicate what I/O devices are expected to be connected to the BASIC Stamp. Listing 2-2 shows a version of the LED Flasher program that includes both types of comments.

You may notice that the `$PBASIC` and `$STAMP` directives begin with an apostrophe. Technically, these lines are treated as comments.

**Book VII
Chapter 2**

Programming
in PBASIC

Listing 2-2: LED Flasher with Comments

```
' LED Flasher Program
' Doug Lowe
' July 10, 2011
'
' This program flashes LEDs connected to pins 0, 2, 4, 6, 8, and 10
' at one-half second intervals.

' {$PBASIC 2.5}
' {$STAMP BS2}

Main:
  HIGH 0        'Turn the LEDs on
  HIGH 2
  HIGH 4
  HIGH 6
  HIGH 8
  HIGH 10
  PAUSE 500     'Wait one-half second
  LOW 0         'Turn the LEDs off
  LOW 2
  LOW 4
  LOW 6
  LOW 8
  LOW 10
  PAUSE 500     'Wait one-half second
  GOTO Main
```

Creating Names

In PBASIC, you can create your own names to use as program labels. As you'll learn later in this chapter, you can also create names for constants and variables. You can also assign a name of your own to I/O pins, which makes it easier to remember what kind of input or output is expected from each pin.

You must follow a few simple rules when you create names in PBASIC:

✦ Names can consist of a combination of upper- and lowercase letters, numbers, and underscore characters (_). Other special characters, such as dollar signs or exclamation marks, aren't allowed. Thus, `Timer_Routine` and `Relay7` are valid names, but `LED$` or `Bang!` aren't.

✦ Names must begin with a letter or an underscore but can't begin with a number. Thus, `Timer1` and `_Timer1` are both valid names, `1Timer` isn't.

✦ Names may be as long as 32 characters.

✦ Names aren't case-sensitive, which is to say that PBASIC doesn't distinguish between upper- and lowercase letters. Thus, PBASIC considers all of the following names to be identical: `TimerCheck`, `timercheck`, `TIMERCHECK`, and `TiMeRcHeCk`.

✦ Actually, nothing in PBASIC is case sensitive, so anything can be written in upper- or lowercase. However, it is a common PBASIC programming convention that keywords such as HIGH and GOTO are written in all caps, while names are written with just the first letter capitalized.

Using Constants

A *constant* is a name that has been assigned a value. This allows you to use the constant name in your program rather than the value itself. Later, if you decide to change the value, you don't have to hunt through the program to find every occurrence of the constant. Instead, you simply change the line that defines the constant.

Here's a statement that creates a constant named Delay and assigns the value 500 to it:

```
Delay   CON 500
```

The CON keyword indicates that Delay is a constant whose assigned value is 500.

To use a constant, just substitute the name of the constant wherever you would use the value. For example, this line pauses the program for the value assigned to the Delay constant:

```
PAUSE Delay
```

Listing 2-3 shows a version of the LED Flasher program that uses a constant to determine how fast the LEDs should flash.

Listing 2-3:   The LED Flasher Program with a Constant

```
' LED Flasher Program
' Doug Lowe
' July 10, 2011
'
' This program flashes LEDs connected to pins 0, 2, 4, 6, 8, and 10
' at one-half second intervals.
'
' This version of the program uses a constant
' for the time interval.

' {$PBASIC 2.5}
' {$STAMP BS2}

Delay CON 500
```

(continued)

Listing 2-3 *(continued)*

```
Main:
  HIGH 0
  HIGH 2
  HIGH 4
  HIGH 6
  HIGH 8
  HIGH 10
  PAUSE Delay
  LOW 0
  LOW 2
  LOW 4
  LOW 6
  LOW 8
  LOW 10
  PAUSE Delay
  GOTO Main
```

Assigning Names to I/O Pins

As you already know, you can use the `HIGH` and `LOW` statements to set the output status of an I/O pin. For example, the following statement sets pin 6 to `HIGH`:

```
HIGH 6
```

Here, the number 6 indicates that pin 6 should be set to `HIGH`.

The problem with using just the pin number to identify which pin you want to control is that you can't tell what kind of device is connected to pin 6 simply by looking at the statement. It could be an LED, but it could also be a motor or a servo or even a pneumatic valve that causes a Frankenstein creature to pop up.

To remedy this situation, PBASIC lets you assign a name to an I/O pin by placing a statement similar to this one near the beginning of your program:

```
Led1 PIN 0
```

Here, the name `Led1` is assigned to pin 0. Now, you can use the name `Led1` in a `HIGH` or `LOW` statement, like this:

```
HIGH Led1
```

This statement sets the I/O pin referenced by the name `Led1` to `HIGH`.

Listing 2-4 shows a version of the LED Flasher program that uses pin names instead of the pin numbers. The real advantage of creating PIN names is that it makes it much easier to change the pin configuration of your project later on. For example, suppose you decide that instead of connecting the six LEDs to pins 0, 2, 4, 6, 8, and 10, you want to connect them to pins 0, 1, 2, 3, 4, and 5. By using pin names, you must change the pin assignments just once when you modify the program, in the PIN statements near the beginning of the program.

Listing 2-4:   The LED Flasher Program with Pin Names

```
' LED Flasher Program
' Doug Lowe
' July 10, 2011
'
' This program flashes LEDs connected to pins 0, 2, 4, 6, 8, and 10
' at one-half second intervals.
'
' This version of the program uses pin names instead of numbers.

' {$PBASIC 2.5}
' {$STAMP BS2}

Led1   PIN 0
Led2   PIN 2
Led3   PIN 4
Led4   PIN 6
Led5   PIN 8
Led6   PIN 10

Main:
   HIGH Led1
   HIGH Led2
   HIGH Led3
   HIGH Led4
   HIGH Led5
   HIGH Led6
   PAUSE 500
   LOW Led1
   LOW Led2
   LOW Led3
   LOW Led4
   LOW Led5
   LOW Led6
   PAUSE 500
   GOTO Main
```

Using Variables

As I mention in the preceding chapter, the BASIC Stamp 2 microprocessor has a whopping 32 bytes of RAM memory that's available for use by your programs. That might not sound like a lot, but it's plenty of RAM for most of the programs you're likely to create for your BASIC Stamp projects.

To use RAM memory in PBASIC, you create variables. A *variable* is simply a name that refers to a location in RAM. Once you've created a variable, you can use the variable name in your program code to set or retrieve the value of the variable, and you can use the variable name in *expressions* that perform simple mathematical calculations on the value of variables.

To create a variable, you list the name you want to use for the variable, followed by the keyword VAR, followed by one of four keywords that indicates the *type* of the variable you're creating. For example, the following creates a variable named Count, using the variable type BYTE:

```
Count VAR BYTE
```

There are four choices for the variable type:

+ **BYTE** — Uses one of the 32 available bytes of RAM and can have a value ranging from 0 to 255. This type of variable is useful for simple counters that don't need to exceed the value 255. For example, if you're creating a timer that will count down 60 seconds, a BYTE variable will do the trick.

+ **WORD** — Uses two of the 32 available bytes and can have a value ranging from 0 to 65,535. You need to use a WORD variable whenever the value to be stored in the variable is greater than 255. For example, a WORD variable is ideal for holding the length of a delay used by the PAUSE statement.

+ **NIB** — If you have a very small counter whose value will never exceed 15, you can use a NIB variable, which requires only one-half of one byte of RAM. Thus, you can in theory create as many as 64 different NIB variables in a BASIC Stamp 2's 32 bytes of RAM.

+ **BIT** — Uses just one binary bit. Thus, the BASIC Stamp can squeeze up to eight BIT variables in each of its 32 bytes of available RAM. BIT variables are mostly used to keep track of whether some event has occurred. For example, you could set up a BIT variable to remember whether a user has pressed an input button. The value 0 would indicate that the button hasn't yet been pressed; the value 1 would mean that the button has been pressed.

Once you've created a variable, you can use it in an *assignment statement* to assign it a value. An assignment statement consists of a variable name followed by an equals sign, followed by the value to be assigned. For example, this assignment statement assigns the value 500 to a variable named Delay:

```
Delay = 500
```

The value on the right side of the equals sign can be an arithmetic calculation. For example:

```
Delay = 500 + 10
```

In this example, the value 510 is assigned to the variable named `Delay`.

There's not a lot of point in doing arithmetic using only numerals. After all, you could just do the calculation yourself. Thus, the previous example could be written like this:

```
Delay = 510
```

The real power of variable assignments happens when you use variables on the right side of the equals sign. For example, the following statement increases the value of the `Delay` variable by 10:

```
Delay = Delay + 10
```

In this example, the previous value of `Delay` is increased by 10. For example, if the `Delay` variable's value was 150 before this statement executed, it will be 160 after.

Listing 2-5 shows a program that uses a variable to change the speed at which the LEDs flash each time the `GOTO` statement causes the program to loop. As you can see, a variable named `Delay` is used to provide the number of milliseconds that the `PAUSE` statement should pause. Each time through the loop, the value of the `Delay` variable is increased by 10. Thus, the LEDs flash very fast when the program first starts, but the flashing gets progressively slower as the program loops.

Listing 2-5: The LED Flasher Program with a Variable

```
' LED Flasher Program
' Doug Lowe
' July 10, 2011
'
' This program flashes LEDs connected to pins 0, 2, 4, 6, 8, and 10
' at one-half second intervals.
'
' This version of the program uses a variable delay.

' {$PBASIC 2.5}
' {$STAMP BS2}
```

(continued)

Listing 2-5 *(continued)*

```
Led1  PIN  0
Led2  PIN  2
Led3  PIN  4
Led4  PIN  6
Led5  PIN  8
Led6  PIN  10

Delay VAR Word
Delay = 10

Main:
  HIGH Led1
  HIGH Led2
  HIGH Led3
  HIGH Led4
  HIGH Led5
  HIGH Led6
  PAUSE Delay
  LOW Led1
  LOW Led2
  LOW Led3
  LOW Led4
  LOW Led5
  LOW Led6
  PAUSE Delay
  Delay = Delay + 10
  GOTO Main
```

One final note about using variables: PBASIC lets you use a variable in a HIGH or LOW statement to indicate which pin should be controlled. For example:

```
Led VAR BYTE
Led = 0
HIGH Led
```

This sequence of statements creates a variable named Led, assigns the value 0 to it, and then uses it in a HIGH statement. The result is that I/O pin 0 is set to HIGH.

Doing Math

PBASIC lets you perform addition, subtraction, multiplication, and division using the symbols (called *operators*) +, -, *, and /. Here's an example of an assignment that uses all four of these symbols:

```
X VAR BYTE
X = 10 * 3 / 2 + 5
```

In this example, the value 20 will be assigned to the variable X. (10 × 3 = 30, 30 / 2 = 15, and 15 + 5 = 20.)

Here are a few things you need to know about how PBASIC does math:

✦ Unlike most programming languages, PBASIC performs mathematical operations strictly on a left-to-right basis. For example, consider the following assignment:

```
X = 10 + 3 * 2
```

Most programming languages would first multiply the 3 by the 2, giving a result of 6, and then add the 6 to the 10, giving the final result 16. That's because multiplication is ordinarily done before addition. But PBASIC calculates the expression left to right, so it first adds 10 and 3, giving the result 13, and then multiplies the 13 by 2, giving the result 26.

✦ You can use parenthesis to force PBASIC to calculate a certain part of the formula first. For example:

```
X = 10 + (3 * 2)
```

Here, PBASIC first does the calculation inside the parenthesis, giving a result of 6. It then adds the 6 to the 10 to give the final result, 16.

✦ When PBASIC does division, it discards the remainder and returns the result as a whole number. For example:

```
X = 8 / 3
```

This statement assigns the value 2 to X. That's because 8 divided by 3 is 2 with a remainder of 2. PBASIC discards the remainder and returns the result 2.

Using If Statements

An IF statement lets you add conditional testing to your programs. In other words, it lets you execute certain statements only if a particular condition is met. This type of conditional processing is an important part of any but the most trivial of programs.

Every IF statement must include a *conditional expression* that lays out a logical test to determine whether the condition is true or false. For example:

```
X = 5
```

This condition is true if the value of the variable X is 5. If X has any other value, the condition is false.

You can use less-than or greater-than signs in a conditional expression, like these:

```
Led < 10
Speed > 1000
```

Here, the first expression is true if the value of Led is less than 10. The second expression is true if the value of Speed is greater than 1,000.

In its simplest form, the IF statement causes the program to jump to a label if a condition is true. For example:

```
IF Led < 11 THEN Main
```

Here, the program jumps to the Main label if the value of the Led variable is less than 11.

Listing 2-6 shows a program that flashes the LEDs in sequence. This program uses a variable named Led to represent the output pin to be used. On each pass through the loop, it adds 2 to the Led variable to determine the next LED to be fired. Then, an IF statement is used to loop back to the Main label if the Led variable is less than 11. This sets up the basic loop that first flashes the LED on pin 0, then the LED on pin 2, and then pins 4, 6, and 8, and finally 10.

After the program flashes the LED in pin 10, the program adds 2 to the Led variable, setting this variable to 12. Then, the conditional expression in the IF statement (X < 11) tests false instead of true, so the IF statement doesn't skip to the Main label at this point. Instead, the statement after the IF statement is executed, which resets the Led variable to zero. Then, a GOTO statement sends the program back to the Main label, where the first LED is flashed again.

Listing 2-6:   The LED Flasher Program with an IF statement

```
' LED Flasher Program
' Doug Lowe
' July 10, 2011
'
' This program flashes LEDs connected to pins 0, 2, 4, 6, 8, and 10
' in sequence.
'
' This version of the program uses a simple IF statement.

' {$PBASIC 2.5}
' {$STAMP BS2}

Speed VAR BYTE
Led VAR BYTE
```

```
Speed = 50
Led = 0

Main:
  HIGH Led
  PAUSE Speed
  LOW Led
  PAUSE Speed
  Led = Led + 2
  IF Led < 11 THEN Main
  Led = 0
  GOTO Main
```

A second and more useful form of the IF statement lets you list one or more statements that should be executed if the condition is true. For example:

```
IF Led < 10 THEN
    Led = Led + 2
ENDIF
```

In this example, 2 is added to the Led variable if the value of the Led variable is less than 10.

You can place as many statements as you want between the IF and ENDIF statements. For example:

```
IF Led < 10 THEN
    Speed = Speed + 10
    Led = Led + 2
ENDIF
```

**Book VII
Chapter 2**

**Programming
in PBASIC**

Here, the Speed variable is also increased if the condition expression is true.

The main difference between the IF statement with ENDIF and an IF statement without ENDIF is that without the ENDIF, the statement that's executed if the IF condition is true must be on the same line as the IF and THEN keywords. If the THEN keyword is the last word on a line, PBASIC assumes that you will use an ENDIF to mark the end of the list of statements to be executed if the IF condition is true. If you forget to include the ENDIF statement, the program won't work properly.

One last trick that the IF statement lets you do is list statements that you want to execute if the condition *isn't* true. You do that by using an ELSE statement along with the IF statement. For example:

```
IF Led < 10 THEN
    Led = Led + 2
ELSE
    Led = 0
ENDIF
```

Here, Led is increased by 2 if its current value is less than 10. But if the current value of Led isn't less than 10, the Led variable is reset to 0.

Listing 2-7 shows a version of the LED Flasher program that uses an IF-THEN-ELSE statement to flash the LEDs in sequence.

Listing 2-7:   LED Flasher with an IF-THEN-ELSE statement

```
' LED Flasher Program
' Doug Lowe
' July 10, 2011
'
' This program flashes LEDs connected to pins 0, 2, 4, 6, 8, and 10
' in sequence.
'
' This version of the program uses an IF-THEN-ELSE statement.

' {$PBASIC 2.5}
' {$STAMP BS2}

Speed VAR BYTE
Led VAR BYTE

Speed = 50
Led = 0

Main:
  HIGH Led
  PAUSE Speed
  LOW Led
  PAUSE Speed
  IF Led < 10 THEN
    Led = Led + 2
  ELSE
    Led = 0
  ENDIF
  GOTO Main
```

Using DO Loops

The DO loop is a special PBASIC statement that performs essentially the same function as a label and a GOTO statement. For example, consider the following:

```
Main:
  HIGH 0
  PAUSE 500
  LOW 0
  PAUSE 500
  GOTO Main
```

The same function can be accomplished without the `Main` label or the `GOTO` statement by placing the lines that turn the LED on and off between `DO` and `LOOP` statements, like this:

```
DO
   HIGH 0
   PAUSE 500
   LOW 0
   PAUSE 500
LOOP
```

The lines between the `DO` and `LOOP` statements will be executed over and over again indefinitely.

Listing 2-8 shows the LED Flashing program implemented with a simple `DO` loop instead of a label and a `GOTO` statement.

Listing 2-8:   LED Flasher with a DO loop

```
' LED Flasher Program
' Doug Lowe
' July 10, 2011
'
' This program flashes LEDs connected to pins 0, 2, 4, 6, 8, and 10
' in sequence.
'
' This version of the program uses a DO loop.

' {$PBASIC 2.5}
' {$STAMP BS2}

Speed VAR BYTE
Led VAR BYTE

Speed = 50
Led = 0

DO
   HIGH Led
   PAUSE Speed
   LOW Led
   PAUSE Speed
   IF Led < 10 THEN
     Led = Led + 2
   ELSE
     Led = 0
   ENDIF
LOOP
```

You can add a conditional test to the LOOP statement to make the loop conditional. For example:

```
Led = 0
DO
   HIGH Led
   PAUSE 500
   LOW Led
   PAUSE 500
   Led = Led + 2
LOOP UNTIL Led > 10
```

This code will flash the LEDs on pins 0, 2, 4, 6, 8, and 10. After the LED on pin 10 is flashed, the next-to-last line sets the Led variable to 12. Then, the LOOP UNTIL statement sees that Led is greater than 10, so it stops looping.

Instead of the word UNTIL, you can use the word WHILE to mark the condition in a DO loop. There's a substantial difference between UNTIL and WHILE, and the difference is just as the words suggest. When you use the word UNTIL, the loop will execute until the condition tests true. When you use the word WHILE, the loop will execute until the condition tests false.

Note that you can also include the condition test on the DO statement or on the LOOP statement. If you place the condition test on the DO statement, the condition is tested *before* each execution of the loop. If you place it on the LOOP statement, the condition is tested *after* the completion of each loop. It's common to place WHILE tests on the DO statement and UNTIL tests on the LOOP statement. For example:

```
Led = 0
DO WHILE Led < 11
   HIGH Led
   PAUSE 500
   LOW Led
   PAUSE 500
   Led = Led + 2
LOOP
```

Here, the value of Led is tested prior to each execution of the loop. The loop is executed as long as Led is less than 11.

DO loops can be *nested*, which simply means that one DO loop can contain another DO loop. Note that when DO loops are nested, the inner loop must have a conditional test. Otherwise, it will loop forever, and the outer loop will never have a chance to complete. Listing 2-9 shows a program that uses two nested DO loops to flash the LEDs in sequence. The innermost DO loop flashes the six LEDs once. It uses an UNTIL condition to stop the loop after the last LED has flashed. The outermost DO loop continues endlessly, causing the flashing sequence to continue indefinitely.

Listing 2-9:   The LED Flasher Program with Nested DO Loops

```
' LED Flasher Program
' Doug Lowe
' July 10, 2011
'
' This program flashes LEDs connected to pins 0, 2, 4, 6, 8, and 10
' in sequence.
'
' This version of the program uses nested DO loops.

' {$PBASIC 2.5}
' {$STAMP BS2}

Speed VAR BYTE
Led VAR BYTE

Speed = 50

DO
  Led = 0
  DO
    HIGH Led
    PAUSE Speed
    LOW Led
    PAUSE Speed
    Led = Led + 2
  LOOP UNTIL Led > 10
LOOP
```

**Book VII
Chapter 2**

**Programming
in PBASIC**

Using FOR Loops

A FOR loop is a special type of looping statement that automatically keeps a counter variable. FOR loops are ideal when you want to execute a loop a certain number of times or when you want to perform an action on multiple I/O pins. Thus, a FOR loop is the ideal way to implement the LED Flasher program.

The basic structure of a FOR loop looks like this:

```
FOR counter = start-value TO end-value
  Statements...
NEXT
```

Here's an example that flashes the LED on pin 0 ten times:

```
X VAR BYTE
FOR X = 1 TO 10
  HIGH 0
  PAUSE 500
  LOW 0
  PAUSE 500
NEXT
```

In this example, the loop is executed ten times. The value of the variable X is increased by 1 each time through the loop.

In the preceding example, the program didn't actually use the counter variable. That's common in FOR loops; sometimes the only purpose for the counter variable is to control how many times the loop is executed. But you can use the counter variable within the loop. For example, here's a loop that makes every I/O pin on the Stamp HIGH for one-tenth of a second:

```
IO_Pin VAR BYTE
FOR IO_Pin = 0 TO 15
  HIGH IO_Pin
  PAUSE 100
  LOW IO_Pin
NEXT
```

Normally, the counter variable is increased by one on each pass through the loop. You can use the STEP keyword to specify a different step value if you want. When you use the STEP keyword, the basic structure of the FOR statement looks like this:

```
FOR counter = start-value TO end-value STEP step-value
  Statements...
NEXT
```

For example, you could flash LEDs on just the even-numbered pins like this:

```
Led VAR Byte
FOR Led = 0 TO 10 STEP 2
  HIGH Led
  PAUSE 100
  LOW Led
NEXT
```

Another interesting feature of FOR loops is that they can count backward. All you have to do is specify a start value that's larger than the end value, like this:

```
Led VAR Byte
FOR Led = 10 TO 0 STEP 2
  HIGH Led
  PAUSE 100
  LOW Led
NEXT
```

Listing 2-10 shows a version of the LED Flasher program that uses a pair of FOR loops to flash the LEDs first in one direction, and then in the opposite direction. This creates an effect similar to the spooky electronic eyes on the evil Cylons in the old TV series *Battlestar Galactica*. The first FOR loop flashes the LEDs on pins 0, 2, 4, 6, and 8. Then, the second FOR loop flashes the LEDs on pins 10, 8, 6, 4, and 2. Both FOR loops are contained within a DO loop that keeps the LEDs bouncing back and forth indefinitely.

Listing 2-10:   The LED Flasher Program with FOR Loops

```
' LED Flasher Program
' Doug Lowe
' July 10, 2011
'
' This program flashes LEDs connected to pins 0, 2, 4, 6, 8, and 10
' back and forth, like Cylon eyes.
'
' This version of the program uses FOR loops.

' {$STAMP BS2}
' {$PBASIC 2.5}

Led VAR Byte

Main:
  FOR Led = 0 TO 8 STEP 2
    HIGH Led
    PAUSE 100
    LOW Led
  NEXT
  FOR Led = 10 TO 2 STEP 2
    HIGH Led
    PAUSE 100
    LOW Led
  NEXT
  GOTO Main
```

Like DO loops, FOR loops can be nested. When FOR loops are nested, the innermost loop(s) complete their entire cycle each time through the outer loop. In other words, if a FOR loop that repeats ten times is placed within an outer loop that repeats ten times, the statements within the innermost loop will execute a total of 100 times — ten times for each of the ten repetitions of the outer loop.

**Book VII
Chapter 2**

**Programming
in PBASIC**

Listing 2-11 shows a variation on the Cylon eyes program shown in Listing 2-10. This one uses an outer FOR loop that varies the delay time for the PAUSE statements. The result is that the LEDs sweep very fast at first, but slow by 10 ms on each repetition of the outer loop until the delay reaches one second per LED.

Listing 2-11:   The LED Flasher Program with Nested FOR Loops

```
' LED Flasher Program
' Doug Lowe
' July 10, 2011
'
' This program flashes LEDs connected to pins 0, 2, 4, 6, 8, and 10
' back and forth, like Cylon eyes.
'
' This version of the program uses nested FOR-NEXT loops to slow the
' sweeping motion of the LEDs.

' {$STAMP BS2}
' {$PBASIC 2.5}

Led VAR Byte
Speed VAR Word

FOR Speed = 10 TO 1000 STEP 10
  FOR Led = 0 TO 8 STEP 2
    HIGH Led
    PAUSE Speed
    LOW Led
  NEXT
  FOR Led = 10 TO 2 STEP 2
    HIGH Led
    PAUSE Speed
    LOW Led
  NEXT
NEXT
```

Chapter 3: More PBASIC Programming Tricks

In This Chapter

✔ Reading the status of pushbuttons

✔ Using a potentiometer as input

✔ Generating random numbers

✔ Creating subroutines with the GOSUB command

In this chapter, you learn some additional PBASIC programming techniques that will become invaluable in your BASIC Stamp projects. Specifically, you learn how to handle input data in the form of pushbuttons, how to generate random numbers that will make your programs more interesting by adding a degree of randomness, how to read the value of a potentiometer, and how to organize your program into subroutines using the GOSUB command.

Using a Pushbutton with a BASIC Stamp

In Chapter 2 of this minibook, you learn how to connect an LED to a BASIC Stamp I/O pin and turn the LED on or off by using the HIGH and LOW commands in a PBASIC program. Those commands are designed to use BASIC Stamp I/O pins as output pins by setting the status of an I/O pin to HIGH or LOW so that external circuitry (such as an LED) can react to the pin's status.

But what if you want to use an I/O pin as an input instead of an output? In other words, what if you want the BASIC Stamp to react to the status of an external circuit instead of the other way around? The easiest way to do that is to connect a pushbutton to an I/O pin. Then, you can add commands to your PBASIC program to detect whether the pushbutton is pressed.

There are two ways to connect a pushbutton to a BASIC Stamp I/O pin:

♦ **Active-high:** This connection places +5 V on the I/O pin when the pushbutton is pressed. When the button is released, the I/O pin sees 0 V.

♦ **Active-low:** This connection sees +5 V when the pushbutton is not pressed. When you press the pushbutton, the +5 V is removed, and the I/O pin sees no voltage.

Figure 3-1 shows examples of both active-high and active-low pushbuttons. In the active-high circuit, the I/O pin is connected to ground through R1 and R2 when the pushbutton isn't pressed. Thus, the voltage at the I/O pin is 0. When the pushbutton is pressed, the I/O pin is connected to Vdd (+5 V) through R1, causing the I/O pin to see +5 V. As a result, the I/O pin is LOW when the button isn't pressed and HIGH when the button is pressed.

In the active-low circuit, the I/O pin is connected to Vdd (+5 V) through R1 and R2, causing the I/O pin to go HIGH. But when the button is pressed, the current from Vdd is shorted to ground through R2, causing the voltage at the I/O pin to drop to zero. Thus, the I/O pin is HIGH when the button isn't pressed and LOW when the button is pressed.

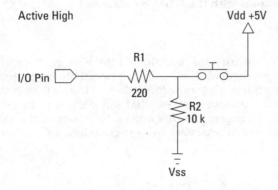

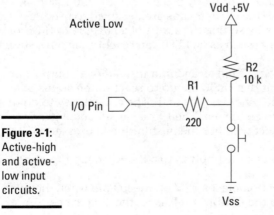

Figure 3-1:
Active-high
and active-
low input
circuits.

Note that in both circuits, R1 is connected directly to the I/O pin to prevent excessive current flow when the switch is pressed. Without this resistor, the pin would be connected directly to Vdd (+5 V) or Vss (ground) when the button is pressed, which could damage the BASIC Stamp.

In an active-high circuit, R2 is called a *pull-down* resistor because it pulls the current from the I/O pin down to zero when the pushbutton isn't depressed. In an active-low circuit, R2 is called a *pull-up resistor* because it pulls the voltage at the I/O pin up to Vdd (+5 V) when the pushbutton isn't depressed.

Checking the Status of a Switch in PBASIC

Once you've connected a switch to a BASIC Stamp I/O pin, you need to know how to determine whether the switch is open or closed from a PBASIC program. The easiest way to do that is to first assign a name to the pin you want to test using the PIN directive. For example, if an active-high input button is connected to pin 14, you can assign it a name like this:

```
Button1    PIN 14
```

Here, the name Button1 is assigned to pin 14.

Then, to determine whether the pushbutton is pressed, you can use an IF statement like this:

```
IF Button1 = 1 THEN
   HIGH Led1
ENDIF
```

Here, the output pin designated as Led1 is made HIGH when the button is pressed.

If you want Led1 to be HIGH *only* when Button1 is pressed, use this code:

```
IF Button1 = 1 THEN
   HIGH Led1
ELSE
   LOW Led1
ENDIF
```

Here, Led1 is made HIGH if the button is pressed and LOW if the button isn't pressed.

You can put the whole thing in a loop to repeatedly test the status of the button and turn the LED on and off accordingly:

```
DO
  IF Button1 = 1 THEN
    HIGH Led1
  ELSE
    LOW Led1
  ENDIF
LOOP
```

Listing 3-1 shows an interesting program that works with a BASIC Stamp that has a pushbutton switch connected to pin 14 and LEDs connected to pins 0 and 2. The program flashes the LED connected to pin 2 on and off at half-second intervals until the pushbutton switch is depressed. Then, it flashes the LED on pin 0.

Listing 3-1:   The Pushbutton Program

```
' Pushbutton Program
' Doug Lowe
' July 13, 2011


' {$STAMP BS2}
' {$PBASIC 2.5}

Led1 PIN 0
Led2 PIN 2
BUTTON1 PIN 14

DO

  IF BUTTON1 = 1 THEN
    LOW Led2
    HIGH Led1
    PAUSE 100
    LOW Led1
    PAUSE 100
  ELSE
    LOW Led1
    HIGH Led2
    PAUSE 100
    LOW Led2
    PAUSE 100
  ENDIF

  PAUSE 100

LOOP
```

Project 3-1 shows how to build a simple circuit you can use to test the program in Listing 3-1, and Figure 3-2 shows the completed circuit.

Figure 3-2:
A circuit for testing an active-high pushbutton switch (Project 3-1).

Randomizing Your Programs

Many computer-controlled applications require a degree of randomness to their operation. A classic example is the *Indiana Jones* ride at Disneyland and Disney World. Every time you go on this ride, the adventure is slightly different. At the start, there are three doors that your vehicle can drive through; the exact door chosen for your ride is determined randomly. Many other details of the ride are randomly varied in an effort to make the adventure slightly different every time you ride it.

You can add a bit of randomness to your own BASIC Stamp programs by using the RANDOM command. This command scrambles the bits of a variable. Although you can use any type of variable with a RANDOM command, you'll almost always want to use a BYTE or WORD variable. When you use a Byte variable, the RANDOM command generates a random number between 0 and 255. When you use a Word variable, the random value will be between 0 and 65,535.

Project 3-1: A Pushbutton-Controlled LED Flasher

In this project, you connect two LEDs to a BASIC Stamp HomeWork board or Board of Education to test the programs presented in this chapter.

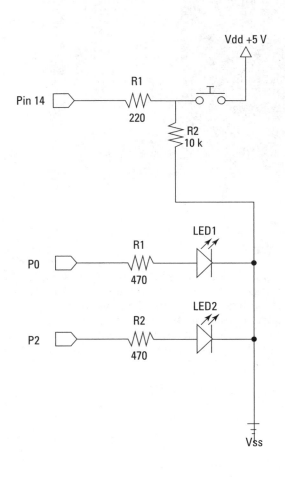

Parts List

1 Computer with BASIC Stamp Editor software installed
1 BASIC Stamp HomeWork board or Board of Education
1 9 V Battery
1 USB cable
1 USB to serial adapter
2 LEDs
2 470 Ω resistors (yellow-violet-brown)
1 220 Ω resistors (red-red-brown)
1 10 kΩ resistor (brown-black-orange)
5 Jumper wires

Switch Test Program

```
'{$STAMP BS2}
'{$PBASIC 2.5}

Led1 PIN 0
Led2 PIN 2
Button1 PIN 14


DO
  IF Button1 = 1 THEN
    HIGH Led1
    LOW Led2
  ELSE
    LOW Led1
    HIGH Led2
  ENDIF
LOOP
```

Steps

1. **Connect the two LEDs to the output pins as follows:**

Pin (Cathode)	Breadboard Hole
P0	A17
P2	A15

2. **Connect the resistors across the gap in the center of the breadboard as follows:**

Resistor	From	To
R1	E17	F17
R2	E15	F15

3. **Use the jumper wires to connect the resistors to ground.**

   I17 to I15

   J15 to Vss

4. **Open the BASIC Stamp Editor.**

5. **Connect your BASIC Stamp to the computer and identify it in the Stamp Editor.**

   For more information about how to do this, refer to Chapter 1 of this minibook.

6. **Type the Switch Test program (see nearby listing) into the Editor program window, and then save the file.**

7. **Choose Run⇨Run.**

   If you prefer, you can click the Run button in the toolbar or press F9.

   When you choose the Run command, the program is downloaded to the BASIC Stamp. Once the program is downloaded, it automatically starts to run on the Stamp.

8. **Press the button.**

When you run the program, the LED connected to pin 2 lights up. When you push the button, the LED on pin 2 goes dark, and the LED on pin 0 lights.

9. **Now try the programs in Listing 3-1.**

   This program is similar to the Switch Test program but adds the complication that the LEDs are flashing on and off at half-second intervals.

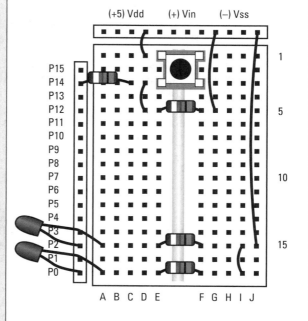

Here's an example of a simple way to use the RANDOM command to add a random pause:

```
Result VAR Word
RANDOM Result
PAUSE Result
```

This sequence of code creates a Word variable named Result, randomizes the variable, and then pauses for the number of milliseconds indicated by the Result variable.

In most cases, the value returned by the RANDOM command isn't really the random number you're looking for. Usually, you want to determine a random number that falls within a range of numbers. For example, if you're writing a dice program and want to simulate the roll of a single die, you'll need to generate a random number between 1 and 6. You can easily reduce the value of a full Word variable to a number between 1 and 6 by using a simple mathematical calculation that involves a special type of division operation called *modulus division.*

You may recall from Chapter 2 of this minibook that when PBASIC does division, it keeps the integer portion of the quotient and discards the remainder. For example, 10 / 3 = 3; the remainder (1) is simply discarded.

Modulus division, which is represented by two slashes (//) instead of one, throws away the quotient and keeps only the remainder. Thus, 10 // 3 = 1 because the remainder of 10 / 3 = 1.

You can put modulus division to good use when working with random numbers. For example:

```
Result VAR Word
Die    VAR Byte
RANDOM Result
Die = Result // 6
```

Well, actually, the preceding calculation isn't quite right. I include it to point out a common pitfall that happens if you forget that the remainder might happen to be 0. The above calculation actually returns a random number that falls between 0 and 6.

To find a random number that falls between 1 and 6, you should calculate the modulus division by 5, not by 6, and then add 1 to the result. For example:

```
Result VAR Word
Die    VAR Byte
RANDOM Result
Die = Result // 5 + 1
```

The preceding calculation returns a random number between 1 and 6.

Listing 3-2 shows a sample program that lights LED1 (pin 0) until the push-button on pin 14 is pressed. Then, it lights LED2 (pin 2) for a random period of time between 1 and 10 seconds. It uses the following equation to calculate the number of seconds to pause:

```
Seconds = Result // 9 + 1
```

It then multiplies the seconds by 1,000 to convert to milliseconds as required by the PAUSE command.

Listing 3-2:   The Random Program

```
' Random Program
' Doug Lowe
' July 10, 2011
'
' This program turns on the LED at pin 2 for a random number
' of seconds between 1 and 10 when the pushbutton on pin 14
' is pressed.

' {$STAMP BS2}
' {$PBASIC 2.5}

Result      VAR Word
Seconds     VAR Byte
Button1     PIN 14
Led1        PIN 0
Led2        PIN 2

DO
  HIGH Led1
  IF Button1 = 1 THEN
    RANDOM Result
    Seconds = Result // 9 + 1
    LOW Led1
    HIGH Led2
    PAUSE Seconds * 1000
    HIGH Led1
    LOW Led2
  ENDIF
LOOP
```

Each time you press the pushbutton in this program, LED2 lights up for a different length of time, between 1 and 10 seconds.

It turns out that the RANDOM command isn't really all that random. The starting value of the variable used by the RANDOM command is called the *seed value*. The RANDOM command applies a complex mathematical calculation to the seed value to determine a result that appears to be random. However, the result isn't actually random: Given a particular seed value, the RANDOM command will always return the same result.

For example, if you use a Word variable whose initial value is 0, the RANDOM command will always change the variable's value to 64992. If you apply the RANDOM command to the same variable again, the result will always be 9,072. The sequence of numbers generated from a given initial value is distributed randomly across the entire range of possible values (for example, 0 to 65,535), but the sequence is the same every time.

Thus, the program in Listing 3-2 always generates the same sequence of random delays. In particular, the sequence for the first ten button presses will always be as follows:

+ **First press:** 4 seconds
+ **Second press:** 1 second
+ **Third press:** 4 seconds
+ **Fourth press:** 1 second
+ **Fifth press:** 3 seconds
+ **Sixth press:** 5 seconds
+ **Seventh press:** 4 seconds
+ **Eighth press:** 1 second
+ **Ninth press:** 1 second
+ **Tenth press:** 4 seconds

This sequence appears fairly random, but every time you reset the program and start over, the sequence will be identical.

The easiest way around this is to make the initial value of the variable fed to the RANDOM command dependent on some external event, such as the press of a button. You can easily do that by creating a loop that counts while it waits for the user to press the button. Because your program has no way to determine exactly when the user will press the button, the number used to start the random number generator will be truly random.

Listing 3-3 shows an improved version of the random program that uses this technique to create a truly random delay. It turns out that only one additional line of code is needed to properly randomize the delay:

```
Result = Result + 1
```

By adding 1 to the Result variable each pass through the loop, the seed value for the RANDOM command will be unpredictable, so a true random value will be generated.

Listing 3-3:   An Improved Version of the Random Program

```
' Improved Random Program
' Doug Lowe
' July 10, 2011
'
' This program turns on the LED at pin 2 for a random number
' of seconds between 1 and 10 when the pushbutton on pin 14
' is pressed.
'
' This version of the program uses a counter to create a truly random number.

' {$STAMP BS2}
' {$PBASIC 2.5}

Result      VAR Word
Seconds     VAR Byte
Button1     PIN 14
Led1        PIN 0
Led2        PIN 2

DO
  HIGH Led1
  Result = Result + 1
  IF Button1 = 1 THEN
    RANDOM Result
    Seconds = Result // 9 + 1
    LOW Led1
    HIGH Led2
    PAUSE Seconds * 1000
    HIGH Led1
    LOW Led2
  ENDIF
LOOP
```

Reading a Value from a Potentiometer

As you know, a *potentiometer* (often called a *pot*) is simply a variable resistor with a knob you can turn to vary the resistance. Pots of various types are often used as input devices for BASIC Stamp projects. For example, you might use a simple pot to control the speed of a pair of flashing LEDs: As you turn the pot's knob, the rate at which the LEDs flash changes.

Although the most common type of pot has a mechanical knob to vary its resistance, many pots use some other means to vary their resistance. For example, a joystick uses pots that are connected to a moveable stick. One of the pots measures the motion of the stick in the *x*-axis; the other measures the motion of the stick in the *y*-axis. You might also connect a pot to the hinge on a door. Then, the resistance of the pot would indicate not only whether the door is opened or closed, but also the door's angle if it's partially opened.

The BASIC Stamp doesn't have the ability to directly read the value of a pot connected to one of its I/O pins. However, by cleverly combining the pot with a small capacitor, you can measure how long the capacitor takes to discharge. With this knowledge, you can calculate the resistance of the POT by using the resistor-capacitor time calculations that are presented way back in Book II, Chapter 3. If you want to review how resistor-capacitor (RC) circuits work and how to do the time calculations, please refer to that chapter. For the purposes of this chapter, you don't need to perform the time calculations yourself. However, it wouldn't hurt to brush up on how RC circuits work.

Figure 3-3 shows a typical RC circuit connected to a pin on a BASIC Stamp. Here, a 10 kΩ pot is placed in parallel with a 0.1 μF capacitor. In addition, a 220 Ω resistor is placed in series with the pot. This is done to protect the BASIC Stamp from damage that might be caused by excess current if you turn the pot's knob so that the resistance of the pot drops to zero.

The capacitor in this circuit is small enough (0.1 μF) that the circuit will charge and discharge very fast — within about a millisecond or so, depending on where the pot knob is set. Thus, your program won't be delayed significantly while it waits for the capacitor to discharge so it can determine the resistance of the pot.

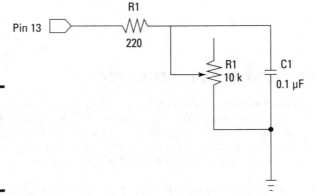

Figure 3-3:
Connecting
a pot to
a BASIC
Stamp I/O
pin.

So given the circuit shown in Figure 3-3, how would you go about measuring the resistance of the pot? The answer requires a clever bit of programming: First, you set pin 13 to HIGH, which charges the capacitor. Then, you set up a loop to monitor the input status of pin 13. Each time you check the status of pin 13, you add one to a counter. When the capacitor has discharged, pin 13 will go LOW. When pin 13 is LOW, the loop ends, and the counter indicates how long it took to discharge the capacitor. Knowing the size of the capacitor and the length of time it took to discharge the capacitor, you can calculate the resistance of the pot.

Fortunately, PBASIC includes a command called RCTIME that does all of this automatically. All you have to do is tell the RCTIME command what pin the RC circuit is on, whether you want to measure how long it takes the RC circuit to charge or discharge, and the name of a variable to store the resulting time calculation in.

Here's how to use the RCTIME command to determine how long it takes an RC circuit on pin 13 to discharge, storing the answer in a variable named Timer:

```
RCTIME 13, 1, Timer
```

This RCTIME command sets the variable named Timer to a value that indicates how long it took the RC circuit to discharge. Immediately before this command, you should set the I/O pin (in this case, pin 13) to HIGH to charge the capacitor. You'll also need to pause for a short time (usually, 1 ms is enough) to allow the circuit to charge.

Although you can use this technique to calculate the actual resistance of a pot, you don't usually have to know the exact resistance. Instead, it's usually sufficient to know that the counter increases when the resistance of the pot increases, and it decreases when the resistance of the pot decreases.

For the circuit shown in Figure 3-3, the RCTIME command calculates time values ranging from about 12 when the resistance of the pot is near 0 to about 54 when the resistance of the pot is at its maximum (10 kΩ).

Listing 3-4 shows a simple program that alternately flashes LEDs connected to pins 0 and 2. The rate at which the LEDs flash is set by a pot in an RC circuit on pin 13. As you can see, the program simply multiples the time value calculated by the RCTIME command by ten to determine how long the program should pause between flashes. As you turn the pot's knob to decrease the resistance of the pot, the LEDs flash at a faster rate.

Listing 3-4: An LED Flashing Program That Uses a Pot

```
' Potentiomter LED Flashing Program
' Doug Lowe
' July 10, 2011
'
' This program flashes LEDs connected to pins 0 and 2
' at a rate determined by an RC circuit on pin 13.

' {$STAMP BS2}
' {$PBASIC 2.5}

Time VAR Word
Led1 PIN 0
Led2 PIN 2
Pot  PIN 13
DO
'  HIGH Pot
   RCTIME Pot, 1, Time
   HIGH Led1
   LOW  Led2
   PAUSE Time * 10
   LOW Led1
   HIGH Led2
   PAUSE Time * 10
LOOP
```

Project 3-2 shows how to build a circuit that includes a 10 kΩ potentiometer and a capacitor so that you can test the code in Listing 3-4. Figure 3-4 shows the completed circuit.

Figure 3-4:
A circuit that uses a potentiometer to control flashing LEDs (Project 3-2).

Using Subroutines and the GOSUB Command

A *subroutine* is a section of a program that can be called upon from any location in the program. When the subroutine finishes, control of the program jumps back to the location from which the subroutine was called. Subroutines are useful because they let you separate long portions of your program from the program's main loop, which simplifies the main program loop to make it easier to understand.

Another benefit of subroutines is that they can make your program smaller. Suppose you're writing a program that needs to perform some complicated calculation several times. If you place the complicated calculation in a subroutine, you can call the subroutine from several places in the program. That way, you have to write the code that performs the complicated calculation only once. Without subroutines, you would have to duplicate the complicated code each time you need to perform the calculation.

To create and use subroutines, you need to use two PBASIC commands. The first is GOSUB, which calls the subroutine. You typically use the GOSUB command within your program's main loop whenever you want to call the subroutine. The second command is RETURN, which is always the last command in the subroutine. RETURN jumps back to the command that immediately follows the GOSUB command.

To create a subroutine, you start with a label and end with a RETURN command. Between them, you write whatever commands you want to execute when the subroutine is called.

Here's an example of a subroutine that generates a random number between 1 and 999 and saves it in a variable named Rnd:

```
GetRandom:
  RANDOM Rnd
  Rnd = Rnd // 999 + 1
  RETURN
```

To call this subroutine, you would simply use a GOSUB command like this:

```
GOSUB GetRandom
```

This GOSUB command transfers control to the GetRandom label. Then, when the GetRandom subroutine reaches its RETURN command, control jumps back to the command immediately following the GOSUB command.

Project 3-2:
Using a Potentiometer to Control Flashing LEDs

In this project, you create a BASIC Stamp circuit that uses a potentiometer to control the rate at which two LEDs alternately flash.

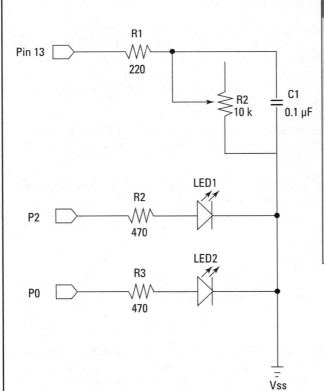

Parts List

1 Computer with BASIC Stamp Editor software installed

1 BASIC Stamp HomeWork board or Board of Education

1 9V Battery

1 USB cable

1 USB to serial adapter

2 LEDs

2 470Ω resistors (yellow-violet-brown)

1 220 Ω resistor (red-red-brown)

1 10 kΩ potentiometer

1 0.1µF capacitor

4 Jumper wires

Steps

1. **Connect the two LEDs to the output pins as follows:**

Pin (Cathode)	Breadboard Hole
P0	A17
P2	A15

2. **Connect the resistors across the gap in the center of the breadboard as follows:**

Resistor	From	To
R1	B4	PIN 13
R2	E17	F17
R3	E15	F15

3. **Insert the jumper wires.**

   F1 to C4

   H3 to Vss

   I17 to I15

   J15 to Vss

4. **Insert the capacitor.**

   One lead goes in G1, the other in Vss.

5. **Insert the potentiometer.**

   The potentiometer has three leads, which should go in D4, F3, and F5.

6. **Open the BASIC Stamp Editor.**

7. **Connect your BASIC Stamp to the computer and identify it in the Stamp Editor.**

   For more information about how to do this, refer to Chapter 1 of this minibook.

8. **Type the Potentiometer LED Flashing program into the Editor program window, and then save the file.**

   The Potentiometer LED Flashing program is shown in Listing 3-4.

9. **Choose Run⇨Run.**

   If you prefer, you can click the Run button in the toolbar or press F9.

   When you choose the Run command, the program is downloaded to the BASIC Stamp. Once the program is downloaded, it automatically starts to run on the Stamp.

10. **Turn the potentiometer knob.**

    The rate at which the LEDs flash will vary as you increase or decrease the resistance of the potentiometer.

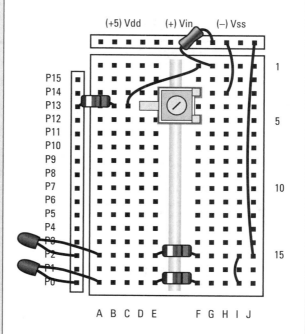

Listing 3-5 shows a complete program that uses a subroutine to get a random number between 1 and 1,000 and uses the random number to cause the LED on pin 0 to blink at random intervals. You can run this program on any Basic Stamp circuit that has an LED on pin 0, including the circuits you built for Projects 3-1 or 3-2 in this chapter.

Listing 3-5:   Using a Subroutine to Blink an LED

```
' LED Blinker Program
' Doug Lowe
' July 10, 2011
'
' This program blinks the LED on pin 0 randomly.

' {$STAMP BS2}
' {$PBASIC 2.5}

Rnd    VAR Word
Led1   PIN 0

DO
  GOSUB GetRandom
  HIGH Led1
  PAUSE Rnd
  LOW Led1
  PAUSE 100
LOOP

GetRandom:
  RANDOM Rnd
  Rnd = Rnd // 999 + 1
  RETURN
```

When you use a subroutine, it's vital that you prevent your program from accidentally "falling into" your subroutine and executing it when you didn't intend it to be executed. For example, suppose that the program in Listing 3-5 used a FOR-NEXT loop instead of a DO loop because you wanted to blink the LED only 100 times. Here's an example of how *not* to write that program:

```
FOR Counter = 1 TO 100
  GOSUB GetRandom
  HIGH Led1
  PAUSE Rnd
  LOW Led1
  PAUSE 100
NEXT
```

```
GetRandom:
  RANDOM Rnd
  Rnd = Rnd // 999 + 1
  RETURN
```

Do you see why? After the FOR-NEXT loop blinks the LED 100 times, the program will continue with the next command after the FOR-NEXT loop, which is the subroutine!

To prevent that from happening, you can use yet another PBASIC command, END, which simply tells the BASIC Stamp that you have reached the end of your program, so it should stop executing commands. You would place the END command after the NEXT command, like this:

```
FOR Counter = 1 TO 100
  GOSUB GetRandom
  HIGH Led1
  PAUSE Rnd
  LOW Led1
  PAUSE 100
NEXT
END

GetRandom:
  RANDOM Rnd
  Rnd = Rnd // 999 + 1
  RETURN
```

Then, the program will stop after the FOR-NEXT loop finishes.

Note that most BASIC Stamp programs don't require an END command. That's because most BASIC Stamp programs are written so that they loop continuously as long as power is applied to the Stamp. Even in the case of programs that loop indefinitely, however, you must be careful to make sure that your subroutines appear after the program's main loop. That way, your subroutines will be executed only when you explicitly call them with a GOSUB command.

Chapter 4: Adding Sound and Motion to Your BASIC Stamp Projects

In This Chapter

✓ Making sound with a piezo speaker

✓ Using the FREQOUT command to create frequencies

✓ Making music

✓ Creating motion with a servo

✓ Using the PULSOUT command to generate servo pulses

*I*n this chapter, you learn how to work with devices that add sound and motion to your BASIC Stamp projects. To create sound, you can add a piezo speaker to create audible output tones. This is useful in situations where your BASIC Stamp program might need to get someone's attention or when you want to create a sound effect. To create motion, you can add a very useful device called a *servo*, which lets you control mechanical motion with a BASIC Stamp program.

Using a Piezo Speaker with a BASIC Stamp

The BASIC Stamp Activity Kit comes with a small piezoelectric speaker, which you can connect directly to an I/O pin to create beautiful music. Well, the music probably won't be so beautiful, but you can coax the BASIC Stamp into emitting a variety of squeaks, burps, and squelches that resemble musical notes. And you can create interesting sound effects like police sirens or chirping crickets. Figure 4-1 shows this handy little speaker.

Figure 4-1:
The piezo-
electric
speaker that
comes with
the BASIC
Stamp
Activity Kit.

If you didn't purchase the BASIC Stamp Activity Kit, you can order the piezo speaker directly from the Parallax website (www.parallax.com) for $1.95.

Note that the piezo speaker is polarized, so when you connect it to an I/O pin, be sure to connect the + terminal to the I/O pin and the other terminal to Vss (ground), as shown in the schematic in Figure 4-2.

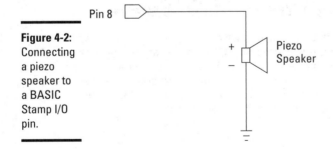

Figure 4-2:
Connecting
a piezo
speaker to
a BASIC
Stamp I/O
pin.

Using the FREQOUT command

Programming a piezo speaker is remarkably simple. PBASIC includes a command called FREQOUT that sends a frequency of your choice to an output pin. Thus, you can create an audible tone on a piezo speaker by using the FREQOUT command, using the following syntax:

```
FREQOUT pin, duration, frequency
```

Here's how that syntax works:

+ *pin* is simply the pin number that you want to send the frequency to.

+ *duration* is simply the length of time in milliseconds you want the frequency to play. And

+ *frequency* is the frequency in hertz that you want to generate.

For example, the following command generates a 2,000 Hz frequency for five seconds on pin 8:

```
FREQOUT 8, 5000, 2000
```

You can easily create a beeping sound by alternately sending short bursts of a frequency to the speaker followed by a brief pause. For example:

```
DO
  FREQOUT 8,250, 1500
  PAUSE 250
LOOP
```

This code repeatedly sends a 1,500 Hz signal for a quarter of a second, and then pauses for a quarter of a second. The result is a *beep-beep-beep* sound.

Testing the piezo speaker

Project 4-1 shows how to build a simple circuit that connects a piezo speaker to a BASIC Stamp so you can create audible output; two pushbuttons vary the sound output. Figure 4-3 shows the circuit.

The piezo speaker here is actually very quiet. This is normal; the piezo speaker draws just 1 mA, and so can't make a lot of noise. The speaker is loudest with frequencies between 4,500 and 5,500 Hz.

Figure 4-3:
A piezo speaker connected to a BASIC Stamp (Project 4-1).

Playing with sound effects

With creative use of the FREQOUT command, PAUSE commands, and FOR-NEXT loops, you can create some interesting and at times annoying sound effects. The idea is to use short durations in the FREQOUT command and use FOR-NEXT loops or some other means to vary the frequency. You can also use PAUSE commands between tones to create beeping or clicking effects.

The best way to learn what kinds of sound effects are possible with the FREQOUT command is to experiment. Listings 4-1, 4-2, and 4-3 give three sample programs you can run using the circuit you created in Project 4-1. Use these programs as starting points for your own experiments.

The program in Listing 4-1 plays two different beeping sounds when you press one of the pushbuttons. If you press Switch1 (on pin 14), a 5,000 Hz tone beeps twice per second. If you press Switch2 (on pin 10), a 5,000 Hz tone beeps five times per second.

Listing 4-1: Generating Two Different Types of Beeping Sounds

```
' Sound Program
' Doug Lowe
' July 15, 2011
'
' This program creates fast and slow beeping sounds.
' A piezo speaker must be connected to pin 0.
' The normally open pushbutton switches must be connect to pins 10 and 14.

' {$STAMP BS2}
' {$PBASIC 2.5}

Speaker     PIN 0
Switch1     PIN 10
Switch2     PIN 14
Frequency   VAR Word
Time        VAR Word

DO
  IF Switch1 = 1 THEN
    FREQOUT Speaker,250, 5000
    PAUSE 250
  ELSEIF Switch2 = 1 THEN
    FREQOUT Speaker,100, 5000
    PAUSE 100
  ENDIF
LOOP
```

Listing 4-2 shows how you can use FREQOUT within a FOR-NEXT loop to create a continuously rising or falling tone, much like a police siren. The program varies the frequency from 3,000 to 5,000 Hz. When you press either of the pushbuttons, the rate at which the pitch rises and falls changes.

The rate at which the pitch rises or falls is governed by a variable named Time. Each time through the FOR-NEXT loop, the program calls a subroutine named GetTime, which checks the status of the pushbutton switches and changes the Time variable if either of the switches is down. That's how the program changes the rate of the pitch change when the buttons are pressed.

**Book VII
Chapter 4**

**Adding Sound and
Motion to Your BASIC
Stamp Projects**

Project 4-1: Creating Sound with a Piezo Speaker

In this project, you connect a piezo speaker to a BASIC Stamp HomeWork board or Board of Education to run programs that generate sound. The circuit also includes two pushbuttons (on pins 10 and 14) that you can use to vary the sound produced by the Stamp program.

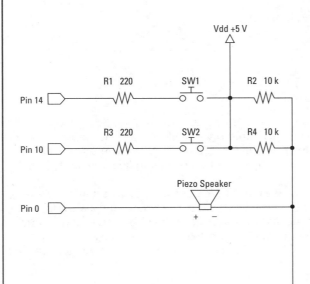

Parts List

1 Computer with BASIC Stamp Editor software installed
1 BASIC Stamp HomeWork board or Board of Education
1 9 V Battery
1 USB cable
1 USB to serial adapter
1 Piezo speaker
2 Normally open push buttons
2 220 Ω resistors
2 10 kΩ resistors
6 Jumper wires

Beeper Program

```
' {$STAMP BS2.5}
' {$PBASIC 2.5}

DO
  FREQOUT 0,250,1500
  PAUSE 250
LOOP
```

Steps

1. **Insert the piezo speaker.**
   The positive lead goes in E17, the negative lead in F17.

2. **Insert the two pushbuttons.**
   SW1 goes in E1, F1, E3, and F3.
   SW2 goes in E5, F5, E7, and F7.

3. **Insert the resistors.**
   R1 (220Ω)   Pin 14 to B3.
   R2 (10KΩ)   E4 to F4
   R3 (220Ω)   Pin 10B7.
   R4 (10KΩ   E8 to F8

4. **Insert the jumper wires.**
   Vdd to D1
   Vss to G4
   C1 to C5
   D3 to D4
   H4 to H8
   D7 to D8
   J8 to J17
   Pin 0 to A17

5. **Open the BASIC Stamp Editor.**

6. **Connect your BASIC Stamp to the computer and identify it in the Stamp Editor.**
   For more information about how to do this, refer to Chapter 1 of this minibook.

7. **Type the Beeper program into the Editor program window, and then save the file.**
   The Beeper program is shown in the accompanying listing.

8. **Choose Run⇨Run.**

If you prefer, you can click the Run button in the toolbar or press F9.

When you choose the Run command, the program is downloaded to the BASIC Stamp. Once the program is downloaded, it automatically starts to run on the Stamp.

9. **Now try the programs in Listings 4-1 and 4-2.**
   These programs vary the tone in different ways to demonstrate the versatility of the FREQOUT command.

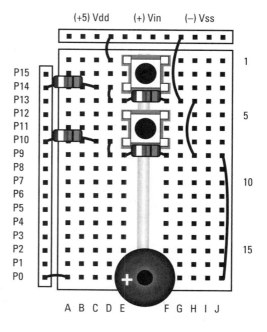

Listing 4-2:   Generating a Siren Effect

```
' Siren Effect Program
' Doug Lowe
' July 15, 2011
'
' This program generates a rising and falling pitch similar to a police siren.
' The rate at which the pitch rises and falls changes if you press either
' of the two pushbuttons.

' {$STAMP BS2}
' {$PBASIC 2.5}

Speaker   PIN 0
Switch1   PIN 10
Switch2   PIN 14
Frequency VAR Word
Time      VAR Word


DO

    FOR Frequency = 3000 TO 5000 STEP 15
       GOSUB SetTime
       FREQOUT 0, Time, Frequency
    NEXT
    FOR Frequency = 5000 TO 3000 STEP 15
       GOSUB SetTime
       FREQOUT 0, Time, Frequency
    NEXT

LOOP

SetTime:
  Time = 15
  IF Switch1 = 1 THEN
    Time = 5
  ENDIF
  IF Switch2 = 1 THEN
    Time = 2
  ENDIF
  RETURN
```

Listing 4-3 shows a program that plays two songs on the piezo speaker: "Mary Had a Little Lamb" and the familiar Happy Birthday song. The former is played when you press SW1; the latter is played when you press SW2.

To simplify the code that generates the musical notes, the program defines several constants that represent the frequency for each of the notes required by the songs. For example, the constant NoteC6 is 1046, the frequency in Hz of C in the sixth octave of a piano keyboard. The constants span two full octaves, which is plenty of range for the songs to be played. Both songs are played in the key of C, so no flats or sharps are required. (If this musical stuff makes no sense to you, don't worry about it — this is an electronics book, not a music book.)

The program also sets up constants for the duration of a quarter note, half note, and whole note. The constants make it easy to specify a particular pitch for a particular duration in a FREQOUT command. Thus, playing a melody is simply a matter of writing a sequence of FREQOUT commands to play the correct notes for the correct durations in the correct order. That's precisely what the subroutines labeled Mary and Birthday do.

Listing 4-3: Making Music with a BASIC Stamp

```
' Song Program
' Doug Lowe
' July 15, 2011
'
' This program plays one of two songs on the piezo speaker
' on pin 0.
' If SW1 on pin 14 is pressed, the program plays "Mary Had a Little Lamb."
' If SW2 on pin 10 is pressed, the program plays "Good Morning to All."

' {$STAMP BS2}
' {$PBASIC 2.5}

SW1       PIN 14
SW2       PIN 10
Speaker   PIN 0

NoteC6    CON 1046
NoteD6    CON 1175
NoteE6    CON 1318
NoteF6    CON 1370
NoteG6    CON 1568
NoteA6    CON 1760
NoteB6    CON 1975
NoteC7    CON 2093
NoteD7    CON 2349
NoteE7    CON 2637
NoteF7    CON 2794
NoteG7    CON 3136
NoteA7    CON 3520
NoteB7    CON 3951
NoteC8    CON 4186

Whole     CON 1000
Half      CON 500
Quarter   CON 250


DO
  IF SW1 = 1 THEN
    GOSUB Mary
  ENDIF
  IF SW2 = 1 THEN
    GOSUB Morning
  ENDIF
LOOP
```

(continued)

Listing 4-3 *(continued)*

```
Mary:
  FREQOUT Speaker, Quarter, NoteE7     ' Mar-
  FREQOUT Speaker, Quarter, NoteD7     ' y
  FREQOUT Speaker, Quarter, NoteC7     ' Had
  FREQOUT Speaker, Quarter, NoteD7     ' a
  FREQOUT Speaker, Quarter, NoteE7     ' Lit-
  FREQOUT Speaker, Quarter, NoteE7     ' tle
  FREQOUT Speaker, Quarter, NoteE7     ' Lamb
  PAUSE Quarter
  FREQOUT Speaker, Quarter, NoteD7     ' Lit-
  FREQOUT Speaker, Quarter, NoteD7     ' tle
  FREQOUT Speaker, Quarter, NoteD7     ' Lamb
  PAUSE Quarter
  FREQOUT Speaker, Quarter, NoteE7     ' Lit-
  FREQOUT Speaker, Quarter, NoteG7     ' tle
  FREQOUT Speaker, Quarter, NoteG7     ' Lamb
  PAUSE Quarter
  FREQOUT Speaker, Quarter, NoteE7     ' Mar-
  FREQOUT Speaker, Quarter, NoteD7     ' y
  FREQOUT Speaker, Quarter, NoteC7     ' Had
  FREQOUT Speaker, Quarter, NoteD7     ' a
  FREQOUT Speaker, Quarter, NoteE7     ' Lit-
  FREQOUT Speaker, Quarter, NoteE7     ' tle
  FREQOUT Speaker, Quarter, NoteE7     ' Lamb
  FREQOUT Speaker, Quarter, NoteE7     ' Its
  FREQOUT Speaker, Quarter, NoteD7     ' Fleece
  FREQOUT Speaker, Quarter, NoteD7     ' Was
  FREQOUT Speaker, Quarter, NoteE7     ' White
  FREQOUT Speaker, Quarter, NoteD7     ' As
  FREQOUT Speaker, Quarter, NoteC7     ' Snow
  PAUSE Half
  RETURN

Morning:
  FREQOUT Speaker, Half, NoteC7        ' Good
  FREQOUT Speaker, Half, NoteD7        ' Morn-
  FREQOUT Speaker, Half, NoteC7        ' ing
  FREQOUT Speaker, Half, NoteF7        ' To
  FREQOUT Speaker, Whole, NoteE7       ' You
  FREQOUT Speaker, Half, NoteC7        ' Good
  FREQOUT Speaker, Half, NoteD7        ' Morn-
  FREQOUT Speaker, Half, NoteC7        ' ing
  FREQOUT Speaker, Half, NoteG7        ' To
  FREQOUT Speaker, Whole, NoteF7       ' You
  FREQOUT Speaker, Half, NoteC7        ' Good
  FREQOUT Speaker, Half, NoteC8        ' Morn-
  FREQOUT Speaker, Half, NoteA7        ' ing
  FREQOUT Speaker, Half, NoteF7        ' Dear
  FREQOUT Speaker, Half, NoteE7        ' Child-
  FREQOUT Speaker, Whole, NoteD7       ' ren
  FREQOUT Speaker, Half, NoteB7        ' Good
  FREQOUT Speaker, Half, NoteA7        ' Morn-
  FREQOUT Speaker, Half, NoteF7        ' ing
  FREQOUT Speaker, Half, NoteG7        ' To
  FREQOUT Speaker, Whole, NoteF7       ' All
  RETURN
```

Using a Servo with a BASIC Stamp

A *servo* is a special type of motor that is designed to rotate to a particular position and hold that position until told to rotate to a different position. Hobby servos are frequently used in radio-controlled vehicles such as airplanes, boats, and cars, but there are many other uses for servos. For example, I often use them in Halloween props to add movement such as eyes or a mouth.

The BASIC Stamp Activity Kit comes with a servo that you can use to learn how to write programs that control servos. You can also purchase servos directly from Parallax (www.parallax.com) or from most hobby stores. Figure 4-4 shows a typical hobby servo.

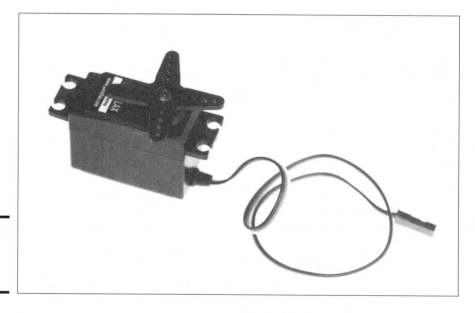

Figure 4-4:
A typical hobby servo.

Connecting a servo to a BASIC Stamp

Servos use a special three-conductor cable to provide power and a *control signal* that tells the servo what position it should move to and hold. The cable's three wires are colored red, black, and white and have the following functions:

✦ **Red:** Supplies the voltage required to operate the servo. For more servos, this voltage can be anywhere between +4 V and +9 V. On a BASIC Stamp HomeWork board, you should connect this to one of the Vdd pins.

✦ **Black:** The ground connection. On the BASIC Stamp HomeWork board, you should connect it to a Vss pin.

✦ **White:** The control wire; it connects to one of the BASIC Stamp's I/O pins.

Figure 4-5 shows how these wires should be connected in a BASIC Stamp circuit.

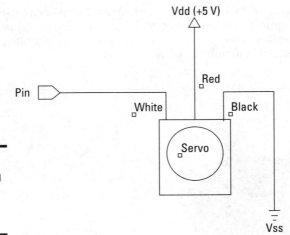

Figure 4-5:
Connecting
a servo to
a BASIC
Stamp.

The control wire controls the position of the servo by sending the servo a series of pulses approximately 20 ms apart. The length of each one of these pulses determines the position that the servo rotates to and holds.

Most hobby servos have a range of motion of 180° — that is, one-half of a complete revolution. The complete range of pulse durations is 0.5 ms to 2.5 ms, where 0.5 ms pulses move the servo to its minimum position (0°), and 2.5ms pulses move the servo to its maximum position (180°). To hold the servo at the center point of this range (90°), the pulses should be 1.5 ms in duration.

To connect a servo to a BASIC Stamp HomeWork Board, you must use a 3-pin header, which comes with the BASIC Stamp Activity Kit and is pictured in Figure 4-6. This header consists of three pins that you can plug in to the solderless breadboard. Then, you can plug the servo cable into the adapter.

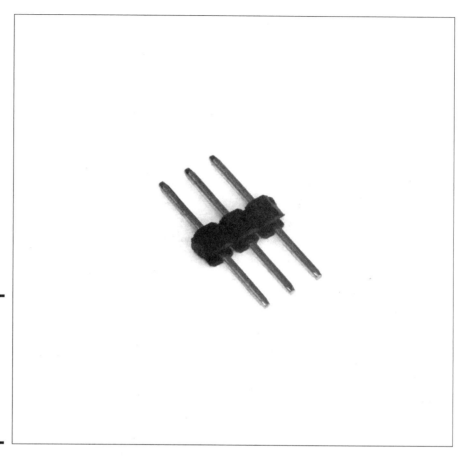

Figure 4-6:
3-pin
header for
connecting
the servo
to a BASIC
Stamp
HomeWork
Board.

Programming a servo in PBASIC

The easiest way to control a servo from a BASIC Stamp microcontroller is to use the PULSOUT command. This command sends a pulse of any duration you specify to an I/O pin of your choosing. The syntax of this command is as follows:

```
PULSOUT pin, duration
```

You specify the duration in units of two microseconds. A microsecond is one-millionth of a second. There are one thousand microseconds in a millisecond. Thus, to send a 1.5 ms pulse with the PULSOUT command, you must specify 750 as the duration, like this:

```
PULSOUT 0,750
```

Here, a 1.5 ms pulse is sent to pin 0.

To make your servo-programming life easier, Table 4-1 lists the duration values you should use for a typical hobby servo for various angles.

Table 4-1	PULSOUT Duration Values for Servo Control		
Angle	*Duration*	*Angle*	*Duration*
0	250	95	778
5	278	100	806
10	306	105	833
15	333	110	861
20	361	115	889
25	389	120	917
30	417	125	944
35	444	130	972
40	472	135	1000
45	500	140	1028
50	528	145	1056
55	556	150	1083
60	583	155	1111
65	611	160	1139
70	639	165	1167
75	667	170	1194
80	694	175	1222
85	722	180	1250
90	750		

For example, to move the servo on pin 0 to 75°, use this command:

```
PULSOUT 0,667
```

Remember that to hold its position, a servo needs a constant stream of pulses approximately 20 ms apart. Thus, PULSOUT commands are usually contained in either DO loops or FOR-NEXT loops. For example, here's a bit of code that keeps the servo on pin 0 at 45° indefinitely:

```
DO
  PULSOUT 0,500
  PAUSE 20
LOOP
```

Listing 4-4 shows a complete program that moves the servo to 45° when SW1 (a pushbutton on pin 14) is pressed and 135° when SW2 (a pushbutton on pin 10) is pressed.

Listing 4-4:   A Servo Control Program

```
' Servo Control Program
' Doug Lowe
' July 15, 2011
'
' This program moves a servo to one of two when SW1 is
    pressed
' and returns the servo to center position when SW2 is
    pressed.

' {$STAMP BS2}
' {$PBASIC 2.5}


Servo PIN 0
SW1    PIN 14
SW2    PIN 10

Position VAR Word
Position = 500
DO
  IF SW1 = 1 THEN
    Position = 500
  ENDIF
  IF SW2 = 1 THEN
    Position = 1000
  ENDIF
  PULSOUT Servo, Position
  PAUSE 20
LOOP
```

Building a servo project

Project 4-2 shows you how to build a complete circuit that uses a servo as well as two pushbuttons. This circuit is capable of running the program that was shown in Listing 4-4. Figure 4-7 shows the assembled project.

Project 4-2: Using a Servo with a BASIC Stamp

In this project, you connect a servo to a BASIC Stamp HomeWork Board. The circuit also includes two pushbuttons (on pins 10 and 14) that you can use to control the action of the servo.

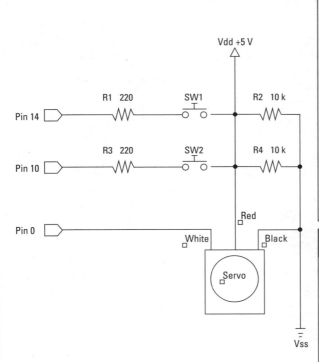

Parts List

1 Computer with BASIC Stamp Editor software installed
1 BASIC Stamp HomeWork board or Board of Education
1 9 V Battery
1 USB cable
1 USB to serial adapter
1 Hobby servo
1 3-pin male-to-male header
2 Normally open pushbuttons
2 220 Ω resistors
2 10 kΩ resistors
8 Jumper wires

Beeper Program

```
'  {$STAMP BS2}
'  {$PBASIC 2.5}

X VAR Byte
DO
  FOR X = 1 To 200
    PULSOUT 0, 350
    PAUSE 10
  NEXT
  FOR X = 1 To 200
    PULSOUT 0, 1150
  PAUSE 10
  NEXT
LOOP
```

Steps

1. **Insert the 3-pin header.**

   The three pins go in G15, G16, and G17.

2. **Insert the two pushbuttons.**

   SW1 goes in E1, F1, E3, and F3.
   SW2 goes in E5, F5, E7, and F7.

3. **Insert the resistors.**

R1 (220 Ω)	Pin 14 to B3
R2 (10 kΩ)	E4 to F4
R3 (220 Ω)	Pin 10B7
R4 (10 kΩ)	E8 to F8

4. **Insert the jumper wires.**

   Vdd to D1
   Vss to G4
   C1 to C5
   D3 to D4
   H4 to H8
   D7 to D8
   J8 to J15
   Pin 0 to A17
   E17 to F17

5. **Connect the servo.**

   Plug the 3-pin servo connector into the 3-pin male-to-male header. Connect the white wire to G15.

6. **Open the BASIC Stamp Editor.**

7. **Connect your BASIC Stamp to the computer and identify it in the Stamp Editor.**

   For more information about how to do this, refer to Chapter 1 of this minibook.

8. **Type the Beeper program into the Editor program window, and then save the file.**

   The Beeper program is shown in the accompanying listing.

9. **Choose Run⇨Run.**

   If you prefer, you can click the Run button in the toolbar or press F9. When you choose the Run command, the program is downloaded to the BASIC Stamp. Once the program is downloaded, it automatically starts to run on the Stamp. The servo should alternate from about 20° to about 165° once per second.

10. **Now try the program in Listings 4-4.**

    This program uses the switches to control the servo's action.

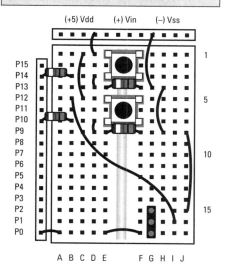

Figure 4-7:
A BASIC
Stamp
project that
controls
a servo
(Project 4-2).

For projects that require multiple servos or a lot of other work besides managing the servo, you should consider using Parallax's Propeller chip instead of a BASIC Stamp. The Propeller processor is designed for programs that have to do several things at once, such as managing servos. The propeller costs a little more than a BASIC Stamp and its programming language is a little more complicated, but in the end, controlling multiple servos is much easier with a Propeller than with a BASIC Stamp.

Book VIII

Special Effects

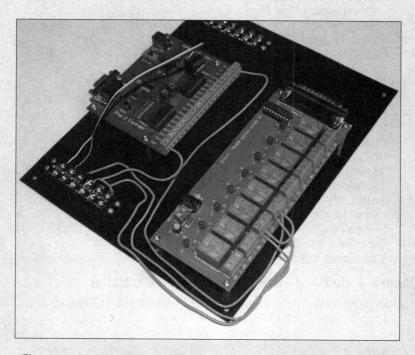

The completed wiring for the judge's console terminal in the Quiz-O-Matic! project.

Contents at a Glance

Chapter 1: Building a Quiz-O-Matic!

In This Chapter

✔ **Introducing the microprocessor-based Quiz-O-Matic!**

✔ **Building the Quiz-O-Matic!**

✔ **Programming the Quiz-O-Matic!**

Game shows are as popular on television today as they ever have been. A new generation of game shows have sprung up in recent years, challenging contestants to become millionaires by answering trivia questions, competing against fifth graders in basic knowledge questions, or remembering the lyrics to classic rock songs. But still, the king of the game shows is the classic quiz game, whose name shall go unmentioned lest I find myself in jeopardy of legal action. The game show which I refer to not by name first entered our living rooms way back in 1964, and it's still among the most popular game shows today.

In the classic quiz show format, the host reads a question to three contestants, who are then given a few seconds to punch a button indicating that they think they know the answer. The first to push his or her button gets a chance to answer the question within another short period of time. If the contestant is right, points are awarded. If the contest is incorrect, points are deducted. Sounds simple enough.

In this chapter, I show you a prototype of an electronic system that handles the punch-the-button and answer-the-question-within-the-allotted-time part of the game show. This system doesn't display the questions or the answers, nor does it keep score. That function can easily be handled with a laptop computer and a program such as Microsoft PowerPoint.

Instead, the system described in this chapter handles the task of determining which contestant pushes his or her button first. It gives a satisfying chime and lights up a lamp when a player presses his or her button. But if no one presses a button within a few seconds, it gives an annoying buzz.

For lack of a better name, I refer to the game show controller described in this chapter as the Quiz-O-Matic! (Yes, with an exclamation mark. I use the exclamation mark as part of the name as a subtle homage to the quintessential quiz show, which once again I don't want to mention by name. If you want to guess which show I'm talking about, remember to phrase your answer in the form of a question.)

You can use the Quiz-O-Matic! at home if you and your buddies like to play quiz games, but where I expect it will come in most useful is in schools and youth groups, where the quiz show format is often used as a teaching tool. For example, a friend of mine who is a fourth-grade teacher often uses the quiz show format to teach history, math, or science. The kids find the game more engaging when they can actually press a button rather than simply raise their hands when they think they know the answer.

Being Safe with Line Voltage

As with many other chapters in this book, I need to start this chapter with a warning: This project involves the use of line voltage, which can be very dangerous – indeed deadly – if not properly secured. If you decide to make a Quiz-O-Matic! of your own, you must be very careful when you are working with the portions of the project that use line voltage.

This chapter shows a prototype of the Quiz-O-Matic! that is designed to use only in the safe confines of a workbench environment, and only by someone who knows how to be very careful around line voltage circuitry. Above all, you must be constantly on your toes whenever the Quiz-O-Matic! is plugged in. Do not under any circumstances touch *any* part of the Quiz-O-Matic main board circuitry when it is plugged in.

If you intend to use the Quiz-O-Matic anywhere other than in the safe confines of your workbench, you must add some important additional safety features. In particular:

✦ Build a solid enclosure for the main controller board so that you can contain all of the line-voltage wiring.

✦ Use proper strain relief on all line-voltage cords that exit the enclosure you build so that they don't inadvertently pull out.

✦ Use grommets at the point where the line voltage cords exit the enclosure to prevent the insulation from wearing off.

✦ Add a low-amperage fuse (1A should be sufficient) in-line with the line-voltage power circuit.

✦ Add an On/Off switch that can cut power to the entire Quiz-O-Matic! circuit.

✦ If any part of your enclosure is made of metal, make sure the metal is properly grounded.

Looking at the Quiz-O-Matic!

The Quiz-O-Matic! is easily the most complicated project in this book. Figure 1-1 shows a block diagram of its various components. ***Note:*** This is a simple block diagram, not an actual schematic, so this diagram doesn't use formal schematic symbols or show complete circuits. It simply shows how the various pieces that make up the Quiz-O-Matic! work together.

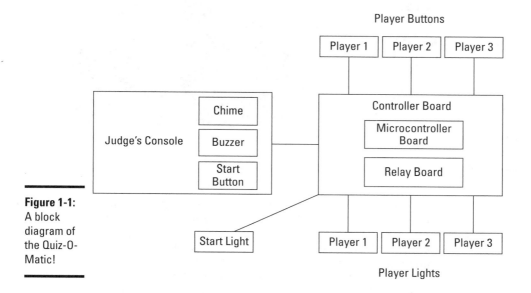

Figure 1-1:
A block diagram of the Quiz-O-Matic!

At the heart of the Quiz-O-Matic! is a Basic Stamp microcontroller and a relay board, which are together mounted on a Plexiglas panel to form the main controller board. The controller module also includes several wiring terminals and line-voltage receptacles made from indoor extension cords that allow you to connect the other components. It also includes a 12 V power supply that powers the BASIC Stamp and relay board, as well as a 120 VAC plug to plug the entire system into a wall outlet.

The main controller board contains line-voltage wiring which is potentially dangerous. Be extremely careful when working with this board. Be very careful when you assemble it that none of the line voltage wires are loose or exposed. And never touch any part of the main controller board when it is plugged in.

Figure 1-2 shows the assembled main controller board.

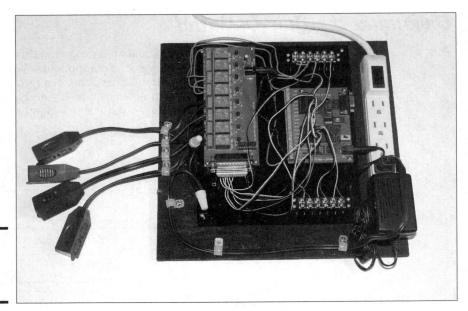

Figure 1-2:
The main
controller
board.

The judge's console is a second Plexiglas panel that contains three components: a doorbell chime, buzzer, and doorbell button. All three of these components can be purchased at any hardware store. The judge's console is connected to the main controller board by a ten-foot length of six-conductor thermostat wire, which can also be purchased at most hardware stores. This allows you to locate the judge's console some distance from the main controller board. Figure 1-3 shows the assembled judge's console.

Each of the players holds one of the three player buttons (shown in Figure 1-4), which are made from doorbell buttons (again purchased at a hardware store) and small project boxes purchased from RadioShack. The boxes are just the right size to hold in your hand with your thumb resting on the button, ready to press quickly when you think you know the answer. The buttons are connected to the main controller board using 18-gauge, two-conductor stranded wire, once again available at most hardware stores.

Finally, there are the four lights — the start light and the three player lights. These are standard line-voltage (120 VAC) lights. You connect them to the main controller board by plugging them into the four extension cord receptacles that dangle from the relay card. The extension cords can be as long as you want, so you can position the lights anywhere in the room you wish. You can also use any 120 VAC lamp that you wish. To keep the cost down, I use simple 4 W night lights that can be purchased for $1 from most dollar stores.

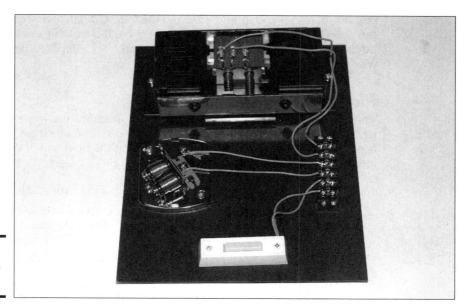

Figure 1-3:
The judge's
console.

Figure 1-4:
One of the
three player
buttons.

The 120 VAC lights and the wiring that connects them to the relay board
are potentially dangerous. Thus, you must be careful when you assemble
the wiring to make sure there are no exposed conductors that could pose a
shock or fire hazard. Carefully inspect the main controller board every time

you use it. And if you intend on using the Quiz-O-Matic! anywhere other than the safe confines of your own workbench, build a suitable enclosure so that no one can accidentally touch any of the circuitry. Also, be sure to unplug it when it isn't in use.

Playing the game

The Quiz-O-Matic! is designed to play a simple quiz show–type format in which three contestants vie for the right to answer a question read by a host. In addition to the three contestants and the host, you'll also need a judge. Here's how the game works:

1. Each player takes one of the player buttons in his or her hand.

2. The judge takes position near the judge's console with his finger on the button. Ideally, the button and the judge's hand shouldn't be visible to the players.

3. The host reads the question out loud so that all three contestants can hear the question.

4. When the host finishes reading the question, the judge presses the start button turning on the start light.

5. When the players see the start light come on, they have approximately five seconds to press their buttons. The first player to press his or her button within the five seconds is the only one who's allowed to answer the question. That player's light will come on, indicating which player pressed the button first.

6. If a player presses the button before the start light comes on, that player is penalized by having to wait two full seconds before he or she can press the button again. This is to prevent players from pressing their buttons before the question has been read.

7. If no player presses his or her button before the five seconds elapses, the buzzer sounds. The host can then read the answer and move on to the next question.

8. When the player whose light comes on answers the question correctly, the judge presses the judge's button again. The chime sounds three times to indicate that the player has answered the question correctly. All the lights then go out, and the host can move on to the next question.

9. If the player doesn't answer the question correctly within ten seconds (that is, if the judge doesn't press the judge's button to indicate a correct answer), a buzzer sounds to indicate that the player didn't answer correctly.

    Note: If the player answers incorrectly, the judge should simply not press the judge's button. Once the ten seconds have passed, the buzzer will sound.

10. Whether the player answers correctly, incorrectly, or no player presses his or her button, all the lights eventually go out. This is the signal that the host can begin the next question. Then, the entire process starts all over again.

Configuring the controller

The program that controls the game controller uses two timing intervals: the amount of time that the players have to press their buttons and the amount of time the winning player has to answer the question. The default values for these intervals are five and ten seconds. However, when power is first applied to the Quiz-O-Matic!, the controller goes through a configuration routine that lets you change these timings.

The configuration routine works like this:

1. When power is applied to the controller, the controller sounds the chime once to indicate that it's waiting for you to set the time allowed for the players to press their buttons.

2. The default time to wait for the players to press their buttons is five seconds, but you can add additional seconds by pressing the Player 1 button. Each time you press the Player 1 button, one second is added. The buzzer sounds to let you know that the button press has been received.

For example, if you want to give your players eight seconds to press their buttons, press the Player 1 button three times after the first chime when you power up the controller.

3. When you've finished adding seconds, press the judge's button. The chime will sound again to indicate that you can now add time to the interval allotted for the winning player to provide the correct answer. The default is 10 seconds.

4. Press the Player 1 button to add additional time in increments of five seconds. For example, if you want to allow 20 seconds to answer the question, press the Player 1 button twice. This adds 10 seconds to the initial default interval of 10 seconds.

5. Press the judge's button again to indicate that you are finished configuring the controller. The chime will sound three times to indicate that you may begin playing the game.

What the Quiz-O-Matic! does not do

Before I get into the details of how to build the Quiz-O-Matic!, I want to point out a few of the things that the Quiz-O-Matic! *doesn't* do. It's designed to handle just that part of the game that determines which of the three contestants gets the chance to answer the question to win points, but it *doesn't*

Book VIII Chapter 1

Building a Quiz-O-Matic!

display the questions on a screen or determine whether a player answers correctly or incorrectly. Nor does it keep score.

If you need to display the questions on a computer screen or projector, I suggest you use a presentation program such as Microsoft PowerPoint 2010. In fact, I would be remiss if I didn't take the opportunity for a shameless plug and encourage you to run out to your local bookstore right now and buy a copy of my book, *PowerPoint 2010 For Dummies*. It will teach you more than you need to know to figure out how to create a nice display of quiz game questions.

As for keeping score, you'll need to devise your own clever system for doing that. I suggest using a decidedly low-tech but highly reliable system: a pencil and piece of paper.

Looking at the Parts List for the Quiz-O-Matic!

The Quiz-O-Matic! is easily the most complicated project presented anywhere in this book, so it not surprisingly has a long shopping list. The complete list of parts you'll need to build the Quiz-O-Matic! is compiled in Table 1-1.

Remember that if you decide to build your own Quiz-O-Matic! circuit, the instructions provided in this chapter are for building a prototype workbench version, not a version that can be safely used away from your workbench. If you want to build a Quiz-O-Matic! you can use anywhere, you'll need to purchase additional parts to build an enclosure to safely contain the circuitry so that there is no exposed line voltage wiring or other safety hazards.

You can find many of the parts you'll need at your local hardware store. A few of the pieces are most easily obtained at RadioShack or another electronics part supplier. You might even find a few important elements — such as inexpensive lights and extension cords — at a discount overstock store such as Big Lots or even at a dollar store.

Two of the most important (and expensive) components must be obtained over the Internet. The first is the microcontroller module. The Quiz-O-Matic! uses a controller board called a Prop-2 controller that's made by a company called EFX-TEK. You can purchase this controller board for about $100 from their website at www.efx-tek.com. While you're there, you should also pick up a 12 V power supply.

If you haven't done any Basic Stamp programming, you'll also need the free Parallax Basic Stamp program editor and an interface cable to connect the Prop-2 to your computer. You can order the interface cable from the EFX-TEK website. You can also find a link to download the program editor. (For more information, refer to Book VII of Chapter 1.)

While you're at the EFX-TEK website, you should also order the servo extension cables. RadioShack doesn't stock them, and although you may be able to find them at a local hobby store, you can probably get them cheaper from EFX-TEK. You can also get the 2.1 mm pigtail power cord for a few bucks from the EFX-TEK online store.

If you're new to BASIC Stamp programming, I suggest you purchase the Prop-2 Starter Kit from EFX-TEK. For just a few dollars more than the cost of the controller alone, it includes a power supply and the programming interface cable.

The second item you'll need to order online is the relay board controller, called the Kit-74 PC Parallel Port Relay Board. There are several online stores that sell this board; just search for *Kit-74* and you'll find plenty of sources. The price from most sources is under $40. Be aware that there are two versions of the Kit-74 board: unassembled or assembled. If you want to spend an hour or so soldering the kit together, get the unassembled version. The assembled version is usually just a few dollars more.

Before you begin assembly of the Quiz-O-Matic!, spread some newspaper and paint one side of the two Plexiglas sheets with the black paint. Sand and paint the 12 x 12" plywood as well.

Table 1-1   Parts List for the Quiz-O-Matic!

Quantity	Description
1	EFX-TEK Prop-2 controller board
1	12 V power supply (from EFX-TEK)
1	2.1 mm pigtail cable (to power Kit-74; purchase from EFX-TEK)
10	14", 3-pin, servo-style extension cables (from EFX-TEK)
1	Kit-74 PC Parallel Port Relay Board (assembled)
1	DB25 female connector (solder-style)
2	6-position, dual-row barrier strip (RadioShack part # 274-659)
8	20 mm PC board standoffs (RadioShack part # 276-195 provides four standoffs, so get two packages)
12	2-56 1/2" machine screws and nuts
1	6-outlet electrical power strip
3	3 x 2 x 1" project box (RadioShack part #270-1801)
3	Round doorbell pushbuttons (not lighted)
1	Rectangular doorbell pushbutton (not lighted)
1	12 V doorbell chime

(continued)

Table 1-1 *(continued)*

Quantity	Description
1	12 V doorbell buzzer
1	Roll of 20-gauge stranded hookup wire (about 8')
4	6' indoor nongrounded extension cords
4	4 W, 120 VAC night lights
2	Yellow twist-on wire connectors (capable of handling 5, 16-gauge wires)
10	⅜" plastic cable clamps (to secure line-voltage cables to plywood base)
10	1/2" sheet-metal screws (to mount the plastic cable clamps)
10'	6-conductor thermostat wire (More conductors are acceptable, but at least six are required)
30'	18-gauge, 2-conductor stranded wire
6	#8 vinyl-insulated crimp-on spade terminals
2	8 x 10" sheets of 3/16"-thick Plexiglas (or other similar acrylic material)
1	12 x 12" sheet of 1/2"-thick plywood
1	Can of black spray paint (to paint Plexiglas and board)

Building the Main Controller Board

The first step in building the Quiz-O-Matic! is to build the main controller board, which consists of the following elements mounted on an 8 x 10" piece of Plexiglas:

+ The Prop-2 controller, mounted on four, 20 mm standoffs.

+ The Kit-74 PC Parallel Port Relay Board, also mounted on four, 20 mm standoffs.

+ Two of the 6-position, dual-row barrier strips, used to connect the judge's console and the player buttons. (You'll connect the player lamps and the start lamp directly to the Kit-74 board.)

Once you've assembled the main controller board, you'll mount the 8 x10" Plexiglas board to a 12" square piece of plywood. The plywood board provides space to mount the 6-outlet electrical power strip and provides a good base on which to mount the plastic cable clamps that secure the wiring used for the player and start lamps.

The following sections describe the steps for assembling the main controller board.

Mounting the components

Start the assembly of the main controller board by mounting the Prop-2 controller, the Kit-74 board, and the two, 6-position dual-row barrier strips. Figure 1-5 is a guide for placing these components. You'll need to drill four corner holes in the Plexiglas for each component. Select a drill bit that's appropriate for the mounting screws you have, and use the components themselves to determine the exact drilling location.

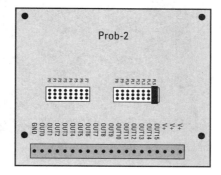

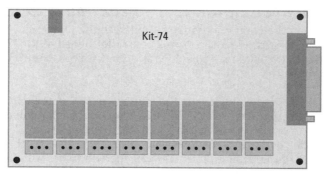

Figure 1-5:
Locating the components on the main controller board.

Plexiglas is a bit brittle, so it's easy to accidentally crack it when drilling. To avoid this, place the Plexiglas flat on a piece of scrap wood and hold the Plexiglas firmly while you drill so that it doesn't flex or turn as you drill.

Once you've drilled the holes, use the standoffs to mount the Prop-2 and the Kit-74. Then, use the machine screws and nuts to mount the two barrier strips. Figure 1-6 shows how the board should appear once the components are mounted.

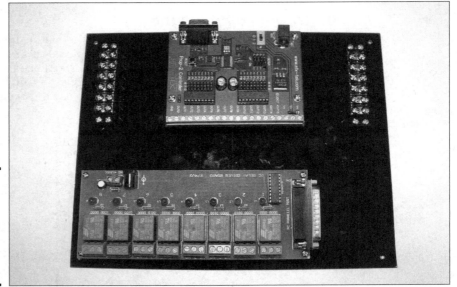

Figure 1-6: Positioning the components on the main controller board.

Connecting the Prop-2 controller to the relay board

When the components are mounted to the main controller board, the next step is to connect the Prop-2 controller board to the Kit-74 relay board. To do that, you'll need to solder a total of seven wires to the female DB25 connector.

Six of these wires will be the white wire of the 3-pin servo-style extension cables. These wires will be used to connect six of the eight relays to I/O ports on the Prop-2 card. To create these wires, you'll cut the connector off of one end of the wire and discard it. Then, peel off the white wire, strip back a bit of the insulation, and solder the end of the white wire to the appropriate pin terminal on the back of the DB25 connector.

The seventh wire is simply a 20-gauge hookup wire that will serve as a common ground connector for the six white wires. The use of this common ground wire eliminates the need to solder the black wires from the six, 3-pin extension cables.

For more information about how to work with a DB25 connector (and the Kit-74 in general), you may want to refer to Book VI, Chapter 4.

Figure 1-7 shows how DB25 connector pins are numbered. Be sure to use the diagram at the top of this figure since you'll be working with a female connector, not a male connector.

DB25 Parallel Port Pins (female)

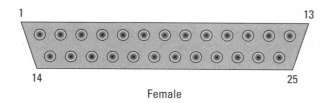

Female

Figure 1-7: Pin numbering for DB25 connectors.

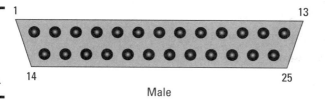

Male

The seven wires connect to the following pins:

Pin	*Description*	*Connects to*
2	White wire from 3-pin extension cable	Prop-2 P0
3	White wire from 3-pin extension cable	Prop-2 P1
4	White wire from 3-pin extension cable	Prop-2 P2
5	White wire from 3-pin extension cable	Prop-2 P3
6	White wire from 3-pin extension cable	Prop-2 P4
7	White wire from 3-pin extension cable	Prop-2 P5
25	20-gauge hookup wire	Prop-2 ground connector

Here are the detailed steps for assembling this connector:

1. **Heat up your soldering iron.**

2. Mount the DB25 connector in a hobby vice or use a third-hand tool to support the connector so that you can easily access the solder connections on the back of the connector.

3. Cut the 3-pin connector off one end of one of the extension cables.

4. Separate the white conductor from the red and black conductors and strip about ¼" of insulation from the end of the white conductor.

5. Peel the red and black conductors back about 2" from the stripped end of the white conductor and cut the loose ends of the red and black conductors off.

6. Twist the stranded wires revealed when you stripped the white wire so that there are no loose strands.

7. Carefully lay the twisted strands into the solder connector for pin 2.

8. Solder the wire to the solder connector.

9. Repeat Steps 3 through 8, soldering the white wires to pins 3, 4, 5, 6, and 7.

10. Cut a 10" length of 20-gauge hookup wire and strip about ¼" of insulation from each end.

11. Solder one end of the hookup wire to pin 25.

12. You're done!

Figure 1-8 shows the assembled connector.

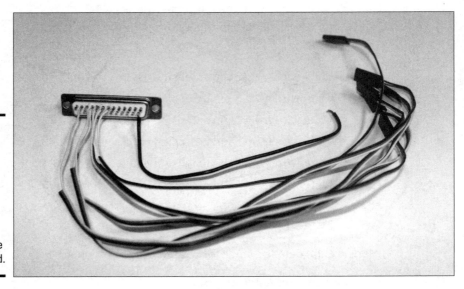

Figure 1-8:
The assembled DB25 connector used to connect the Prop-2 board to the Kit-74 board.

Once you've completed the connector, connect it to the Prop-2 and Kit-74 boards as follows:

1. **Plug the DB25 connector into the Kit-74 board.**

2. **Plug the 3-pin connector from pin 2 into the 3-pin connector on the Prop-2 board labeled P0.**

   Be sure to plug the extension cable's connector into the Prop-2 connector correctly: The pins are labeled W, R, and B to indicate how to orient the white, red, and black.

Figure 1-9 shows how the main controller board should appear with the Prop-2 and Kit-74 boards connected.

Figure 1-9: The main controller board with the Prop-2 and Kit-74 boards connected.

Completing the line-voltage wiring

Four of the six relays used on the Kit-74 control the lights that indicate when the players can press their buttons and which player pressed his or her button first. The start lamp is controlled by relay 1, which is connected to pin 0 on the Prop-2 controller. The player lamps for players 1, 2, and 3 are on relays 2, 3, and 4. They're connected to pins 1, 2, and 3.

Yes, I know these numberings are a little confusing. There's not much that I can do about it, however. The I/O pins on a Basic Stamp processor are numbered starting with 0, while the relays on the Kit-74 board are numbered starting with 1. And to make matters worse, relay 1 is controlled by pin 2 on the DB25 connector. Thus, the start lamp is controlled by I/O pin 0 on the Prop-2 board, which is connected to pin 2 on the DB25 connector, which activates relay 1 on the Kit-74. Sigh.

The player and start lights run on 120 VAC line voltage, so you must be very careful when assembling this part of the circuit to make sure that your connections are secure and there are no bare wires. And under no circumstances should you work on this wiring while the Quiz-O-Matic! is plugged in!

There are several different ways to wire the relays in a Kit-74 to handle line voltage. I find the easiest is to use ordinary indoor extension cords. Simply cut off the female end of the cord, leaving 8–10" of wire. You'll also use the male end of one of the cords, cut to a length of about 18". Finally, you'll need four, 5" lengths of wire peeled from the left over extension wire. Strip the loose ends of all the wires, connect them properly, and secure them with cable ties so they don't move around and wiggle loose.

By "connect them properly," I mean by following the wiring diagram shown in Figure 1-10. In a nutshell, one wire from each of the four female extension cords connects directly to one wire from the male cord. The other end of the four female cords connects to the other end of the male cord through one of the Kit-74 relay connections.

The connections are made using two twist-on wire connectors, also known as *wire nuts*. These come in various sizes; make sure you get connectors that can accommodate a total of five, 16-gauge wires. To make a proper connection with a wire nut, first gather all the wires that you need to connect so that the stripped ends of the insulation are together. Give the big mass of bare wire a twist with your fingers to make sure that there aren't any loose strands. Then, firmly twist the connector on to the bare wires. You'll have to apply plenty of downward pressure as you twist. Don't stop until all of the bare strands are completely covered by the connector.

If you examine the terminals on the Kit-74 board, you'll see that each relay has three connectors, labeled NC, C, and NO. These stand for Normally Closed, Common, and Normally Open. The four, 5" wires will connect to the Common terminal for relays 1 through 4. These four wires will then be connected to one wire from the male extension cord and secured with one of the wire nuts. One wire from each of the female extension cords will connect to the NO terminal for relays 1 through 4. The other wire from each female extension cord will be connected to the remaining wire from the male cord using the other wire nut.

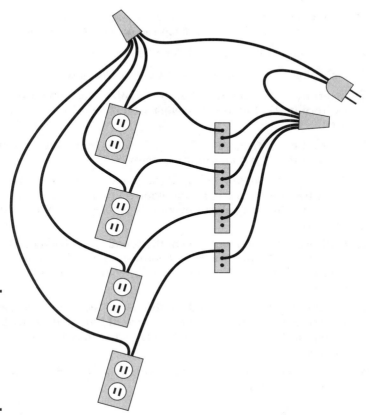

Figure 1-10:
Wiring
diagram for
the Quiz-
O-Matic!'s
line-voltage
wiring.

Here's the complete procedure to follow for making these connections:

1. **Cut the female end of each of the four extension cords approximately 10 inches from the connector.**

2. **Cut the male end of one of the extension cords approximately 18 inches from the connector.**

3. **Use the tip of your wire cutters to snip between the two conductors on the free ends of the wires you just cut, and then peel the conductors apart approximately two inches.**

4. **Cut two lengths of wire approximately five inches long. Then peel and separate the two conductors of these two lengths so that you have a total of four, single-conductor wires.**

5. **Use a small flat-head screwdriver to loosen the NO and C connections for relays 1 through 4 on the Kit-74 board.**

**Book VIII
Chapter 1**

**Building a
Quiz-O-Matic!**

6. Strip about ⁵⁄₁₆" of insulation from one wire of each of the female cords and about ½" of insulation from the other wire of each of these cords.

7. Do the same for the four, 5" lengths of wire.

8. Strip about ½" of insulation from both wires of the mail extension cord.

9. Insert the ⁵⁄₁₆" stripped end of the female extension cords into the NO terminals of relays 1 through 4 on the Kit-74 board and tighten securely.

10. Insert the ⁵⁄₁₆" stripped end of the 5" wires into the C terminals of relays 1 through 4 and tighten securely.

11. Gather the ½" stripped ends of the female extension cords along with one wire of the male cord and connect all five wires using one of the wire nuts. Be sure the connection is secure and no bare strands are exposed.

12. Gather the ½" stripped ends of the female extension cords along with one wire of the male cord and connect all five wires using one of the wire nuts. Be sure the connection is secure and no bare strands are exposed.

13. Gather the ½" stripped ends of the four 5" wires and the remaining wire from the male cord and connect all five wires using the other wire nut. Be sure the connection is secure and no bare strands are exposed.

14. Use labels or a light-colored marking pen to mark the female outlets *Start, 1, 2,* and *3.*

    Mark the outlets as follows:

Relay	*Mark*
1	Start
2	1
3	2
4	3

15. Carefully review your work.

    This part of the circuit carries line voltage that could potentially kill you or someone else. Therefore, you are not finished with this part of the assembly until you have carefully inspected every one of the connections to make sure that all of your connections are solid and that there are no exposed wires.

That's it! The line-voltage wiring is now assembled. It should resemble the wiring shown in Figure 1-11.

Figure 1-11: The main controller board with the line voltage wiring in place.

Wiring the judge's console terminal

The barrier strip on the left side of the Prop-2 controller board is used to connect the judge's console. This barrier strip has six connections, two each for the start button, the buzzer, and the doorbell chime.

Although the terminals on the barrier strip aren't actually numbered, number them 1 through 6 starting at the top of the Plexiglas board. Thus, terminals 1 and 2 are used for the start button, 3 and 4 are used for the buzzer, and 5 and 6 are used for the doorbell.

Figure 1-12 shows the wiring diagram to follow for wiring up this section of the main controller board.

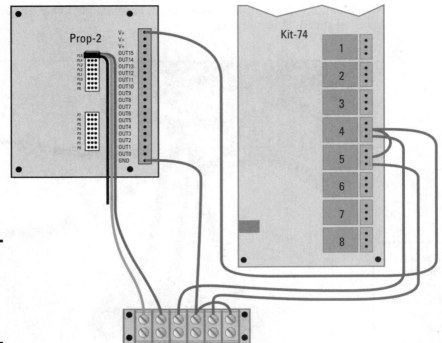

Figure 1-12:
Wiring
diagram for
the judge's
console
terminal.

To assemble this wiring, gather a roll of 20-gauge hookup wire, a 3-pin servo-style extension cable, wire cutters, wire strippers, and a screwdriver. For Steps 1 through 6, you'll need to measure and cut the hookup wire and strip about ⅜" of insulation from each end. Here are the steps:

1. **Use a short length of wire to connect terminals 4 and 6 on the barrier strip.**

2. **Connect terminal 5 to the NO terminal of relay 5 on the Kit-74 board.**

3. **Connect terminal 3 to the NO terminal of relay 4 on the Kit-74 board.**

4. **Connect terminal 4 to the GND connection on the Prop-2 controller board.**

5. **Connect the first V+ connection on the Prop-2 controller board to the C terminal of relay 4 on the Kit-74.**

6. **Use a short length of wire to connect the C terminal of relay 4 to the C terminal of relay 5.**

7. **Cut one end off the 3-pin extension cable. Peel the three wires to separate them, and then cut the black wire about 2" back. Strip about ½" of insulation from the white and red wires, and then connect the white wire to terminal 1 and the red wire to terminal 2.**

8. **Plug the other end of the 3-pin extension cable into the 3-pin header labeled *P15*. Make sure you orient the wire correctly; see the labels *W R B* next to the header pins on the Prop-2 controller for the correct orientation.**

Figure 1-13 shows how the wiring for the judge's console terminal appears when it's complete.

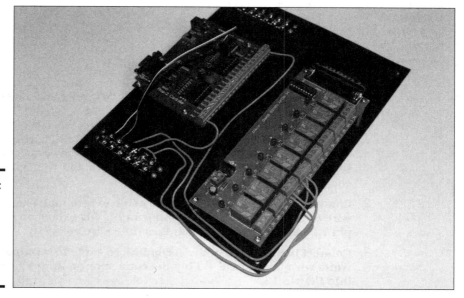

Figure 1-13:
The completed wiring for the judge's console terminal.

Wiring the player button terminal

The three player buttons will be connected to the main controller board via the other six-terminal barrier strip. Two terminals are used for each button, and the terminals are connected to the red and white wires of I/O pins 12, 13, and 14 on the Prop-2 controller via 3-pin, servo-style extension cables. Figure 1-14 shows the wiring diagram for the three player buttons.

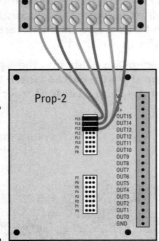

Figure 1-14:
The wiring diagram for the terminal used to connect the three player buttons.

To wire up the player button terminal, grab the three remaining servo-style extension cables and follow these steps:

1. **On each of the three servo cables, cut one end off, leaving at least six inches of wire. Cut the black wire off close to the connector, and then separate the white and red wires and strip about ½" insulation from each wire.**

2. **Connect the red wire of one servo cable to terminal 1 and the white wire to terminal 2. Plug the connector on this extension cable into I/O pin 12, being careful to orient the white wire correctly.**

3. **Connect the red wire of the second servo cable to terminal 3 and the white wire to terminal 4. Plug the connector on this extension cable into I/O pin 13.**

4. **Connect the red wire of the third servo cable to terminal 5 and the white wire to terminal 6. Plug the connector on this extension cable into I/O pin 14.**

Figure 1-15 shows how the wiring for the player button terminal appears when it's complete.

Figure 1-15:
The completed wiring for the player button terminal.

Finishing the main controller board

Once you've finished assembling the components and wiring on the Plexiglas base of the main controller board, you can mount it on the plywood base and attach the rest of the components. Here are the steps:

1. **Mount the 6-outlet power strip as close to the top edge of the board as you can.**

Use two wood screws to mount the strip.

2. **Drill a hole in each corner of the Plexiglas main controller board, and then use wood screws to mount the board to the plywood base as close as possible to the power strip.**

Drill carefully, as Plexiglas is prone to crack.

3. **Plug the male end of the extension cord into one of the outlets on the power strip.**

4. **Plug the 12 V power supply into one of the outlets on the power strip, and then plug the 2.1 mm power supply connector into the Prop-2's power connector.**

**Book VIII
Chapter 1**

**Building a
Quiz-O-Matic!**

When positioning the 12 V power supply (and the male end of the extension cord in the preceding step, be sure to leave the area immediately in front of the DB9 programming connector on the Prop-2 controller clear. You'll need room to plug the USB programming adapter in when it comes time to program Quiz-O-Matic!

5. **Use the plastic cable clamps to secure the line-voltage wiring. Use one clamp to secure each of the female extension cord connectors and use several clamps to secure the male extension cord on its route from the Kit-74 to the power outlet strip.**

6. **Connect the positive lead of the 2.1 mm pigtail connector to one of the V+ terminals on the Prop-2 controller and connect the negative lead to the GND terminal. Then plug the 2.1 mm connector into the power connector on the Kit-74. This connection supplies power for the Kit-74 from the Prop-2 controller.**

Congratulations! You're now finished with the main controller board of the Quiz-O-Matic! Figure 1-16 shows what the controller board should look like once it's finished.

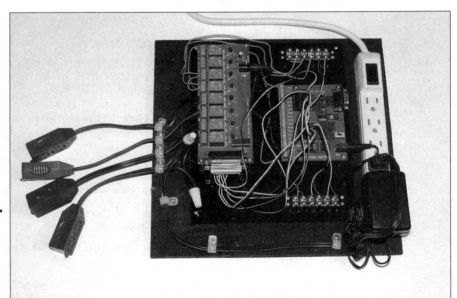

Figure 1-16: The finished main controller board.

Building the Judges Console

Once the main controller board of the Quiz-O-Matic! is complete, the next step is to build the judge's console. Like the main controller board, the judge's console consists of components mounted on an 8 x 10" sheet of Plexiglas. Before you assemble the judge's console, you should paint one side of the Plexiglas black to give it a professional look.

Figure 1-17 shows the layout of the components you'll mount to the judge's console. This figure also shows the wiring.

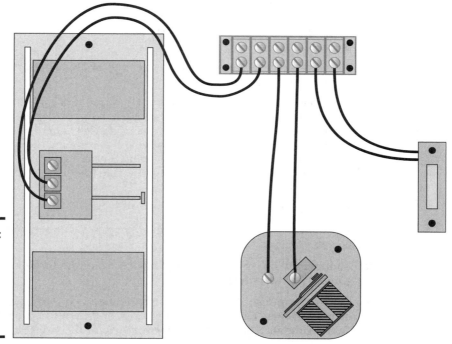

Figure 1-17: The component layout and wiring for the judge's console.

Here are the steps to assemble the judge's console:

1. **Mount the six-position, dual-row barrier strip.**

   You'll need to drill four mounting holes and use four, ½" 2-56 machine screws and nuts to mount the strip.

2. **Mount the doorbell chime.**

   You need two machine screws and nuts to mount the chime. Because doorbell chimes vary, you'll have to check yours first to determine which size mounting screws you'll need.

   Note that you'll probably need to pry off the cover of the doorbell chime to gain access to the mounting holes as well as the wiring terminals.

When the judge's console is finished, you can reattach the cover if you want, or you can leave the chime's inner mechanism exposed.

3. **Mount the buzzer.**

You'll need to pry the cover off the buzzer to get to the mounting holes and wiring terminals. You'll also need a pair of machine screws and nuts to mount the buzzer.

4. **Drill a ³⁄₁₆" hole through one side of the doorbell button. Then pass two, 4" lengths of hookup wire, stripped ½" on each end, and connect these wires to the screw terminals on the back side of the doorbell button.**

Depending on the button, you may need to pry off a back panel to expose the screw terminals.

Note that if the doorbell button is lit (that is, if it has a light that illuminates so you can see it in the dark), you'll have to remove the light. Usually, the light is a small circuit board that's connected between the two screw terminals. You can remove it by removing the screws, pulling the circuit board off, and then replacing the screws. If you leave the light in place, the Prop-2 controller won't be able to detect when the judge's button is pressed.

5. **Mount the doorbell button.**

You'll need some long (⅞" or so) machine screws and nuts to mount the button.

6. **Connect the other two ends of the wires from the button to terminals 1 and 2 on the barrier strip.**

7. **Cut two lengths of hookup wire approximately 6" long, string ½" of insulation from each end of both wires, and then use the wires to connect terminals 3 and 4 on the barrier strip to the two terminals on the buzzer.**

8. **Cut two lengths of hookup wire approximately 10" long, string ½" of insulation from each end of both wires, and then use the wires to connect terminals 3 and 4 on the barrier strip to the doorbell chime.**

Most doorbell chimes have three terminals, labeled *Front, Rear,* and *Trans.* You should use the Trans and Rear connections for the Quiz-O-Matic! judge's console.

9. **Congratulate yourself.**

The judge's console is done! Figure 1-18 shows how the judge's console appears when the assembly is complete.

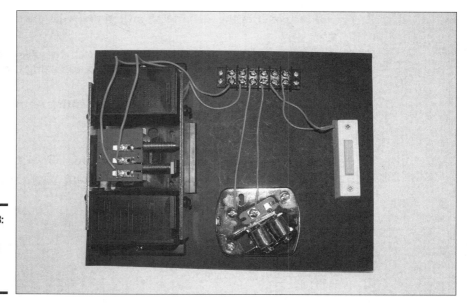

Figure 1-18:
The assembled judge's console.

Building the Player Buttons

The final pieces to assemble for the Quiz-O-Matic! are the three player buttons. For each player button, you need a round unlighted doorbell button; a 3 x 2 x 1" project box (available at RadioShack); 10' of 18-gauge, two-conductor stranded wire; and six, #8 vinyl-insulated crimp-on spade terminals.

Here's the assembly procedure:

1. **Separate the ends of the 2-conductor wire and strip about ½" of insulation from both ends of both wires.**

2. **Use a wire crimper to attach the six spade terminals to both conductors on one end of the wire.**

3. **Drill ¼" holes in the center of the plastic top of the project box and in the center of one of the short ends of the box.**

4. **Insert the wire through the end of the box and tie a knot leaving about 4" of free wire past the knot.**

The knot will act as a strain relief so the wire doesn't come loose when the player pulls on the cord.

5. **Feed the wire through the hole in the project box top.**

6. **Connect the wires to the two screw terminals in the bottom of the pushbutton switch.**

7. Pry off the cover of the pushbutton switch to reveal the mounting holes.

8. Center the pushbutton switch over the hole in the project box top and then mount the pushbutton to the project box top using the mounting screws that came with the pushbutton.

9. Mount the lid to the project box using the mounting screws that came with the box.

10. Replace the switch cover.

11. Repeat Steps 1 through 10 for the two remaining player buttons.

12. Use a printed label or a white marker to label the boxes 1, 2, and 3.

Figure 1-19 shows one of the assembled player buttons.

Figure 1-19: One of the player buttons.

Connecting the parts

Whew! Almost done. The final remaining step is to connect the judge's console and the player buttons to the main controller console and plug in the four lights.

To connect the player buttons, simply connect the two spade terminals on each button to the corresponding terminals on the player button barrier strip — the one located at the top right of the main controller board. Numbering from the top down, player 1's button should connect to terminals 1 and 2, player 2's button to terminals 3 and 4, and player 3's button to terminals 5 and 6.

Connecting the judge's console is a bit trickier, but is still pretty simple. The console is connected to the main controller board via a 10' cable, which gives you plenty of flexibility to place the judge who will operate the start button well away from the main controller board.

Here are the steps for connecting the judge's console:

1. **Cut away about two inches of the outer insulation from each end of the 10' length of 6-conductor thermostat wire.**

   Inside this insulation, you'll find six wires, each a different color. *Note:* If you're using wire with more than six conductors, you can safely ignore the extra colors. To avoid confusion, cut the unneeded conductors short so they won't be in the way.

2. **Strip about half an inch of insulation from each of these six wires.**

3. **Starting at the top of the barrier strip on the judge's console, connect one of the wires to each of the six terminals on the barrier strip.**

   At this point, it doesn't matter which color you connect to each terminal.

4. **Connect the six wires to the barrier strip on the main controller board, being sure to match the colors in the same order as you connected them on the judge's console.**

   In other words, if you connected the red conductor to terminal 1 on the judge's console, you must connect the red conductor to terminal 1 on the main controller console.

5. **You're done!**

Once the player buttons and judge's console are connected, the last step is to connect four lamps to the four female extension cord outlets that dangle from the relay board. For testing purposes, I simply plug four inexpensive night-lights into the outlets. If you're actually playing a game, you can use any incandescent lights you want, and you can use extension cords to locate the lights anywhere in the room.

Figure 1-20 shows the Quiz-O-Matic! all connected and ready to test.

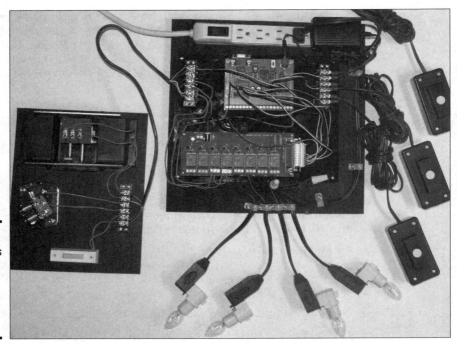

Figure 1-20:
Everything is connected; the Quiz-O-Matic! is ready to test.

Testing the Quiz-O-Matic!

Having completed the assembly of the Quiz-O-Matic!, you're probably dying to know if it works. Before you download an actual quiz show program that will be pretty complicated, it's worthwhile to test it with a very simple program that activates all the outputs and then tests all the input buttons, just to make sure all the Quiz-O-Matic!'s hardware is working as it should.

Toward that end, Listing 1-1 shows a simple test program that turns all four lights on and off for one second, and then rings the chime and sounds the buzzer. Once all of this happens, the program waits for you to press the buttons. When you press any of the buttons, the corresponding light turns on for one second, and then turns back off.

Listing 1-1: A Test Program for the Quiz-O-Matic!

```
' {$STAMP BS2}
' {$PBASIC 2.5}
'
```

```
' =====================================================================
'
'   GameShowTest.bs2
'
'   Tests the GameShow controller.
'
' =====================================================================
'
' ----- BS2 Pin Assignments ------------------------------------------
'
'   P0      Output    Kit74 Relay 1    Start Lamp
'   P1      Output    Kit74 Relay 2    Player 1 Lamp
'   P2      Output    Kit74 Relay 3    Player 2 Lamp
'   P3      Output    Kit74 Relay 4    Player 3 Lamp
'   P4      Output    Kit74 Relay 5    Buzzer
'   P5      Output    Kit74 Relay 6    Chime
'   P6                not used
'   P7                not used
'   P8                not used
'   P9                not used
'   P10               not used
'   P11               not used
'   P12     Input     Player 1 Button
'   P13     Input     Player 2 Button
'   P14     Input     Player 3 Button
'   P15     Input     Start Button
'
'
' ----- PIN symbols --------------------------------------------------
'
StartLamp       PIN     0
Player1Lamp     PIN     1
Player2Lamp     PIN     2
Player3Lamp     PIN     3
Buzzer          PIN     4
Chime           PIN     5
Player1Button   PIN     12
Player2Button   PIN     13
Player3Button   PIN     14
StartButton     PIN     15
'
' ----- Constants ----------------------------------------------------
'

LampOn          CON     1
LampOff         CON     0

BuzzerOn        CON     1
BuzzerOff       CON     0

ChimeOn         CON     1
ChimeOff        CON     0

SwitchOn        CON     1
SwitchOff       CON     0

OutputPin       CON     1

' ----- Initialization -----------------------------------------------

Reset:
  OUTS = $0000                              ' clear all outputs
```

(continued)

Listing 1-1 *(continued)*

```
DIR0 = OutputPin
DIR1 = OutputPin
DIR2 = OutputPin
DIR3 = OutputPin
DIR4 = OutputPin
DIR5 = OutputPin

Main:

'    Test the outputs

     StartLamp = LampOn
     PAUSE 1000
     StartLamp = LampOff

     Player1Lamp = LampOn
     PAUSE 1000
     Player1Lamp = LampOff

     Player2Lamp = LampOn
     PAUSE 1000
     Player2Lamp = LampOff

     Player3Lamp = LampOn
     PAUSE 1000
     Player3Lamp = LampOff

     Chime = ChimeOn
     PAUSE 20
     Chime = ChimeOff
     PAUSE 1000

     Buzzer = BuzzerOn
     PAUSE 500
     Buzzer = BuzzerOff

'    Test the inputs

InputTest:

     IF Player1Button = SwitchOn THEN
         Player1Lamp = LampOn
         PAUSE 1000
         Player1Lamp = LampOff
         GOTO InputTest
     ENDIF

     IF Player2Button = SwitchOn THEN
         Player2Lamp = LampOn
         PAUSE 1000
         Player2Lamp = LampOff
         GOTO InputTest
     ENDIF

     IF Player3Button = SwitchOn THEN
         Player3Lamp = LampOn
         PAUSE 1000
```

```
        Player3Lamp = LampOff
        GOTO InputTest
ENDIF

IF StartButton = SwitchOn THEN
    StartLamp = LampOn
    PAUSE 1000
    StartLamp = LampOff
    GOTO InputTest
ENDIF

GOTO InputTest
```

To download this program into the Quiz-O-Matic!, follow these steps:

1. **Fire up the Parallax Basic Stamp Editor.**

   If you don't have the program editor, you can download it from the Parallax website at `www.parallax.com`.

2. **Copy the program into the editor and save the file.**

   Or if you prefer, you can download the program from this book's companion website.

3. **Plug the Quiz-O-Matic! into a wall outlet, and then turn on the main switch on the electrical outlet strip on the main controller board.**

4. **Turn the power switch on the Prop-2 controller to the 2 position.**

   At this point, you should see an LED light up on the Prop-2 controller. If it doesn't, recheck your wiring to make sure that the 12 V power supply is plugged into the outlet strip and into the Prop-2 controller.

5. **Connect the serial adapter cable to your computer and the Prop-2 controller.**

   This cable allows you to download programs from your computer to the Basic Stamp microprocessor.

6. **Click the Run button in the toolbar in the Parallax Basic Stamp Editor.**

   This connects to the Prop-2 controller, downloads the program, and then runs the program.

7. **Verify that the lights flash in sequence, the chime rings, and the buzzer sounds.**

8. **Press each of the four buttons to verify that the lights come on when the buttons are pressed.**

9. **Celebrate your achievement! The Quiz-O-Matic! works!**

In case you encounter difficulties when testing the Quiz-O-Matic!, here are some tips:

✦ **If the Parallax Basic Stamp Editor can't connect to the Prop-2, the most likely problem is that the driver for the adapter isn't correctly installed.** Reread the documentation that came with the adapter and follow the steps recommended for correcting the connection problem.

✦ **If nothing seems to happen when you run the program, check to make sure the Kit-74 relay board has power.** It needs to be connected via the pigtail connector to the V+ and GND connectors on the Prop-2 board. Verify that the polarity of your connection is correct and also that the power switch on the Prop-2 board is in the 2 position.

✦ **If the Quiz-O-Matic! doesn't respond to the buttons, verify that the buttons are wired properly and make sure that the buttons aren't lit.** Lit buttons have a lamp that draws current across the two terminals of the switch. The lamp circuit can confuse the Prop-2 controller, making it believe that a player has pressed the button.

Programming the Quiz-O-Matic!

Once you've established that all the parts are working, the final step in building the Quiz-O-Matic! is to download the program that will allow it to play its games. Listing 1-2 shows the complete Quiz-O-Matic! program. This program is pretty long, so instead of keying it in yourself, you'll probably want to download it from this book's companion website. Once you've obtained this program, you can download it into the Quiz-O-Matic! by following the instructions that were in the previous section.

Before you delve deep into this program to try to understand how it works, I suggest you review the sections "Playing the game" and "Configuring the controller" earlier in this chapter. The program in Listing 1-2 implements the rules described in those sections, so having a solid understanding of those rules will go a long way toward helping you understand how the program works.

Listing 1-2: The game show program for the Quiz-O-Matic!

```
' {$STAMP BS2}
' {$PBASIC 2.5}
'
' ====================================================================
'   Quiz-O-Matic!.bs2
'
'   Quiz-O-Matic! GameShow controller software.
'
'
' ====================================================================
'
' ----- BS2 Pin Assignments -------------------------------------------
'
'    P0     Output    Kit74 Relay 1      Start Lamp
'    P1     Output    Kit74 Relay 2      Player 1 Lamp
```

```
'    P2      Output    Kit74 Relay 3    Player 2 Lamp
'    P3      Output    Kit74 Relay 4    Player 3 Lamp
'    P4      Output    Kit74 Relay 5    Buzzer
'    P5      Output    Kit74 Relay 6    Chime
'    P6                not used
'    P7                not used
'    P8                not used
'    P9                not used
'    P10               not used
'    P11               not used
'    P12     Input     Player 1 Button
'    P13     Input     Player 2 Button
'    P14     Input     Player 3 Button
'    P15     Input     Start Button
'
'
' ----- PIN symbols ------------------------------------------------
'
StartLamp       PIN     0
Player1Lamp     PIN     1
Player2Lamp     PIN     2
Player3Lamp     PIN     3
Buzzer          PIN     4
Chime           PIN     5
Player1Button   PIN     12
Player2Button   PIN     13
Player3Button   PIN     14
StartButton     PIN     15
'
'
' ----- Constants --------------------------------------------------
'
LampOn          CON     1
LampOff         CON     0

BuzzerOn        CON     1
BuzzerOff       CON     0

ChimeOn         CON     1
ChimeOff        CON     0

SwitchOn        CON     1
SwitchOff       CON     0

PenaltyOn       CON     1
PenaltyOff      CON     0

OutputPin       CON     1

OneChimeSecond  CON     200
OneAnswerSecond CON     350
'
'
' ----- Variables --------------------------------------------------
'
chimeTimerMax        VAR     Word
chimeTimer           VAR     Word
chimeSeconds         VAR     Byte

answerTimerMax       VAR     Word
answerTimer          VAR     Word
```

(continued)

**Book VIII
Chapter 1**

**Building a
Quiz-O-Matic!**

Listing 1-2 *(continued)*

```
answerSeconds          VAR     Byte

penaltyTimerMax        VAR     Word

Player1Penalty         VAR     Bit
Player2Penalty         VAR     Bit
Player3Penalty         VAR     Bit
'
'
' ----- Initialization -----------------------------------------------
'
Reset:
  OUTS = $0000                    ' clear all outputs
  DIR0 = OutputPin                ' configures I/O pins 0 through 5
  DIR1 = OutputPin                ' for use as outputs
  DIR2 = OutputPin
  DIR3 = OutputPin
  DIR4 = OutputPin
  DIR5 = OutputPin

  GOSUB Setup


Main:

' Turn off all lamps

    Player1Lamp = LampOff
    Player2Lamp = LampOff
    Player3Lamp = LampOff
    StartLamp   = LampOff

' Reset timers

    ChimeTimer = 0
    AnswerTimer = 0

' Reset penalties

    Player1Penalty = PenaltyOff
    Player2Penalty = PenaltyOff
    Player3Penalty = PenaltyOff

' Wait for Start button to be pressed

WaitForStart:
    IF StartButton = SwitchOff THEN
      IF Player1Button = SwitchON THEN
          Player1Penalty = PenaltyOn
      ENDIF
      IF Player2Button = SwitchON THEN
          Player2Penalty = PenaltyOn
      ENDIF
      IF Player3Button = SwitchON THEN
          Player3Penalty = PenaltyOn
      ENDIF

      GOTO WaitForStart
    ENDIF
```

```
        StartLamp = LampOn

' Wait for Player to chime in

WaitForPlayer:

    IF Player1Penalty = PenaltyOff THEN
      IF Player1Button = SwitchOn THEN
        Player1Lamp = LampOn
        GOSUB SoundChime
        GOTO WaitForTurnOver
      ENDIF
    ENDIF

    IF Player2Penalty = PenaltyOff THEN
      IF Player2Button = SwitchOn THEN
        Player2Lamp = LampOn
        GOSUB SoundChime
        GOTO WaitForTurnOver
      ENDIF
    ENDIF

    IF Player3Penalty = PenaltyOff THEN
      IF Player3Button = SwitchOn THEN
        Player3Lamp = LampOn
        GOSUB SoundChime
        GOTO WaitForTurnOver
      ENDIF
    ENDIF

    chimeTimer = chimeTimer + 1

    IF chimeTimer = PenaltyTimerMax THEN
        Player1Penalty = PenaltyOff
        Player2Penalty = PenaltyOff
        Player3Penalty = PenaltyOff
    ENDIF

    IF chimeTimer >= chimeTimerMax THEN
      GOSUB SingleBuzzer
      GOTO Main
    ENDIF

    GOTO WaitForPlayer

' Wait for judge to press Start button

WaitForTurnOver:
    IF StartButton = SwitchOff THEN
      AnswerTimer = AnswerTimer + 1
      IF AnswerTimer >= AnswerTimerMax THEN
        GOSUB DoubleBuzzer
        GOTO Main
      ENDIF
      PAUSE 1
      GOTO WaitForTurnOver
    ENDIF
```

(continued)

Listing 1-2 *(continued)*

```
    PAUSE 500
    GOTO Main


' ----- Subroutines -------------------------------------------------

SoundChime:
    Chime = ChimeOn
    PAUSE 20
    Chime = ChimeOff
    RETURN

SingleBuzzer:
    Buzzer = BuzzerOn
    PAUSE 200
    Buzzer = BuzzerOff
    RETURN

DoubleBuzzer:
    Buzzer = BuzzerOn
    PAUSE 50
    Buzzer = BuzzerOff
    PAUSE 100
    Buzzer = BuzzerOn
    PAUSE 100
    Buzzer = BuzzerOff
    RETURN


Setup:

    , Get the number of seconds for the player chime timer
    , Minimum is 5 seconds
    , Each press of the Player 1 button adds 1 second

    GOSUB SoundChime
    chimeSeconds = 5
    DO
      IF Player1Button = SwitchOn THEN
        chimeSeconds = chimeSeconds + 1
        GOSUB SingleBuzzer
      ENDIF
    LOOP UNTIL StartButton = SwitchOn
    PAUSE 250

    , Get the number of seconds for the answer timer
    , Minimum is 10 seconds
    , Each press of the Player 1 button adds 5 seconds

    GOSUB SoundChime
    answerSeconds = 10
    DO
      IF Player1Button = SwitchOn THEN
        answerSeconds = answerSeconds + 5
        GOSUB SingleBuzzer
      ENDIF
    LOOP UNTIL StartButton = SwitchOn
    PAUSE 250

    chimeTimerMax = OneChimeSecond * chimeSeconds
```

```
chimeTimer = 0
penaltyTimerMax = OneChimeSecond * 2

answerTimerMax = OneAnswerSecond * answerSeconds
answerTimer = 0


GOSUB SoundChime
PAUSE 500
GOSUB SoundChime
PAUSE 500
GOSUB SoundChime

RETURN
```

Because this is an electronics book, not a computer programming book, I don't describe the operation of this program in detail. The program is filled with comments, so you shouldn't have much trouble following its logic. I would, however, like to point out a few of the more interesting details:

✦ The I/O pins on a Basic Stamp are configured by default to work as inputs. The DIR assignments in the initialization section of the program are required to establish pins 0 through 5 as outputs.

✦ The timing aspects of the program are a bit tricky because a Basic Stamp microprocessor doesn't have a real-time clock. So the sections of the program that wait for players to press buttons increment counters that are tested against constants whose values I determined by trial and error. In other words, I used a stopwatch to determine how long the program waited for a player to press a button before sounding the buzzer. Then, I adjusted the counter constant until I found a value that came close to the five-second interval I was looking for.

✦ Technically speaking, the portion of the code that determines which player pushes his or her button first isn't 100 percent accurate. Each of the IF statements that check to see whether a player's button has been pressed takes a few milliseconds to process. So it's remotely possible that player 2 could press his or her button a few milliseconds before player 1, and the program would award the question to player 1.

It would be possible to program the game to be more fair in this regard, but doing so would require advanced programming techniques that are a bit beyond the reach of this book. I think you'll agree that the program is complicated enough as it is! If you'd like an additional challenge, look into this problem and see if you can devise a way of detecting which player pressed first with no margin of error.

Chapter 2: Building a Color Organ

In This Chapter

- ✔ **Creating a fake thunderstorm**
- ✔ **Powering lights with a rickety old generator**
- ✔ **Making a skeleton's heart beat**

*O*ne of the great things about Disneyland is that sometimes the long line you have to wait in to go on a particular ride is almost as good as the ride itself. One of the best examples of this is the famous *Indiana Jones Adventure: Temple of the Forbidden Eye*. Just outside of an ancient temple, you pass by a rickety steam-powered generator that is barely running. The clickity-clickity sound of the generator alternately grows louder and softer as the generator sputters and threatens. Once inside the temple, you pass through narrow tunnels and creepy caverns that are lit overhead by lights that appear to be powered by the rickety generator. The lights flicker and dim, then grow brighter for a moment, then flicker and dim again in sync with the laboring generator.

In this chapter, you learn how to build an electronic circuit you can use to create this creepy lighting effect. I've used this circuit for many different purposes, including lighting the narrow passageways in my own haunted tomb at Halloween. I've also used the same circuit to create a thunderstorm in my front yard to add the right ambiance to my haunted Halloween grave-yard. And the same circuit creates a spooky red heartbeat in the chest of a plastic skeleton that stands watch over the scene.

The circuit is called a *color organ*, and its operation is simple: It converts the volume of an audio input into an output voltage that gets higher as the sound source gets louder. If you connect a light to the output, the light will glow brighter when the audio input is louder and dimmer when the input is quieter.

As with most of the chapters in this minibook, I need to start with a warning that this circuit requires that you work with line-level voltages (120 VAC), so it's potentially dangerous. The circuit is designed with safeguards, but you must be careful to not bypass them. You should inspect the project every time you use it to make sure none of the wiring has come loose or frayed, and you must never work on the circuit while it's plugged in.

Examining the Color Organ Project

The completed color organ described in this chapter is shown in Figure 2-1. The project is housed in a 2 x 3 x 6" project box that includes a power plug that connects the circuit to a wall outlet, a power outlet that you can connect a light to (maximum 120 W), an RCA-style audio input connector that you can connect a sound source to (up to 60 W), a knob for adjusting sensitivity, and a power switch.

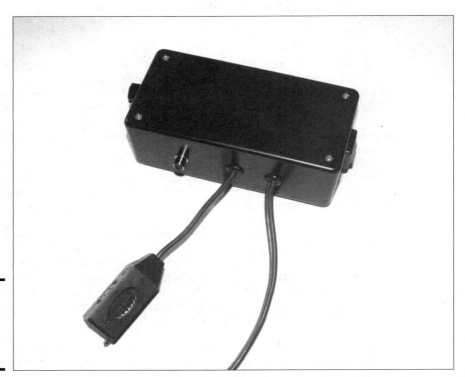

Figure 2-1: The completed color organ project.

To keep the project simple, you can build the electronics using an inexpensive kit — specifically, the Velleman MK110 Simple One Channel Light Organ kit. This kit is widely available on the Internet for under $10; just search for *Velleman MK110*, and you'll find several sources.

Figure 2-2 shows how the color organ project can be hooked up to create light that varies in brightness with a sound source. You'll need a source for the sound, such as a boom box or other sound system with external speaker outputs that you can tap into.

Note that the diagram in Figure 2-2 doesn't show the details of how to connect the sound system to the color organ and the speakers. The easiest way to connect the color organ to the sound system is to simply replace one of the speakers with the color organ. That way, the color organ responds to one of the stereo channels while the speaker plays the sound coming from the other stereo channel. For this type of hookup, you'll need just a single cable that has an RCA connector on one end (to plug into the color organ) and the proper connector on the other end to connect to the sound system's speaker output.

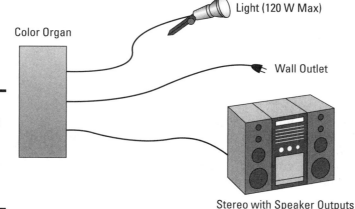

Figure 2-2:
Connecting the color organ to a light and a sound source.

You'll also need a suitable recording for your sound source. For example, if you want to use the color organ to create a thunderstorm effect, you'll need a recording of thunder. To make a red light flash in sync with a heartbeat, you'll need a recording of a heartbeat. The Internet is rife with such sound effects, so you shouldn't have too much trouble locating and downloading a sound effect to meet your needs. If you want to customize the sound effect, you can download a free audio editor called Audacity from `http://audacity.sourceforge.net`.

**Book VIII
Chapter 2**

Building a Color Organ

Understanding How the Color Organ Works

There are several different ways to design a color organ circuit. Most of them rely on a special type of electronic component called a *triac*, which is essentially a transistor that's designed to work with alternating current. It has three terminals. Two are anodes, called A1 and A2, and the third is a gate. A voltage at the gate — either positive or negative — allows the anodes to conduct. The anodes are connected to the line load, and the gate voltage is derived from the audio input.

The audio input isn't connected directly to the triac gate, however. Instead, most color organs use one of two techniques to isolate the audio input from the line-voltage side of the circuit. One method is to use a transformer. The other is to use an *optoisolator,* which is a single component that consists of an infrared LED and a photodiode or other light-sensitive semiconductor. Voltage on the LED causes the LED to emit light, which is detected by the photodiode and passed on to the output circuit.

The Velleman MK110 kit uses an optoisolator triac, in which the photosensitive semiconductor is actually a triac whose gate is stimulated by light rather than by voltage. The optoisolator is an integrated circuit in a 6-pin DIP package.

Figure 2-3 shows a simplified schematic diagram for the circuit used by the Velleman MK110 kit. As you can see, the audio input is applied to the LED side of the optoisolator, controlled by a potentiometer, which lets you adjust the sensitivity of the circuit. The output from the optoisolator is applied to the gate of the triac, whose anodes are connected across the line voltage circuit. Thus, the volume of the audio input directly controls the voltage of the output circuit.

Figure 2-3:
The schematic diagram for the color organ circuit.

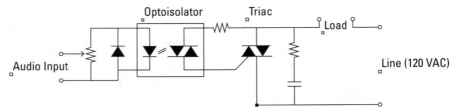

Getting What You Need to Build the Color Organ

Other than the Velleman kit itself, most of the materials you need to build the color organ project can be purchased at your local RadioShack store or any other supplier of electronic components. The table below lists all of the materials you'll need.

Quantity	Description
1	Velleman MK110 Simple Onee Channel Light Organ kit
1	2 x 3 x 6" project box (RadioShack part 270-1805)
1	20 mm PC board standoffs (package of 4, RadioShack part 276-195)
1	RCA-style phono jack (RadioShack part 274-0346)
1	¾" control knob (RadioShack part 274-415)
1	Chassis-type fuse holder for 1-¼" x ¼" fuses (RadioShack part 270-0739)
1	1 A, 250 V, fast-acting 1-¼" x ¼" fuse (RadioShack part 270-1005)
1	⅜" 4-40 machine screw and nut (for mounting the fuse holder)
1	SPST rocker switch (RadioShack part 275-694)
2	⅜" grommets (RadioShack part 64-3025)
2	Screw-on wire connectors (RadioShack part 64-3057)
5"	20-gauge stranded hook-up wire
1	Indoor extension cord

Assembling the Color Organ

Once you have gathered all the materials for the color organ, you're ready to assemble the project. You'll need the following tools:

+ Soldering iron, preferably with both 20 and 40 W settings

+ Solder

Use thicker solder for the line-voltage wires and thin solder for assembling the MK110 kit.

+ Magnifying goggles
+ Phillips screwdriver
+ Small flat-edge jewelers screwdriver
+ Wire cutters
+ Wire strippers
+ Pliers
+ Hobby vise
+ Drill with ⅛", ⁵⁄₃₂", ¼", ⁵⁄₁₆", ⅜", and ¾" bits

Here are the steps for constructing this project:

1. Assemble the Velleman MK110 kit.

The kit comes with simple but accurate instructions. Basically, you just mount and solder all the components onto the circuit board. Pay especial attention to the color codes for the resistors and the orientation of the diode.

Figure 2-4 shows the completed MK110 kit.

Figure 2-4:
The assembled Velleman MK110 kit.

When I build this kit, I find it best to mount the circuit board in a good hobby vise and use an alligator clip or masking tape to hold the components in place while soldering.

2. **Drill all the mounting holes in the project box except the hole for the sensitivity control on the left side of the box.**

   Figure 2-5 shows the orientation of the approximate location of the mounting holes.

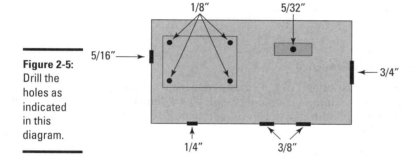

Figure 2-5: Drill the holes as indicated in this diagram.

   Use the assembled circuit board to determine the exact drilling locations for the four holes that will mount the circuit board. The position of the other holes isn't critical, with the exception of the hole for the potentiometer knob. Don't drill that hole until Step 4.

3. **Mount the four standoffs in the four MK110 circuit board mounting holes.**

   Use four of the machine screws that came with the standoffs.

4. **Drill the hole for the circuit board's potentiometer.**

   Set the circuit board on top of the four standoffs to determine the exact location for this hole.

5. **Insert the two rubber grommets into the two 3/8" holes.**

   The grommets are difficult to squeeze into the hole, but work at it and you'll get them in. If necessary, use the small edge of a flat screwdriver to push the rubber edges into the holes.

In the steps that follow, you assemble all the parts into the project box. Use Figure 2-6 as a guide for the proper placement of each of the parts.

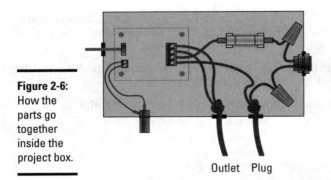

Figure 2-6:
How the
parts go
together
inside the
project box.

Outlet   Plug

1. Cut the extension cord.

First cut the outlet end of the extension cord, leaving about 12" of wire attached to the outlet. Then cut the plug end, leaving about three or four feet of wire attached to the plug. You'll have a few feet of wire left over; set this wire aside for later.

2. Push the power cords through the grommets and tie a knot inside the box.

It will be a tight squeeze, but the cords will fit. Pull about a foot of the cord with the plug attached through the hole nearest the switch. Then, tie it in a knot, cinch the knot down tight, and pull the plug so that the knot is snug against the grommet. The knot acts as a strain relief.

Repeat the same process with the cord that's attached to the outlet, passing it through the other grommet, tying a tight knot, and pulling the knot up against the grommet.

When both power cords are in place, separate the two wires of each cord inside the project box and strip about ⅜" of insulation from each wire.

3. Cut two, 1.5" lengths of extension cord wire and solder them to the switch terminals.

You'll need to strip about ⅜" of insulation from each end of both wires. Put your soldering iron on its High setting and use thick solder. Set the switch aside when the solder sets.

4. Cut two 1.5" lengths of extension cord wire and solder them to the terminals on the fuse holder.

Again, you'll need to strip about ⅜" of insulation from each end of both wires and solder with high heat.

5. Cut two, 2.5" lengths of the hookup wire and strip ⅜" of insulation from the ends.

6. **Solder one of the hookup wires to the center terminal of the RCA-style phono jack and the other wire to the ground terminal.**

At this point, you're done with the soldering iron, so you can turn it off.

7. **Mount the RCA-style phono jack in the ¼" hole in the project box.**

To mount the jack, you'll first have to remove the nut, the ground terminal, and the lock washer from the jack. Then, pass the wire connected to the center terminal of the phono jack through the ¼" hole, and then insert the threaded end of the phono jack into the hole. Slip the lock washer, the ground terminal, and the nut over the wire connected to the center terminal, and then thread them onto the threaded part of the jack. Tighten with needle-nose pliers.

In the next few steps, you attach wires to the MK110 circuit card. Do *not* mount the circuit board to the standoffs quite yet. You'll have an easier time connecting the wires if the circuit board is loose. After the wires are all connected, you mount the board.

1. **Connect the separated wires of the cord that's attached to the outlet to the two terminals marked *Load* on the back of the MK110 circuit board.**

Use a small, flat screwdriver to tighten the terminals. Make sure the wires are securely connected.

2. **Connect one of the wires attached to the fuse holder to one of the Mains terminals at the back of the MK110 circuit board.**

3. **Connect one of the extension cord wires that's attached to the plug to the other Mains terminal at the back of the MK110 circuit board.**

4. **Connect the two hook-up wires from the phono jack to the input terminals at the front of the MK110 circuit board.**

The input terminals are labeled *LS* on the board. You'll need a very small flat-blade screwdriver to tighten these terminals.

5. **Mount the MK110 circuit board on the standoffs.**

To mount the board, you'll need to tilt it a bit to slide the shaft of the potentiometer through the ⁵⁄₁₆" hole. Once the shaft is through, set the board down on the standoffs and secure it with the remaining four machine screws that came with the standoffs.

6. **Use the ³⁄₈" 4-40 machine screw and nut to mount the fuse holder.**

Slide the machine screw through the ⁵⁄₃₂" hole in the bottom of the project box. Then pass the machine screw through the hole in the center of the fuse holder and attach the nut. Tighten with a screwdriver.

7. Mount the switch.

To mount the switch, first remove the plastic nut on the threaded end of the switch. Then, pass the wires and the threaded end of the switch through the ¾" hole in the side of the project box. Finally, slip the nut over the wires and tighten it onto the switch.

8. Connect the switch to the power cord and the fuse.

Use one of the screw-on wire connectors to connect one of the switch wires to the unconnected wire on the fuse holder. Then, use the other wire connector to connect the other switch wire to the unconnected wire that leads to the power plug.

9. Insert the fuse in the fuse holder.

Guess what — you're almost done! Figure 2-7 shows the project with all the parts assembled.

Figure 2-7: The color organ is almost finished.

10. Attach the knob to the potentiometer shaft protruding from the box.

Use a small, flat screwdriver to tighten the set screw on the knob.

11. Place the lid over the project box and secure it with the provided screws.

Now you really are done!

Using the Color Organ

Having finished the color organ, it's time to put it to use making interesting sound and lighting effects. To do that, you'll have to connect both lights and a sound system to the color organ.

The procedure for actually using the color organ is quite simple:

1. **Connect a light to the color organ's female extension cord connector.**
2. **Connect a speaker-level audio input to the RCA connector.**
3. **Plug the male extension cord connector into a power outlet.**
4. **Turn on the color organ.**
5. **Play the sound.**
6. **Turn the knob on the color organ to adjust the sensitivity.**
7. **If the light never comes on, try increasing the output volume on the stereo.**

The color organ can handle just 120 W on the output circuit, so you need to be careful not to overload the circuit. You can use a single 100 W flood light or a couple of 60 W lamps. Or you can use a few strings of Christmas lights or other low-wattage lights.

As I mention earlier in this chapter, the easiest way to connect the color organ to the sound system is to simply replace one of the speakers with the color organ. The type of cable you'll need to do that depends on how the speakers connect to the sound system. If the speakers connect with simple post connectors, you'll need a cable with bare wire on one end and an RCA plug on the other end. If the speakers connect with RCA connectors, you'll need a cable with RCA plugs on both ends.

When you use the color organ in this way, it's important to realize that the color organ will respond to one channel of the stereo recording while the speaker plays the other channel. In most cases, you won't notice much difference. However, in some recordings, the sound on the left channel is very different from the sound on the right channel. This can affect the quality of the sound, and it can also prevent the light from flashing in perfect sync with the sound, since the light is responding to a different sound source from the one heard through the speakers.

In some cases, you can use this to improve the effect you're trying to achieve with the color organ. For example, in an actual thunderstorm, the lightning flashes well before the thunder is heard. To reproduce this effect, all you need is a sound recording of a thunderstorm in which the thunder is heard on the left channel before it's heard on the right channel. Then, if you connect the color organ to the left channel and the speaker to the right channel, the light will flash before the sound is heard.

Have fun with your color organ!

Chapter 3: Animating Holiday Lights

In This Chapter

✔ Decorating your house with computer-controlled flashy holiday lights

✔ Looking at Light-O-Rama controllers

✔ Programming a light sequence with Light-O-Rama's Sequence Editor software

*H*ave you seen those crazy animated light shows some people put up for the holidays, in which thousands of lights (or maybe tens of thousands or in some cases hundreds of thousands) are sequenced in time with music to create dazzling displays? Usually the lights are sequenced to cool holiday music by Mannheim Steamroller or Trans-Siberian Orchestra. Often the music is broadcast over a low-wattage FM transmitter, so when you drive by the house to see the show, you can just tune your radio to the local broadcast to hear the music. If you're not sure what I'm talking about, just hop onto YouTube and search for *Christmas lights.* You'll find plenty of videos.

If you've ever wanted to create such a display yourself, this chapter will show you how. Although you can design the circuitry to control light shows like this from scratch, the easiest way to build a light show is to buy an inexpensive lighting controller. That way, you can focus your energies on the design of the show rather than on the design of the lighting controller.

There are several companies that sell lighting controllers for holiday displays. My favorite is Light-O-Rama, based in South Glen Falls, New York. What I like most about Light-O-Rama is that you can purchase preassembled lighting controllers, or you can purchase kits and assemble the controllers yourself. So if you're handy with a soldering iron (and I assume that you are, or you wouldn't be reading this book), you can save a few hundred dollars on a basic sequencer by building the circuit yourself.

For complete information about Light-O-Rama and their products, point your browser to www.lightorama.com.

Light-O-Rama controllers are useful for much more than just holiday displays; they're also useful for school events, carnivals, amusement parks, store and museum displays, theatrical productions, malls and shopping centers, and so on. In fact, the Light-O-Rama products shown in this chapter were donated by the good folks at Light-O-Rama for use at the Fresno Chaffee Zoo.

Introducing the ShowTime PC Controller

Light-O-Rama's most popular lighting controller for residential use is called the *ShowTime PC* controller, shown in Figure 3-1. This controller (model number CTB16PC) offers the following basic features:

✦ **Sixteen separate *channels* of light:** Each channel is a separate 120 VAC circuit that can power up to 8 amps of lights. The power cords dangling from the bottom of the unit are where you plug the lights in; each cord connects the lights for one of the 16 channels. (The total current load for the controller should not exceed 30 amps.)

In most cases, you won't plug your lights directly into the power cords dangling from the bottom of the lighting controller. Instead, you'll use extension cords to reach from the lighting cords to where the lights are actually placed. If you plan on making a holiday lighting display, you'll need a lot of extension cords.

Sixteen channels is enough to get started with Light-O-Rama displays, but once you've set up your first show, you'll wish you had additional channels. Most Light-O-Rama setups include two, three, or four controllers for a total of 32, 48, or 64 channels.

✦ **The ability to turn the lights on and off, separate for each channel:** The controller also has a few other special effects, such as fade up, fade down, different intensity levels, twinkling, and shimmering.

These special effects are what make the Light-O-Rama controller special. You can find much less expensive ways to simply turn lights on and off. For example, two Kit-74 relay controllers connected to a computer can turn 16 channels of lights on and off for less than half the price of a ShowTime PC controller. However, the Kit-74 can't fade the lights up or down, and the ability to gradually fade the lights adds a lot to the impact of the light show.

✦ **Computer control of the light show:** You simply connect the ShowTime PC controller to the computer via a special USB adapter cable. Then, you run special software available from Light-O-Rama to control the show.

Although the ShowTime PC controller is mounted in a weatherproof container, your computer isn't. As a result, you'll want to place the computer inside your house or garage, then use a long cable to reach the ShowTime PC controller.

✦ **Light show control without a computer:** If you don't want to drive the show from a computer, you can purchase a special device called an *MP3 Director* that lets you control the show without a computer.

✦ **Expandability:** You can connect as many as 240 ShowTime PC controllers together to create a massive show with as many as 3,840 separate light circuits. You'd need a small nuclear reactor to power it all, but it's possible.

Figure 3-1:
The basic ShowTime PC controller from Light-O-Rama.

If you want to get started with Light-O-Rama, I suggest you purchase one of their starter kits. You can purchase a completely preassembled kit including a 16-channel ShowTime PC controller, USB adapter, and software for under $400, or you can purchase built-it-yourself kits for considerably less. These kits require you to solder the components to the board, but assembling the board yourself will give you the satisfaction of knowing you built it yourself.

Looking at a Basic Light-O-Rama Setup

Figure 3-2 shows a basic setup for using a single, 16-channel ShowTime PC controller to control up to 16 separate strands of lights. The following paragraphs describe each of the elements in this setup:

✦ **ShowTime PC controller:** Usually located outside, close to where your lights are. It comes with a weatherproof enclosure so you can place it outside.

Although it isn't shown in the figure, the ShowTime PC controller requires a source of AC power. As a result, you should locate the controller near an electrical outlet.

The controller has two electrical plugs; each provides power to half of the 16 light channels. If you plug both of these cords into the same electrical outlet, the total lighting capacity of the controller will be limited by the maximum amperage rating of the outlet you plug the cords into (typically 15 A). However, you can plug the two cords into separate outlets to double the lighting capacity of the controller to 30 A, provided that the two outlets you plug the controller into are located on separate electrical circuits.

✦ **Lights:** Connected to the 16 power cords that hang from the bottom of the controller. For more information about choosing lights, see the section "Choosing Lights for Your Display" later in this chapter.

✦ **Computer:** Runs the ShowTime software that controls the lights. You'll want to place the computer in a secure location that isn't exposed to weather.

Unfortunately, Light-O-Rama's software isn't free. If you purchase one of Light-O-Rama's starter kits, the software is included in the purchase price, but if you purchase a do-it-yourself kit and assemble the circuit board yourself, you'll have to buy the software separately. (The cost is under $50.)

For more about using the Light-O-Rama software, see the section "Using the Light-O-Rama Software" later in this chapter.

✦ **USB adapter:** Required to connect the computer to the ShowTime PC controller. For more information about the USB adapter, see the section "Connecting the Controller to a Computer" later in this chapter.

✦ **Sound system:** Plays or broadcasts the sound that is synchronized to the lights. The sound system connects to the computer's headphone output and either amplifies it for speakers or broadcasts it so it can be heard on an FM radio.

If you want to play the sound through speakers, you can use any amplifier that has an input jack and is powerful enough to play the sound at the volume you desire. I have a guitar amplifier that I usually use, but I have also used a boom box.

If you want to broadcast the music so that people can listen to it on their car radios as they drive by your house, you can purchase a low-power FM broadcaster from many sources on the Internet. Light-O-Rama sells one for about $125.

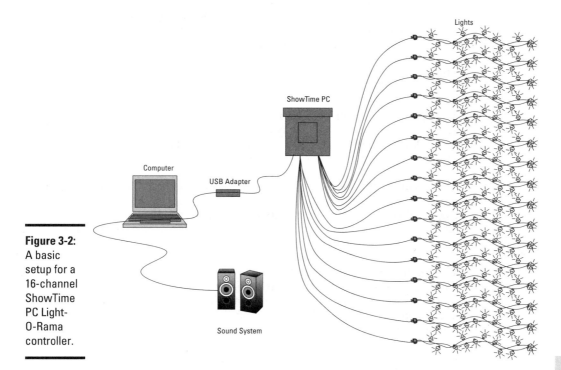

Figure 3-2:
A basic
setup for a
16-channel
ShowTime
PC Light-
O-Rama
controller.

Understanding Channels and Sequences

In a ShowTime PC controller or other similar holiday lighting controller, lights are controlled via *channels*. Each channel is a separate electrical circuit to which you can connect one or more strings of lights. It's important to keep in mind that the controller can't control the individual lights that are connected to a channel; all the lights on a single channel operate together as a single unit.

A *sequence* is simply a recorded program that activates the lights connected to each of the controller's channels in a particular order, most often synchronized with music. The art of designing a good holiday light display lies in creating clever sequences that squeeze the most impact out of a limited number of channels.

When planning a light show, one of the first steps is to determine how many channels your show requires and what lighting elements will be controlled by each channel. For example, you might run several strings of icicle lights across the roofline of your house and connect those lights to one of the channels. Then, the controller can turn the icicle lights on or off together as a unit. Or you might put colored lights on a shrub connected to its own channel. Then, the controller can turn the shrub on or off.

There's no reason you can't physically overlap lights that are connected to different channels. For example, you might put a string of green lights connected to one channel on a shrub and place a string of red lights on the same shrub but connected to a different channel. Then, the controller can make the shrub glow green, glow red, or glow green and red at the same time. Or the controller might make the shrub alternately flash green and red, in rhythm with the music of course.

There's also nothing preventing you from connecting different colors or kinds of lights in different areas of your yard to the same controller. For example, you might use red lights on one shrub and green lights on another shrub on the same channel. Then, the controller can turn both shrubs on or off — one green, the other red.

You might even put a string of green lights and a string of red lights on both shrubs, but connect the green string on the first shrub and the red string on the second shrub to one channel and the red string on the first shrub and the green string on the second shrub to another channel. Then, you can create a sequence that alternately flashes the shrub green and red.

Keep in mind that the controller can do more than just turn lights on and off. For example, the controller can cause each channel to fade up or down at any speed you want. If you have green and red lights on a single shrub, you can create a sequence that gradually changes the shrub from green to red, then back to green.

You can also use channels to create the effect of motion in your light show. As a simple example, suppose you set up several strings of light radiating from a central point, like spokes on a wheel. If the lights on each spoke are connected to several channels, you can create the illusion that the wheel is spinning by activating each spoke in sequence.

The possibilities are limited only by your creativity — and the number of channels you have available to you. More channels are better because having more channels lets you create more complicated sequences. If you start with a 16-channel controller, it won't be long before you want to expand it to 32 channels so you can create more creative sequences.

Choosing Lights for Your Display

There are many different types of lights you can use with a Light-O-Rama or other lighting controller. Take a trip to any department store during the holiday season and you'll find a wide variety of lights to choose from, or you can purchase lights from online sources such as www.1000bulbs.com or www.christmaslightsetc.com.

The best time to buy lights for your display is the day after Christmas, when most retail stores mark them down at least 50 percent.

The following paragraphs describe the most common types of lights used in holiday displays:

+ **Incandescent minilights:** Strings of 100 or more lights in various colors. These lights are readily available in white, red, green, blue, violet, and yellow. You can also find multicolored strings, which contain a mix of colors. If you hunt around, you may also find other colors. During the Halloween season, you can find orange and purple minilights.

   Minilights are designed to run on 2.5 V AC. They're wired in series with 50 lights on each run, which enables them to operate on 120 VAC. The drawback of this arrangement is that because the lights are wired in series, if one of the lights pops out of its socket or comes loose, continuity is broken for the entire run, and all 50 lights will go out.

   You can connect strings of minilights end-to-end on a single channel, but you shouldn't connect more than five strings of 100 lights together end-to-end or you'll blow the fuses that are built in to the plugs.

+ **Net lights:** These are minilights woven together into a meshlike net that can be spread over a shrub or other small bush. Usually, each net has 300 lights.

+ **LED lights:** Usually the same size as minilights but made with LEDs rather than incandescent light bulbs. LEDs consume much less power than incandescent lights; in fact, an entire string of LED lights consumes about the same amount of power as a single minilight, and you can connect as many as 20 strings together end-to-end. Thus, you can connect more LEDs to a single channel. However, LED lights are a lot more expensive than incandescent minilights.

✦ **C7 lights:** Incandescent lights that are the same size as night lights. Strings of C7 lights are usually made with 25 sockets per string. (The designation *C7* refers to the size of the base.)

C7 lights operate on 120 VAC, so they're wired together in parallel. If one goes out or if you remove one, the rest of the string will still light up. You can connect C7 strings end-to-end, but you should connect no more than three strings together to avoid overloading the circuit and blowing out the fuses.

✦ **C9 lights:** Similar to C7 lights but bigger.

✦ **Rope lights:** Strings of incandescent minilights or LEDs enclosed in a flexible transparent tube. You can purchase short lengths (typically 18') of rope light at retail stores, or you can buy 150" commercial-grade spools of rope light online from sources such as `www.1000bulbs.com` or`www.christmaslightsetc.com`.

✦ **Blow-mold figures:** Sculptures made of translucent plastic with a light bulb inside. Most blow-mold figures have a single incandescent light bulb inside, typically 40 or 60 W. If you have several blow-mold figures in your display, you can connect them all to a single channel on your controller to turn them on or off together, or, if you have enough channels at your disposal, you can connect each to a separate channel to control each individually. (See why you need more channels?)

✦ **Wire-frame sculptures:** Sculptures made from thick wires that are bent and welded into shapes such as deer, Christmas trees, presents, and so on. They are then lit by strings of incandescent minilights or LED lights.

✦ **Megatree:** A popular type of light sculpture that is simple to make. All you need is a tall pole and several strings of lights. Attach one end of each string of lights to the top of the pole, and then use stakes to plant the other end of the light strings into the ground, forming a ring around the base of the pole. The more strings of light you use, the better. If you use alternating colors of lights and connect each color to a separate channel, you can create some interesting animation effects.

Designing Your Layout

One of the most important steps for creating a holiday light display is designing the layout for your lights. To do that, first start with a diagram of your yard (or whatever space you'll be using for your display). It doesn't have to be exactly to scale, but it will help if the proportions are relatively accurate. If you want, grab a tape measure, make some rough measurements, and sketch out your space on a piece of graph paper. Then, sketch in where you'd like to place your lights and other decorations such as wire sculptures, blow-mold figures, or a megatree.

Once you've placed your lighting elements where you want them, start assigning channel numbers to each element. If you have a 16-channel controller, assign each lighting element a number from 1 to 16.

Assembling the ShowTime PC Controller

If you purchase the ShowTime PC controller from Light-O-Rama as a do-it-yourself kit, you'll have to assemble it yourself. That means you'll have to solder all the components to the main circuit board, install the board into the weatherproof container, and connect all the various cords that are required for the unit to work.

I'm not going to repeat the assembly instructions that come with the kit here because the instructions that come with the kit are excellent. But I would like to offer a few tips that may help ease the process:

✦ **Allow plenty of time.** Even if you're skilled with a soldering iron, the circuit has a lot of components, and it may well take you two or three evenings to completely assemble the kit.

✦ **Read through the entire assembly instructions** before you start. That way, there won't be any surprises.

✦ **Clear off your workbench.** There are a million little parts in the kit, so you need plenty of space to spread your stuff out.

✦ **Organize the parts.** Get some small containers to hold all the various parts and help keep them organized. I like to use small disposable plastic bowls.

✦ **Secure parts on the board before you solder.** Get some blue painter's tape to hold parts to the board while you solder them in place.

✦ **Turn off the TV.** The last thing you need while you're assembling a complicated kit is distraction.

✦ **Use lighted magnifying goggles.** They make your work a lot easier.

✦ **Secure the circuit board.** A third-hand tool or, better yet, a hobby vice will come in handy to hold the circuit board while you solder.

✦ **Use the proper soldering tools and techniques.** Make sure you use a low-wattage soldering iron (25 W) and fine-gauge solder for the sensitive parts, especially the integrated circuits. The ICs are soldered directly to the circuit board and can be damaged if you heat them too much.

Figure 3-3 shows the circuit board being assembled.

Figure 3-3:
Assembling the ShowTime PC circuit board.

After you've assembled the circuit board, you can install it in the weatherproof plastic box and connect the power cords. Again, Light-O-Rama provides excellent illustrated instructions for this task, so I won't duplicate them here.

One other bit of advice I will offer, however, is to pick up a set of numbered electrician's stickers so that you can properly label each of the 16 power cords. You can find these stickers at any home-improvement store that stocks electrical items. If you don't label the cords as you install them, you won't know which cord corresponds to each of the 16 channels.

Connecting the Controller to a Computer

The ShowTime PC controller communicates with the outside world using a digital protocol called RS-485. Most computers don't have an RS-485 port but do have a USB port. The USB adapter converts the USB signals to RS-485 signals so that the computer can communicate with the ShowTime PC controller. Figure 3-4 shows the USB adapter.

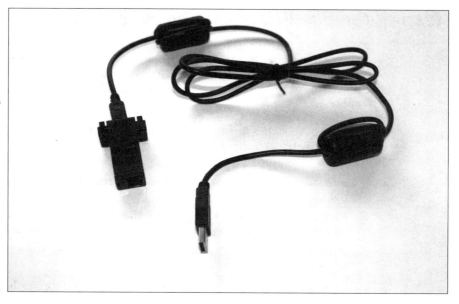

Figure 3-4:
A USB adapter converts the RS-485 protocol required by the ShowTime PC controller to USB.

The USB adapter comes with a short USB cable used to connect the adapter to the USB port on a computer as well as a Cat5 network cable used to connect the USB adapter to the ShowTime PC controller. Cat5 cable is the same type of cable used by most computer networks. The Cat5 cable can be as long as 4,000'. A cable that long might be suitable for the likes of Disneyland, but you'll probably want to keep your controller within a few dozen feet of the computer.

Light-O-Rama also sells a serial adapter that converts RS-485 signals to the standard serial port protocol that was once commonplace on PCs. Most computers don't come with serial ports anymore, so you'll probably use the USB adapter instead. (Note that if you use the serial adapter, the maximum length of the Cat5 cable that connects the adapter to the controller is 100'.

It's very important that you follow the instructions that come with the USB adapter *before* you plug the adapter in to a USB port on your computer. Specifically, you should first run the driver installation software from the CD that came with the USB adapter. Plug the adapter into a USB port only after you've run the driver installation program. If you plug the adapter into a USB port before you run the driver program, there's a very good chance that Windows will install the wrong driver for the adapter, and that driver will be associated with the adapter, making it difficult to get the correct driver installed.

**Book VIII
Chapter 3**

Animating Holiday Lights

Testing the ShowTime PC Controller

When you've finished assembling your ShowTime PC controller, you should test it to make sure it's operating properly. You must first install the Light-O-Rama ShowTime Software Suite and the device driver for the USB adapter. Then, follow these steps to connect your ShowTime PC Controller to your computer:

1. **Plug both of the power supply cords from the ShowTime PC controller into electrical outlets.**

   The Status LED on the ShowTime PC controller's circuit board flashes, indicating that the power is on but the device isn't connected to a computer.

2. **Plug one end of the USB cable supplied with the USB adapter into the adapter, and then plug the other end into your computer.**

3. **Plug the Cat5 network cable that's attached to the ShowTime PC controller into the USB adapter.**

   The Status LED in the ShowTime PC controller continues blinking.

4. **Plug a light into one or more of the power cords.**

   I recommend using inexpensive night-lights, which you can usually find at any dollar store for — you guessed it, $1. Since they're so inexpensive, you may want to purchase 16 of them and plug one into each of the channels. Otherwise, just get a few and plug them into the first few channels.

Once you have connected the controller to the computer, you can use the Light-O-Rama software to test the connection. Here are the steps:

1. **On the computer, run the Light-O-Rama Hardware program.**

   To run the program, click the Start button, choose All Programs, choose Light-O-Rama, and then choose Light-O-Rama Hardware. The Hardware program springs to life and displays the screen shown in Figure 3-5.

2. **Click the Auto Configure button near the top left of the Hardware program's window.**

   The Hardware program searches all the available COM ports on your computer until it finds the one that's associated with your USB adapter.

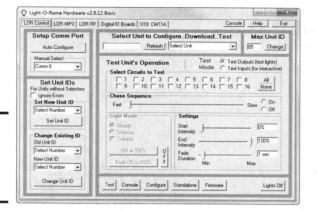

Figure 3-5:
The Light-
O-Rama
Hardware
program.

3. **Change the maximum number of units to be searched to 5.**

To do so, click the Change button located in the Max Units section of the window (near the top right). This brings up a dialog box that has a slider control; drag this slider control all the way to the left, and then click Save.

This step saves you time in the next step, which searches for any available controllers. It takes a few seconds to look for controllers at every possible address, so changing the maximum from 240 to 5 can save you several minutes of your life you would never get back as you waited for the program to search for controllers that don't exist.

4. **Click the Refresh button to look for ShowTime PC controllers.**

Assuming you completed Step 7, the Hardware program finds your controller within a few seconds. The display changes slightly to indicate that the controller has been found.

5. **Click the Console button located near the bottom of the window.**

This action brings up the fun console window shown in Figure 3-6.

6. **For each channel that you connected a light to, drag the corresponding slider up and down and verify that the light turns on and off.**

If a channel doesn't work, verify that the light bulb isn't burned out, that you've connected the light to the correct power cord, and that the light is turned on if it has a switch.

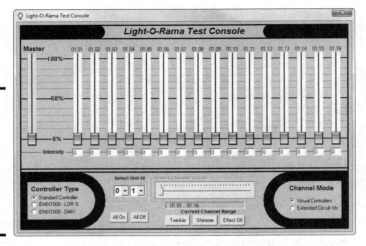

Figure 3-6:
The Hardware program's Console lets you test your controller channels.

7. **Move the lights to other cords and repeat Step 6 to test all channels.**

   You'll want to verify that all 16 of your controller's channels are operating correctly.

8. **Use the Twinkle and Shimmer buttons to verify that the twinkle and shimmer functions are working.**

   When you're tired of the twinkling and shimmering, click the All Off button.

9. **You're done!**

Assuming all the lights came on, you've successfully tested your Light-O-Rama controller and are now ready to start creating sequences. (If any of the channels don't work, carefully double-check your wiring to make sure everything is connected properly. If you get stuck, contact Light-O-Rama's technical support department; they can help you get your controller working.)

Using the Light-O-Rama Sequence Editor

The meat of the Light-O-Rama Software Suite is the Sequence Editor, which lets you create light sequences that are synchronized with music. A *sequence* is a file that stores the information needed to synchronize lights with one selection of music. The music itself is stored in a separate audio file in the MP3 format.

You can also use the Sequence Editor to play a sequence. When it plays a sequence, the MP3 file is loaded and played through your computer's speakers or other audio output. In addition, the Sequence Editor sends instructions to the ShowTime PC controller to activate the lights in sequence with the music.

In Light-O-Rama, a *sequence* is usually associated with an individual song. Two or more sequences can be combined to create a *show*, which lets you cycle through a set of songs so that your audience doesn't have to watch just one song over and over again. You can also set up a *schedule*, which automatically starts and stops shows on particular times and days. Thus, by setting up sequences, shows, and a schedule, you can completely automate every aspect of your light show.

In the remainder of this chapter, I introduce you to some of the basic concepts for working with the Sequence Editor to create simple sequences. You should realize, however, that you can purchase preprogrammed sequences from Light-O-Rama's website for about $30 per song. Programming even a simple sequence for a single song can easily take 10 hours or more, and programming complicated sequences can quickly become a full-time job. So even at $30 per song, purchasing preprogrammed sequences can be worthwhile.

Understanding Sequences

In Light-O-Rama, a sequence is represented as a grid that's somewhat similar to the grid in a spreadsheet program. For example, Figure 3-7 shows part of a very simple sequence in which the lights on channels 1, 3, and 5 alternately turn on and off every half second.

Each row in the grid represents one of the channels available in the controller. Thus, a grid for a sequence played on a 16-channel controller will have 16 rows, one for each channel. Initially, these channels are named according to the control unit number (01 in the figure) and the channel number (1 through 16 in the figure). However, you can easily change the row names to something more meaningful, such as *Red Bush* or *Santa's Hat.*

Each column in the grid represents a time interval. When you first set up a sequence, you can designate the time intervals to use for the grid. For the sequence in Figure 3-7, I specified that each column should be one tenth of a second.

**Book VIII
Chapter 3**

Animating Holiday Lights

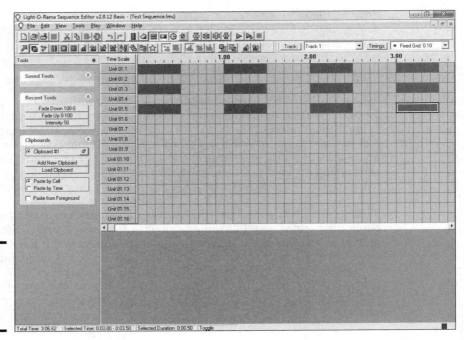

Figure 3-7:
A simple
Light-O-
Rama
sequence.

When a grid square is filled dark, the corresponding channel is turned on.
When a grid square is empty, the corresponding channel is turned off. Thus,
by following the grid cells from left to right, you can see that channels 1, 3, and
5 are on for 0.5 seconds, off for 0.5 seconds, on for 0.5 seconds, and so on.

Light-O-Rama lets you create these two distinct types of sequences:

✦ **Musical:** Lights are synchronized with music in this sequence. Musical
sequences always have an MP3 file associated with them, and Light-O-
Rama can play only one musical sequence at a time.

✦ **Animation:** This sequence doesn't have a music file associated with it.
Instead of synchronizing lights with music, animation sequences are
used to create simple light animations, such as a waving Santa Claus or a
snowman tossing a snowball. Light-O-Rama can play multiple animation
sequences simultaneously. Thus, you can have your snowman throw his
snowballs at the waving Santa Claus if you want.

Creating a Musical Sequence

To create a musical sequence using the Sequence Editor, follow these steps:

1. **In the Sequence Editor, choose File⇨New.**

   This brings up the dialog box shown in Figure 3-8.

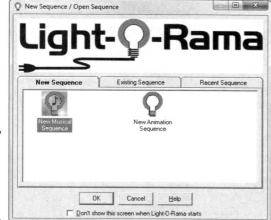

Figure 3-8: Choosing the type of sequence to create.

2. **Select New Musical Sequence, and then click OK.**

   You're prompted to select an MP3 file to use as the musical basis for your sequence.

3. **Locate and select the MP3 file to use for your sequence, and then click Open.**

   The dialog box shown in Figure 3-9 then appears.

4. **Stare at all the options for a while, and then change a few of them if you want.**

   You should familiarize yourself with the range of options that are available for creating new musical sequences. For your first few sequences, I suggest you choose the following options:

   - *Number of Channels Used:* Set this to the number of channels in your controller. The default is 8. If you have a 16-channel controller, change this value to 16.

   - *Initial Timing:* Change this value from the default choice (Use the Tapper Wizard) to A Tenth of a Second. The Tapper Wizard is very useful, but you should create a few simple sequences without it before you try out the Tapper Wizard.

Book VIII
Chapter 3

Animating Holiday
Lights

5. Click OK.

The Sequence Editor creates a blank sequence, as shown in Figure 3-10.

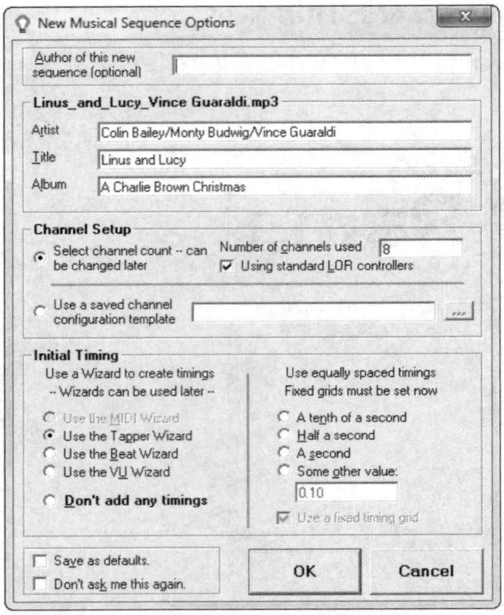

Figure 3-9:
Filling in
the musical
sequence
options.

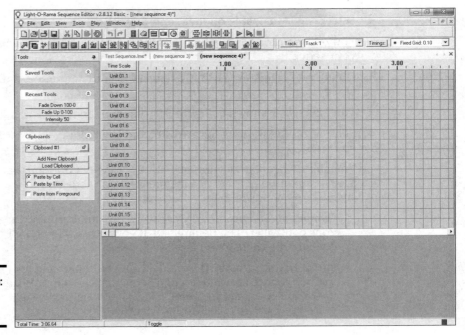

Figure 3-9:
Filling in
the musical
sequence
options.

Figure 3-10:
A blank
sequence.

6. Fill in the sequence.

To do that, you use the buttons in the Tools toolbar to turn channels on and off in various ways. Here's the rundown on what the most popular of these buttons do:

Button	Name	What It Does
	On	Turns the light on.
	Off	Turns the light off.
	Toggle	Toggles the light: Turns it on if it's off or off if it's on.
	Set Intensity	Sets the light to any intensity you choose.
	Fade Up	Gradually fades the light from off to on. Click this tool and then drag over a range of cells to indicate how quickly to fade the light.
	Fade Down	Gradually fades the light from on to off. Click this tool and then drag over a range of cells to indicate how quickly to fade the light.
	Intelligent Fade	If you drag from left to right, fades the light up. If you drag from right to left, fades the light down.
	Twinkle	Causes the light to slowly twinkle.
	Shimmer	Causes the light to quickly shimmer.

7. To test a sequence, click the Play button in the toolbar.

When you've heard enough, click the Stop button to stop the playback.

8. Use the File⇨Save command to save your sequence.

As you work on your sequence, be sure to save it often so that if something goes wrong you won't lose much work.

Here are some additional tips and tidbits to ponder as you create your sequences:

✦ You can use Undo (Ctrl+Z) to undo any mistakes you make.

✦ If you select a tool, then drag over a range of cells, the tool will be applied to all cells within the range. Note that if you use the Toggle tool while dragging over a range of cells, each cell within the range will be individually changed from on to off or off to on.

✦ When you use one of the fade tools, the coloration of the cells involved in the fade indicates how the lights will fade.

✦ You can change the name of each row and the color it's rendered in by clicking the row button in the left margin of the row. This brings up the Channel Settings dialog box shown in Figure 3-11. Although it isn't a requirement, changing the color of the row to match the color of lights displayed by the channel can help you better visualize the show, and changing the name of the channel makes it a lot easier to keep track of how the channels are used in your show.

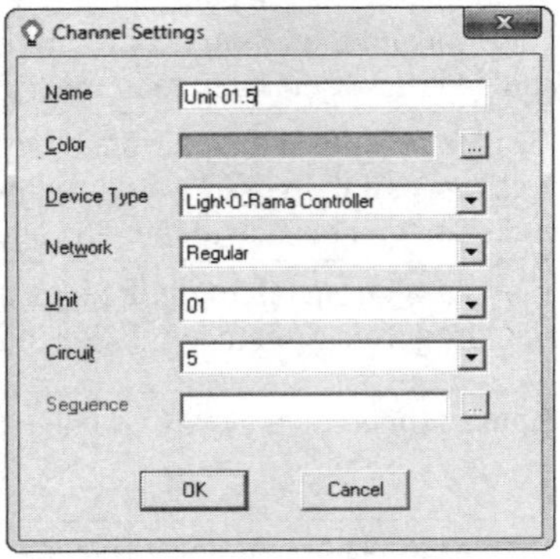

Figure 3-11: Changing the name or color of a row.

✦ Another way to apply effects is to right-click a cell or range of cells. When you do, a pop-up menu listing all the various effects appears.

✦ You can use the View⇨Zoom commands to zoom in or out to get finer control or see more of the big picture.

✦ If it helps, you can make the wave file visible by choosing View⇨ WaveForm. Figure 3-12 shows a sequence with the waveform visible.

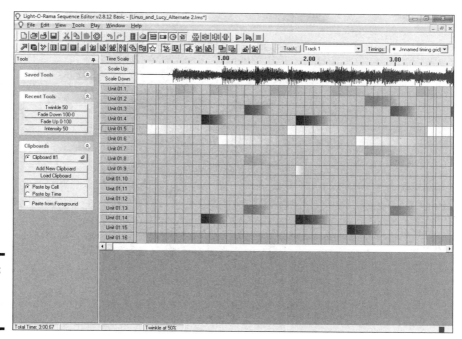

Figure 3-12:
Displaying
the
waveform.

Visualizing Your Show

To help you picture how your show will appear when it is run, you can use the View⇨Animation command. This brings up the window shown in Figure 3-13, which lets you draw a crude representation of your lights. When you run the show within the Sequence Editor, the lights you draw in this window are turned on or off to simulate the appearance of the actual show.

Before you start drawing lights onto the Animation window, you should first import a photograph of your yard. To do so, you'll obviously first need to take a picture of your yard and upload it to your computer. Then, click the Select button, browse to the picture, and select it. You're asked to indicate the size of the grid you want to superimpose on the picture. Then, the picture is added to the Animation window, as shown in Figure 3-14.

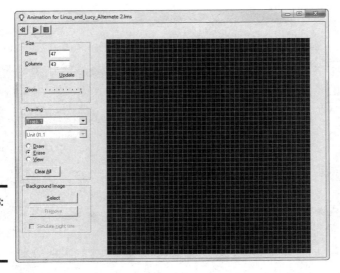

Figure 3-13:
The
Animation
window.

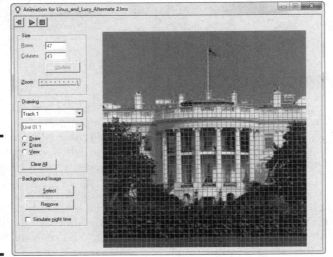

Figure 3-14:
The
Animation
window
with a
background
picture.

To color in your lights, first select the channel you want to color in from the drop-down to the left of the picture. Then, select the Draw radio button and draw your lights on the picture roughly where they will appear in your yard. Draw the lights for each channel, one at a time. If you make a mistake, select the Erase radio button, and then click over your mistake to erase it. Figure 3-15 shows a picture of a house with some lights drawn on it.

Figure 3-15:
The
Animation
window
with lights
drawn in.

Chapter 4: Building an Animatronic Prop Controller

In This Chapter

✔ Considering control requirements for animatronic props

✔ Looking at microcontroller products from EFX-TEK

✔ Examining programming requirements for relay control

✔ Investigating programming requirements for sound control

✔ Responding to a motion detector

✔ A complete program for a scary jack-in-the-box prop

*A*nimatronics refers to the technology used to create automated puppets that move by themselves. The movements are provided by mechanical means such as motors, servos, compressed air, or pneumatics. Animatronic puppets are usually controlled by microprocessors.

My fascination with animatronics started at a very young age, when I first visited Disneyland as a child and was mesmerized by three of the most amazing things I had ever seen: *Enchanted Tiki Room*, filled with marvelous robotic birds that sang and told jokes; *Pirates of the Caribbean*, with its swashbuckling pirates; and of course *The Haunted Mansion*, with its 999 ghosts and room for one more.

The Haunted Mansion was the *pièce de résistance*. I have been fascinated with all things Halloween ever since and tinkered on and off over the years with building my own animatronic ghosts and goblins.

In this chapter, I show you how to build a microprocessor-based controller system that I've used in several of my best Halloween props. I've used this exact circuit to create a Frankenstein creature that pops up, a mummy that pops up, and a giant jack-in-the-box, but you can use this circuit to control just about any kind of animatronic creation you can imagine.

Note that Halloween isn't the only application for animatronics. You can use them in theatrical productions, school fundraisers, or for promotional or marketing purposes. For example, if you own a flower shop, you could create animatronic flowers for your storefront display, or if you own a tuxedo store, you could create an animatronic dancer dressed in your finest tuxedo to draw customers in. The only limit is your imagination.

Animatronics should be built and used with care, as they can be dangerous. You must ensure that no one who views your animatronic will be exposed to any danger, such as loose electrical connections or faulty compressed-air fittings. Make sure you place your finished animatronic in a secure area where no one will be able to touch it, and make sure that all electrical and compressed-air fittings are completely secure.

If your animatronic prop uses compressed air, be especially careful! This chapter doesn't show you how to work with compressed air because that's a subject for an entire book. If you want to work with compressed air, always use professionally designed and manufactured components such as air cylinders and pop-up mechanisms. Don't try to build your own compressed air cylinders out of PVC pipe, bicycle pumps, or any other material that isn't specifically designed to work with compressed air. And beware of inexpensive compressed-air devices available on eBay or other online sites. A good, safe compressed air mechanism should cost you several hundred dollars. If it's selling online for $89, it's probably not safe.

Looking at the Requirements of Animatronic Prop Control

Whether it's used in a haunted house, trade show exhibit, store display, or museum, a typical animatronic prop requires the following elements to achieve the desired effect:

✦ Most props require some sort of trigger to initiate the prop's routine. The trigger can be as simple as a button that you or a customer pushes to start the routine, or it could be a timer that causes the prop to start up automatically every five or ten minutes. A third option is to use a motion detector that causes the prop to start when motion is detected near the prop.

✦ All props require a way to control the prop's *actuators,* which are the components that cause the prop to move. The most common types of actuators are electric motors, servos, and compressed air cylinders. If you're using an electric motor, the prop controller usually turns the motor on or off directly by applying current to the motor. If you're using servos, the prop controller must send timing pulses to the servo to control the servo's position. And if you're using compressed air cylinders, the prop controller opens and closes electric valves that allow air into the cylinders.

✦ Most props require lighting effects that enable the viewer to see the prop in action. In some cases, this is as simple as one or more lights that come on when the prop starts its routine and go off when the routine ends. In other cases, a variety of lights are used during the routine to highlight different parts of the prop during different parts of the show.

✦ Some props use sound that is synchronized with the prop's movements. For example, an animated face might speak dialog, a scary creature might scream when it suddenly moves, or the prop may use music to enhance its routine.

The prop controller I show you how to build in this chapter meets all of these needs. It's based on a BASIC Stamp microprocessor, which provides more than enough computing power to control lights and sound and activate a few pneumatic valves and electric motors.

Instead of building the controller from a BASIC Stamp module from Parallax, the controller shown in this chapter uses components from a company named EFX-TEK, which specializes in microprocessor components for animatronic prop control. The prop controller uses three products from EFX-TEK: a Prop-1 controller board, which includes a BASIC Stamp micro-processor; an RC-4 relay board that can hold up to four relays for switching 120 VAC circuits; and an AP-16+ sound board, which can be programmed to play sounds saved on a standard SD memory card.

For more information about these components and how to use them, see the section "Building the Prop Controller" later in this chapter.

Examining a Typical Animatronic Prop

Figure 4-1 shows a typical animatronic prop that I made a few years ago to use in a Halloween haunted house. The prop is a large jack-in-the-box that benignly plays "Pop Goes the Weasel" while a crank on one side of the prop slowly spins until an unsuspecting visitor approaches it. Then — all in one frightening moment — the music stops, the lid flies open, a large clown head pops up, a bright light comes on to ensure that you can see the clown, and a ghastly voice screams, "Trick or treat!" A moment later, the light goes off, and the clown retracts back into the box. A few seconds after that, the crank starts spinning and the music plays again while the jack-in-the-box waits for another victim.

Figure 4-1:
The jack-in-the-box prop.

The basic pop-up operation of the jack-in-the-box is accomplished with a compressed air mechanism that can be controlled with an electrically-activated valve. When the valve is closed, the mechanism is retracted inside the box. When the valve opens, compressed air flows into an air cylinder that extends the mechanism upwards. The clown head is mounted on top of the mechanism. The lid of the jack-in-the-box is positioned such that the clown's head pushes the lid open when the mechanism extends. Gravity takes care of closing the lid when the mechanism retracts, and the clown goes back into the box.

Note that the point of this chapter is to show you how to build the electronic part of this prop — the prop controller that controls the prop's action. This chapter doesn't show you how to build the prop itself. The prop controller is generic enough that it can control just about any prop that includes simple actions such as pneumatic pop-ups, motors, lights, and sound. In fact I've used this very same design for half a dozen different Halloween props.

Nevertheless, it's beneficial to see how the prop controller works in a specific prop, so the following paragraphs describe in general terms how I built the jack-in-the-box prop. This will give you a general idea of how it operates so that you can see how the prop controller works with it.

✦ The most important component of this prop is the pneumatic lifting mechanism. I purchased it for about $250 from an online distributor of Halloween prop mechanisms. If you search the web, you'll find several distributors that make safe, professional-quality compressed-air mechanisms. Just search for *Halloween prop lifter* and you'll find several suppliers. You can also find them for sale on eBay.

Figure 4-2 shows a close-up photograph of the lift mechanism.

The lift mechanism I purchased included the welded steel frame, a pneumatic cylinder that provides the motion needed to lift the frame, and an electrical valve that can be connected to a compressed air hose. When the valve is connected to a source of compressed air and voltage is applied, the valve opens to allow compressed air into the cylinder.

Do not, under any circumstances, attempt to cut corners by building your own lifter mechanism unless you are an experienced metalworker. The Internet has plans for similar devices made out of sprinkler pipe or bicycle pumps, but putting compressed air into plastic or metal parts that were not designed to handle compressed air is an excellent way to maim or even kill someone.

✦ The prop is housed in a 22" square box I constructed from ⅝" plywood, reinforced inside with 2 x 2" lumber. I was able to cut all six sides and the lid out of a single 4 x 8' sheet of plywood.

✦ To create the lid, I used a sabre saw to cut out the opening. Then, I cut a separate piece of plywood about 1" larger than the opening to use as the lid. I used a pair of small gate hinges to connect the lid to the box.

✦ I mounted the lift mechanism directly to the bottom of the box. Depending on the design of the mechanism you use, you may have to mount the lift on the back of the box instead of the bottom.

**Book VIII
Chapter 4**

Building an Animatronic Prop Controller

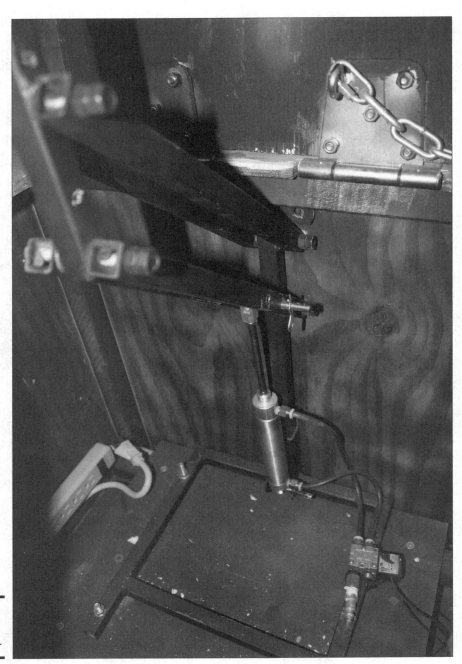

Figure 4-2:
The lift
mechanism.

- I drilled two, 1" diameter holes in the back of the box to pass the electrical power cord and the compressed air hose into the box.

- The clown consists of a small foam wig head that I bought at a thrift store and a mask that I purchased at a seasonal Halloween costume shop.

- I used a surplus automotive windshield wiper motor to turn the crank. You can find them for sale on the Internet for about $15. I fabricated the crank itself from a few pieces of scrap metal I had lying around.

- The prop controller is mounted inside the box, secured to one of the side panels with four screws.

- To trigger the prop, an infrared motion detector called a *PIR* is used. The motion detector is mounted to the outside of the box and connected via a 24" cable to the prop controller. When the motion detector detects nearby motion, the programming in the prop controller triggers the prop to activate the valve so the clown pops up.

- To light up the clown's face when he pops up, I used a 75 W white flood lamp mounted in a plastic flood lamp holder I got from a home improvement store.

- For the sound, I used a set of inexpensive speakers that I had lying around. The speakers are connected to the audio output from the prop controller's sound board. This sound board includes a 20 W amplifier, which is more than loud enough to give a good scare when the clown pops up.

- Figure 4-3 shows a wiring diagram of the prop so you can get an idea of the prop's electrical requirements. As you can see, the prop requires a total of three, 12 VDC power adapters, commonly known as *wall warts.* The first is to power the prop controller, the second provides 12 VDC for the windshield wiper motor, and the third provides 12 VDC for the pneumatic valve.

To simplify the wiring, I used three, six-outlet power strips and an eight-terminal barrier strip, which aren't shown on this diagram. One of the power strips plugs directly into a wall outlet to supply the main 120 VAC power for the prop. The other two power strips are connected to the RC-4 relay control board. Using these power strips allows you to simply plug in the 12 VDC power adapters and the flood lamp holder.

The prop runs on 120 VAC household power, and there are several exposed electrical connections inside the box. Therefore, it's imperative that the inside of the box is inaccessible to humans.

It's especially important that the wiring is secure enough to withstand the vibrations that the box will experience every time the prop is activated. When the valve closes, the cylinder extends, the lid swings open, and the clown pops up his head, the box will be jolted more than a little. To insure that the frequent shaking doesn't cause any electrical connects to come lose, all electrical wires are fastened down with wire holders, and all connections are made with soldered terminals that are tightly screwed to the barrier strip.

That's about it for the jack-in-the-box prop itself. As I've already mentioned, the primary focus of this chapter is building the prop controller that provides the brains for this prop. The prop controller can be used for any other type of prop with similar requirements. And, if your prop is more complex than the jack-in-the-box, you can easily expand the prop controller by including additional relay control boards to control more devices.

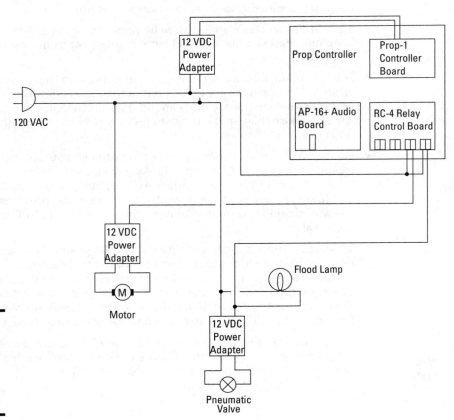

Figure 4-3:
Schematic diagram for the jack-in-the-box prop.

Building the Prop Controller

Figure 4-4 shows the fully assembled prop controller used to control the jack-in-the-box prop. As you can see, the prop controller is made from three circuit boards, which you can purchase from EFX-TEK (www.efx-tek.com). The boards are mounted on an 8 x 10" sheet of Plexiglas, which you can purchase from most hardware stores. Before mounting the components on Plexiglas, spray-paint the back of the Plexiglas black to create a nice, professional-looking board.

Figure 4-4:
The assembled animatronic prop controller.

Here's a complete list of the parts you'll need to build the prop controller:

Quantity	Description
1	EFX-TEK Prop-1 controller
1	EFX-TEK PIR sensor (includes a 14" extension cable for connecting to the Prop-1)
1	EFX-TEK AP-16+ audio player
1	EFX-TEK RC-4 relay control board
2	Crydom D2W203F solid-state relays

Quantity	Description
3	EFX-TEK stand-off kits (each kit includes four, ⅝" stand-offs and 8, 4-40 machine screws)
2	14" servo-style extension cable
2	6" lengths of 16-gauge stranded wire
1	8 x 10" sheet of 3⁄16" Plexiglas
1	Can of black spray paint

All of these components except the wire, Plexiglas, and the paint can be ordered from EFX-TEK at www.efx-tek.com. The total price for all components should be around $275.

The following paragraphs describe the four main components of the prop controller, which are listed first in the parts list:

✦ **Prop-1 controller:** The brains of the prop controller. This inexpensive controller board ($39.95) contains a BASIC Stamp microcontroller and two I/O busses that give you direct access to the BASIC Stamp's eight I/O pins. The first bus is a series of three-pin headers that you can connect standard servo cables to. These connectors provide a TTL interface to the BASIC Stamp and are used for low-current applications such as communicating with other prop-control elements.

The second bus is for high-current applications that can handle up to 500 mA. These outputs can be used to directly drive small motors, relays, solenoid valves, and so on.

The Prop-1 also includes a 5 V regulated power supply and a programming interface so you can connect it directly to your computer to download programs.

✦ **PIR sensor:** An inexpensive (under $10) motion detector that can be connected via a three-pin servo cable to any of the Prop-1's low-current I/O pins. This device is used to trigger the prop's action.

✦ **AP-16+ audio player:** At $129.95, the audio player is the most expensive component of the prop controller. However, sound is an essential element of any good prop, and the AP-16+ is an extremely versatile sound player. It can play sounds in WAV format directly from a micro-SD card and includes a built-in 20 W amplifier, so you can connect speakers directly to the AP-16+ without using a separate amplifier. The AP-16+ also connects to the Prop-1 via a three-pin servo cable. This connection allows the Prop-1 to send commands to the AP-16+ to tell the AP-16+ to play specific sound files on the micro-SD card.

✦ **RC-4 relay control:** This module lets you control up to four line-voltage (120 VAC) circuits by using solid-state relays. It connects to the Prop-1 via a three-pin servo cable, enabling the Prop-1 to send commands to the RC-4 to tell it to turn its relays on or off under program control.

The RC-4 board itself costs $39.95, but it doesn't come with any relays. The relays are an additional $14.95 each, so the cost of this board populated with two relays (as this project requires) is about $70.

I can't say it too many times in this chapter: The prop controller uses line-level voltages that can hurt or even kill you if you aren't careful. Make sure that the line-level wiring is properly secured and enclosed, and *never* work on the controller board when the line-level circuits are plugged in.

Here are the steps to assemble the prop controller:

1. **Paint one side of the Plexiglas with the black spray paint and allow the paint to dry.**

2. **Drill the necessary mounting holes in the Plexiglas.**

   You'll need to drill four mounting holes for each of the three boards that will be mounted to the Plexiglas (Prop-1, RC-4, and AP-16+). Use a ³⁄₁₆" drill bit. Figure 4-5 is a rough drilling guide, but note that this diagram isn't to scale. You'll need to lay your actual components out on the Plexiglas to determine the exact drilling locations.

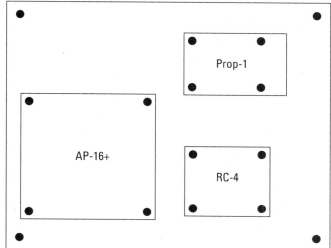

Figure 4-5: Where to mount the circuit boards and drill the mounting holes for the animatronic prop controller.

3. **Mount the Prop-1 controller on the Plexiglas.**

Use four of the stand-offs and eight of the 4/40 machine screws. Position the Prop-1 controller so that the row of screw terminal connectors is on the left side of the controller as oriented in Figure 4-5.

4. **Mount the RC-4 board on the Plexiglas.**

Use four of the stand-offs and eight of the 4-40 machine screws. Position the RC-4 board so that the four screw-terminal connectors are at the bottom of the board as oriented in Figure 4-5.

5. **Mount the AP-16+ board on the Plexiglas.**

Use four of the stand-offs and eight of the 4-40 machine screws. Position the AP-16+ board so that the audio output connectors are at the bottom of the board as oriented in Figure 4-5.

6. **Use the two wires to connect the V+ and GND terminals on the Prop-1 to the V+ and GND terminals on the AP-16+ board.**

These wires are used to provide power for the AP-16+ board. After test-fitting the wires, cut them to the correct length and strip ⅜" of insulation off each end. Then insert each end of the wire into the appropriate screw terminal and tighten the screw to ensure a solid connection.

7. **Use one of the extension cables to connect the P7 header on the Prop-1 controller to the SER header on the RC-4.**

This cable enables the Prop-1 controller to send commands to the RC-4 board to cause the RC-4 board to turn its relays on or off.

Eight, three-pin headers are found near the center of the Prop-1 controller board. These headers provide access to the eight I/O ports of the BASIC Stamp microprocessor. The headers are labeled P0 through P7, so you shouldn't have trouble finding the correct header.

There are two, three-pin headers on the RC-4 board, both labeled SER. These two headers are connected to one another, so it doesn't matter which one you use as they're electrically the same.

The only trick to connecting the extension cable is making sure you get them oriented correctly. The extension cable has three wires: white, red, and black. The cables must be inserted into the header correctly, or the Prop-1 won't be able to communicate with the RC-4.

Orienting the cable correctly on the Prop-1 is easy because the three-pin headers are labeled to indicate which pin is white, which is red, and which is black. On the RC-4, the cable should be inserted into the header with the black wire facing the outside edge of the board and the white wire facing the center of the board.

8. **Use the second extension cable to connect the empty SER header on the RC-4 board to one of the SERIAL headers on the AP-16+ board.**

The purpose of this connection is to enable the Prop-1 controller to communicate with the AP-16+ audio player to tell it when to play its sounds.

As oriented in Figure 4-5, the SERIAL header pins are located near the top-right corner of the AP-16+ board. The pins on the AP-16+ headers are labeled W, R, and B, so you should have no trouble orienting the wire colors. As with the RC-4, there are two SERIAL headers on the AP-16+; it doesn't matter which of them you use.

9. **Use the third extension cable to connect the PIR sensor to the P6 header on the Prop-1 controller.**

The P6 header is right next to the P7 header.

10. **Insert the two Crydom solid-state relays in the K1 and K2 positions on the RC-4 board.**

Note that if your prop needs to control more than two 120 VAC circuits, you can add up to four Crydom relays to the RC-4 board. And you can add additional RC-4 boards if you need to control more than four circuits.

Programming the Prop-1 Controller

At the heart of a Prop-1 controller is a Parallax BASIC Stamp microcontroller. As a result, you can program the Prop-1 controller using the same PBASIC programming language you use to program a BASIC Stamp. The techniques for programming in PBASIC are presented in detail in Book VII. As a result, I won't review those details here. You'll need to be familiar with the concepts presented in Book VII before you attempt to program the Prop-1 controller.

However, realize that the Prop-1 controller uses a simpler form of the BASIC Stamp microprocessor, known as the BASIC Stamp 1. Book VII covers the more advanced version of the BASIC Stamp, called the BASIC Stamp 2. The BASIC Stamp 1 uses a simpler version of PBASIC than the BASIC Stamp. Here are some of the differences you'll need to be aware of:

✦ The BASIC Stamp 1 has less memory than the BASIC Stamp 2. Specifically, the Stamp 1 has just 256 bytes of program memory, which means that the longest program for a Prop-1 is around 80 to 100 statements. In addition, you're limited to just 14 bytes of variable memory, which means your programs can have just a few variables.

For most simple animatronic props, these limitations aren't a problem. If your prop is more complicated, EFX-TEK offers a Prop-2 controller, which uses a BASIC Stamp 2 microcontroller. It's about $60 more than the Prop-1 controller, however.

✦ Some of the programming instructions covered in Book VII don't work on the BASIC Stamp 1. The biggest limitation is that the IF statement is much simpler: The BASIC Stamp 1 IF statement can test for a condition and branch to a label if the condition is true, like this:

```
IF PIN0 = 0 THEN Main
```

Here, the program branches to the label named `Main` if the value of the variable `PIN0` is zero. The more complicated variants of the IF statement covered in Book VII, Chapter 2 aren't allowed.

✦ Because there's such a limited amount of variable memory, variables have preassigned names on the BASIC Stamp 1.

There are two types of variables on the BASIC Stamp 1: 8-bit variables called Byte variables and 16-bit variables called Word variables.

The Byte variables are named B0 through B13.

The Word variables are named W0 through W6.

Note that the Word variables occupy the same 14 bytes of memory that the Byte variables occupy. Thus, W0 occupies the same two bytes as B0 and B1; W1 occupies the same two bytes as B2 and B3; and so on.

✦ You can use the SYMBOL statement to assign a more meaningful name to one of the prenamed variables. For example:

```
SYMBOL Counter = B0
```

This statement assigns the symbol `Counter` to the byte variable B0. Thereafter, you can use the name `Counter` to refer to this variable.

✦ The SYMBOL statement can also give meaningful names to I/O pins, as in this example:

```
SYMBOL Pir = PIN6
```

In this example, the name `Pir` is used to refer to I/O pin 6.

Other than these variations, programming the BASIC Stamp 1 microcontroller is similar to programming its more advanced counterpart.

Sending Commands to the RC-4 or AP-16+ Modules

I don't cover the PBASIC command called SEROUT in Book VII, but you'll need to be familiar with it to use EFX-TEK's RC-4 and AP-16+ modules. The SEROUT command sends data to an external device that's connected to the

Prop-1 controller via one of the I/O pins. It uses a very simple serial communications protocol to accomplish this. Fortunately, you don't need to worry about the details of how the serial communications protocol works to use the SEROUT command to talk to an RC-4 or AP-16+ module. But you do need to know how to use the SEROUT command.

The SEROUT command has a simple syntax:

```
SEROUT pin, mode, (data)
```

The first parameter of the SEROUT command, *pin*, is simply the pin number that the data is to be sent to. Since the RC-4 and AP-16+ modules are connected to PIN7 in the animatronic prop controller, you'll simply specify PIN7 for this parameter.

The second parameter indicates the communication mode that will be used to send the data. For all EFX-TEK I/O modules, you should specify the built-in constant OT2400 for this parameter.

In case you're interested, OT2400 means that the transmission speed is 2400 bits per second, and the output polarity is open-drain driven high — which sounds pretty technical to me, so I wouldn't worry about it. Just use OT2400 for this parameter.

The third parameter, *(data)*, is the one that actually specifies the data to be sent to the external device. You must enclose the data in parentheses, and within the parentheses you can list as many string constants or variables as you want. For example:

```
SEROUT PIN7, OT2400, ("Hello World!")
```

This statement sends the string constant "Hello World!" to the device on PIN7.

Here's an example that accomplishes the same thing but with three string constants instead of one:

```
SEROUT PIN7, OT2400, ("Hello ", "World", "!")
```

And here's an example that uses a variable:

```
SYMBOL Bam = B0
Bam = "!"
SEROUT PIN7, OT2400, ("Hello, World", Bam)
```

You might reasonably ask why you would use several string constants or variables instead of just one. The answer to that excellent question will become apparent when you read the following two sections about programming the RC-4 and AP-16+ modules.

It's a good idea to assign a symbol to the I/O pin to which the RC-4 and AP-16+ modules are attached, and to the communication mode. For example:

```
SYMBOL Sio = 7
SEROUT Sio, OT2400, ("Hello, World!")
```

That way, if you decide to change the I/O pin to which the external device is attached, you have to make only one change to your program — the symbol definition — instead of multiple changes in each SEROUT statement.

Though not as important, it's also a good idea to use a symbol for the mode parameter, like this:

```
SYMBOL Sio = 7
SYMBOL Baud = OT2400
SEROUT Sio, Baud, ("Hello, World!")
```

Why? Because someday you may decide to move your program over to a Prop-2 controller. When you do, you'll discover that the BASIC Stamp 2 SEROUT command has additional mode settings that enable faster and more reliable serial communications. You'll be able to change the serial communications mode for all of your program's SEROUT commands simply by changing the SYMBOL statement.

Programming the RC-4 Relay Control Module

The RC-4 relay control module supports up to four high-voltage solid-state relays that can turn line-voltage circuits on and off. EFX-TEK sells the board and relays separately, so you can equip it with just the number of relays your project needs.

A pair of jumper blocks on the RC-4 board lets you give it an address value of 0, 1, 2, or 3. This allows you to daisy-chain up to four RC-4 boards together, allowing you to control up to 16 separate relay circuits. The examples in this section assume that you have removed these two jumpers so that the address of the RC-4 is 0.

To control an RC-4 board from a Prop-1 controller, you use the SEROUT command to send data to the pin that the RC-4 board is connected to. The data that you send must follow this format:

```
"!RC4", address, command {,data}
```

The first data item is the string constant "!RC4". This string constant is a pre-amble that indicates that the data that follows is intended for an RC-4 relay control card. This allows you to chain several different kinds of EFX-TEK cards together and send commands only to the RC-4.

The second data item is the address of the RC-4 card you want to send the data to. The possible values are 0, 1, 2, or 3.

The third data item is a command, which is a single character that repre-sents one of several commands that the RC-4 can respond to. Depending on the command used, you may also need to provide one or two additional bytes of data after the command. The possible commands are:

Command	What it does
X	Resets all relays. No additional data is needed.
R	Sets or resets an individual relay to a particular value. Two data bytes are required: one to indicate which relay to set or reset, and a second to indicate whether the relay should be closed (set) or opened (reset).
S	Sets all relays. The S command is followed by a single byte of binary data that indicates which relays should be open and which should be closed.
V	Causes the RC-4 to send its firmware version informa-tion back to the prop controller. The version informa-tion can be read with a SERIN command.
G	Gets the current status of all relays. Use the SERIN command to read the data from the RC-4 after sending this command.

The three RC-4 commands you'll use most often are X, to reset all relays; R, to set an individual relay; and S, to set each of the four relays to a particular status (on or off). These commands are described in the following sections.

Turning all relays off

To turn all the relays on an RC-4 module off, use the X command. For example:

```
SEROUT PIN7, OT2400, ("!RC4", 0, "X")
```

It's a good idea to use this command near the beginning of your program to make sure that the program starts with all relays off.

Book VIII
Chapter 4

Building an Animatronic Prop Controller

Turning an individual relay on or off

To turn an individual relay on or off, you send an R command to the RC-4. The R command requires that you send five separate data items:

✦ The preamble: "!RC4"

✦ The RC4 board's address

✦ The command "R"

✦ The relay number: 1 through 4

✦ The relays status: 1 to turn the relay on, 0 to turn the relay off

Here's a SEROUT statement to turn the first relay (number 1) on:

```
SEROUT PIN7, OT2400, ("!RC4", 0, "R", 1, 1)
```

And here's a SEROUT statement to turn the first relay off:

```
SEROUT PIN7, OT2400, ("!RC4", 0, "R", 1, 0)
```

Setting all four relays at once

To set the status of all four relays at one time, you can use the S command. This command requires four data items:

✦ The preamble: "!RC4"

✦ The RC4 board's address

✦ The command "S"

✦ A single byte value that represents the status of the four relays

The value that controls the relay status is usually written as a percent sign followed by four binary bits representing the status of each relay — 1 for on and 0 for off. The only trick is that the four binary bits are written in reverse order: The first is relay 4, the second is relay 3, the third is relay 2, and the fourth is relay 1.

For example, to turn relay 1 on and the other three relays off, you would use %0001 as the relay status. To turn relays 1 and 3 on and relays 2 and 4 off, you would use %0101 as the status.

Here's a SEROUT command that sends an S command to turn all four relays on:

```
SEROUT PIN7, OT2400, ("!RC4", 0, "S", %1111)
```

And here's a SEROUT command that turns relays 1 and 3 on and simultaneously turns relays 2 and 4 off:

```
SEROUT PIN7, OT2400, ("!RC4", 0, "S", %0101)
```

The following SEROUT command turns all relays off, which is equivalent to sending an "X" command:

```
SEROUT PIN7, OT2400, ("!RC4", 0, "S", %0000)
```

Using symbols to make RC-4 commands more readable

As I mentioned earlier in this chapter, it's a good idea to use symbols for the I/O pin and output mode. It's also a good idea to use symbols for the RC-4 address, for the relay number, and for the relay status (on or off). That can make your program more readable.

For example, here are some typical symbol declarations:

```
SYMBOL Sio      = 7
SYMBOL Baud     = OT2400
SYMBOL RC4      = 0
SYMBOL Relay1   = 1
SYMBOL Relay2   = 2
SYMBOL Relay3   = 3
SYMBOL Relay4   = 4
SYMBOL RelayOn  = 1
SYMBOL RelayOff = 0
```

With these symbol declarations in place, you can turn on relay 1 with this statement:

```
SEROUT Sio, Baud, ("!RC4", RC4, "R", Relay1, RelayOn)
```

And here's a statement that turns relay 1 off:

```
SEROUT Sio, Baud, ("!RC4", RC4, "R", Relay1, RelayOff)
```

Here's a SEROUT statement that uses the symbols to send an S command that turns relays 2 and 4 on and relays 1 and 3 off:

```
SEROUT Sio, Baud, ("!RC4", RC4, "S",%1010)
```

A sample program for controlling all four RC-4 relays

Listing 4-1 shows a complete PBASIC program that gives the RC-4 card a bit of a workout by performing the following sequence of steps:

1. **Turn off all relays with an X command.**

2. **Pause one second.**

3. **Use a series of S commands to turn on relays 1, 2, 3, and 4 in sequence at one-second intervals.**

   As each relay is turned on, the other three are turned off.

4. **Pause one second, and then turn off all relays.**

5. **Use a series of R commands to turn on relays 1, 2, 3, and 4 in sequence at one-second intervals.**

   As each relay is turned on, the status of the other relays are left unchanged. As a result, this part of the program first turns on relay 1, then adds relay 2, then adds relay 3, and then adds relay 4. At this point, all four relays are on.

6. **Wait one second, and then repeat the entire program.**

You can download this program from this book's companion website.

Listing 4-1:   An RC-4 Control Program

```
' {$STAMP BS1}
' {$PBASIC 1.0}

SYMBOL Sio      = 7
SYMBOL Baud     = OT2400
SYMBOL RC4      = 0
SYMBOL Relay1   = 1
SYMBOL Relay2   = 2
SYMBOL Relay3   = 3
SYMBOL Relay4   = 4
SYMBOL RelayOn  = 1
SYMBOL RelayOff = 0


Main:
  'Turn off all relays
  SEROUT Sio, Baud, ("!RC4", RC4, "X")
  PAUSE 1000

  'Cycle through the relays one at a time
  SEROUT Sio, Baud, ("!RC4", RC4, "S", %0001)
  PAUSE 1000
```

```
SEROUT Sio, Baud, ("!RC4", RC4, "S", %0010)
PAUSE 1000
SEROUT Sio, Baud, ("!RC4", RC4, "S", %0100)
PAUSE 1000
SEROUT Sio, Baud, ("!RC4", RC4, "S", %1000)
PAUSE 1000

'Turn off all relays
SEROUT Sio, Baud, ("!RC4", RC4, "S", %0000)
PAUSE 1000

'Turn on relay 1
SEROUT Sio, Baud, ("!RC4", RC4, "R", Relay1, RelayOn)
PAUSE 1000

'Add relay 2
SEROUT Sio, Baud, ("!RC4", RC4, "R", Relay2, RelayOn)
PAUSE 1000

'Add relay 3
SEROUT Sio, Baud, ("!RC4", RC4, "R", Relay3, RelayOn)
PAUSE 1000

'Add relay 4
SEROUT Sio, Baud, ("!RC4", RC4, "R", Relay4, RelayOn)
PAUSE 1000

'Repeat!
GOTO Main
```

Programming the AP-16+ Audio Player Module

The AP-16+ audio player module is an amazingly versatile way to add sound to your animatronic prop. Like the RC-4, you communicate with the AP-16+ module using serial communication via the SEROUT statement. Some commands cause the AP-16+ module to return data, which you can retrieve with the SERIN statement. But for most props, all you have to do is send a few commands using SEROUT, so you can dispense with the SERIN statement.

The AP-16+ board can play sounds recorded in WAV format stored on a micro-SD card. So before you program your prop, you'll need to prepare a micro-SD card with the files you want the prop to play. You can simply use your computer to copy the files to the micro-SD card; the only restriction is that the filenames must be no longer than eight characters, and the filename extension must be .wav. (The AP-16+ card can't play MP3 files.)

For the jack-in-the-box prop, just two sound files are needed. The file that contains the "Pop Goes the Weasel" song is named weasel.wav. I created this recording myself; you can download it from this book's companion website.

The other file is a horrific scream that is played when the clown pops up. This file is named scream.wav. It is also available on the companion website.

The AP-16+ is an extremely versatile card that's designed to be usable in stand-alone mode — that is, not controlled by a microcontroller. Many of its features are useful primarily in stand-alone mode. For example, you can store files on the micro-SD card with predefined filenames such as SFX*nn*.wav and AUX*nn*.wav, where *nn* is a two-digit number (for example, SFX01.wav or AUX04.wav). The AP-16+ can be configured to automatically play these files in random ways in response to direct trigger inputs on the card. The AP-16+ can also be configured to play an ambient background sound whose filename must be ambient.wav. These are all very useful features, but not relevant to the prop controller presented in this chapter. So although I ignore those features in this section, I suggest you carefully read the AP-16+ documentation supplied by EFX-TEK to see how these features work.

To control an AP-16+ module from a Prop-1 controller, you use the SEROUT command to send data to the pin that the AP-16+ is connected to. The data that you send must follow this format:

```
"!AP16", address, command {,data}
```

The string constant "!AP16" is the preamble that indicates that the command is intended for an AP-16+ module. This allows you to chain different kinds of modules together with the AP-16+, making it possible to control all of the modules used in a project from a single I/O pin.

The second data item is the address of the AP-16+ module you want to send the command to. The possible values are 0, 1, 2, or 3. The address is set by a pair of jumpers on the AP-16+ board. For the examples in this section, I assume the AP-16+ address is 0.

The third data item is the command. Depending on the command used, you may also need to provide additional data after the command. The possible commands are:

Command	What it does
X	Resets the AP-16+, which stops any sound that may be playing.
L	Sets the output volume.
PW	Plays a named file. Additional data provides the name of the file and the number of times to loop the file. (Specify 0 for the loop value to repeat the file indefinitely.)
G	Gets the current status of the AP-16+ card. Use the SERIN command to read a single byte that indicates whether the AP-16+ is currently playing a sound file and, if so, which sound file is being played.

Command	What it does
PS	Plays one of the SFX*nn*.wav files. Additional data provides the SFX number (that is, the *nn*) and the number of times the file should be looped. (Specify 0 for the loop value to repeat the file indefinitely.)
P?	Randomly selects one of the SFX*nn*.wav files to play.
PA	Plays one of the AUX*nn*.wav files. Additional data provides the SFX number (that is, the *nn*) and the number of times the file should be looped. (Specify 0 for the loop value to repeat the file indefinitely.)
S	Changes the playback speed.
V	Gets the AP-16+ version information. The version information can be read with a SERIN command.

The four AP-16+ commands you'll use most are X, to reset the sound player and stop any sound that may be playing; L to set the output volume; PW, to play a specific file; and G, to get the current status of the AP-16+ so you can find out if it has finished playing a sound file. These commands are described in the following sections.

Note that throughout these sections, I assume that the following symbols have been defined near the top of the program:

```
SYMBOL Sio  = 7
SYMBOL Baud = OT2400
SYMBOL AP16 = 0
```

Resetting the AP-16+

To reset the AP-16+, send an X command like this:

```
SEROUT Sio, Baud, ("!AP16", AP16, "X")
```

This causes any WAV file that may be playing to immediately stop. Not only is it a good idea to send this command at the beginning of your program to make sure that the AP-16+ is ready to receive commands, but this command is also the best way to silence the AP-16+ if you want to stop it before the current file finishes playing.

Changing the volume

If you want, you can change the output volume from your program. This is useful if you want to play a file at less than full volume. It's also useful if you want the sound output to come from one side of the room or the other.

To change the output volume, use the L command. The syntax of the L command is this:

```
"!AP16", address, "L", left, right
```

Where *left* and *right* are values between 0 and 100 that indicate the percentage volume for the left and right outputs, respectively.

For example, to set the output volume to 50% for both left and right, use this statement:

```
SEROUT Sio, Baud, ("!AP16", AP16, "L", 50, 50)
```

To make the sound come exclusively from the left channel, use this:

```
SEROUT Sio, Baud, ("!AP16", AP16, "L", 100, 0)
```

To restore the sound to full volume, use this statement:

```
SEROUT Sio, Baud, ("!AP16", AP16, "L", 100, 100)
```

Playing a specific file

To play a specific WAV file, use the PW command. Here's the syntax:

```
"!AP16", address, "PW", name, 13, loops
```

The *name* parameter is the name of the file, without the .wav extension. The 13 is required because of a quirk in the way the AP-16+ programming works. Think of it as a good luck charm!

The *loops* parameter indicates how many times the file should be repeated. Normally, you'll just specify 1 to play the file once. The maximum is 255, but specify 0 if you want the file to keep playing forever — or at least until you reset the AP-16+ with an X command or use another PW command to play a different file.

Here's the command to play a file named scream1.wav just once:

```
SEROUT Sio, Baud, ("!AP16", AP16, "PW", "SCREAM1", 13, 1)
```

Here's a command that plays the file weasel.wav indefinitely:

```
SEROUT Sio, Baud, ("!AP16", AP16, "PW", "WEASEL", 13, 0)
```

Note that whenever you play a sound file, any other sound file that happens to be playing is immediately stopped. Thus, you don't have to wait for one sound file to end to play another.

Waiting for a file to finish playing

When you play a sound file, you will often want your program to wait until the sound file finishes playing. You could do that with a PAUSE command if you know exactly how long the sound is. But what if you aren't sure?

Fortunately, the AP-16+ includes a G command that gets the current status of the card. To use this command, you must first send the G command to the AP-16+ with a SEROUT statement. Then, you must retrieve the status result by using a SERIN statement.

The SERIN statement is pretty straightforward: Its syntax looks like this:

```
SERIN pin, mode, variable
```

The *pin* and *mode* parameters are the same as for the SEROUT statement. The *variable* parameter is simply the name of a variable you have set aside to receive the status result.

Here's how you can use the SEROUT and SERIN statements together to get the status of an AP-16+:

```
SYMBOL ap16Status B0
SEROUT Sio, AP16, ("!AP16", AP16, "G")
SERIN Sio, AP16, ap16Status
```

In this example, the SYMBOL statement equates the name ap16status with the variable B0. Then, the SEROUT statement sends a G command to the AP-16+ and the SERIN statement retrieves the status data in the variable named ap16Status.

The status data includes a variety of useful information, but in nearly every case the only thing you really want to know is whether the AP-16+ is currently playing a sound. You can find that out by testing bit 7 of the status byte. The easiest way to do that is to define a symbol using the Basic Stamp's predefined bit-variables, as in this example:

```
SYMBOL ap16Status  = B0
SYMBOL ap16Playing = BIT7
```

Then, you can simply test the ap16Playing variable, like this:

```
IF ap16Playing = 0 THEN SoundDone
```

This statement will branch to the label SoundDone if the AP-16+ has finished playing the sound.

Here is a complete snippet of code that plays a sound and then locks itself in a loop until the sound is finished playing:

```
SEROUT Sio, Baud, ("!AP16", AP16, "PW", "SCREAM", 13, 1)
StillPlaying:
   SEROUT Sio, Baud, ("!AP16", AP16, "G")
   SERIN Sio, Baud, ap16Status
   IF ap16Playing = 0 THEN SoundDone
   GOTO StillPlaying
SoundDone:
   'The sound is done!
```

A sample AP-16+ program

Listing 4-2 shows a complete PBASIC program that plays sounds on an AP-16+4 card. The program plays the sound `weasel.wav` for 10 seconds, and then plays the `scream.wav` sound until that sound ends. Then it repeats. You can download this program from this book's companion website.

Listing 4-2: An AP-16+ Program

```
' {$STAMP BS1}
' {$PBASIC 1.0}
'
'
' AP-16 Test Program
'
' This program requires an AP-16+ on Pin 7
'
' AP-16+ Sound Files
'
'    WEASEL.WAV
'    SCREAM.WAV
'
' The action of this program is as follows:
'
' 1. Play the file "WEASEL.WAV" for 10 seconds
'
' 2. Play the file "SCREAM.WAV" once
'
' 3. Repeat.


SYMBOL Sio        = 7
SYMBOL Baud       = OT2400

SYMBOL AP16       = 0
SYMBOL ap16Status = B0
SYMBOL ap16Playing = BIT7


'Reset the AP16
   SEROUT Sio, Baud, ("!AP16", AP16, "X")
```

```
Main:

  'Play Pop Goes the Weasel for 10 seconds
  SEROUT Sio, Baud, ("!AP16", AP16, "PW", "WEASEL", 13, 0)
  PAUSE 10000

  'Play the scream
  SEROUT Sio, Baud, ("!AP16", AP16, "PW", "SCREAM", 13, 1)

  'Wait for the scream to end
  StillScreaming:
    SEROUT Sio, Baud, ("!AP16", AP16, "G")
    SERIN Sio, Baud, ap16Status
    IF ap16Playing = 1 THEN StillScreaming

  'Do it again
  GOTO Main
```

Programming the PIR Motion Detector

The PIR sensor uses infrared light to detect motion. It sends a 1 to the micro-controller when such motion is detected.

At first glance, then, you would think that the PIR sensor is pretty easy to use. Just plug it into one of the Basic Stamp's I/O pins and wait for that pin to have a value of 1 to trigger your prop. The code would look something like this, assuming that Pir is a symbol assigned to the I/O pin that the PIR is connected to:

```
WaitForMovement:
  IF Pir = 0 THEN WaitForMovement

MovementDetected:
 'Activate your prop now!
```

Unfortunately, life isn't that easy where PIR motion detection is concerned. For starters, the PIR won't be reliable for a while when power is first applied. Thus, any program that uses a PIR to detect motion should start with an immediate 60-second pause. This gives the PIR time to settle down so it can reliably detect motion.

Second, the PIR will often send very brief false triggers when there's no actual motion present in the room. The usual way to get around this problem is to wait until the PIR's value is 1 for a quarter of a second before triggering your prop.

**Book VIII
Chapter 4**

**Building an
Animatronic Prop
Controller**

You can do this with a tricky bit of code:

```
SYMBOL Pir       = PIN6
SYMBOL waitTimer = B2

PAUSE 60000
Main:
  waitTimer = 0

  WaitForMotion:
    PAUSE 10
    waitTimer = waitTimer + Pir * Pir
    IF waitTimer < 25 THEN WaitForMotion

  'The PIR has detected motion for 0.25 seconds
  'so the prop can now be triggered
```

The cleverness in this bit of code is that the WaitForMotion loop pauses for 10 milliseconds, then adds the value of the PIR input status to the waitTimer variable while at the same time multiplying the result by the status of the PIR input. If the PIR input is 0, multiplying by the input status resets the wait-Timer variable to 0 because any number times zero is zero. So this loop won't exit until the PIR has detected motion consistently for a quarter of a second.

Looking at Complete Jack-In-The-Box Program

Now that you've seen the techniques for programming the individual components of the prop controller, put them together to create a program for an actual prop. Listing 4-3 shows the complete program for the jack-in-the-box prop. I made liberal use of comments throughout this program, so you should have no trouble following its basic operation. (The program is available for download at this book's companion website.)

Listing 4-3: The Jack-In-The-Box Program

```
' {$STAMP BS1}
' {$PBASIC 1.0}


' Jack-In-The-Box Program
'
' The following devices are connected to the prop controller for this prop
'
' Pin 6          PIR Sensor
'
' Pin 7          RC-4   Address %00
'                AP-16+ Address %00
'
' RC-4 Relay 1   Pneumatic Solenoid Valve to open prop
'                Light to illuminate clown
'
```

```
' RC-4 Relay 2     Motor to turn crank handle
'
'
' AP-16+ Sound Files
'
' WEASEL.WAV      Pop Goes the Weasel to be played while motor is running
' SCREAM.WAV      Scream to play when clown pops up


' The action of the prop is as follows:
'
' Turn the crank arm motor and play "Pop Goes the Weasel" until
' motion is detected in the room.
'
' When motion is detected, pop up the clown and play the scream sound.
'
' When the scream sound ends, retract the clown, wait 2 seconds,
' and then resume the crank motor and "Pop Goes the Weasel"
'
' Wait at least 10 seconds before triggering the prop again.



SYMBOL Sio         = 7
SYMBOL Baud        = OT2400

SYMBOL RC4         = 3
SYMBOL RelayValve  = 1
SYMBOL RelayMotor  = 2

SYMBOL RelayOn     = 1
SYMBOL RelayOff    = 0

SYMBOL AP16        = 0
SYMBOL ap16Status  = B0
SYMBOL ap16Playing = BIT7

SYMBOL Pir         = PIN6
SYMBOL waitTimer   = B2



'Reset everything & pause one minute to let the PIR stabilize
  SEROUT Sio, Baud, ("!AP16", AP16, "X")
  SEROUT Sio, Baud, ("!RC4", RC4, "X")
  PAUSE 50000

Main:

  'Play Pop Goes the Weasel in an indefinite loop
  SEROUT Sio, Baud, ("!AP16", AP16, "PW", "WEASEL", 13, 0)

  'Start the motor
  SEROUT Sio, Baud, ("!RC4", RC4, "R", RelayMotor, RelayOn)

  'Wait 10 seconds for the room to clear
  PAUSE 10000
```

**Book VIII
Chapter 4**

**Building an
Animatronic Prop
Controller**

(continued)

Listing 4-3 *(continued)*

```
'Wait for the PIR to trigger for 0.25 seconds
waitTimer = 0
WaitForMotion:
  PAUSE 10
  waitTimer = waitTimer + Pir * Pir
  IF waitTimer < 25 THEN WaitForMotion

'At this point the trigger has been activated

'Stop the motor
SEROUT Sio, Baud, ("!RC4", RC4, "R", RelayMotor, RelayOff)

'Play the scream
SEROUT Sio, Baud, ("!AP16", AP16, "PW", "SCREAM", 13, 1)

'Pop up the clown
SEROUT Sio, Baud, ("!RC4", RC4, "R", RelayValve, RelayOn)

'Wait for the scream to end
StillScreaming:
  SEROUT Sio, Baud, ("!AP16", AP16, "G")
  SERIN Sio, Baud, ap16Status
  IF ap16Playing = 1 THEN StillScreaming

'Retract the clown
SEROUT Sio, Baud, ("!RC4", RC4, "R", RelayValve, RelayOff)

'Wait 2 seconds
PAUSE 2000

'Do it again
GOTO Main
```

Index

H

1

J

K

L

N

S

W